Alternative respiratory pathways in higher plants

Alternative respiratory pathways in higher plants

EDITED BY

Kapuganti Jagadis Gupta

Department of Plant Sciences
University of Oxford
Oxford, UK

Luis A.J. Mur

Institute of Biological
Environmental and Rural Science
Aberystwyth University
Aberystwyth, UK

Bhagyalakshmi Neelwarne

Plant Cell and Biotechnology Department
CSIR-Central Food Technological Research Institute
Mysore, India

WILEY Blackwell

Registered Office
John Wiley & Sons, Ltd, The Atrium, Southern Gate, Chichester, West Sussex, PO19 8SQ, UK

Editorial Offices
9600 Garsington Road, Oxford, OX4 2DQ, UK
The Atrium, Southern Gate, Chichester, West Sussex, PO19 8SQ, UK
111 River Street, Hoboken, NJ 07030-5774, USA

For details of our global editorial offices, for customer services and for information about how to apply for permission to reuse the copyright material in this book please see our website at www.wiley.com/wiley-blackwell.

Library of Congress Cataloging-in-Publication Data:

Gupta, Kapuganti Jagadis
Alternative respiratory pathways in higher plants / Kapuganti Jagadis Gupta, Luis A.J. Mur, and Bhagyalakshmi Neelwarne.
pages cm
Includes bibliographical references and index.
ISBN 978-1-118-79046-5 (cloth)
1. Plants–Respiration. 2. Plant genetics. 3. Plant physiology. I. Mur, Luis A. J.
II. Neelwarne, Bhagyalakshmi. III. Title. IV. Title: Respiratory pathways in higher plants.
QK891.K37 2015
581.3′5–dc23

2014050165

A catalogue record for this book is available from the British Library.

Wiley also publishes its books in a variety of electronic formats. Some content that appears in print may not be available in electronic books.

Cover image: Main cover picture created by Birgit Arnholdt Schmidt and Kapuganti Jagadis Gupta

Set in 9.5/13pt Meridien by SPi Publisher Services, Pondicherry, India

1 2015

Contents

List of contributors

Salvador Abreu
Department of Computer Science, Universidade de Évora, Évora, Portugal

Mary S. Albury
Biochemistry and Molecular Biology, School of Life Sciences, University of Sussex, Falmer, Brighton, East Sussex, UK

Birgit Arnholdt-Schmitt
EU Marie Curie Chair, ICAAM, Universidade de Évora, Évora, Portugal

Natalia V. Bykova
Cereal Research Centre, Agriculture and Agri-Food Canada, Morden, MB, Canada

Maria Doroteia Campos
EU Marie Curie Chair, ICAAM, Universidade de Évora, Évora, Portugal

Hélia G. Cardoso
EU Marie Curie Chair, ICAAM, Universidade de Évora, Évora, Portugal

José Hélio Costa
Department of Biochemistry and Molecular Biology, Federal University of Ceara, Fortaleza, Ceara, Brazil

Marina Cvetkovska
Department of Biological Sciences and Department of Cell and Systems Biology, University of Toronto Scarborough, Toronto, Ontario, Canada

Keshav Dahal
Department of Biological Sciences and Department of Cell and Systems Biology, University of Toronto Scarborough, Toronto, Ontario, Canada

Matthias Döring
INOQ GmbH, Solkau, Schnega, Germany

Padmanabh Dwivedi
Department of Plant Physiology, Institute of Agricultural Sciences, Banaras Hindu University, Varanasi, India

Catherine Elliott
Biochemistry and Molecular Biology, School of Life Sciences, University of Sussex, Falmer, Brighton, East Sussex, UK

Margarida Espada
NemaLab-ICAAM, Departamento de Biologia, Universidade de Évora, Évora, Portugal

Ralph Ewald
Institut für Biowissenschaften, Abteilung Pflanzengenetik, Universität Rostock, Rostock, Germany

António Miguel Frederico
EU Marie Curie Chair, ICAAM, Universidade de Évora, Évora, Portugal

Teresa Gonçalves
Department of Computer Science, University of Évora, Évora, Portugal

Kapuganti Jagadis Gupta
Department of Plant Sciences, University of Oxford, Oxford, UK
Current address:
National Institute of Plant Genome Research, Aruna Asaf Ali Road, New Delhi, India

Lee D. Hansen
Department of Chemistry and Biochemistry, Brigham Young University, Provo, Utah, USA

Renate Horn
Institut für Biowissenschaften, Abteilung Pflanzengenetik, Universität Rostock, Rostock, Germany

Abir U. Igamberdiev
Department of Biology, Memorial University of Newfoundland, St. John's, Newfoundland and Labrador, Canada

Jens Jurgeleit
INOQ GmbH, Solkau, Schnega, Germany

Sarma Rajeev Kumar
Plant Genetic Engineering Laboratory, Department of Biotechnology, Bharathiar University, Coimbatore, India

Wu-Sheng Liang
Institute of Biotechnology, College of Agriculture and Biotechnology, Zhejiang University, Hangzhou, People's Republic of China

Eva Lucic
INOQ GmbH, Solkau, Schnega, Germany

Allison E. McDonald
Department of Biology, Wilfrid Laurier University, Waterloo, Ontario, Canada

Kaveh Mashayekhi
BioTalentum Ltd, Budapest, Hungary

Ben May
Biochemistry and Molecular Biology, School of Life Sciences, University of Sussex, Falmer, Brighton, East Sussex, UK

Dirce Fernandes de Melo
Department of Biochemistry and Molecular Biology, Federal University of Ceara, Fortaleza, Ceara, Brazil

Louis Mercy
INOQ GmbH, Solkau, Schnega, Germany

Anthony L. Moore
Biochemistry and Molecular Biology, School of Life Sciences, University of Sussex, Falmer, Brighton, East Sussex, UK

Manuel Mota
NemaLab-ICAAM, Departamento de Biologia, Universidade de Évora, Évora, Portugal

Luz Muñoz-Sanhueza
EU Marie Curie Chair, ICAAM, Universidade de Évora, Évora, Portugal
Current address:
Department of Plant and Environmental Sciences (IPM), Norwegian University of Life Sciences, Ås, Norway

Luis A.J. Mur
Institute of Biological, Environmental and Rural Science, Aberystwyth University, Aberystwyth, UK

Bhagyalakshmi Neelwarne
Plant Cell and Biotechnology Department, CSIR-Central Food Technological Research Institute, Mysore, India

Tânia Nobre
EU Marie Curie Chair, ICAAM, Universidade de Évora, Évora, Portugal

Carlos Noceda
EU Marie Curie Chair, ICAAM, Universidade de Évora, Évora, Portugal
Current address:
Prometeo Project (SENESCYT), CIBE (ESPOL), Guayaquil, Ecuador

Amaia Nogales
EU Marie Curie Chair, ICAAM, Universidade de Évora, Évora, Portugal

Halley Caixeta Oliveira
Departamento de Biologia Animal e Vegetal, Centro de Ciências Biológicas, Universidade Estadual de Londrina (UEL), Londrina, Paraná, Brazil

Michail Orfanoudakis
Department of Forestry and Management of the Environment and Natural Resources, Forest Soil Lab, Democritus University of Thrace, Orestiada, Greece

Augusto Peixe
Melhoramento e Biotecnologia Vegetal, ICAAM, Universidade de Évora, Évora, Portugal

Alexios Polidoros
Department of Genetics and Plant Breeding, School of Agriculture, Aristotle University of Thessaloniki, Thessaloniki, Greece

Paulo Quaresma
Department of Computer Science, University of Évora, Évora, Portugal

Carla Ragonezi
EU Marie Curie Chair, ICAAM, Universidade de Évora, Évora, Portugal

Anna M. Rychter
Institute of Experimental Plant Biology and Biotechnology, Faculty of Biology, University of Warsaw, Warsaw, Poland

Ione Salgado
Departamento de Biologia Vegetal, Instituto de Biologia, Universidade Estadual de Campinas (UNICAMP), São Paulo, Brazil

Elisete Santos Macedo
EU Marie Curie Chair, ICAAM, Universidade de Évora, Évora, Portugal

Ramalingam Sathishkumar
Plant Genetic Engineering Laboratory, Department of Biotechnology, Bharathiar University, Coimbatore, India

Caroline Schneider
INOQ GmbH, Solkau, Schnega, Germany

Samir Sharma
Department of Biochemistry, University of Lucknow, Lucknow, India

Evangelia Sinapidou
Department of Agricultural Development, Democritus University of Thrace, Orestiada, Greece

Debabrata Sircar
Biotechnology Department, Indian Institute of Technology Roorkee, Uttarakhand, India

Jan T. Svensson
EU Marie Curie Chair, ICAAM, Universidade de Évora, Évora, Portugal
Current address:
Nordic Genetic Resource Center, Alnarp, Sweden

Bożena Szal
Institute of Experimental Plant Biology and Biotechnology, Faculty of Biology, University of Warsaw, Warsaw, Poland

Vera Valadas
EU Marie Curie Chair, ICAAM, Universidade de Évora, Évora, Portugal

Greg C. Vanlerberghe
Department of Biological Sciences and Department of Cell and Systems Biology, University of Toronto Scarborough, Toronto, Ontario, Canada

Cláudia Vicente
NemaLab-ICAAM, Departamento de Biologia, Universidade de Évora, Évora, Portugal

Jia Wang
Department of Biological Sciences and Department of Cell and Systems Biology, University of Toronto Scarborough, Toronto, Ontario, Canada

Luke Young
Biochemistry and Molecular Biology, School of Life Sciences, University of Sussex, Falmer, Brighton, East Sussex, UK

Preface

Respiration is a crucial biochemical process found in all living organisms for meeting their energy demands. A cell adapts to its surroundings and dynamically caters to the energy needs of a wide array of functions. Thus, cells have evolved mechanisms to ingeniously 'switch on' and 'switch off' the different steps of respiratory mechanisms. Among the biochemical processes involved in respiration, three major highly conserved 'classical' pathways are involved; glycolysis, where energy is generated by breaking down glucose; the tricarboxylic acid (TCA) cycle, where the energy is generated in a form that can be used in cellular biochemical reactions; and electron transfer through an electron transport chain to form reducing equivalents leading to the generation ATP. Additionally, plant cells can regulate respiration in a manner deviating from fundamental and generic pathways via so-called alternative respiratory pathways (ARP), which form the focus of this book. While alternative modes of respiration occur in parallel to normal respiration, different sets of regulatory mechanisms are involved in the regulation of genes encoding for the proteins that are involved in alternative pathways. Understanding the regulation of these genes is an important theme in ARP research. Thus, the means through which alternative respiratory processes are regulated to help maintain classical respiration under various stresses or during discrete developmental or ecological conditions, features prominently in ARP publications. Linked to such research are attempts to predict the responses to climate change – changes in temperature, gases, physical vibrations, light, cosmic energy and so on. Even at the shortest and smallest scales, the plant's immediate environment directly influences *in planta* physiological processes – via processes such as respiration – which are ultimately regulated at the genetic level. As a result, on longer and larger spatiotemporal scales, such environmental effects bring about changes in the distribution of plant species and ecosystems. Such changes will in turn also impact on the climate through the exchange of energy and gases among the flora and fauna around them. Equally, a failure to understand and respond to the impacts of climate change on respiration in crops will compromise yield, perturbing food security. Aware of these facts, plant physiologists have focused their research into each aspect of these interactions. A great deal of research has recently been published on how plants display different modes of respiration in different organs by switching over to ARP and on what set of parameters regulate alternative oxidases. To highlight the contribution of ARP to these fundamentally important topics we have brought together scientists with global reputations in the field to produce what we consider to be an important book with relevance to ecology, plant biodiversity and crop production.

This book therefore considers both classical and alternative respiratory pathways in diverse plant species and in different organs of the same plant at different times of its life cycle. Another driving principle has been to consider the potential applications of this knowledge to plant science and agriculture. The sixteen chapters are split into three sections: the first shows how plant respiratory mechanism have developed to thrive by cleverly rationing cellular energy under differing circumstances, while the second section highlights the application of ARP in plant breeding. The book wraps up the third and final section with the description of important protocols that will be useful for newer researchers.

Within Section A, Chapter 1 introduces readers to the basic principles and the principal difference between classic respiration and the alternative respiratory mechanisms. Complex regulatory mechanisms are described indicating the possibility of not only switching from glycolysis to fermentative metabolism but also the utilization of ARP to maintain substrate oxidation while minimizing the production of ATP. Equally, new insights are indicated on how ATP generation can be maintained under hypoxia. Chapter 2 describes the uncoupling pathways of plant mitochondrial electron transport and the mechanisms variously evolved to maintain the energy flux. How the regulatory proteins – the alternative oxidases – are distributed among the plant kingdom is brought into focus in Chapter 3.

Chapters 4 to 9 deal with alternative respiration under endogenous biochemical perturbations that occur due to certain signal molecules and exogenous stress, as well as how mitochondrial metabolism is regulated and cellular energy is balanced. Chapters 10 and 11 specifically address certain issues related to horticultural commodities – ARP in fruit ripening and in bulky storage tissues.

Section B contains subsections 12 to 14 – a package of 12 chapters – that consider how the molecular information on alternative oxidases may be developed as functional markers in plant breeding programmes. In-depth information is provided by the most renowned experts in the field, discussing how alternative oxidase genes also serve to develop phenotyping tools based on calorespirometry. Since alternative respiratory pathways play a role in the generation of heat during flower blooming and fruit ripening – where heat is needed for emitting volatiles – it serves as an excellent tool for calorespirometric measurements of metabolic heat rates and carbon dioxide rates of respiring tissues as functions of temperature. This enables the rapid responses of plant metabolic events to temperature fluctuations to be determined and, therefore, plant adaptability to environmental conditions to be deduced. Investigating such responses often involves cumbersome and expensive experiments which may be avoided by opting for methods such as calorespirometry. This area has great potential for projecting the effects of global warming on the plant kingdom as a whole and for predicting the geographical distribution of different crops and plant species.

Section C, which includes Chapters 15 and 16, provides updated protocols that describe the steps involved in the isolation of mitochondria for different studies, written by the most experienced workers in the field.

This book, with its breadth of information from the classical understanding of plant respiratory mechanisms to the highly specialized physiological changes that occur in plants during ARP, is expected to find a large readership among life science students and researchers in plant science.

Reputed scientists from nine different countries have contributed to this book and to whom we editors are extremely grateful. We owe our heartfelt gratitude to the internal editors and book publishing staff of John Wiley & Sons, Ltd. for their continuous support and timely advice during the course of the preparation of this volume.

K.J. Gupta, L.A.J. Mur and B. Neelwarne

SECTION A
Physiology of plant respiration and involvement of alternative oxidase

Contents

CHAPTER 1

Integrating classical and alternative respiratory pathways

Kapuganti Jagadis Gupta[1,*], Bhagyalakshmi Neelwarne[2] and Luis A.J. Mur[3]

[1] *Department of Plant Sciences, University of Oxford, Oxford, UK*

[2] *Plant Cell and Biotechnology Department, CSIR-Central Food Technological Research Institute, Mysore, India*

[3] *Institute of Biological, Environmental and Rural Science, Aberystwyth University, Aberystwyth, UK*

**Current address: National Institute of Plant Genome Research, Aruna Asaf Ali Road, New Delhi, India*

Introduction

Respiratory pathways are vital for plant carbon and energy metabolism, which is the main use of most assimilated carbohydrates. Most respiratory pathways are very well established, the prominent being glycolysis in cytosol and the tricarboxylic acid (TCA) cycle, which occurs in the matrix of mitochondria coupled with the electron transport chain (ETC) which functions along the inner mitochondrial membrane. Some glycolytic enzymes also associate with the mitochondrial membrane and dynamically support substrate channelling (Giegé *et al.*, 2003; Graham *et al.*, 2007). Despite cross-kingdom commonalities in glycolysis and the TCA cycle, the regulation of respiration is relatively poorly understood (Fernie *et al.*, 2004) which reflects the complexity of respiratory pathways. In plants this complexity encompasses the only possibility of switching from glycolysis to fermentative metabolism but the utilization of alternative pathways in plants allows the maintenance of substrate oxidation while minimizing the production of ATP. Equally, new insights have suggested how ATP generation can be maintained under hypoxia. With this overview, this chapter will integrate such alternative respiratory pathways with components of the classical oxidative-phosphorylative pathways.

Mitochondrial electron transport generates ATP by using the reducing equivalents derived through the operation of the TCA-cycle. The classic operation of the ETC pathway involves the transport of electrons from such as NAD(P)H or succinate to oxygen via four integral membrane oxidoreductase complexes: NADH dehydrogenase (complex I), succinate dehydrogenase (complex II), cytochrome *c* reductase (complex III), cytochrome *c* oxidase (complex IV or COX), linked to a mobile electron transfer protein (cytochrome *c*) and ATP synthase complex (complex V). In complex V, the active extrusion of protons from the inner membrane space to the matrix leads to the generation of ATP (Boekema

Alternative Respiratory Pathways in Higher Plants, First Edition.
Edited by Kapuganti Jagadis Gupta, Luis A.J. Mur and Bhagyalakshmi Neelwarne.

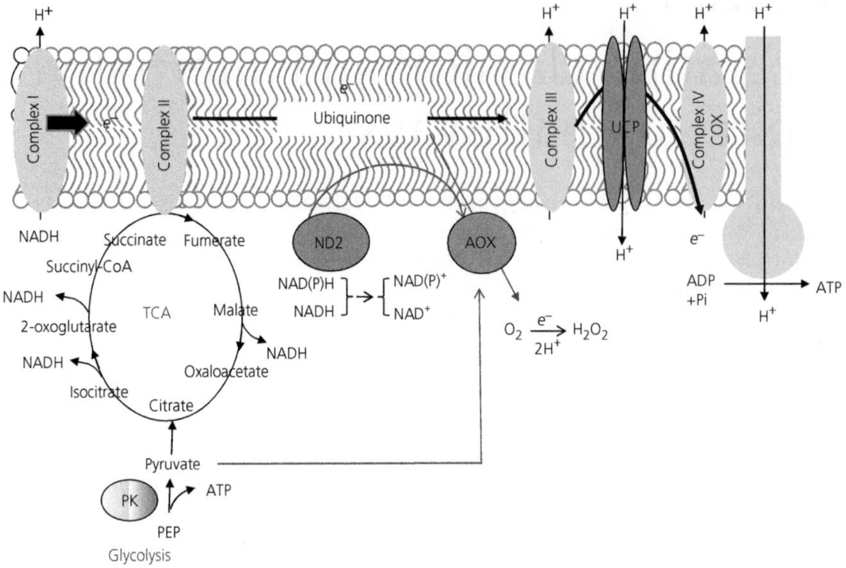

Figure 1.1 Overview of electron transport chain dissipatory mechanisms in plant mitochondria.

and Braun, 2007) (Figure 1.1). Apart from this classical operation of the ETC, mitochondrial complexes interact to form so-called super-complexes or respirosomes (Boekema and Braun, 2007). Complex I, II and IV are involved in the formation of super-complexes with different degrees and configurations. It may be that the formation of super-complexes represents a regulatory mechanism that controls the passage of electrons through the ETC (Eubel *et al.*, 2003). Super-complex formation helps in increasing the stability of individual complexes, in the dense packing of complexes in the membrane and in fine tuning energy metabolism and ATP synthesis (Ramírez-Aguilar *et al.*, 2011).

Currently most research on alternative electron transfer is focused on non-phosphorylating bypass mechanisms: a second oxidase – the alternative oxidase (AOX), an external NAD(P)H dehydrogenases in the first part of ETC, and also plant uncoupling mitochondrial proteins (PUCPs).

Alternative oxidase (AOX)

AOX is located in the inner mitochondrial membrane of all plants and fungi and a limited number of protists. AOX also appears to be present in several prokaryotes and even some animal systems (Chaudhuri and Hill, 1996; McDonald, 2008; McDonald and Vanlerberghe, 2006). Two forms of AOX are present in dicot plants (AOX1 and AOX2) while in monocots there is only one AOX (AOX1) (Considine *et al.*, 2002; Karpova *et al.*, 2002).

AOX are homodimeric proteins orientated towards the inner mitochondrial matrix. AOX diverts electrons from the main respiratory chain at the ubiquinone pool and mediates the four-electron reduction of oxygen to water (Figure 1.1). In comparison to electron transfer by the cytochrome chain (complex III and IV), AOX does not pump H^+, therefore transfer of electrons by AOX does not create a transmembrane potential, and the decline in free energy between ubiquinol and oxygen is dissipated and mostly released as heat (Vanlerberghe *et al.*, 1999). The diversion of electrons to the AOX pathway can reduce ATP generation by up to 60% (Rasmussen *et al.*, 2008). The AOX ATP dissipatory pathway plays an important role when the ETC is inhibited by various stress conditions. ETC inhibition increases NADH/NAD^+ and ATP/ADP ratios and as a consequence the TCA cycle could slow down. In addition to the energetic consequences of this, the number of carbon skeletons being produced will also be limited as the export of citrate supports nitrogen assimilation. Against this, AOX contributes to the maintenance of electron flow and the production of reducing equivalents to help maintain the TCA cycle. Indeed, AOX activation occurs in direct response to stress. A feature of all stress conditions is an increase in the production of reactive oxygen species (ROS): a process that can occur from the over-reduction of cytochrome components through the disruption of the ETC. In response to this, ROS or ROS-induced signals such as salicylic acid, act to induce the transcription of AOX (Vanlerberghe and McIntosh, 1997; Mackenzie and McIntosh, 1999) as also suggested from the observation that the addition of antioxidants leads to the suppression of AOX (Maxwell *et al.*, 2002).

Oxygen, AOX and COX

Once induced by ROS, AOX may function as a negative feedback mechanism to suppress ROS production; a feature that we have named oxygen homeostasis (Gupta *et al.*, 2009). This feedback mechanism is a consequence of large differences in O_2 affinities of the classical and alternative respiratory pathways. The K_m of COX is approximately 0.1 µmol but in AOX it is between 10 and 20 µmol (although the study by Millar *et al.*, 1993 suggested a 10-fold higher AOX affinity for O_2). Given these affinities, COX will maintain respiration whilst AOX reduces the O_2 concentration, thereby decreasing the production of ROS inside the mitochondrion (Puntarulo and Cederbaum, 1988; Skutnik and Rychter, 2009). This is supported by the observations of Ribas-Carbo *et al.* (1995) who used an oxygen isotope discrimination technique to show that the inhibition of AOX by its inhibitor salicylhydroxamic acid (SHAM) did not lead to a decrease in total respiratory rates. This mechanism would be an exception to the 'energy over flow' model proposed by Lambers (1982), who suggested that in certain situations (e.g. excess carbohydrate), non-phosphorylating alternative pathways might contribute significantly to total respiration. Oxygen homeostasis could be of especial relevance in situations where different plant tissues are subjected to fluctuating O_2 concentration due to diffusion gradients, and more so under environmental conditions such as flooding (Rolletschek *et al.*, 2002; Bailey-Serres and Chang, 2005; Schmälzlin *et al.*, 2005; Bailey-Serres and Voesenek, 2008; Rasmusson *et al.*, 2008).

The electron partitioning model of Ribas-Carbo *et al.* (1995) suggests that COX and AOX compete for electron and electron passage but this must be influenced by the stress response of each pathway and particularly if exposed to low partial pressures of O_2 (Po_2). In a study undertaken by the senior author's group, root slices of several species were incubated in a sealed cuvette and the respiratory rate of the tissue was measured until total oxygen was depleted in the vial. Until a partial pressure of 4% Po_2, the decrease in respiratory rate correlated linearly with O_2 concentration; however, at <4% Po_2 level, the respiratory oxygen consumption rate slowed, taking a longer time to consume oxygen, indicating that a more slowly respiring plant would promote survival under the latter condition (Zabalza *et al.*, 2009). This unique phenomenon has been named as the 'adaptive response of plant respiration (ARPR) to hypoxia'. To determine which among the respiratory pathways could be influencing ARPR, each pathway was selectively inhibited in hydroponically grown pea using either KCN (an inhibitor for COX) or SHAM (an inhibitor for AOX). When AOX was the only electron acceptor, O_2 consumption continued without any alteration until all the oxygen was depleted, but when AOX was inhibited, ARPR was still observed. Thus, the COX pathway was found to be responsible for ARPR (Zabalza *et al.*, 2009). Clearly, ARPR is not a consequence of differentially responsive O_2 affinities of the terminal oxidases (see earlier) as it occurs at Po_2 above the K_m of both oxidases. The decline in respiration could not be explained by a depletion of carbohydrates, as respiratory substrates, since when the same root material was immediately reused in experiments, ARPR was still observed. Moreover, oxygen diffusion through the tissue was not limiting at low Po_2 because ARPR was also observed with in single-celled organism *Chlamydomonas* which has a diameter approximately 20 µm (Gupta *et al.*, 2009). The lower Po_2 was not in itself limiting respiratory rates as respiration could be elevated by the prior addition of 10 mM pyruvate prior to assessing ARPR. Taken together, these observations point towards the most likely scenario of the existence of an oxygen sensing mechanism that regulates the rate of mitochondrial oxygen consumption at low Po_2.

Pyruvate kinases, classical respiratory metabolism and AOX

Pyruvate kinase (PK; EC 2.7.1.40) plays a critical role in glycolytic pathway catalyzing the terminal reaction of the glycolytic pathway by converting ADP and phosphoenolpyruvate (PEP) to ATP and pyruvate. As pyruvate regulates both glycolysis and the TCA cycle (Pilkis and Granner, 1992; Teusink *et al.*, 2000), PK represents a crucial respiratory regulatory node. PK exists as tissue-specific isozymes that exhibit significant differences in their physical and kinetic properties (reviewed by Plaxton and Podesta, 2006). This reflects the presence of different PK isozymes in the cytosolic and plastidial compartments in plants; designated as PKc and PKp forms respectively (Plaxton, 1996; Givan, 1999). Transgenic tobacco plants which were deficient in PKc were used to demonstrate its role in regulating development via modulation of carbon sink-source relationships (Knowles *et al.*, 1998; Grodzinski *et al.*,1999). PKc lines exhibited delayed shoot and flower development and this was correlated with poor export of previously fixed $^{14}CO_2$ from leaves in the 'night-time' phase of a light-dark cycle but increased $^{14}CO_2$ release from respiration (Grodzinski *et al.*,1999). Conversely, in another study with *Arabidopsis* seeds, PKp has been shown to play an important role in fatty acid biosynthesis (Andre and Benning, 2007; Andre *et al.*, 2007). PKs also exist as tissue specific isozymes (Turner *et al.*, 2005). The subtle respiratory regulation that these difference in PK isoforms affords is well-illustrated by a classic study of PKc repression and activation in castor seed endosperm (Podesta and Plaxton, 1991). In castor seeds, during aerobic conditions, the allosteric inhibition of endosperm PKc facilitated larger gluconeogenic conversion of stored triacylglycerides to hexose-phosphates assisting in germination. However, under low oxygen PKc became active in order to compensate for ATP depletion that occurs due to hypoxic stress (Podesta and Plaxton, 1991).

A key study also used a transgenic approach to provide greater insight into the role of PKc in carbon metabolism through the coordinated regulation of glycolysis, the TCA cycle, the mitochondrial ETC and also AOX in potato tuber (Oliver *et al.*, 2008). A role for PKc in these respiratory pathways was implied from a series of observations. Firstly, pyruvate addition experiments showed an effect on glycolytic flux and the consequences that altered the dynamics of mitochondrial ETC (Zabalza *et al.*, 2009). The link to AOX was suggested when an increase in AOX activity was seen after pyruvate was added to isolated mitochondria (Millar *et al.*, 2003). This AOX effect was then explained through the interaction of pyruvate to cysteine residue of AOX (Umbach *et al.*, 2006).

Transgenic potato tubers with decreased in PKc levels were generated through an RNA interference (RNAi) gene silencing approach, among which three lines were selected, lines PKC-25, 6 and 15 – where PK activity was reduced to ~40%, 37% and 29% respectively (Oliver *et al.*, 2008). As expected, lowering PKc expression led to a higher PEP to pyruvate ratio in actively growing tubers. This decrease in pyruvate levels correlated with a decrease in the various organic acids involved in the TCA cycle and there was also a decrease in the level of total protein in the tubers. [^{14}C]Glc labelling and feeding experiments showed a slight decrease in carbon partitioning towards organic acid and protein synthesis upon decrease in PKc levels. These results clearly demonstrated that PKc plays a very important role in the regulation of the levels of organic acids in tubers and partitioning the carbon toward the TCA cycle but interestingly total respiration and TCA cycle flux did not alter. One reason could be that residual pyruvate levels are probably enough to maintain the respiratory activity in these tubers. Equally, other enzymes that generate pyruvate such as PKp, PEPC, or PEP phosphatase could be compensating for the loss in PKc. Alternatively; there could be a compensatory change in electron transport through the COX pathway, which is in line with the electron partition model (Ribas-Carbo *et al.*, 1995). This would imply that respiratory metabolism has a high homeostatic ability allowing considerable flexibility in response to changes in metabolite and transcript levels (Nunes-Nesi *et al.*, 2005, 2007; Studart-Guimarães *et al.*, 2007).

The potato RNAi lines also exhibited a suppression of AOX-dependent respiration which could be reversed by external feeding of pyruvate to tuber tissue. Suppression of the AOX pathway would be beneficial in growing tubers, which characteristically have low internal oxygen concentrations and low adenylate energy charge (Geigenberger, 2003). In line with this, PKc silenced plants produced significantly more tubers which also tended to be larger than the control tubers (Oliver *et al.*, 2008). Thus, PKc modulation of pyruvate accumulation would be of great agronomic importance, functioning as a key regulatory step in potato tuber development by influencing the AOX in heterotrophic potato tubers.

NADPH dehydogenases linked to AOX

In addition to complex I (NADH dehydrogenase) there are some additional proteins which can use NADH and NADPH to reduce ubiquinone pool. There are NAD(P)H dehydrogenases. Type II NAD(P)H dehydrogenases (ND2) are membrane-bound proteins that face either the matrix or the inter-membrane side (Figure 1.1). Unlike complex I these are not involved in proton translocation and therefore do not contribute for ATP synthesis. As shown in Figure 1.1 there are at least four types of NADH dehydrogenase proteins; two on the external side of the inner mitochondrial membrane (one oxidizing NADH and one NADPH) and two to the inner face of the inner membrane (similarly one devoted use NADH and other use NADPH) (Rasmusson and Møller, 1991). Substrate specificity for these dehydrogenases is based on pH and calcium. Since various environmental conditions and biotic abiotic stresses influence the dynamics of calcium and pH, which in turn have cascading effects on activities of NADH and NADPH dehydrogenases (Felle, 2005; Dodd *et al.*, 2010). For instance NADPH dehydrogenases are involved in nitric oxide generation under anoxia. In view of these intricate dynamic processes, uncovering the roles of different dehydrogenases has been an area of intense research (Michalecka *et al.*, 2003; Rasmusson *et al.*, 2008). There are reports that specificity for NADPH of the external NADPH dehydrogenase NDB1 at low pH becomes important under hypoxia (Felle, 2005). This leads to oxidation of cytosolic NADH under hypoxia which leads to recycling of NAD^+.

Uncoupling proteins (UCPs)

Plant uncoupling proteins are a class of mitochondrial anion carrier proteins. UCP is a specialized protein that uncouples electron transport from ATP synthesis in mitochondria by acting downstream of complex IV (Figure 1.1). The primary functions of UCPs are to transport protons from the intermembrane space into the mitochondrial matrix. This translocation leads to generation of electrochemical gradient ($\Delta\psi$) (Rial *et al.*, 1983) which is opposite of ATP and this action leads to a decrease in $\Delta\psi$, and the potential energy of the $\Delta\psi$ is dissipated as heat (Vercesi *et al.*, 2006). Therefore UCPs were initially considered as energy wastage proteins. UCP mediates a fatty acid dependent, purine nucleotide-inhibited proton leakage across the inner mitochondrial membrane (Krauss *et al.*, 2005). Therefore, within the context of plant energy-balance rearrangements, UCP may have overlapping functions with other alternative pathway proteins in the ETC like AOX and NAD(P)H dehydrogenases. Due to this, a tight regulation of UCP takes place in mitochondria. UCPs are mainly activated by free fatty acids and activity diminishes by ADP, GDP, ATP and GTP; (Vercesi *et al.*, 1995; Jezek *et al.*, 1996). Various physiological states such as pH, redox status of the ubiquinone pool control UCPs activity (Navet *et al.*, 2005; Borecký *et al.*, 2001). For instance, a decline in pH from 7.1 to 6.3 promotes the inhibitory effect of UCPs (Borecký *et al.*, 2001). It was also found that ROS can increase the activity of UCP. First interaction of ROS with membrane lipids leads to the production of 4-hydroxy-2-trans-nonenal which then activates the proton translocation activity of the UCPs (Smith *et al.*, 2004). UCPs are known to protect plants from high light, drought or heat stress. Supporting evidence in line with this is that the over-expression of Arabidopsis UCP (AtUCP1) in tobacco suppressed drought and salt stress-associated respiration. The AtUCP1 transgenic lines exhibited lower levels of ROS and higher tolerance to drought and salt stress (Begcy *et al.*, 2011). Not only to combat stress, UCPs also facilitate the synthesis of intermediates for amino acid and lipid biosynthesis (Tielens and Van Hellemond, 1998; Sweetlove *et al.*, 2007). This is via increasing metabolic flux during the conditions of excess ATP production by, for example, photosynthetic light reactions. Sweetlove *et al.* (2007) demonstrated that UCPs are involved in the recycling of metabolic intermediates of photorespiration and play important role in maintaining the metabolite flux during the condition of photorespiration.

Electron transfer flavoprotein (ETF)

Besides uncovering pathways which remove excess reducing power and balance the redox poise of the cell, several additional electron donors to the mitochondrial ETC in addition to NADH and NADPH have been uncovered in plants.

Most of them are similarities to well-characterized animal systems (Fe, 1988; Frerman *et al.*, 2001). One of such components is the electron transfer flavoprotein (ETF). ETF was first identified by Crane and Beinert in 1956 based on its capability to transfer electrons to various acceptors from fatty acyl-CoA dehydrogenases. Mammalian ETF is a heterodimer of alpha and beta subunits which are 31 and 27 kDa respectively, each binding to a single flavin adenine dinucleotide (FAD) as a redox responsive co-factor (McKean, Beckmann and Frerman, 1983). This protein is located in mitochondrial matrix and encoded by nuclear genome. ETF is an electron acceptor for at least nine mitochondrial matrix flavoprotein dehydrogenases. These are four straight fatty acyl-CoA dehydrogenases and five dehydrogenases which are involved in the catabolism of amino acids such as glutaryl, isovaleryl short and long chain and choline (reviewed by Roberts *et al.*, 1996 and the literature therein). These donors can be also classified as seven acyl-CoA dehydrogenases and two N-methyl dehydrogenases, isovaleryl-CoA dehydrogenase (IVDH), 2-methyl branched-chain acyl-CoA dehydrogenase, glutaryl-CoA dehydrogenase, sarcosine and dimethylglycine dehydrogenases. ETF donates electrons to flavoprotein:ubiquinone oxidoreductase (ETFQO) which are transferred to the ubiquinone pool (Ishizaki *et al.*, 2005). In mammalian systems the ETF-EFFQO has been shown to link the β-oxidation of fatty acids, choline and various amino acids to respiratory metabolism (Frerman, 1987). As a result mutation in either ETF or ETFQO leads to type II glutaric acidemia disease in humans where the build-up of incomplete processed proteins and fats leads to blood plasma acidosis (Frerman and Goodman, 2001).

Within plant science ETF came into picture when Heazlewood *et al.* (2004) identified the ETF system by liquid chromatography, tandem mass spectrometry mitochondrion proteomic analysis of *Arabidopsis*. Very soon afterwards *Arabidopsis* genes encoding ETFQO were discovered (Ishizaki *et al.*, 2005). It quickly emerged that the ETF-ETFQO system was involved in plant senescence which includes lipid mobilisation. Thus, Buchanan-Wollaston *et al.*, (2005) found the ETF system was transcriptionally induced during dark-induced senescence but this role was unambiguously demonstrated with T-DNA tagged mutants in *Arabidopsis* (Ishizaki *et al.*, 2005, 2006). Both ETF and ETFQO T-DNA mutants exhibited accelerated senescence and early death compared to wild-type during extended darkness. Interestingly, the mutants exhibited altered amino acid metabolism and in particular the accumulation of a leucine catabolism intermediate (Ishizaki *et al.*, 2005, 2006). The ETC complex was induced by oxidative stress following menadione treatment (Lehmann *et al.*, 2009) and it is tempting to suggest that senescence-associated oxidative stress triggers the ETC to contribute towards the energetic demands of cellular catabolism. Indeed, phytol and branched chain amino acid degradation leads to the formation of isovaleryl-CoA which can be oxidized by isovaleryl dehydrogenase (IVDH)

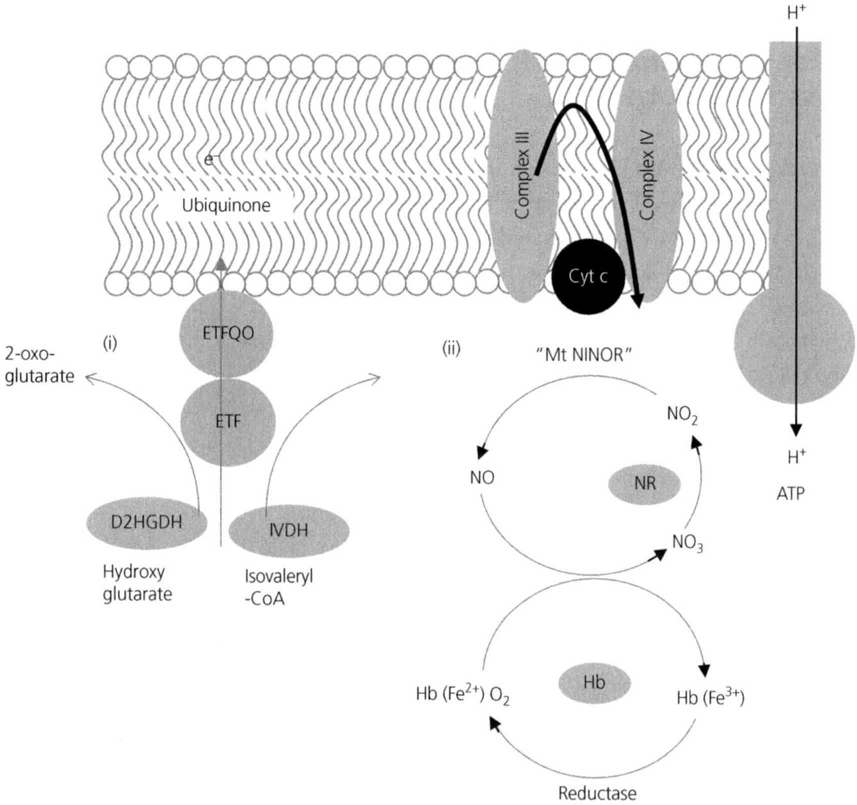

Figure 1.2 Alternative ATP generating mechanisms via operation of ETF/ETFQO and hemoglobin nitric oxide cycle.

leading to a transfer of electrons to the ETF/ETFQO system (Araújo *et al.*, 2010). Similarly, hydroxyglutarate formed via lysine degradation is oxidized by 2-hydroxyglutarate dehydrogenase (D2HGDH) to 2-oxo-glutarate (2-OG) to transfer electrons to ETF (Figure 1.2). The relative importance of each pathway in ETF/ETFQO expression has been investigated using knockout mutants of IVDH and D2HGDH, both enzymes being encoded by single genes (Araújo *et al.*, 2010). Comparing continuous light (24 h light), short day (8 h light/16 h dark) and in cold conditions (13 °C, 16 h light/8 h dark) *ivdh-1* plants exhibited a clearer accelerated senescence than the *d2hgdh1–2* plants. This finding suggests that IVDH is more likely to control the provision of electrons to the ETF/ETFQO complex than D2HGDH. Lysine was found to accumulate in both mutants, implying that this amino acid accumulation is important to flux the electrons through the ETF/ETFQO complex.

Deploying electron dissipatory mechanisms whilst maintaining ATP production under stress situations

Stress imposes certain conditions in plants during which ETC components can become over-reduced to produce ROS and electron dissipation becomes vital but ATP production is still required for energy requirement. Due to the situation of O_2 limitation, hypoxia represents a fascinating interplay between aerobic and anaerobic respiratory metabolism that is discussed here.

Hypoxia is one of the barriers for respiration in bulky tissues (Rolletschek *et al.*, 2003; Rolletschek *et al.*, 2005a, 2005b, 2007) but also plants experience hypoxia that lead to alterations in respiration, for instance during the period of flooding or waterlogging. O_2 depletion occurs where respiration dominates over O_2 availability that result in the depletion of ATP (Zabalza *et al.*, 2009; van Dongen *et al.*, 2009). One major structural change that occurs in certain plant roots is the formation of aerenchyma (Drew *et al.*, 2000). However, this is mediated by ethylene whose biosynthesis is dependent on O_2-requiring ACC oxidase so aerenchyma tend to form only under hypoxic conditions (He *et al.*, 1996). In anoxic conditions aerenchyma formation might takes place only with active photosynthesis which can transfer O_2 to the roots. The metabolic adjustment to low oxygen includes the down-regulation of energy–consuming metabolic pathways (Geigenberger, 2003; van Dongen *et al.*, 2011 that include the down-regulation of storage carbohydrate metabolism (Geigenberger *et al.*, 2000), the metabolic shift from invertase to sucrose synthase pathway (Bologa *et al.*, 2003; Huang *et al.*, 2008), and the inhibition of mitochondrial respiration at near low oxygen to utilize available oxygen for longer time (Gupta *et al.*, 2009; Zabalza *et al.*, 2009). Downregulation of energy inefficient pathways such as AOX pathway also takes place at low oxygen which is a part of the plant survival strategy. When the O_2 concentration decreases below the level of operation of oxidative phosphorylation, plant cells follow various alternative strategies to produce ATP. These include the operation of glycolytic pathway (even in low oxygen situations), which produces two ATP and two pyruvate molecules per unit of hexose utilizing while concomitantly reducing NAD^+ to NADH. However, for the glycolytic pathway to operate NAD^+ must be continuously regenerated from NADH via fermentative pathways. By using pyruvate as substrate, fermentative metabolism either produces ethanol via pyruvate decarboxylase (*Pdc*) and alcohol dehydrogenase (*Adh*) or lactate via lactate dehydrogenase (Tadege *et al.*, 1999). It seems likely that these pathways play role in hypoxic survival as both that *Pdc* and *Adh* are strongly induced in response to this stress (Rahman *et al.*, 2001; Kürsteiner *et al.*, 2003). However lactate and ethanol are potentially cytotoxic, if produced in high concentrations (Figure 1.3).

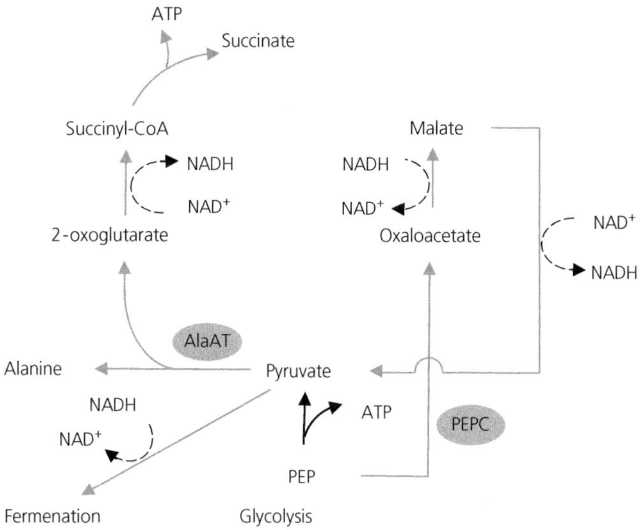

Figure 1.3 Reconfigured TCA metabolism during hypoxia via alanine aminotranferase.

Another important chemical induced at low oxygen is nitric oxide (Planchet *et al.*, 2005). NO production by mitochondria leads to NAD(P)H consumption and the generation of a limited amount of ATP under anoxic conditions (Stiomenova *et al.*, 2007), via operation of haemoglobin (Hb)-NO cycle (Figure 1.2). Non-symbiotic Hbs ($NO+O_2 \rightarrow NO_3^-$) have a high affinity for oxygen; over two orders of magnitude lower than that of COX which allows a limited respiration at very low Po_2. NO oxidation by Hb results in the formation of oxidized ferric metHb [$Hb(Fe^{3+})$] and so the reaction is ($Hb(Fe^{2+})O_2+NO+\rightarrow Hb(Fe^{3+})+NO_3^-$). The Hb is then reduced to its ferrous form [$Hb(Fe^{2+})$] by an associated reductase. NO_3^- is reduced to NO_2^- by nitrate reductase ($NO_3^-+NAD(P)H\rightarrow NO_2^-+NAD(P)^+ + OH^-$) and NO_2^- is reduced back to NO by mitochondrial nitrite NO-reductase activity (Mt NINOR) at complex III and cytochrome *c* oxidase ($2NO_2^-+H^+ + NAD(P)H\rightarrow 2NO+NAD(P)^+ + 2OH^-$) donates electrons to the ETC and also restarts the cycle (Igamberdiev and Hill, 2009). Crucially, the Hb-Mt-NINOR cycle only comes into play when the O_2 concentration falls below 2 μM and so appears to be particularly tailored to confer tolerance during anoxic conditions (Gupta *et al.*, 2005).

Alanine is a metabolite that accumulates at high concentrations at low Po_2 (de Sousa and Sodek, 2003) and under hypoxia, alanine comprised 50% of the soluble amino acid fraction of excised rice roots representing 1.2% of the root dry weight (Reggiani *et al.*, 1988). Recent ^{15}N labelling experiments suggested that while N uptake was reduced, amino acid metabolism was redirected towards alanine and γ-aminobutyric acid synthesis (Oliveira and Sodek, 2013). This substantial production of alanine is driven by alanine aminotransferase (AlaAT) (EC 2.6.1.2) which catalyses the reaction between pyruvate and glutamate to form alanine and 2-oxoglutarate (2-OG). In *Arabidopsis* there are two sequences that code for the AlaAT; with AlaAT-1 likely targeted to the cytosol and AlaAT-2 to the mitochondria (Liepman and Olsen, 2003). Two subclasses of AlaAT have been extensively characterized in soybean plants that were exposed to hypoxic stress with different nitrogen sources. Semi-quantitative PCR expression analysis showed that AlaAT were highly expressed in hypoxic roots and nodules. Reoxygenation caused a decrease in transcript and alanine content without altering the activity of enzyme, possibly suggesting an allosteric control mechanism operating under such conditions. Under NH_4^+ nutrition, the transcript abundance and enzyme activities were found to be higher in comparison to NO_3^- nutrition (Rocha *et al.*, 2010a, 2010b). Further, by using AlaAT T-DNA knockout plants, it was demonstrated that alanine production does not purely depend on these enzymes (Miyashita *et al.*, 2007), and that alanine can also be made by γ-aminobutyric acid transaminase (GABA-T) using pyruvate as co-substrate (Miyashita and Good, 2008). Obviously, the next central question would be on the role of alanine in hypoxia. The active transport of the accumulated alanine to the shoot via the xylem after the flooding period suggests that the recycling of alanine takes place after flooding. This may improve carbon and nitrogen distribution after flooding, conferring faster recovery of the plant (de Sousa and Sodek, 2003). Drew (1997) suggested that the accumulated alanine could improve energy-producing efficiency via the glycolytic flux, thereby assisting plant survival during hypoxic conditions. However, this argument is defeated by the fact that AlaAT-mediated alanine production is not coupled to NAD(P)H to regenerate NAD^+, as is the case with such fermentative pathways. An alternative suggestion that alanine accumulation might serve to buffer the pH in anoxic cells was made by Reggiani (1988). However, the most obvious metabolic role for alanine accumulation is the prevention of excess pyruvate accumulation which could impact on AOX activity (Zabalza *et al.*, 2009). In the absence of AlaAT activity, a pyruvate-driven increase in respiration could deplete internal O_2, instead of the required decrease in O_2 consumption needed for short-term plant survival (Gupta *et al.*, 2009). Therefore, alanine accumulation serves as an indirect survival strategy evolved by plant cells as a response to hypoxic stress (Rocha *et al.*, 2010a, 2010b)

Conclusions

To conclude, this chapter provides an overview, illustrating the functional flexibility of classical and alternative respiratory metabolism that coordinate with high precision to maintain ATP generation under a range of situations that could otherwise lead to an over-reduction in ETC components, and more so during hypoxia. As such it is clear that understanding the pathways and their interactions during various environmental conditions is an essential prerequisite to any appreciation of plant physiology and, thus, topics such as crop yield. The chapters in this book expand many of these themes, which are fundamental to plant biology.

References

Andre, C. and Benning, C. (2007) *Arabidopsis* seedlings deficient in a plastidic pyruvate kinase are unable to utilize seed storage compounds for germination and establishment. *Plant Physiology* **145**: 1670–1680.

Andre, C., Froehlich, J.E., Moll, M.R. and Benning, C. (2007) A heteromeric plastidic pyruvate kinase complex involved in seed oil biosynthesis in *Arabidopsis*. *The Plant Cell* **19**: 2006–2022.

Araújo, W.L., Ishizaki, K., Nunes-Nesi, A., Larson, T.R., Tohge, T., Krahnert, I., *et al.* (2010) Identification of the 2-hydroxyglutarate and isovaleryl-CoA dehydrogenases as alternative electron donors linking lysine catabolism to the electron transport chain of *Arabidopsis* mitochondria. *The Plant Cell* **22**: 1549–1563.

Bailey-Serres, J. and Chang, R. (2005) Sensing and signalling in response to oxygen deprivation in plants and other organisms. *Annals of Botany* **96**: 507–518.

Bailey-Serres, J. and Voesenek, L.A.C.J. (2008) Flooding stress: acclimations and genetic diversity. *Annual Review of Plant Biology* **59**: 313–339.

Boekema, E.J. and Braun, H.-P. (2007) Supramolecular structure of the mitochondrial oxidative phosphorylation system. *Journal of Biological Chemistry* **282**: 1–4.

Bologa, K.L., Fernie, A.R., Leisse, A. *et al.* (2003) A bypass of sucrose synthase leads to low internal oxygen and impaired metabolic performance in growing potato tubers. *Plant Physiology* **132**: 2058–2072.

Borecký, J., Maia, I.G., Costa, A.D., Jezek, P., Chaimovich, H., de Andrade, P.B., Vercesi, A.E. and Arruda, P. (2001) Functional reconstitution of *Arabidopsis thaliana* plant uncoupling mitochondrial protein (AtPUMP1) expressed in *Escherichia coli*. *FEBS Letters* **505**(2): 240–244.

Begcy, K., Mariano, E.D., Mattiello, L. *et al.* (2011) An *Arabidopsis* mitochondrial uncoupling protein confers tolerance to drought and salt stress in transgenic tobacco plants. *PLoS ONE* **6**: e23776.

Buchanan-Wollaston, V. (1997) The molecular biology of leaf senescence. *Journal of Experimental Botany* **48**: 181–199.

Chaudhuri, M. and Hill, G.C. (1996) Cloning, sequencing, and functional activity of the Trypanosoma brucei brucei alternative oxidase. *Molecular and Biochemical Parasitology* **83**: 125–129.

Considine, M.J., Holtzapffel, R.C., Day, D.A. *et al.* (2002) Molecular distinction between alternative oxidase from monocots and dicots. *Plant Physiology* **129**: 949–953.

Crane, F.L. and Beinert, H. (1956) On the mechanism of dehydrogenation of fatty acyl derivatives of coenzyme A. *Journal of Biological Chemistry* **218**: 701–716.

Day, D.A., Millar, A.H., Wiskich, J.T. and Whelan, J. (1994) Regulation of alternative oxidase activity by pyruvate in soybean mitochondria. *Plant Physiology* **106**: 1421–1427.

Dodd, A.N., Kudla, J. and Sanders, D. (2010) The language of calcium signaling. *Annual Review of Plant Biology* **61**: 593–620.

van Dongen, J.T., Fröhlich, A., Ramírez-Aguilar, S.J. *et al.* (2009) Transcript and metabolite profiling of the adaptive response to mild decreases in oxygen concentration in the roots of *Arabidopsis* plants. *Annals of Botany* **103**: 269–280.

van Dongen, J.T., Gupta, K.J., Ramírez-Aguilar, S.J. *et al.* (2011) Regulation of respiration in plants: A role for alternative metabolic pathways. *Journal of Plant Physiology* **168**: 1434–1443.

van Dongen, J.T., Roeb, G.W., Dautzenberg, M. *et al.* (2004) Phloem import and storage metabolism are highly coordinated by the low oxygen concentrations within developing wheat seeds. *Plant Physiology* **135**: 1809–1821.

Drew, M.C. (1997) Oxygen deficiency and root metabolism: Injury and acclimation under hypoxia and anoxia. *Annual Review of Plant Physiology and Plant Molecular Biology* **48**: 223–250.

Drew, M.C., He, C.-J. and Morgan, P.W. (2000) Programmed cell death and aerenchyma formation in roots. *Trends in Plant Science* **5**: 123–127.

Eubel, H., Jänsch, L. and Braun, H.-P. (2003) New insights into the respiratory chain of plant mitochondria. Supercomplexes and a unique composition of complex II. *Plant Physiology* **133**: 274–286.

Fe, F. (1988) Acyl-CoA dehydrogenases, electron transfer flavoprotein and electron transfer flavoprotein dehydrogenase. *Biochemical Society Transactions* **16**: 416–418.

Felle, H.H. (2005) pH Regulation in anoxic plants. *Annals of Botany* **96**: 519–532.

Fernie, A.R., Carrari, F. and Sweetlove, L.J. (2004) Respiratory metabolism: glycolysis, the TCA cycle and mitochondrial electron transport. *Current Opinion in Plant Biology* **7**: 254–261.

Frerman, F.E. (1987) Reaction of electron-transfer flavoprotein ubiquinone oxidoreductase with the mitochondrial respiratory chain. *Biochimica et Biophysica Acta – Bioenergetics* **893**: 161–169.

Frerman, F.E. and Goodman, S.I. (2001) Defects of electron transfer flavoprotein and electron transfer flavoprotein-ubiquinone oxidoreductase: glutaric acidemia type II. In *The Metabolic and Molecular Bases of Inherited Disease* Vol 2 (C.R. Scriver, W.S. Sly, B. Childs *et al.*, eds), pp. 2357–2365. McGraw-Hill, New York.

Geigenberger, P. (2003) Response of plant metabolism to too little oxygen. *Current Opinion in Plant Biology* **6**: 247–256.

Geigenberger, P., Fernie, A.R., Gibon, Y. *et al.* (2000) Metabolic activity decreases as an adaptive response to low internal oxygen in growing potato tubers. *Biological Chemistry* **381**: 723–740.

Giegé, P., Heazlewood, J.L., Roessner-Tunali, U. *et al.* (2003) Enzymes of glycolysis are functionally associated with the mitochondrion in *Arabidopsis* cells. *The Plant Cell* **15**: 2140–2151.

Givan, C.V. (1999) Evolving concepts in plant glycolysis: Two centuries of progress. *Biological Reviews* **74**: 277–309.

Graham, J.W.A., Williams, T.C.R., Morgan, M. *et al.* (2007) Glycolytic enzymes associate dynamically with mitochondria in response to respiratory demand and support substrate channelling. *The Plant Cell* **19**: 3723–3738.

Grodzinski, B., Jiao, J., Knowles, V.L. and Plaxton, W.C. (1999) Photosynthesis and carbon partitioning in transgenic tobacco plants deficient in leaf cytosolic pyruvate kinase. *Plant Physiology* **120**: 887–896.

Gupta, K.J., Zabalza, A. and Van Dongen, J.T. (2009) Regulation of respiration when the oxygen availability changes. *Physiologia Plantarum* **137**: 383–391.

He, C., Finlayson, S.A., Drew, M.C. *et al.* (1996) Ethylene biosynthesis during aerenchyma formation in roots of maize subjected to mechanical impedance and hypoxia. *Plant Physiology* **112**: 1679–1685.

Heazlewood, J.L., Tonti-Filippini, J.S., Gout, A.M. *et al.* (2004) Experimental analysis of the *Arabidopsis* mitochondrial proteome highlights signaling and regulatory components, provides assessment of targeting prediction programs, and indicates plant-specific mitochondrial proteins. *The Plant Cell* **16**: 241–256.

Huang, S., Colmer, T.D. and Millar, A.H. (2008) Does anoxia tolerance involve altering the energy currency towards PPi? *Trends in Plant Science* **13**: 221–227.

Igamberdiev, A.U. and Hill, R.D. (2009) Plant mitochondrial function during anaerobiosis. *Annals of Botany* **103**: 259–268.

Ishizaki, K., Larson, T.R., Schauer, N. *et al.* (2005) The critical role of *Arabidopsis* electron-transfer flavoprotein:ubiquinone oxidoreductase during dark-induced starvation. *The Plant Cell* **17**: 2587–2600.

Ishizaki, K., Schauer, N., Larson, T.R. *et al.* (2006) The mitochondrial electron transfer flavoprotein complex is essential for survival of *Arabidopsis* in extended darkness. *The Plant Journal* **47**: 751–760.

Ježek, P., Hanuš, J., Semrad, C. and Garlid, K.D. (1996) Photoactivated azido fatty acid irreversibly inhibits anion and proton transport through the mitochondrial uncoupling protein. *Journal of Biological Chemistry* **271**: 6199–6205.

Karpova, O.V., Kuzmin, E.V., Elthon, T.E. and Newton, K.J. (2002) Differential expression of alternative oxidase genes in maize mitochondrial mutants. *The Plant Cell* **14**: 3271–3284.

Knowles, V.L., McHugh, S.G., Hu, Z. *et al.* (1998) Altered growth of transgenic tobacco lacking leaf cytosolic pyruvate kinase. *Plant Physiology* **116**: 45–51.

Krauss, S., Zhang, C.Y. and Lowell, B.B. (2005) The mitochondrial uncoupling-protein homologues. *Nature Reviews: Molecular Cell Biology* **6**: 248–261.

Kürsteiner, O., Dupuis, I. and Kuhlemeier, C. (2003) The pyruvate decarboxylase1 gene of *Arabidopsis* is required during anoxia but not other environmental stresses. *Plant Physiology* **132**: 968–978.

Lambers, H. (1982) Cyanide-resistant respiration: A non-phosphorylating electron transport pathway acting as an energy overflow. *Physiologia Plantarum* **55**: 478–485.

Lehmann, M., Schwarzländer, M., Obata, T. *et al.* (2009) The metabolic response of *Arabidopsis* roots to oxidative stress is distinct from that of heterotrophic cells in culture and highlights a complex relationship between the levels of transcripts, metabolites, and flux. *Molecular Plant* **2**: 390–406.

Liepman, A.H. and Olsen, L.J. (2003) Alanine aminotransferase homologs catalyze the glutamate:glyoxylate aminotransferase reaction in peroxisomes of *Arabidopsis*. *Plant Physiology* **131**: 215–227.

Mackenzie, S. and McIntosh, L. (1999) Higher plant mitochondria. *The Plant Cell* **11**: 571–585.

Maxwell, D.P., Nickels, R. and McIntosh, L. (2002) Evidence of mitochondrial involvement in the transduction of signals required for the induction of genes associated with pathogen attack and senescence. *The Plant Journal* **29**: 269–279.

McDonald, A.E. (2008) Alternative oxidase: an inter-kingdom perspective on the function and regulation of this broadly distributed 'cyanide-resistant' terminal oxidase. *Functional Plant Biology* **35**: 535–552.

McDonald, A.E. and Vanlerberghe, G.C. (2006) Origins, evolutionary history, and taxonomic distribution of alternative oxidase and plastoquinol terminal oxidase. *Comparative Biochemistry and Physiology Part D: Genomics and Proteomics* **1**: 357–364.

McKean, M.C., Beckmann, J.D. and Frerman, F.E. (1983) Subunit structure of electron transfer flavoprotein. *Journal of Biological Chemistry* **258**: 1866–1870.

Michalecka, A.M., Svensson, Å.S., Johansson, F.I. *et al.* (2003) *Arabidopsis* genes encoding mitochondrial type II NAD(P)H dehydrogenases have different evolutionary origin and show distinct responses to light. *Plant Physiology* **133**: 642–652.

Navet R., Douette P., Puttine-Marique F. *et al.* (2005) Regulation of uncoupling protein activity in phosphorylating potato tuber mitochondria. *FEBS Letters* **579**(20): 4437–4442.

Millar, A.H., Mittova, V., Kiddle, G. *et al.* (2003) Control of ascorbate acid synthesis by respiration and its implications for stress responses. *Plant Physiology* **133**: 443–447.

Millar, A.H., Wiskich, J.T., Whelan, J. and Day, D.A. (1993) Organic acid activation of the alterNatlve oxidase of plant mitochondria. *FEBS Letters* **329**: 259–262.

Miyashita, Y., and Good, A.G. (2008) Contribution of the GABA shunt in hypoxia-induced alanine accumulation in roots of *Arabidopsis thaliana*. *Plant and Cell Physiology* **49**: 92–102.

Miyashita, Y., Dolferus, R., Ismond, K.P. and Good, A.G. (2007) Alanine aminotransferase catalyses the breakdown of alanine after hypoxia in *Arabidopsis thaliana. The Plant Journal* **49**: 1108–1121.

Nunes-Nesi, A., Carrari, F., Gibon, Y. *et al.* (2007) Deficiency of mitochondrial fumarase activity in tomato plants impairs photosynthesis via an effect on stomatal function. *The Plant Journal* **50**: 1093–1106.

Nunes-Nesi, A., Carrari, F., Lytovchenko, A. *et al.* (2005) Enhanced photosynthetic performance and growth as a consequence of decreasing mitochondrial malate dehydrogenase activity in transgenic tomato plants. *Plant Physiology* **137**: 611–622.

Oliveia, H.C. and Sodek, L. (2013) Effect of oxygen deficiency on nitrogen assimilation and amino acid metabolism by soybean root segments. *Amino Acids* **44**: 743–755.

Oliver, S.N., Lunn, J.E., Urbanczyk-Wochniak, E. *et al.* (2008) Decreased expression of cytosolic pyruvate kinase in potato tubers leads to a decline in pyruvate resulting in an *in vivo* repression of the alternative oxidase. *Plant Physiology* **148**: 1640–1654.

Pilkis, S.J. and Granner, D.K. (1992) Molecular physiology of the regulation of hepatic gluconeogenesis and glycolysis. *Annual Review of Physiology* **54**: 885–909.

Planchet, E., Sonoda, M., Gupta, K.J. and Kaiser, W.M. (2005) Nitric oxide emission from tobacco leaves and cell suspensions: rate limiting factors and evidence for the involvement of mitochondrial electron transport. *The Plant Journal* **41**: 732–743.

Plaxton, W.C. (1996) The organization and regulation of plant glycolysis. *Annual Review of Plant Physiology and Plant Molecular Biology* **47**: 185–214.

Plaxton, W.C. and Podesta, F.E. (2006) The functional organization and control of plant respiration. *Critical Review of Plant Science* **25**: 159–198.

Podesta, F.E. and Plaxton, W.C. (1991) Kinetic and regulatory properties of cytosolic pyruvate kinase from germinating castor oil seeds. *Biochemical Journal* **279**: 495–501.

Puntarulo, S. and Cederbaum, A.I. (1988) Effect of oxygen concentration on microsomal oxidation of ethanol and generation of oxygen radicals. *Biochemical Journal* **251**: 787–794.

Rahman, M., Grover, A., Peacock, W.J. *et al.* (2001) Effects of manipulation of pyruvate decarboxylase and alcohol dehydrogenase levels on the submergence tolerance of rice. *Functional Plant Biology* **28**: 1231–1241.

Ramírez-Aguilar, S.J., Keuthe, M., Rocha, M. *et al.* (2011) The composition of plant mitochondrial super complexes changes with oxygen availability. *Journal of Biological Chemistry* **286**: 43045–43053.

Rasmusson, A.G., Geisler, D.A. and Møller, I.M. (2008) The multiplicity of dehydrogenases in the electron transport chain of plant mitochondria. *Mitochondrion* **8**: 47–60.

Rasmusson, A.G. and Møller, I.M. (1991) NAD(P)H dehydrogenases on the inner surface of the inner mitochondrial membrane studied using inside-out submitochondrial particles. *Physiologia Plantarum* **83**: 357–365.

Reggiani, R., Cantú, C.A., Brambilla, I. and Bertani, A. (1988) Accumulation and interconversion of amino acids in rice roots under anoxia. *Plant Cell Physiology* **29**: 981–987.

Rial, E., Poustie, A. and Nicholls, D.G. (1983) Brown-adipose-tissue mitochondria: The regulation of the 32000-Mr uncoupling protein by fatty acids and purine nucleotides. *European Journal Biochemistry* **137**: 197–203.

Ribas-Carbo, M., Berry, J.A., Yakir, D. *et al.* (1995) Electron partitioning between the cytochrome and alternative pathways in plant mitochondria. *Plant Physiology* **109**: 829–837.

Roberts, D.L., Frerman, F.E. and Kim, J.-J.P. (1996) Three-dimensional structure of human electron transfer flavoprotein to 2.1-Å resolution. *PNAS* **93**: 14355–14360.

Rolletschek, H., Borisjuk, L., Koschorreck, M. *et al.* (2002) Legume embryos develop in a hypoxic environment. *Journal of Experimental Botany* **53**: 1099–1107.

Rolletschek, H., Borisjuk, L., Sánchez-García, A. *et al.* (2007) Temperature-dependent endogenous oxygen concentration regulates microsomal oleate desaturase in developing sunflower seeds. *Journal of Experimental Botany* **58**: 3171–3181.

Rolletschek, H., Koch, K., Wobus, U. and Borisjuk, L. (2005a) Positional cues for the starch/lipid balance in maize kernels and resource partitioning to the embryo. *The Plant Journal* **42**: 69–83.

Rolletschek, H., Radchuk, R., Klukas, C. *et al.* (2005b) Evidence of a key role for photosynthetic oxygen release in oil storage in developing soybean seeds. *New Phytologist* **167**: 777–786.
Rolletschek, H., Weber, H. and Borisjuk, L. (2003) Energy status and its control on embryogenesis of legumes. Embryo photosynthesis contributes to oxygen supply and is coupled to biosynthetic fluxes. *Plant Physiology* **132**: 1196–1206.
Rocha, M., Licausi, F., Araujo, W. L. *et al.* (2010a). Glycolysis and the tricarboxylic acid cycle are linked by alanine aminotransferase during hypoxia induced by waterlogging of *Lotus japonicus*. *Plant Physiology* **152**(3): 1501–1513. DOI:10.1104/pp.109.150045.
Rocha, M., Sodek, L., Licausi, F. *et al.* (2010b). Analysis of alanine aminotransferase in various organs of soybean (*Glycine max*) and in dependence of different nitrogen fertilisers during hypoxic stress. *Amino Acids* **39**(4): 1043–1053. DOI:10.1007/s00726-010-0596-1.
Schmälzlin, E., van Dongen, J.T., Klimant, I. *et al.* (2005) An optical multifrequency phase-modulation method using microbeads for measuring intracellular oxygen concentrations in plants. *Biophysical Journal* **89**: 1339–1345.
Skutnik, M. and Rychter, A.M. (2009) Differential response of antioxidant systems in leaves and roots of barley subjected to anoxia and post-anoxia. *Journal of Plant Physiology* **166**: 926–937.
Smith, A.M., Ratcliffe, R.G. and Sweetlove, L.J. (2004) Activation and function of mitochondrial uncoupling protein in plants. *Journal of Biological Chemistry* **279**(50): 51944–51952.
de Sousa, C.A.F. and Sodek, L. (2002) The metabolic response of plants to oxygen deficiency. *Brazilian Journal of Plant Physiology* **14**: 83–94.
de Sousa, C.A.F. and Sodek, L. (2003) Alanine metabolism and alanine aminotransferase activity in soybean (*Glycine max*) during hypoxia of the root system and subsequent return to normoxia. *Environmental and Experimental Botany* **50**: 1–8.
Stoimenova, M., Igamberdiev, A.U., Gupta, K.J. *et al.* (2007) Nitrite-driven anaerobic ATP synthesis in barley and rice root mitochondria. *Planta* **226**: 465–474.
Studart-Guimarães, C., Fait, A., Nunes-Nesi, A. *et al.* (2007) Reduced expression of succinyl-coenzyme a ligase can be compensated for by up-regulation of the γ-aminobutyrate shunt in illuminated tomato leaves. *Plant Physiology* **145**: 626–639.
Sweetlove, L.J., Fait, A., Nunes-Nesi, A. *et al.* (2007) The mitochondrion: an integration point of cellular metabolism and signalling. *Critical Reviews in Plant Sciences* **26**: 17–43.
Tadege, M., Dupuis, I. and Kuhlemeier, C. (1999) Ethanolic fermentation: new functions for an old pathway. *Trends in Plant Science* **4**: 320–325.
Teusink, B., Passarge, J., Reijenga, C.A. *et al.* (2000) Can yeast glycolysis be understood in terms of *in vitro* kinetics of the constituent enzymes? Testing biochemistry. *European Journal of Biochemistry* **267**: 5313–5329.
Tielens, A.G. and Van Hellemond, J.J. (1998) The electron transport chain in anaerobically functioning eukaryotes. *Biochimica et Biophysica Acta – Bioenergetics* **1365**: 71–78.
Turner, W.L., Knowles, V.L. and Plaxton, W.C. (2005) Cytosolic pyruvate kinase: Subunit composition, activity, and amount in developing castor and soybean seeds, and biochemical characterization of the purified castor seed enzyme. *Planta* **222**: 1051–1062.
Umbach, A.L., Ng, V.S. and Siedow, J.N. (2006) Regulation of plant alternative oxidase activity: A tale of two cysteines. *Biochimica et Biophysica Acta — Bioenergetics* **1757**: 135–142.
Vanlerberghe, G.C., and McIntosh, L. (1997) Alternative oxidase: From gene to function. *Annual Review of Plant Physiology and Plant Molecular Biology* **48**: 703–734.
Vanlerberghe, G.C., Yip, J. and Parsons, H.L. (1999) *In organello* and *in vivo* evidence of the importance of the regulatory sulfhydryl/disulfide system and pyruvate for alternative oxidase activity in tobacco. *Plant Physiology* **121**: 793–803.
Vercesi, A.E., Borecky, J., Maia Ide, G. *et al.* (2006) Plant uncoupling mitochondrial proteins. *Annual Review of Plant Biology* **57**: 383–404.
Vercesi, A.E., Martins, I.S., Silva, M.A.P. *et al.* (1995) PUMPing plants. *Nature* **375**: 24.
Zabalza, A., van Dongen, J.T., Froehlich, A. *et al.* (2009) Regulation of respiration and fermentation to control the plant internal oxygen concentration. *Plant Physiology* **149**: 1087–1098.

CHAPTER 2

Non-coupled pathways of plant mitochondrial electron transport and the maintenance of photorespiratory flux

Abir U. Igamberdiev[1] and Natalia V. Bykova[2]
[1] *Department of Biology, Memorial University of Newfoundland, St. John's, Newfoundland and Labrador, Canada*
[2] *Cereal Research Centre, Agriculture and Agri-Food Canada, Morden, MB, Canada*

Introduction: Carbon fluxes through plant mitochondria in the light

Mitochondria of C_3 plant leaves in the light support oxidation of the high fluxes of photorespiratory glycine, which is synthesized in the glycolate pathway as a consequence of the oxygenase reaction of ribulose-1,5-bisphosphate carboxylase/oxygenase (Rubisco), while the oxidation of respiratory substrates, although partially inhibited, proceeds simultaneously with the photorespiratory glycine oxidation (Gardeström, Igamberdiev and Raghavendra, 2002). In the absence of photorespiration under saturating light conditions, the intensity of photosynthesis can reach values up to ~5000 nmol (O_2 evolved or CO_2 consumed) mg^{-1} (chlorophyll (Chl)) min^{-1} at 25 °C (calculated from Edwards and Walker, 1983). The rate of respiration in the darkness is about 10 times lower than the photosynthetic rate in the post-illumination period and further 2.5 times lower during the prolonged darkness (Byrd *et al.*, 1992). This allows estimation of respiration rates at ~500 and 200 nmol (O_2 consumed) mg^{-1} (Chl) min^{-1} correspondingly in post-illumination period and in prolonged darkness. Taking the mitochondrial volume of 4 μl mg^{-1} Chl (with very little variation for barley, spinach and potato as determined by Winter *et al.*, 1993, 1994; Leidreiter *et al.*, 1995), the maximum respiratory rates of 120 nmol (O_2 consumed) mg^{-1} (mitochondrial protein) min^{-1} after illumination and near 50 nmol mg^{-1} min^{-1} during prolonged darkness can be calculated approximately. In the light, respiration is usually inhibited and its rate will be lower but there is a controversy in the literature about the rate of this inhibition (Atkin *et al.*, 1998;

Alternative Respiratory Pathways in Higher Plants, First Edition.
Edited by Kapuganti Jagadis Gupta, Luis A.J. Mur and Bhagyalakshmi Neelwarne.

Pinelli and Loreto, 2003; Tcherkez *et al.*, 2005). An average rate value of 50% compared to that in the dark can be estimated (Atkin *et al.*, 2000a, 2000b).

Photorespiratory flux depends on CO_2 concentration near the centres of carboxylation. This parameter cannot be estimated precisely because of its dynamic nature. The depletion of CO_2 at the Rubisco site is balanced by its photorespiratory release, therefore the CO_2 concentration oscillates in photosynthetic cells (Roussel *et al.*, 2007; Roussel and Igamberdiev, 2011). The depletion of CO_2 in the proximity of Rubisco is avoided through a CO_2 active supply from the bicarbonate pool and through CO_2 pumping from the thylakoid lumen due to the carbonic anhydrase activity of photosystem II (Igamberdiev and Roussel, 2012). At current atmospheric CO_2 levels (approaching 400 ppm), the average apparent concentration is approximately 200 ppm in the chloroplast stroma and may decrease under drought or in xeromorphic plants down to 150–90 ppm (Di Marco *et al.*, 1990), i.e. to the values closer to the compensation point. The ratio of rates of oxygenation and carboxylation of ribulose-1,5-bisphosphate (V_o/V_c) is estimated as 20–25% of the assimilation rate at 400–300 ppm CO_2 increasing up to 50% under CO_2 depletion (glacial CO_2 levels or closing stomata under drought) (Sharkey, 1988). The latter, according to the stoichiometry of photorespiratory pathway, will correspond to the value of 1200 nmol (CO_2 evolved) mg^{-1} (mitochondrial protein) min^{-1} or 600 nmol O_2 consumed $mg^{-1} min^{-1}$ in the glycine decarboxylase reaction (assuming that all NADH is oxidized in the electron transport chain). This does not include O_2 consumption in the Rubisco oxygenase and glycolate oxidase reactions. It is possible that the oxygenation rate of Rubisco is even higher (André, 2011a, 2011b) resulting correspondingly in higher rates of glycine oxidation.

The glycine decarboxylase complex (GDC) contains four component proteins (P, H, T and L) with a stoichiometry 2P:27H:9T:1L and has a total molecular mass of 1.3 MDa (Oliver, 1994). Together with serine hydroxymethyltransferase (SHMT), GDC converts glycine to serine with the concomitant release of CO_2 and NH_3 (Figure 2.1). The GDC protein components are encoded in nuclear genes and expressed with N-terminal presequences targeting them to the mitochondria. The expression of genes encoding GDC proteins is controlled in a similar way as that of the small subunit of Rubisco (Oliver, 1994).

The concentration of GDC in the mitochondrial matrix of C_3 plant leaves is nearly 50% of the total mitochondrial protein content (Douce *et al.*, 1994) and the impairment of GDC even by 30–50% leads to the accumulation of glycine, increased susceptibility to drought and formate production (Wingler *et al.*, 1999a; Wingler *et al.*, 1999b; Heineke *et al.*, 2001), which suggests that GDC operates at a subsaturating substrate level in the conditions of the current CO_2 content in the ambient air (Bykova *et al.*, 2005). This indirectly indicates that the observed high GDC concentration is needed for the maintenance of photorespiratory flux through mitochondria, with the intensity determined by atmospheric O_2/CO_2 ratio. The GDC concentration of 0.2 mM in the leaf tissue (Douce *et al.*, 1994) means that its concentration in the mitochondrial matrix, assuming that

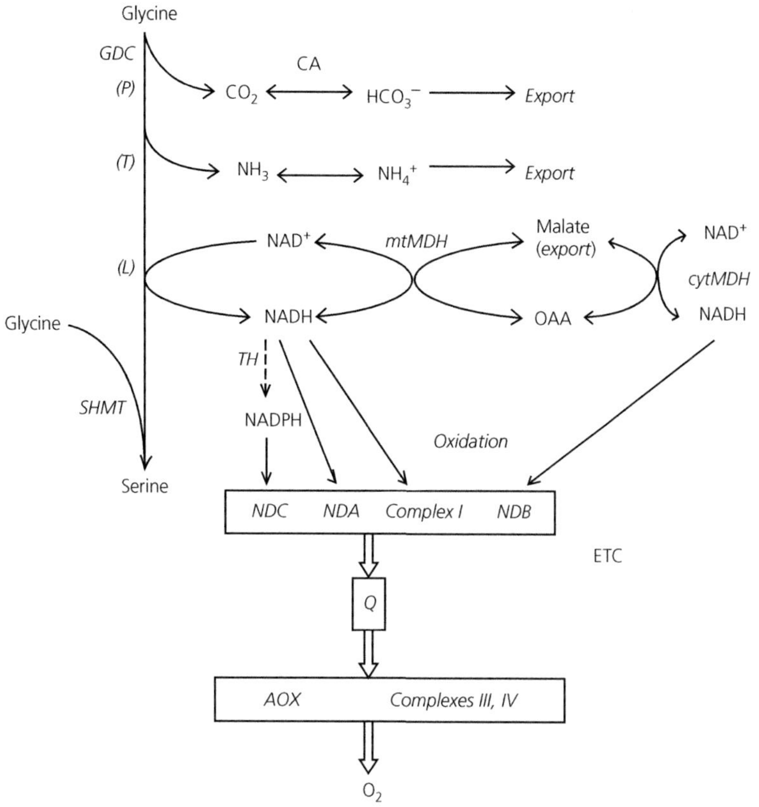

Figure 2.1 The scheme of the glycine decarboxylase complex (GDC) reactions catalysed by its different proteins, with links to metabolic processes. P-protein is involved in decarboxylation; T-protein – in release of ammonia; L-protein – in NAD^+ reduction. CO_2 is equilibrated by carbonic anhydrase (CA) with bicarbonate (HCO_3^-) which is exported to the cytosol. NH_3 is protonated to NH_4^+ which is exported and used in chloroplast. NADH is equilibrated by the mitochondrial malate dehydrogenase (mtMDH), malate is exported to the cytosol where it is equilibrated by cytosolic malate dehydrogenase (cytMDH). Cytosolic NADH can be oxidized by external rotenone-insensitive dehydrogenases (NDB), mitochondrial NADH – by complex I and internal rotenone-insensitive dehydrogenase (NDA). Mitochondrial NADPH (formed in the non-proton-pumping transhydrogenase reaction, TH) is oxidized by internal rotenone-insensitive dehydrogenase (NDC). The electrons from the ubiquinone pool are transported to O_2 either via the cytochrome pathway (complexes III and IV) or via alternative oxidase (AOX). Other abbreviations: OAA, oxaloacetate; ETC, electron transport chain; SHMT, serine hydroxymethyltransferase.

mitochondria occupy ~5% of the cell volume (Winter *et al.*, 1993, 1994), of at least 20 times higher, i.e. 4 mM, is comparable to the concentration of Rubisco in chloroplasts (Pickersgill, 1986). The link between high GDC concentration in mitochondria, equilibration of NADH concentration by malate dehydrogenase (MDH), and engagement of the non-coupled pathways of mitochondrial electron transport will be discussed later in this chapter.

Activation of glycine oxidation by malate

The rates for oxidation of glycine in isolated mitochondria are usually around 200 nmol (O_2 consumed) mg^{-1} (mitochondrial protein) min^{-1}. For isocitrate oxidation (a 'bottleneck' in the tricarboxylic acid (TCA) cycle) the rates are not higher than 50 nmol $mg^{-1} min^{-1}$ (Day and Wiskich 1977). In the presence of malate, the rate of glycine oxidation increases to 500–600 nmol (O_2 consumed) mg^{-1} (mitochondrial protein) min^{-1} (Wiskich *et al.*, 1990; Bykova *et al.*, 1998; Bykova and Møller, 2001), and it is close to the maximum photorespiratory flux possible *in vivo*.

The observed activation of glycine oxidation by malate is related to the buffering role of MDH operating at its equilibrium (Hagedorn *et al.*, 2004) and decreasing NADH concentration, which rises in the course of glycine oxidation to the levels corresponding to that equilibrium. Also, oxaloacetate (OAA) formation during malate oxidation causes the recycling of NADH formed in the GDC reaction and facilitates NADH oxidation in the mitochondrial respiratory chain (Wiskich and Dry, 1985; Wiskich *et al.*, 1990). The initial hypothesis (Wiskich *et al.*, 1990) assumes the existence of separate metabolons of MDH, one serving for oxidation of photorespiratory NADH via OAA reduction and another participating in the oxidation of malate in the TCA cycle. However, there is no experimental evidence of separate MDH pools in mitochondria. Metabolon organization of electron transport chain proteins in supercomplexes – respirasomes (Krause *et al.*, 2004; Dudkina *et al.*, 2006) – is confirmed and it is also possible for the TCA cycle enzymes (Vélot *et al.*, 1997), but there is no evidence confirming its relevance to the oxidation of photorespiratory glycine.

The model of NADH recycling from glycine oxidation was presented by Wiskich *et al.* (1990) in two versions. One involves the recycling of malate within one mitochondrion, and another assumes the existence of two mitochondrial subpopulations, oxidizing either glycine or TCA cycle substrates. Although the existence of such subpopulations has not been confirmed, leaf mitochondria are indeed not uniform across the leaf blade (Tobin *et al.*, 1989), and they are more enriched by GDC at the upper surface of the leaf, while the TCA cycle enzymes exhibit higher concentrations at the lower surface (reviewed in Igamberdiev *et al.*, 2014). This may result in the fluxes of malate between the cells. A similar mechanism exists in C_3–C_4 intermediate plants where GDC is located in the bundle sheath cells while mitochondria enriched with the TCA cycle enzymes are present in the mesophyll (Ueno *et al.*, 2003). Intercellular and intertissue operation of the malate valve is a possibility that could be investigated for a better understanding of the interactions between respiratory and photorespiratory metabolism in leaves.

Accepting the possibility of the proposed spatial separation between glycine and malate oxidations, we should however emphasize that for the effective recycling of photorespiratory NADH there is actually no need for such separation for the removal of NADH from the GDC active site. A high activity and turnover rate of the MDH reaction (exceeding that of GDC by several orders of magnitude) will result in the effective equilibration of NADH, NAD^+, malate and OAA in accordance with kinetic equilibrium of MDH and relieve the inhibition of GDC by NADH (with K_i 15 μM) (Oliver, 1994). Therefore, the kinetic compartmentation (based on a big difference in the catalytic constants of MDH and GDC) can effectively substitute for the proposed spatial compartmentation. Nevertheless, NADH concentration in the matrix tends to increase during photorespiration (Igamberdiev *et al.*, 2001a) and this will involve the non-coupled pathways of the mitochondrial electron transport, which we will discuss later.

Because of high MDH activity, malate buffers glycine oxidation in a way that immediately re-establishes MDH equilibrium after each turnover of GDC. The data obtained on the knockout single and double mutants with the high expression of mMDH1 and low expression of mMDH2, the mitochondrial isoforms of MDH, showed changes in other mitochondrial NAD-linked dehydrogenases, indicating a reorganization of these enzymes in the mitochondrial matrix (Tomaz *et al.*, 2010). The slow-growing double mutant exhibited elevated whole leaf respiration rate in the dark and light, which indicates that mMDH uses NADH to reduce oxaloacetate to malate, which is in turn then exported to the cytosol, rather than driving mitochondrial respiration. Increased respiratory rate in leaves can account in part for the low net CO_2 assimilation and slow growth rate of double mutants lacking both mitochondrial MDHs. It was also shown that the loss of mMDH affected photorespiration, as evidenced by a lower post-illumination burst, alterations in CO_2 assimilation/intercellular CO_2 curves at low CO_2, and the light-dependent elevated concentration of photorespiratory metabolites. This directly supports the role of mitochondrial MDH in the equilibration of NADH from the GDC reaction.

Oscillations of respiratory and photorespiratory fluxes

After the light is turned off, the two major oscillations are observed, one linked to oxidation of the remnant of photorespiratory glycine (post-illumination burst, PIB), and the other to oxidation of mainly malate (light-enhanced dark respiration, LEDR) (Igamberdiev *et al.*, 2001b). LEDR comes later and is preceded by PIB (in photorespiratory conditions). Therefore we observe two separate fluxes after illumination, one related to photorespiration and the other to respiration. LEDR occurs due to the inhibition of respiration by the light (Atkin *et al.*, 1998) causing accumulation of organic acids, mainly malate (Hill and Bryce, 1992; Igamberdiev *et al.*, 2001b). The lag-phase between the light turned off and the start of the PIB peak is usually considered to be between 10 and 15 s. In a CO_2-free atmosphere it is shorter being only ~4 s with the peak at 5–6 s (Laisk and Sumberg, 1994). This corresponds well to the duration of photorespiration-related CO_2 oscillations observed in tobacco leaves (Roussel *et al.*, 2007). The CO_2 evolved during PIB is partially refixed, the process being relatively slow in the dark (30–60 s) and most likely involving PEP carboxylase (Laisk and Sumberg, 1994). Oscillations of CO_2 during PIB when the leaf was previously exposed to high CO_2 (2000 ppm) were observed reflecting a balance between refixation and CO_2 release in the malic enzyme reaction (Laisk and Sumberg, 1994). These oscillations are short (1–3 s) and reflect fast exchange between the carboxylation reaction (cytosolic enzyme PEP carboxylase in the darkness) and the mitochondrial decarboxylation (via NAD-malic enzyme). In the light, refixation of CO_2 could occur mainly at the Rubisco site since its activity is much higher in C_3 plants than the activity of PEP carboxylase.

Rubisco under high light and limiting CO_2 conditions represents the single limitation of photosynthesis in the Calvin cycle (von Caemmerer, 2000). It is highly regulated by several mechanisms including carbamylation of lysine, chaperone-like activity of Rubisco activase, binding of activators and inhibitors, and positive and negative cooperativity (Andrews and Lorimer, 1987). The oxygenase reaction of Rubisco is inhibited by CO_2. As a result, a feedback should exist between photorespiratory CO_2 release in mitochondria and CO_2 assimilation in chloroplasts. The presence of an enzyme in concentrations comparable to those of a substrate and a product may represent a possible source of sustained oscillations in metabolic networks (Ryde-Pettersson, 1992). When two substrates (CO_2 and O_2 in the case of Rubisco) are competing to bind a macromolecule, and when the flow of one substrate is controlled by a feedback mechanism (photorespiration releasing CO_2), sustained oscillations are generated (Ngo and Roussel, 1997). Rubisco concentration in chloroplasts is millimolar, while CO_2 is in a micromolar range (Pickersgill, 1986). In this case, oscillations can occur due to the depletion of the substrate if a simple feedback mechanism exists. Such a feedback will result in the oscillatory phenomena occurring in the leaf system (Ivlev, 1989; Roussel *et al.*, 2007; Roussel and Igamberdiev, 2011).

Ivlev (1989) suggested a hypothesis that CO_2 assimilation has a discrete pattern. He introduced the idea that Rubisco depletes CO_2 concentration near centres of carboxylation thus initiating oxygenation, during which a part of assimilated CO_2 is released. The feedback of photorespiration to photosynthetic electron transport has been modelled by Kukushkin and Soldatova (1996) and observed experimentally in green alga *Bryopsis maxima* (Satoh and Katoh, 1983). Measurement of the internal CO_2 concentration (C_i) in tobacco leaves using a fast-response CO_2 exchange system (Roussel *et al.*, 2007) showed that in the light under conditions of high photorespiration, the C_i oscillations are observed. The oscillations have the range of C_i varying by 2–4 $\mu L\,L^{-1}$ in substomatal cavities with a period of a few seconds. The statistical properties of the time series of the observed oscillations are stationary and the attractor reconstruction shows that they exhibit a stochastic (not chaotic) behaviour. It was proposed that the CO_2

concentration in photosynthetic plant cells might be pulsed with the period in the order of a few seconds (Roussel *et al.*, 2007). This may indicate that in the mitochondria of green tissues in the light, a continuous switching occurs between the oxidation of glycine and the TCA cycle substrates.

In addition to the kinetic and possible spatial compartmentation between oxidations of glycine and malate, we can assume the existence of temporal separation between oxidation of glycine and malate. The sequence of temporal events could be as follows: glycine oxidation increases NADH concentration in mitochondria, this leads to GDC inhibition (K_i 15 μM) and activation of OAA reduction to malate, which further results in NADH consumption. In this system we should observe oscillations of glycine and malate oxidation in mitochondria. Although such oscillations have not been shown (no such experiments have been performed to date), the oscillations are often observed in biological systems in relation to changes in energy state. In pollen tubes during growth, NAD(P) oscillates with the period in the order of 10 s (Cárdenas *et al.*, 2006). Citrate concentration oscillates in isolated respiring animal mitochondria with the period of tens seconds (MacDonald *et al.*, 2003). It is possible that the switch between glycine and malate oxidation takes place *in vitro*. Generation of NADH inhibits GDC, then malate produces OAA and it is oxidized in the futile cycle. This system may be complicated by the participation of malic enzyme (which is less sensitive to high NADH) that may form pyruvate upon the increase of NADH by GDC. The oscillations can also be linked to periodic changes of pH in mitochondria and to selective engagement of non-coupled rotenone-insensitive dehydrogenases (with lower affinity to NADH than complex I) when NADH concentration transiently increases.

NADH and NADPH dehydrogenases in the mitochondrial membranes

Under conditions of low NADH production in mitochondria (usually in non-photosynthetic tissues under non-stress conditions) the operation of complex I and of the cytochrome pathway fulfils major energetic demands of the cell. In the light, when ATP is intensively formed photosynthetically in chloroplasts, the functions of mitochondria are changed and a high carbon flux through mitochondria is also provided by operation of the pathways non-coupled to ATP synthesis (Gardeström *et al.*, 2002). These pathways include rotenone-insensitive NADH and NADPH dehydrogenases (ND; type II) in the inner and outer side of the inner mitochondrial membrane and the cyanide-resistant alternative ubiquinol oxidase (AOX). The NADH dehydrogenase on the outer mitochondrial membrane may also be connected to the electron transport chain of mitochondria (Møller and Lin, 1986) but this is not yet confirmed. Another uncoupling mechanism is related to the function of the uncoupling protein (UCP), which is regulated via change of redox level and reactive oxygen species (Jezek, Costa and Vercesi, 1996). Coordinated switch of NDs, AOX and UCP may result in high flexibility in establishment of redox and energy balance in mitochondria of photosynthetic cells.

Three distinct activities of NAD(P)H dehydrogenases on the inner side of the inner mitochondrial membrane are responsible for oxidation of reducing equivalents formed in the mitochondrial matrix (Møller, 2001). The K_m of complex I to NADH is about 7 µM, the K_m of the rotenone-resistant NADH dehydrogenase (NDA) is 80 µM, while the K_m of Ca^{2+}-dependent NADPH dehydrogenase (NDC) is 25 µM (Møller, 2001). There are also two distinct activities of the external dehydrogenases (NDB), one NADH- and the other NADPH-dependent, in the inner mitochondrial membrane; both are Ca^{2+} dependent. According to the genetic data, seven genes of type II NAD(P)H dehydrogenase are found in *Arabidopsis* (Geisler *et al.*, 2007). These include four genes (*ndb*) for the NDB proteins (external Ca^{2+}-dependent NADH and NADPH dehydrogenases), two genes (*nda*) for the NDA proteins (internal Ca^{2+}-independent NADH dehydrogenase), and one gene (*ndc*) for the NDC protein (internal Ca^{2+}-dependent NADPH dehydrogenase).

Elhafez *et al.* (2006) showed that, according to the microarray data, the gene for internal rotenone-insensitive dehydrogenase *nda1* is clustered closest to the gene encoding the P-subunit of glycine decarboxylase. NDA1, NDB2, NDC and AOX were up-regulated in the light in a similar manner while NDA1 also exhibited a diurnal light-dependent regulation. Induction of both internal NADH dehydrogenases and AOX suggests that a complete non-proton-pumping respiratory chain is specifically activated in the light, accommodating the increased levels of matrix NADH generated by glycine oxidation (Svensson and Rasmusson, 2001; Rasmusson and Escobar, 2007). Photoreceptor-mediated transcriptional control of NDA1 involves an I-box flanked by two GT-1 elements localized to a 99-bp region of the *nda1* promoter, the arrangement similar to the promoters of photosynthesis-associated genes (Escobar *et al.*, 2004). The gene of internal rotenone-insensitive NADPH dehydrogenase (*ndc1*) affiliates in phylogenetic analysis with corresponding cyanobacterial genes suggesting that this gene entered the eukaryotic cell via the chloroplast progenitor (Michalecka *et al.*, 2003). The genes encoding NDA1 and NDC are shown to be activated via phytochrome A (Elhafez *et al.*, 2006). Light-activation of *nda1* and its absence for *nda2* corresponds to different localization of corresponding translated proteins (NDA1 mainly in leaves and NDA2 in non-photosynthetic tissues, e.g. roots) (Rasmusson *et al.*, 2008). While type II NADH and NADPH dehydrogenases increase their expression in the light, the expression of succinate dehydrogenase in the light is down-regulated (Popov *et al.*, 2010).

Increase of the mitochondrial capacity in the light via engagement of rotenone-insensitive dehydrogenases

Plant mitochondria contain, according to our estimations, a 1.5–2.0 mM pool of NAD and NADH and a 0.15–0.2 mM pool of NADP and NADPH (Igamberdiev *et al.*, 2001a). Determination of the reduction level of NAD and NADP gives only approximate results due to unavoidable uncertainties in experiments with cycling enzymes; however, our estimations show that under photorespiratory conditions NAD is about three times and NADP is 1.5–2 times more reduced than in non-photorespiratory conditions. In the darkness, the reduction level is even lower than in the light in non-photorespiratory conditions (Igamberdiev *et al.*, 2001a). The investigation of engagement of different dehydrogenases during glycine plus malate oxidation shows that the capacity of the electron transport chain increases at least two times or more (Igamberdiev *et al.*, 1997; Bykova *et al.*, 1998; Bykova and Møller, 2001). This can be directly related to the increase in NADH and NADPH levels during photorespiration. Glycine oxidation raises the NADH level in mitochondria more than the oxidation of other substrates. The NADH/NAD^+ ratio in mitochondria increases under photorespiratory conditions, being 0.2–0.3 compared to 0.05–0.07 under non-photorespiratory conditions (Wigge *et al.*, 1993; Igamberdiev and Gardeström, 2003). This corresponds to an NADH concentration of approximately 0.4 mM under photorespiratory conditions and 0.15 mM under non-photorespiratory conditions (Igamberdiev and Gardeström, 2003). However such concentrations will be inhibiting for GDC, which has a K_i value for NADH of 15 μM. The real concentrations of free NADH will be lower, especially in green tissues where most NADH is bound (Møller, 2001).

Under maximal rates of glycine plus malate oxidation, the near 50% rate of internal NAD(P)H oxidation is confined to complex I, almost the same rate is achieved by the rotenone-insensitive NADH dehydrogenase (NDA) and a lower activity (which can reach near 15% of complex I activity) belongs to the NADPH dehydrogenase (NDC) (Bykova *et al.*, 1998). Since the kinetic data of glycine oxidation show a high involvement of NDA, the real concentration of free NADH should rise to near saturation level for this dehydrogenase, i.e. above its K_m (80 μM). The engagement of this dehydrogenase is much lower under oxidation of the TCA cycle substrates (less than 20%) (Bykova *et al.*, 1998; Bykova and Møller, 2001), and under non-photorespiratory conditions, free NADH concentration is not higher than 20–30 μM. As in the case of other enzymes present in high concentration (Hanson and Schnell, 2008; Igamberdiev and Roussel, 2012), the values of K_m and V_{max} give only an approximate evaluation of the capacity of corresponding enzymes. More important is estimation of parameters of flux and effective delivery (e.g. via pumping) of substrates to these enzymes. In this regard, the buffering role of mitochondrial MDH, which is involved in fast equilibration of NADH and NAD^+, represents an important mechanism for efficient saturation of coupled and non-coupled dehydrogenases.

The involvement of NDC in glycine oxidation can be achieved via the mechanism of transhydrogenation between NADH and $NADP^+$ (Bykova and Møller, 2001). Plant mitochondria lack proton-translocating transhydrogenase (Bykova *et al.*, 1999) and possess two non-energy-linked transhydrogenase activities, one belonging to the side reaction of complex I and the other to soluble (possibly weakly attached to the membrane) transhydrogenase-like enzyme. This type of transhydrogenase in bacteria establishes a mass action ratio of pyridine nucleotides close to 1, which is the equilibrium value. In higher plant mitochondria *in situ* according to our estimations, the mass action ratio changes from lower levels in darkness and in high CO_2 (1–3) to higher levels in limiting CO_2 (photorespiratory conditions) (Igamberdiev and Gardeström, 2003). Besides providing engagement of the internal NADPH dehydrogenase, the increased NADPH levels inside mitochondria will facilitate reduction of glutathione (Noctor *et al.*, 2007), activate AOX (possibly via the thioredoxin system) and affect isocitrate oxidation (Igamberdiev and Gardeström, 2003). The absence of proton-translocating transhydrogenase means that the redox equilibration of pyridine nucleotides is not linked to the generation of proton potential and hence it does not contribute to ATP synthesis. The transhydrogenation reaction in plant mitochondria is coupled with highly active dehydrogenases (e.g. MDH). It can also involve the participation of NAD- and NADP-dependent isocitrate dehydrogenases (Igamberdiev and Gardeström, 2003) and the side activity (with NADP) of MDH (Scheibe and Stitt, 1988).

Summarizing the role of NADPH in the mitochondrial matrix during photorespiration, we can conclude that when the concentration of NADH increases, it enters into the transhydrogenation reaction with $NADP^+$ thus forming NADPH. The consequence of this process will be the activation of additional oxidation flow via the internal NADPH dehydrogenase of the electron transport chain. Its capacity (up to 15% of the total capacity for NAD(P)H oxidation) (Bykova *et al.*, 1999) provides an additional power to increase flux through the electron transport chain. The rise of NADPH also contributes to the activation of AOX (Vanlerberghe *et al.*, 1995), so the total flux through electron transport chain can increase even much more. It also stimulates the reverse reaction of NADP-dependent isocitrate dehydrogenase, leading to citrate efflux from mitochondria (Igamberdiev and Gardeström, 2003) and to the activation of the AOX gene (Vanlerberghe and McIntosh, 1996).

Physiological role of alternative oxidase

Since the electron transport through AOX is not coupled to ATP production, this results in a very flexible coupling between electron transport and oxidative phosphorylation. For a long time AOX was regarded as a more or less passive overflow (Lambers, 1982) or slippage (Tomashek and Brusilow, 2000) mechanism. Progress in the understanding of AOX regulation has changed this view (Ribas-Carbo *et al.*, 1997). It is now clear that AOX can play a very active role in the regulation of coupling between electron transport and oxidative phosphorylation and thus in the energy- and redox-balance in the cell. According to the estimations (Wigge *et al.*, 1993; Igamberdiev *et al.*, 2001a; Igamberdiev and Gardeström, 2003), NADH/NAD^+ and NADPH/$NADP^+$ ratios increase by approximately three times in photorespiratory conditions. The pyruvate dehydrogenase complex is inhibited by the photorespiratory ammonium and high redox level, resulting in the accumulation of pyruvate, which increases to concentration levels that activate AOX. The half-saturation of AOX is approximately 0.1 mM for pyruvate and higher for other keto acids (Millar *et al.*, 1993). Also, NADPH via the thioredoxin system activates AOX by reducing its disulfide bond, and citrate is accumulated in the light activating the AOX gene (Vanlerberghe and McIntosh, 1996). Thus AOX exists in the light in a fully activated state. In this condition it is fully regulated by the QH_2/Q ratio and by the availability of O_2, affinity to which is lower than that of cytochrome oxidase.

Inside mitochondria, besides the involvement of internal rotenone-insensitive dehydrogenase, which will increase the ubiquinone reduction level, increased NADH will displace the MDH reaction toward formation of malate. The latter will participate in malate shuttle and also enter the NAD-malic enzyme reaction, which is relatively insensitive to higher NADH levels (Pascal *et al.*, 1990). This will lead to the formation of pyruvate, which together with a higher reduction level of ubiquinone activates alternative oxidase, resulting in saturation of all paths of mitochondrial electron transport. Light induction of the internal NADH and NADPH dehydrogenases and AOX suggests that the complete non-proton-pumping respiratory chain is specifically activated in the light, accommodating the increased levels of matrix NADH generated by glycine oxidation (Svensson and Rasmusson, 2001; Escobar *et al.*, 2004; Rasmusson and Escobar, 2007).

Light-dependent regulation of AOX may also clarify the previously described stimulation of its gene expression by accumulating citrate (Vanlerberghe and McIntosh, 1996; Gupta *et al.*, 2012). According to Finnegan *et al.* (1997), the soybean AOX is encoded by a multigene family (*AOX*) with three known members. The relative abundance of *AOX2* transcripts and the corresponding AOX2 protein is light-controlled. *AOX2* has promoter regions associated with phytochrome regulation that support this observation (Thirkettle-Watts *et al.*, 2003). The activation of NDs and AOX in the light prevents further elevation of reduction level of ubiquinone, thus protecting cells from the increased formation of the superoxide anion.

Equilibration of adenylates in the intermembrane space of mitochondria

Although living systems operate far from the equilibrium, non-equilibrium fluxes should be stable, which can be achieved at certain values of metabolic rates (Igamberdiev, 1999; Igamberdiev and Kleczkowski, 2009). This can be reached via continuous and rapid equilibrium processes that contribute to the balance of the fluxes of load and consumption in major metabolic components, e.g. ATP (Stucki, 1980). In the light, the maintenance of fluxes of photosynthetic CO_2 assimilation, photorespiration and respiration is balanced not only via engagement of the non-coupled pathways of electron transport but also via another essentially energy wasting process, which consists of the conversion of a fraction of the synthesized ATP to ADP in the mitochondrial intermembrane space (Igamberdiev and Kleczkowski, 2003). Plant mitochondria contain a high activity of adenylate kinase (AK) in the intermembrane space that exceeds the activity of ATP synthase more than four times (Roberts *et al.*, 1997). AK functions as an engine that prevents depletion of ADP when it is taken for ATP synthesis, so that the ATP/ADP ratio in the intermembrane space of mitochondria is equilibrated and maintained at $ATP_{free}/ADP_{free} = 1$, according to the stoichiometry of the adenylate transporter that takes free adenylates (Igamberdiev and Kleczkowski, 2003).

Generation of the membrane potential drives ATP synthesis and provides a continuous exchange of ATP_{free} and ADP_{free} across the inner mitochondrial membrane. In this way, mitochondria can interact (via cytosol) with chloroplasts and other organelles. The active AK in the intermembrane space also allows Mg^{2+} and other cations such as Mn^{2+} and K^+ to be at appreciable levels under high respiratory rates. This is possible because the AK equilibrium is established between free and Mg-bound adenylates. ADP is exhausted outside mitochondria and regenerated by AK from ATP and AMP. AK establishes a link between the ratios of free and Mg-bound adenylates, the concentration of Mg^{2+} and the inner membrane potentials of mitochondria and chloroplasts (Igamberdiev and Kleczkowski, 2001, 2003, 2006).

With the contribution of AK, the synthesis of ATP in mitochondria is optimized by the two aforementioned processes, i.e. via the buffering by AK and via the uncoupling by the alternative dehydrogenases/oxidase and the uncoupling protein. Animal mitochondria use both AK and creatine kinase for equilibration of adenylates, while plant mitochondria use only AK, but they have far more variable mechanisms to finely regulate the degree of coupling via engagement of the numerous non-coupled pathways of mitochondrial electron transport. These mechanisms optimize the operation of mitochondria, particularly in the light. Depending on the supply of NAD(P)H, this optimization takes place in a time-dependent manner. The degree of coupling can easily change and the intensity of ADP load can also fluctuate. The increase in the ATP/ADP ratio also regulates the functional state of proteins and the activity of mitochondrial enzymes via their reversible phosphorylation (Bykova *et al.*, 2003a; Bykova *et al.*, 2003b; Ito *et al.*, 2009; Taylor *et al.*, 2011).

Bicarbonate pool and refixation of photorespiratory carbon

Photorespiratory release of CO_2 is not necessarily a release of a fixed carbon. The carbon released in photorespiration is mostly refixed. The photorespiratory 'loss' of CO_2 really counts the carbon that is not entered in the leaf due to the proposed mechanism involving a lower capacity for the CO_2 pump upon intensification of the mitochondrial CO_2 release. Thus, the CO_2 loss is attributed mainly not to a released carbon but to a non-fixed carbon due to photorespiration. The refixation capacity is very high and it also involves the operation of chloroplastic, cytosolic and mitochondrial isoforms of carbonic anhydrase (Riazunnisa *et al.*, 2006). Some estimations give a value of more than 80% CO_2 refixed after photorespiratory release in ambient air conditions (Loreto *et al.*, 1999). The observed inhibition of respiration in the light may also be explained in part by more efficient refixation of CO_2 in photosynthetic tissue (Pinelli and Loreto, 2003).

The lack of complex I results in the increase of photorespiration due to decreased mesophyll conductance to CO_2 (Priault *et al.*, 2006). This can be linked to the lack of γ-carbonic anhydrase activity associated with the mitochondrial complex I. The carbonic anhydrase subunits form a matrix-exposed domain attached to the membrane arm of complex I (Sunderhaus *et al.*, 2006). They comprise the γ-type of carbonic anhydrase, which is almost insensitive to ethoxyzolamide and has similarity to corresponding carbonic anhydrases of cyanobacteria (Parisi *et al.*, 2004). The complex I-linked carbonic anhydrase subunits (number of three to five) are involved in an intracellular carbon transport system in higher plants that resembles the carbon concentration system of cyanobacteria (Dudkina *et al.*, 2006). In the matrix fraction, a β-type carbonic anhydrase was found, isolated and characterized in *Chlamydomonas* (Eriksson *et al.*, 1996). It is low CO_2-inducible and therefore its function is related to photorespiration (Eriksson *et al.*, 1998). The analysis of CA genes in *Arabidopsis* showed that one β-type CA is targeted to mitochondria (Fabre *et al.*, 2007). The presence of a carbonic anhydrase-based carbon concentration mechanism in C_3 plants was originally postulated by Fridlyand and Kaler (1987, 1988). It acquired some approval only after the discovery of the mitochondrial carbonic anhydrase. CO_2 released during respiration and photorespiration rapidly comes to equilibrium with bicarbonate, facilitating its solubility in the cytosol and enhancing its assimilation in the chloroplast.

If most of the photorespiratory CO_2 is refixed, this introduces the question of why C_3 plants are not as efficient as C_4 plants in carbon fixation. This can be explained by the way that upon photorespiratory CO_2 release, the pumping capacity for CO_2 is decreased and the plant cell cannot take much CO_2 before it is depleted by Rubisco. This means that we observe not the photorespiratory CO_2

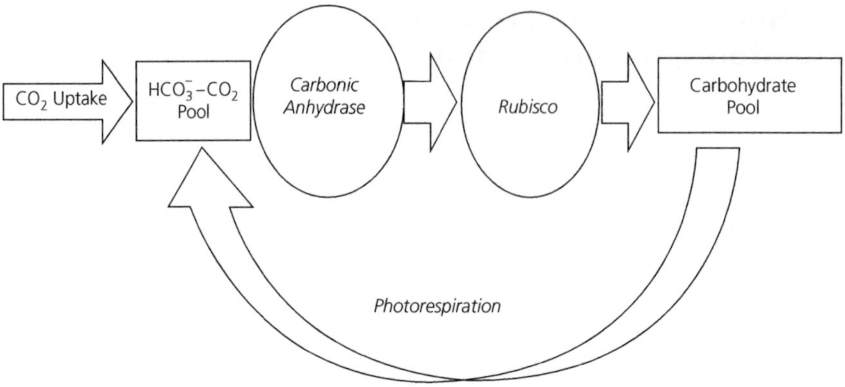

Figure 2.2 General scheme showing joint operation of Rubisco, carbonic anhydrases and photorespiration. The source of CO_2 is a bicarbonate pool fed from the atmosphere and buffered by the carbonic anhydrase serving as a feed-forward pump for Rubisco. The latter is an engine producing carbohydrates and at the same time generating a feedback (photorespiration) to feed the bicarbonate pool in conditions of insufficient CO_2 supply.

loss but the alternating pumping capacity of C_3 plants caused by photorespiration. In other words, photorespiration makes the CO_2 pumping capacity less efficient and this is often perceived as photorespiratory CO_2 loss. The joint operation of Rubisco, carbonic anhydrase and photorespiration in CO_2 equilibrium in the photosynthetic cell is presented in Figure 2.2.

Malate and citrate valves

The increase in the reduction level of NADH/NAD and NADPH/NADP and an increased ATP/ADP ratio in mitochondria in the light due to photorespiration have important consequences for the operation of the TCA cycle (Figure 2.3). The increased redox level is mainly due to the high oxidation rate of glycine and the transhydrogenation reaction forming NADPH (Bykova *et al.*, 1998, 1999; Bykova and Møller, 2001). The TCA cycle is reorganized in the light in such way that it turns from being the main source of energy in the cell to become a flexible mechanism that enables the cell to sustain the photosynthetic process, both through the production of carbon skeletons and by contributing to the redox homeostasis of the cell. When the redox level of mitochondria rises, they export citrate to the cytosol, and while the redox level decreases, the complete TCA cycle is activated (Igamberdiev and Gardeström, 2003). A partial TCA-cycle operates in the light to supply carbon skeletons for biosynthetic purposes (Chen and Gadal, 1990; Hanning and Heldt, 1993; Igamberdiev and Gardeström, 2003; Fernie *et al.*, 2004). Citrate has been suggested as the main exported product of such a partial cycle (Hanning and Heldt, 1993). Recent measurements of subcellular pyridine nucleotide redox status, and the kinetic properties of the key enzymes

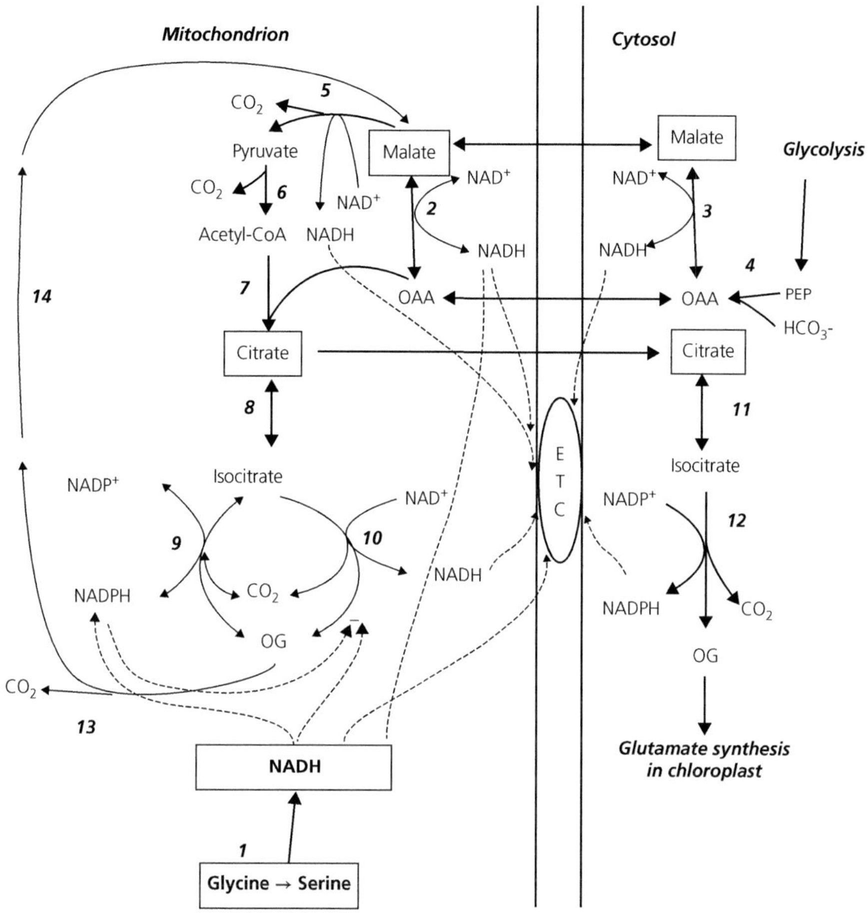

Figure 2.3 Operation of malate and citrate valves during glycine oxidation. The reaction catalyzed by GDC (1) raises NADH in mitochondria, which directs the reaction of mitochondrial malate dehydrogenase (2) toward malate. Malate is exported to cytosol where it is equilibrated with oxaloacetate (OAA) by the cytosolic malate dehydrogenase (3). OAA is formed in the cytosol as a product of glycolysis when PEP enters the reaction catalyzed by PEP-carboxylase and can be transported to mitochondria (4). At elevated NADH, malate in mitochondria can be converted to pyruvate by NAD-malic enzyme, which is relatively insensitive to high redox levels (5). Pyruvate is decarboxylated by the pyruvate dehydrogenase complex (6) with formation of acetyl-CoA. The latter, via condensation with OAA, forms citrate in the citrate synthase reaction (7), which is in equilibrium with isocitrate due to the aconitase reaction (8). Isocitrate oxidation is inhibited at elevated NADH (shown by the 'minus' sign) due to displacing the equilibrium of NADP-dependent isocitrate dehydrogenase (9) into reverse reaction and to inhibition by NADH of NAD-dependent isocitrate dehydrogenase (10). This results in the export of citrate to cytosol, where it is converted to isocitrate by cytosolic aconitase (11) and then to 2-oxoglutarate (OG) by cytosolic NADP-isocitrate dehydrogenase (12). OG is used for glutamate biosynthesis in chloroplasts. 2-oxoglutarate dehydrogenase reaction (13) and the subsequent reactions up to malate formation (14) of the TCA cycle are inhibited in the light.

involved, also support the conclusion that citrate is the exported compound (Igamberdiev and Gardeström, 2003). Two isocitrate dehydrogenases in mitochondria – one irreversible and NAD-dependent and another reversible and NADPH-dependent – represent a system sensitive to changes in the mitochondrial redox state. In photorespiratory conditions, when the $NADH/NAD^+$ and $NADPH/NADP^+$ ratios are high, only the reverse reaction of NADP-isocitrate dehydrogenase may take place and NAD-isocitrate dehydrogenase is inhibited (Igamberdiev and Gardeström, 2003). PDC and isocitrate oxidation will be important steps for the control of this flux. Citrate export from mitochondria may also be important for maintaining the cytosolic NADPH/NADP ratio in the light.

There are two major redox valves that transport redox equivalents in the photosynthetic plant cell. The malate valve, driven by NADPH formed by photosynthetic electron transport in chloroplasts and by NADH formed in the GDC reaction in mitochondria, prevents over-reduction of the chloroplasts and mitochondria and increases the $NADH/NAD^+$ ratio in different cellular compartments. Another valve, the citrate valve, driven by the increased reduction level in mitochondria linked to photorespiratory glycine oxidation, reduces NADP pools and supplies 2-oxoglutarate for glutamate biosynthesis. The active operation of the citrate valve corresponds to the transition from the complete to the partial TCA-cycle in plant mitochondria. The partial TCA-cycle maintains the operation of the citrate valve, supplying the anabolic reduction power (NADPH) via oxidation of isocitrate in the cytosol. In photorespiratory conditions, a part of the NADPH pool in the cytosol is used for the reduction of glyoxylate and hydroxypyruvate exported from peroxisomes (Krömer and Heldt, 1991). NADPH/NADP turnover may be provided by the participation of the cytosolic NADP-isocitrate dehydrogenase and NADPH-dependent hydroxypyruvate and glyoxylate reductases (Igamberdiev and Kleczkowski, 2000; Igamberdiev and Gardeström, 2003). Operation of the modified TCA-cycle and the citrate valve also maintains the concentrations of 2-oxoglutarate, OAA and pyruvate in the cytosol and mitochondria, which is important for nitrogen assimilation in the light. Studies with a barley mutant deficient in mitochondrial GDC showed that, in photorespiratory conditions, the chloroplasts and mitochondria were over-reduced and over-energized (Igamberdiev *et al.*, 2001a). This gives support to a function for photorespiration as an effective redox transfer mechanism from chloroplasts and mitochondria in which the GDC reaction represents the main engine for transporting redox equivalents, ATP and carbon from mitochondria to the cytosol.

Conclusion

We presented here the arguments in support of the role of non-coupled pathways of the mitochondrial electron transport and of the reactions associated with these pathways in the maintenance of high photorespiratory flux. High rates of glycine oxidation are possible via the kinetic mechanisms leading to the increased capacity of the mitochondrial electron transport chain under intensive glycine oxidation via switching to non-coupled pathways. Rising NADH due to the GDC reaction not only engages the non-coupled pathways but also results in the intensification of the malate and citrate mitochondrial valves.

References

André, M.J. (2011a) Modelling $^{18}O_2$ and $^{16}O_2$ unidirectional fluxes in plants: I. regulation of pre-industrial atmosphere. *Biosystems* **103**: 239–251.

André, M.J. (2011b) Modelling $^{18}O_2$ and $^{16}O_2$ unidirectional fluxes in plants: II. analysis of Rubisco evolution. *Biosystems* **103**: 252–264.

Andrews, T.J. and Lorimer, G.H. (1987) Rubisco: Structure, mechanisms and prospects for improvement. In *The Biochemistry of Plants* (M.D. Hatch and N.K. Boardman, eds), pp 132–219. Academic Press, New York.

Atkin, O.K., Evans, J.R. and Siebke, K. (1998) Relationship between the inhibition of leaf respiration by light and enhancement of leaf dark respiration following light treatment. *Australian Journal of Plant Physiology* **25**: 437–443.

Atkin, O.K., Millar, A.H., Gardeström, P. and Day, D.A. (2000a) Photosynthesis, carbohydrate metabolism and respiration in leaves of higher plants. In *Photosynthesis: Physiology and Metabolism* (R.C. Leegood, T.D. Sharkey and S. von Caemmerer, eds), pp. 153–175. Kluwer, Dordrecht.

Atkin, O.K., Evans, J.R., Ball, M.C. *et al.* (2000b) Leaf respiration of snow gum in the light and dark. Interactions between temperature and irradiance. *Plant Physiology* **122**: 915–923.

Bykova, N.V. and Møller, I.M. (2001) Involvement of matrix NADP turnover in the oxidation of NAD^+-linked substrates by pea leaf mitochondria. *Physiologia Plantarum* **111**: 448–456.

Bykova, N.V., Igamberdiev, A.U. and Møller, I.M. (1998) Contribution of respiratory NAD(P)H dehydrogenases to glycine and glycine plus malate oxidation by pea leaf mitochondria. In *Plant Mitochondria: From Gene to Function* (I.M. Møller, P. Gardeström, K. Glimelius and E. Glaser, eds), pp. 347–351. Backhuys Publishers, Leiden, The Netherlands.

Bykova, N.V., Rasmusson, A.G., Igamberdiev, A.U. *et al.* (1999) Two separate transhydrogenase activities are present in plant mitochondria. *Biochemical and Biophysical Research Communications* **265**: 106–111.

Bykova, N.V., Egsgaard, H. and Møller, I.M. (2003a) Identification of 14 new phosphoproteins involved in important mitochondrial processes. *FEBS Letters* **540**: 141–146.

Bykova, N.V., Stensballe, A., Egsgaard, H. *et al.* (2003b) Phosphorylation of formate dehydrogenase in potato tuber mitochondria. *Journal of Biological Chemistry* **278**: 26021–26030.

Bykova, N.V., Keerberg, O., Pärnik, T. *et al.* (2005) Interaction between photorespiration and respiration in transgenic potato plants with antisense reduction in glycine decarboxylase. *Planta* **222**: 130–140.

Byrd, G.T., Sage, R.F. and Brown, R.H. (1992) A comparison of dark respiration between C_3 and C_4 plants. *Plant Physiology* **100**: 191–198.

Cárdenas, L., McKenna, S.T., Kunkel, J.G and Hepler, P.K. (2006) NAD(P)H oscillates in pollen tubes and is correlated with tip growth. *Plant Physiology* **142**: 1460–1468.

Chen, R.-D. and Gadal, P. (1990) Do the mitochondria provide the 2-oxoglutarate needed for glutamate synthesis in higher plant chloroplasts? *Plant Physiology and Biochemistry* **28**: 141–145.

Day, D.A. and Wiskich, J.T. (1977) Factors limiting respiration by isolated cauliflower mitochondria. *Phytochemistry* **16**: 1499–1502.

Di Marco, G., Manes, F., Tricoli, D. and Vitale, E. (1990) Fluorescence parameters measured concurrently with net photosynthesis to investigate chloroplastic CO_2 concentration in leaves of *Quercus ilex* L. *Journal of Plant Physiology* **136**: 538–543.

Douce, R., Bourguignon, J., Macherel, D. and Neuburger, M. (1994) The glycine decarboxylase system in higher plant mitochondria: structure, function and biogenesis. *Biochemical Society Transactions* **22**: 184–188.

Dudkina, N.V., Heinemeyer, J., Sunderhaus, S. *et al.* (2006) Respiratory chain supercomplexes in the plant mitochondrial membrane. *Trends in Plant Science* **11**: 232–240.

Edwards, G., Walker, D.A. (1983) *C3, C4: Mechanisms and Cellular and Environmental Regulation of Photosynthesis*. University of California Press, Berkeley.

Elhafez, D., Murcha, M.W., Clifton, R. *et al.* (2006) Characterization of mitochondrial alternative NAD(P)H dehydrogenases in *Arabidopsis*: intraorganelle location and expression. *Plant and Cell Physiology* **47**: 43–54.

Eriksson, M., Karlsson, J., Ramazanov, Z. *et al.* (1996) Discovery of an algal mitochondrial carbonic anhydrase: Molecular cloning and characterization of a low-CO_2-induced polypeptide in *Chlamydomonas reinhardtii*. *Proceedings of the National Academy of Sciences of the United States of America* **93**: 12031–12034.

Eriksson, M., Villand, P., Gardeström, P. and Samuelsson, G. (1998) Induction and regulation of expression of a low-CO_2-induced mitochondrial carbonic anhydrase in *Chlamydomonas reinhardtii*. *Plant Physiology* **116**: 637–641.

Escobar MA, Franklin KA, Svensson AS, Salter MG, Whitelam GC, Rasmusson AG (2004) Light regulation of the Arabidopsis respiratory chain. Multiple discrete photoreceptor responses contribute to induction of type II NAD(P)H dehydrogenase genes. *Plant Physiology* **136**: 2710–2721.

Fabre, N., Reiter, I.M., Becuwe-Linka, N. *et al.* (2007) Characterization and expression analysis of genes encoding alpha and beta carbonic anhydrases in Arabidopsis. *Plant, Cell and Environment* **30**: 617–629.

Fernie, A.R., Carrari, F. and Sweetlove, L.J. (2004) Respiratory metabolism: glycolysis, the TCA cycle and mitochondrial electron transport. *Current Opinion in Plant Biology* **7**: 254–261.

Finnegan, P.M., Whelan, J., Millar, A.H. *et al.* (1997) Differential expression of the multigene family encoding the soybean mitochondrial alternative oxidase. *Plant Physiology* **114**: 455–466.

Fridlyand, L.E. and Kaler, V.L. (1987) Possible CO_2 concentrating mechanism in chloroplasts of C_3 plants – role of carbonic anhydrase. *General Physiology and Biophysics* **6**: 617–636.

Fridlyand, L.E. and Kaler, VL. (1988) Criteria indicating the existence of hypothetical CO_2 concentrating mechanisms in C_3 plant and estimation of their efficiency. *Soviet Plant Physiology* **35**: 333–340.

Gardeström, P., Igamberdiev, A.U. and Raghavendra, A.S. (2002) Mitochondrial functions in the light and significance to carbon-nitrogen interactions. In: *Photosynthetic Nitrogen Assimilation and Associated Carbon and Respiratory Metabolism* (C.H. Foyer and G. Noctor, eds), pp. 151–172. Kluwer, Dordrecht, The Netherlands.

Geisler, D.A., Broselid, C., Hederstedt, L. and Rasmusson, A.G. (2007) Ca^{2+}-binding and Ca^{2+}-independent respiratory NADH and NADPH dehydrogenases of *Arabidopsis thaliana*. *Journal of Biological Chemistry* **282**: 28455–28464.

Gupta, K.J., Shah, J.K., Brotman, Y. *et al.* (2012) Inhibition of aconitase by nitric oxide leads to induction of the alternative oxidase and to a shift of metabolism towards biosynthesis of amino acids. *Journal of Experimental Botany* **63**: 1773–1784.

Hagedorn, P.H., Flyvbjerg, H. and Møller, I.M. (2004) Modelling NADH turnover in plant mitochondria. *Physiologia Plantarum* **120**: 370–385.

Hanning, I. and Heldt, H.W. (1993) On the function of mitochondrial metabolism during photosynthesis in spinach leaves (*Spinacia oleracea* L.). Partitioning between respiration and export of redox equivalents and precursors for nitrate assimilation products. *Plant Physiology* **103**: 1147–1154.

Hanson, S.M. and Schnell, S. (2008) Reactant stationary approximation in enzyme kinetics. *Journal of Physical Chemistry A* **112**: 8654–8658.

Heineke, D., Bykova, N., Gardeström, P. and Bauwe, H. (2001) Metabolic response of potato plants to an antisense reduction of the P-protein of glycine decarboxylase. *Planta* **212**: 880–887.

Hill, S.A. and Bryce, J.H. (1992) Malate metabolism and light-enhanced dark respiration in barley mesophyll protoplasts. In *Molecular, Biochemical and Physiological Aspects of Plant Respiration* (H. Lambers and L.H.W. van der Plas, eds), pp. 221–230. SPB Academic Publishing, The Hague, The Netherlands.

Igamberdiev, A.U., Bykova, N.V. and Gardeström, P. (1997) Involvement of cyanide-resistant and rotenone-insensitive pathways of mitochondrial electron transport during oxidation of glycine in higher plants. *FEBS Letters* **412**: 265–269.

Igamberdiev, A.U. (1999) Foundations of metabolic organization: coherence as a basis of computational properties in metabolic networks. *Biosystems* **50**: 1–16.

Igamberdiev AU, Kleczkowski LA. (2000) Capacity for NADPH/NADP turnover in the cytosol of barley seed endosperm: The role of NADPH-dependent hydroxypyruvate reductase. *Plant Physiology and Biochemistry* **38**: 747–753.

Igamberdiev, A.U., Bykova, N.V., Lea, P.J. and Gardeström, P. (2001a) The role of photorespiration in redox and energy balance of photosynthetic plant cells: A study with a barley mutant deficient in glycine decarboxylase. *Physiologia Plantarum* **111**: 427–438.

Igamberdiev, A.U., Romanowska, E. and Gardeström, P. (2001b) Photorespiratory flux and mitochondrial contribution to energy and redox balance of barley leaf protoplasts in the light and during light–dark transitions. *Journal of Plant Physiology* **158**: 1325–1332.

Igamberdiev, A.U. and Kleczkowski, L.A. (2001) Implications of adenylate kinase-governed equilibrium of adenylates on contents of free magnesium in plant cells and compartments. *Biochemical Journal* **360**: 225–231.

Igamberdiev AU, Gardeström P. (2003) Regulation of NAD- and NADP-dependent isocitrate dehydrogenases by reduction levels of pyridine nucleotides in mitochondria and cytosol of pea leaves. *Biochimica et Biophysica Acta – Bioenergetics* **1606**: 117–125.

Igamberdiev, A.U. and Kleczkowski, L.A. (2003) Membrane potential, adenylate levels and Mg^{2+} are interconnected via adenylate kinase equilibrium in plant cells. *Biochimica et Biophysica Acta – Bioenergetics* **1607**: 111–119.

Igamberdiev, A.U. and Kleczkowski, L.A. (2006) Equilibration of adenylates in the mitochondrial intermembrane space maintains respiration and regulates cytosolic metabolism. *Journal of Experimental Botany* **57**: 2133–2141.

Igamberdiev, A.U. and Kleczkowski, L.A. (2009) Metabolic systems maintain stable non-equilibrium via thermodynamic buffering. *Bioessays* **31**: 1091–1099.

Igamberdiev, A.U. and Roussel, M.R. (2012) Feedforward non-Michaelis-Menten mechanism for CO_2 uptake by Rubisco: contribution of carbonic anhydrases and photorespiration to optimization of photosynthetic carbon assimilation. *Biosystems* **107**: 158–166.

Igamberdiev, A.U., Eprintsev, A.T., Fedorin, D.N. and Popov, V.N. (2014) Phytochrome-mediated regulation of plant respiration and photorespiration. *Plant, Cell and Environment* **37**: 290–299.

Ito, J., Taylor, N.L., Castleden, I. *et al.* (2009) A survey of the *Arabidopsis thaliana* mitochondrial phosphoproteome. *Proteomics* **9**: 4229–4240.

Ivlev, A.A. (1989) On the discreteness of CO_2 assimilation in the light by C_3 plants. *Biofizika* **34**: 887–891.

Jezek, P., Costa, A.D. and Vercesi, A.E. (1996) Evidence for anion-translocating plant uncoupling mitochondrial protein in potato mitochondria. *Journal of Biological Chemistry* **271**: 32743–32748.

Krause, F., Reifschneider, N.H., Vocke, D. *et al.* (2004) "Respirasome"-like supercomplexes in green leaf mitochondria of spinach. *Journal of Biological Chemistry* **279**: 48369–48375.

Krömer, S. and Heldt, H.W. (1991) Respiration of pea leaf mitochondria and redox transfer between the mitochondrial and extramitochondrial compartment. *Biochimica et Biophysica Acta* **1057**: 42–50.

Kukushkin, A.K. and Soldatova, E.A. (1996) The effect of photorespiration on the induction of delayed millisecond luminescence of higher plants photosystem II. Theoretical investigation. *Biofizika* **41**: 440–444.

Laisk, A. and Sumberg, A. (1994) Partitioning of the leaf CO_2 exchange into components using CO_2 exchange and fluorescence measurements. *Plant Physiology* **106**: 689–695.

Lambers, H. (1982) Cyanide-resistant respiration: a non-phosphorylating electron transport pathway acting as an energy overflow. *Physiologia Plantarum* **55**: 478–485.

Leidreiter, K., Kruse, A., Heineke, D. *et al.* (1995) Subcellular volumes and metabolite concentrations in potato (*Solanum tuberosum* cv. Désirée) leaves. *Botanica Acta* **108**: 439–444.

Loreto, F., Delfine, S. and Di Marco, G. (1999) Estimation of photorespiratory carbon dioxide recycling during photosynthesis. *Australian Journal of Plant Physiology* **26**: 733–736.

MacDonald, M.J., Fahien, L.A., Buss, J.D. *et al.* (2003) Citrate oscillates in liver and pancreatic beta cell mitochondria and in INS-l insulinoma cells. *Journal of Biological Chemistry* **278**: 51894–51900.

Michalecka, A.M., Svensson, A.S., Johansson, F.I. *et al.* (2003) *Arabidopsis* genes encoding mitochondrial type II NAD(P)H dehydrogenases have different evolutionary origin and show distinct responses to light. *Plant Physiology* **133**: 642–652.

Millar, A.H., Wiskich, J.T., Whelan, J. and Day, D.A. (1993) Organic acid activation of the alternative oxidase of plant mitochondria. *FEBS Letters* **329**: 259–262.

Møller, I.M. (2001) Plant mitochondria and oxidative stress: electron transport, NADPH turnover, and metabolism of reactive oxygen species. *Annual Review of Plant Physiology and Plant Molecular Biology* **52**: 561–591.

Møller, I.M. and Lin, W. (1986) Membrane-bound NAD(P)H dehydrogenases in higher plant cells. *Annual Review of Plant Physiology* **37**: 309–334.

Ngo, L.G. and Roussel, M.R. (1997) A new class of biochemical oscillator models based on competitive binding. *European Journal of Biochemistry* **245**: 182–190.

Noctor, G., De Paepe, R. and Foyer, C.H. (2007) Mitochondrial redox biology and homeostasis in plants. *Trends in Plant Science* **12**: 125–134.

Oliver, D.J. (1994) The glycine decarboxylase complex from plant mitochondria. *Annual Review of Plant Physiology and Plant Molecular Biology* **45**: 323–337.

Parisi, G., Perales, M., Fornasari, M.S. *et al.* (2004) Gamma carbonic anhydrases in plant mitochondria. *Plant Molecular Biology* **55**: 193–207.

Pascal, N., Dumas, R. and Douce, R. (1990) Comparison of the kinetic behavior toward pyridine nucleotides of NAD-linked dehydrogenases from plant mitochondria. *Plant Physiology* **94**: 189–193.

Pickersgil, l R.P. (1986) An upper limit to the active site concentration of ribulose bisphosphate carboxylase in chloroplasts. *Biochemical Journal* **236**: 311.

Pinelli, P. and Loreto, F. (2003) $^{12}CO_2$ emission from different metabolic pathways measured in illuminated and darkened C_3 and C_4 leaves at low, atmospheric and elevated CO_2 concentration. *Journal of Experimental Botany* **54**: 1761–1769.

Popov, V.N., Eprintsev, A.T., Fedorin, D.N. and Igamberdiev, A.U. (2010) Succinate dehydrogenase in *Arabidopsis thaliana* is regulated by light via phytochrome A. *FEBS Letters* **584**: 199–202.

Priault, P., Tcherkez, G., Cornic, G. *et al.* (2006) The lack of mitochondrial complex I in a CMSII mutant of *Nicotiana sylvestris* increases photorespiration through an increased internal resistance to CO_2 diffusion. *Journal of Experimental Botany* **57**: 3195–3207.

Rasmusson, A.G. and Escobar, M.A. (2007) Light and diurnal regulation of plant respiratory gene expression. *Physiologia Plantarum* **129**: 57–67.

Rasmusson, A.G., Geisler, D.A. and Møller, I.M. (2008) The multiplicity of dehydrogenases in the electron transport chain of plant mitochondria. *Mitochondrion* **8**: 47–60.

Riazunnisa, K., Padmavathi, L., Bauwe, H. and Raghavendra, A.S. (2006) Markedly low requirement of added CO_2 for photosynthesis by mesophyll protoplasts of pea (*Pisum sativum*): possible roles of photorespiratory CO_2 and carbonic anhydrase. *Physiologia Plantarum* **128**: 763–772.

Ribas-Carbo, M., Lennon, A.M., Robinson, S.A. *et al.* (1997) The regulation of electron partitioning between the cytochrome and alternative pathways in soybean cotyledon and root mitochondria. *Plant Physiology* **113**: 903–911.

Roberts, J., Aubert, S., Gout, E. *et al.* (1997) Cooperation and competition between adenylate kinase, nucleoside diphosphokinase, electron transport, and ATP synthase in plant mitochondria studied by ^{31}P-nuclear magnetic resonance. *Plant Physiology* **113**: 191–199.

Roussel, M.R., Ivlev, A.A. and Igamberdiev, A.U. (2007) Oscillations of the internal CO_2 concentration in tobacco leaves transferred to low CO_2. *Journal of Plant Physiology* **164**: 1188–1196.

Roussel, M.R. and Igamberdiev, A.U. (2011) Dynamics and mechanisms of oscillatory photosynthesis. *Biosystems* **103**: 230–238.

Ryde-Pettersson, U. (1992) Oscillations in coupled enzymic reactions at high concentration of enzyme. *Biochimica et Biophysica Acta* **1119**: 118–122.

Satoh, K. and Katoh, S. (1983) Induction kinetics of millisecond-delayed luminescence in intact *Bryopsis* chloroplasts. *Plant and Cell Physiology* **24**: 953–962.

Scheibe, R. and Stitt, M. (1988) Comparison of NADP-malate dehydrogenase activation, QA reduction and O_2 evolution in spinach leaves. *Plant Physiology and Biochemistry* **26**: 473–481.

Sharkey, T.D. (1988) Estimating the rate of photorespiration in leaves. *Physiologia Plantarum* **73**: 147–152.

Stucki, J.W. (1980) The thermodynamic-buffer enzymes. *European Journal of Biochemistry* **109**: 257–267.

Sunderhaus, S., Dudkina, N.V., Jansch, L. *et al.* (2006) Carbonic anhydrase subunits form a matrix-exposed domain attached to the membrane arm of mitochondrial Complex I in plants. *Journal of Biological Chemistry* **281**: 6482–6488.

Svensson, A.S. and Rasmusson, A.G. (2001) Light-dependent gene expression for proteins in the respiratory chain of potato leaves. *Plant Journal* **28**: 73–82.

Taylor, N.L., Heazlewood, J.L. and Millar, A.H. (2011) The *Arabidopsis thaliana* 2-D gel mitochondrial proteome: Refining the value of reference maps for assessing protein abundance, contaminants and post-translational modifications. *Proteomics* **11**: 1720–1733.

Tcherkez, G., Cornic, G., Bligny, R. *et al.* (2005) *In vivo* respiratory metabolism of illuminated leaves. *Plant Physiology* **138**: 1596–1606.

Thirkettle-Watts, D., McCabe, T.C., Clifton, R. *et al.* (2003) Analysis of the alternative oxidase promoters from soybean. *Plant Physiology* **133**: 1158–1169.

Tobin, A.K., Thorpe, J.R., Hylton, C.M. and Rawsthorne, S. (1989) Spatial and temporal influences on the cell-specific distribution of glycine decarboxylase in leaves of wheat (*Triticum aestivum* L.) and pea (*Pisum sativum* L.). *Plant Physiology* **91**: 1219–1225.

Tomashek, J.J. and Brusilow, W.S. (2000) Stoichiometry of energy coupling by proton-translocating ATPases: A history of variability. *Journal of Bioenergetics and Biomembranes* **32**: 493–500.

Tomaz, T., Bagard, M., Pracharoenwattana, I. *et al.* (2010) Mitochondrial malate dehydrogenase lowers leaf respiration and alters photorespiration and plant growth in *Arabidopsis*. *Plant Physiology* **154**: 1143–1157.

Ueno, O., Bang, S.W., Wada, Y. *et al.* (2003) Structural and biochemical dissection of photorespiration in hybrids differing in genome constitution between *Diplotaxis tenuifolia* (C_3–C_4) and radish (C_3). *Plant Physiology* **132**: 1550–1559.

Vanlerberghe, G.C., Day, D.A., Wiskich, J.T. *et al.* (1995) Alternative oxidase activity in tobacco leaf mitochondria – dependence on tricarboxylic-acid cycle-mediated redox regulation and pyruvate activation. *Plant Physiology* **109**: 353–361.

Vanlerberghe, G.C. and McIntosh, L. (1996) Signals regulating the expression of the nuclear gene encoding alternative oxidase of plant mitochondria. *Plant Physiology* **111**: 589–595.

Vélot, C., Mixon, M.B., Teige, M. and Srere, P.A. (1997) Model of a quinary structure between Krebs TCA cycle enzymes: a model for the metabolon. *Biochemistry* **36**: 14271–14276.

von Caemmerer, S. (2000) *Biochemical Models of Leaf Photosynthesis.* CSIRO Publishing, Collingwood, Australia.

Wigge, B., Krömer, S. and Gardeström, P. (1993) The redox levels and subcellular distribution of pyridine nucleotides in illuminated barley leaf protoplasts studied by rapid fractionation. *Physiologia Plantarum* **88**: 10–18.

Wingler, A., Lea, P.J. and Leegood, R.C. (1999a) Photorespiratory metabolism of glyoxylate and formate in glycine-accumulating mutants of barley and *Amaranthus edulis. Planta* **207**: 518–526.

Wingler, A., Quick, W.P., Bungard, R.A. *et al.* (1999b) The role of photorespiration during drought stress: an analysis utilizing barley mutants with reduced activities of photorespiratory enzymes. *Plant, Cell and Environment* **22**: 361–373.

Winter, H., Robinson, D.G. and Heldt, H.W. (1993) Subcellular volumes and metabolite concentrations in barley leaves. *Planta* **191**: 180–190.

Winter H., Robinson DG., Heldt HW. (1994) Subcellular volumes and metabolite concentrations in spinach leaves. *Planta* **193**: 530–535.

Wiskich, J.T. and Dry, I.B. (1985) The tricarboxylic acid cycle in plant mitochondria: its operation and regulation. In *Encyclopedia of Plant Physiology,* Vol. 18, *Higher Plant Cell Respiration* (R. Douce and D.A. Day, eds), pp. 281–313. Springer-Verlag, Berlin.

Wiskich, J.T., Bryce, J.H., Day, D.A. and Dry, I.B. (1990) Evidence for metabolic domains within the matrix compartment of pea leaf mitochondria. Implications for photorespiratory metabolism. *Plant Physiology* **93**: 611–616.

CHAPTER 3

Taxonomic distribution of alternative oxidase in plants

Allison E. McDonald

Department of Biology, Wilfrid Laurier University, Waterloo, Ontario, Canada

What is alternative oxidase?

A great deal of work during the past several decades has increased our understanding of the alternative oxidase (AOX). It has been conclusively shown that AOX is a quinol oxidase that is present in the respiratory electron transport systems (ETS) of many organisms. AOX introduces a branch-point at the quinol pool where electrons can either be transferred to complex III or AOX. If electrons are passed to AOX, complexes III and IV are bypassed and fewer protons are pumped across the inner mitochondrial membrane. This means that fewer ATP will be synthesized per oxygen consumed. This fact indicates that AOX is therefore energetically wasteful and so much effort has been spent in an attempt to determine why such a pathway has been retained over evolutionary time by many organisms.

Historical investigations of AOX in plants

AOX was first discovered in plants due to the interesting observation of thermogenesis (i.e. heat generation) in the reproductive tissues of members of the Araceae family (Church, 1908). In describing *Arum maculatum* (the cuckoo pint), Church describes the 'unpleasant odour' given off by the plant and the fact that the smell attracted several different fly species (Church, 1908). In several experiments he describes the heating of the spadix tissue to 25–29 °C and that prior to heating the tissue was full of starch granules and after heating starch reserves were vastly depleted (Church, 1908). He comments that the chamber of the plant exhibits low oxygen and high CO_2 during the heating event and suggests that the flies attracted by the smell serve to pollinate this species (Church, 1908).

Alternative Respiratory Pathways in Higher Plants, First Edition.
Edited by Kapuganti Jagadis Gupta, Luis A.J. Mur and Bhagyalakshmi Neelwarne.

It had been observed that some physiological mechanism was responsible for the heating of the spadix to temperatures that were quite warm even compared to ambient temperatures (Church, 1908). In 1950, James and Beevers investigated respiration in the spadices of various *Arum* species. They observed that during the different developmental stages of the spadix, respiration was very insensitive to cyanide, and attributed this to a non-cytochrome terminal oxidase with a low affinity for oxygen (James and Beevers, 1950). The experiments of Bendall and Bonner in 1971 confirmed the presence of this second oxidase in the respiratory electron transport system and indicated that this 'alternative oxidase' might be a non-heme iron protein. Storey identified ubiquinol in 1976 as the point at which the electron transport system branches between the cytochrome and alternative pathways, but it was not until several years later that the protein responsible for alternative oxidase respiration was identified through protein purification (Elthon and McIntosh, 1987). The development of monoclonal antibodies to AOX (i.e. AOA) with broad cross-reactivity to AOXs from a wide variety of plants has been an exceptionally useful tool for the AOX community (Elthon *et al.*, 1989). Shortly thereafter, the first AOX cDNA was cloned from *Sauromatum guttatum* (Rhoads and McIntosh, 1991). Since that time, most research on AOX has understandably therefore occurred in plants. Specifically, most AOX work has taken place in angiosperm plants (i.e. flowering plants).

Taxonomic distribution of alternative oxidase in all domains of life

In addition to its presence in plants, several early studies identified cyanide-resistant respiration, likely attributable to AOX, in the fungi *Neurospora crassa* (Lambowitz and Slayman, 1971) and *Aspergillus nidulans* (Turner and Rowlands, 1976), the soil amoeba *Acanthamoeba castellanii* (Edwards and Lloyd, 1978) and various kinetoplastids (Hill and Cross, 1973). Later work revealed the presence of AOX in the protist *Dictyostelium discoideum* (Jarmuszkiewicz *et al.*, 2002). Recent work on the AOX using the tools of bioinformatics has revealed the presence of AOX in bacteria and animals for the first time (McDonald *et al.*, 2003; McDonald and Vanlerberghe, 2004).

AOX exists in some prokaryotes and many eukaryotic lineages, but the evolutionary relationship of AOX proteins between these groups has not been fully explored. Based on the current knowledge of the taxonomic distribution of AOX, a theory has been put forward for a prokaryotic origin of AOX and its spread to multiple eukaryotic lineages via the endosymbiotic event that led to mitochondria (McDonald and Vanlerberghe, 2006). It has been hypothesized that the original function of oxidases such as AOX may have been to allow cells to survive exposure to oxygen (Gomes *et al.*, 2001). Oxygen levels in Earth's atmosphere rapidly rose once the iron in the oceans could no longer bind the large amounts of oxygen produced by oxygenic photosynthesis (Farquhar *et al.*, 2011); AOX may have been initially important due to its ability to utilize oxygen, an initially toxic molecule, and convert it into water.

Taxonomic distribution of alternative oxidase in plants

Because it was first discovered in plants, it is understandable that the most AOX research has occurred within the plant kingdom. One generally defining feature of plants is the presence of plastids; more specifically, the chloroplast. Based on current understanding of endosymbiotic theory, it is believed that the first primary endosymbiotic event led to the generation of a eukaryotic cell containing mitochondria (Gray *et al.*, 2001). A later, second primary endosymbiotic event involving a cyanobacterium and a eukaryotic cell containing mitochondria is thought to have led to the three classical primary plastid lineages; the green lineage (leading to plants), the red lineage (which includes red algae and various protists), and the glaucocystophytes (Archibald, 2006).

This chapter will only concern itself with the green lineage (i.e. Viridiplantae) which includes the Chlorophyta (green algae) and the Streptophyta (including streptophyte algae and embryophytes) (Becker and Marin, 2009). With this definition in place, AOX can be investigated in all plants and not just angiosperms. It is also worth making the point that explicit definitions of plant groups are important. Many biological studies use the term 'plant-specific' when referring to genes or proteins that have only been investigated in angiosperms (Becker and Marin, 2009). In such cases, the terms 'embryophyte or spermatophyte-specific' are more accurate (Becker and Marin, 2009), especially as the genome from a streptophyte algae has not yet been sequenced.

Chlorophyte algae

The green lineage is commonly divided into chlorophytes and streptophytes; the timing of this split is debatable (Becker, 2013), but molecular clock methods estimate the date to be approximately 725–1200 million years ago (Becker and Marin, 2009). Recent work examining the presence of AOX in chlorophytes using a bioinformatics approach identified sequences in members of the Chlorophyceae, Mamiellophyceae, Prasinophyceae, Trebouxiophyceae and Ulvophyceae (Neimanis *et al.*, 2013; Table 3.1). An AOX sequence was not found in any members of the Pedinophyceae (Neimanis *et al.*, 2013; Table 3.1). Previous work using the AOX inhibitors SHAM and nPG suggests that the green alga *Chlorella* sp. contains AOX (Eriksen and Lewitus, 1999). Interestingly, *Polytomella* sp., a member of the Chlorophyceae, appears to have experienced a secondary loss of its AOX gene (Reyes-Prieto *et al.*, 2002). The only chlorophyte in which

Table 3.1 The presence of AOX in major plant groups

Plant group	Molecular evidence for the presence of AOX?
Chlorophyta	
Chlorophyceae	Yes
Mamiellophyceae	Yes
Pedinophyceae	No
Prasinophyceae	Yes
Trebouxiophyceae	Yes
Ulvophyceae	Yes
Streptophyta	
Chlorokybophyceae	Yes
Klebsormidiophyceae	Yes
Mesostigmatophyceae	No
Zygnemophyceae	Yes
Streptophytina	
Charophyceae	Yes
Coleochaetophyceae	Yes
Embryophyta	
Anthocerotophyta	No
Bryophyta	Yes
Marchantiophyta	Yes
Tracheophyta	
Lycopodiophyta	Yes
Euphyllophyta	
Moniliformopses	
Equisetopsida	No
Marattiopsida	No
Ophioglossopsida	No
Polypodiopsida	Yes
Psilotopsida	No
Spermatophyta	
Coniferophyta	Yes
Cycadophyta	No
Ginkophyta	No
Gnetophyta	No
Magnoliophyta	Yes

AOX has been characterized is the green alga *Chlamydomonas reinhardtii* (Dinant *et al.*, 2001). *C. reinhardtii* contains two AOX genes and AOX1 expression is strongly up-regulated by nitrate (Baurain *et al.*, 2003).

Streptophyte algae

Although the mitochondrial or plastid genomes have been sequenced for several streptophyte algae (Turmel *et al.*, 2007; Lemieux *et al.*, 2007), coverage of the nuclear genomes of these organisms is poor and is limited to a few EST projects (Timme *et al.*, 2012). AOX sequences have been identified in the EST data of eight species of streptophyte algae that include representatives of the Charophyceae, Chlorokybophyceae, Coleochaetophyceae, Klebsormidiophyceae, and Zygnemophyceae (Table 3.1).

The evidence indicates that streptophyte algae were able to colonize terrestrial environments primarily because they had successfully exploited freshwater as opposed to marine niches, unlike the chlorophytes (Becker and Marin, 2009). Recent work using DNA sequences supports the hypothesis that land plants, Coleochaetales and Zygnematales are monophyletic (Laurin-Lemay *et al.*, 2012). Recent work indicates that the Zygnemophyceae are the closest living relative to land plants (Timme *et al.*, 2012).

Land plants

AOX first attracted attention in thermogenic plant species, but subsequent research has indicated that it is widespread in land plants. Most of this research has taken place in angiosperms. Within the bryophytes (non-vascular seedless plants), an AOX sequence has been identified in the model moss *Physcomitrella patens* (Neimanis *et al.*, 2013; Table 3.1). Molecular biology experiments have confirmed the presence of the AOX gene and the expression of mRNA in this species (Neimanis *et al.*, 2013). A 35 kDa protein that cross-reacted with an AOX1/2 antibody was detected in Western blots in isolated moss mitochondria (Lang *et al.*, 2011).

Significantly, the genome of *P. patens* has been fully sequenced and this species contains only one AOX gene. This is in significant contrast to all angiosperm plants investigated to date which contain an AOX multigene family. Evolution of embryophytes has been described as utilizing protein family expansion and later differential expression as opposed to large changes in sequence (Becker and Marin, 2009). Bioinformatics also detected putative AOX sequences in the liverwort *Marchantia polymorpha*; however AOX was not detected in hornworts, likely due to the low availability of sequence data (Neimanis *et al.*, 2013; Table 3.1).

Within the vascular seedless plants (Tracheophyta), bioinformatics analyses found AOX sequences in several species of *Selanginella* and ferns (Neimanis *et al.*, 2013; Table 3.1). No sequence data are available to search for AOX within the horsetails or whiskferns (Neimanis *et al.*, 2013; Table 3.1). Within the gymnosperms, AOX was detected within the gnetophyte *Ephedra distachya* and several species of conifers, but no data are available for Ginkophyta or Cycadophyta (Neimanis *et al.*, 2013; Table 3.1). Within the conifers, reports of cyanide-resistant respiration exist for *Picea glauca* root mitochondria (white spruce) (Johnson-Flanagan and Owens, 1986; Weger and Guy, 1991) and *Araucaria angustifolia* mitochondria (Mariano *et al.*, 2008). Cyanide-resistant respiration has also been observed in purified mitochondria of *Picea abies* and *Abies cephalonica* (Petrussa *et al.*, 2008).

In angiosperms, early studies of AOX occurred in various thermogenic species (generally members of Araceae) and potato tubers (Van der Plas and Wagner, 1980). Early work focused on characterizing the AOX enzyme and identifying the protein responsible. More recent work has focused on identifying the genes encoding AOX and better understanding enzyme structure and regulation. Several recent papers have also put forward novel ideas about the physiological role of the enzyme.

Recent functional hypotheses based on studies of AOX in multiple plants

Some major themes have recently started to take shape about the possible physiological function(s) of AOX when work in several plant systems has been surveyed. The first is the concept that AOX may be serving to influence how a plant is utilizing and partitioning carbon by shifting to more anabolic metabolism (Mathy *et al.*, 2010), the synthesis of particular amino acids (Gupta *et al.*, 2012), phenolic metabolism (Sircar *et al.*, 2012), and adaptive phenylpropanoid and lignin metabolism (Macedo *et al.*, 2012). These changes in carbon allocation may underlie the phenomenon of AOX's role in allowing plants to resist abiotic environmental stresses such as cold (Li *et al.*, 2012) or salt stress (Mhadhbi *et al.*, 2013) or biotic stresses due to pathogens (Zhu *et al.*, 2012) or herbivores (Zhang *et al.*, 2012). It is becoming increasingly clear that the over- or under-expression of AOX genes in many different plants leads to a retooling of metabolism (i.e. cellular reprogramming) within the organism (Arnholdt-Schmitt *et al.*, 2006). Naturally caused perturbations in homeostasis have led to the hypothesis that AOX may allow plants to effectively react to these metabolic fluctuations (Rasmusson *et al.*, 2009). AOX is perfectly positioned at the nexus of carbon metabolism and energy production to convey metabolic flexibility to organisms which have it.

The second theme gaining attention is the protection of cells and mitochondria from the excess generation of reactive oxygen species (ROS). This effect was first observed in transgenic tobacco cells expressing altered levels of AOX (Maxwell *et al.*, 1999). Recent studies have expanded these findings to other members of the Viridiplantae and have investigated the effects on key cellular processes such as photosynthesis (Mathy *et al.*, 2010; Zhang *et al.*, 2011). In addition to ROS, AOX has now been found to have a role in the generation of nitric oxide in tobacco leaf mitochondria (Cvetkovska and Vanlerberghe, 2012). This indicates that AOX may serve to affect cellular signalling pathways by contributing to NO levels.

The third theme revolves around plant reproduction. Our attention was first called to AOX in thermogenic plants, but it has been known for some time that AOX levels rapidly increase during the climacteric stage of fruit development in some species, such as mango (Cruz-Hernández and Gómez-Lim, 1995). AOX is starting to be linked to fruit ripening processes and the hormones that are involved (e.g. ethylene) (Xu *et al.*, 2012). One very interesting hypothesis is that AOX is involved in temperature-dependent seed ejection in the dwarf mistletoe (Friedman *et al.*, 2013).

Where should efforts be focused next?

Bioinformatics and molecular biology tools should be used in the future to fill in some of the gaps in our knowledge. For example, it would be useful to obtain information on AOX in groups where no data are available including the Pedinophyceae within the Chlorophyta and the Mesostigmatophyceae within the streptophyte algae (Table 3.1). Little work has been done to date on the AOXs of algae, with the exception of *C. reinhardtii,* and some of these species may be very amenable to laboratory research. Within the non-vascular seedless plants, we hypothesize that AOX will be present in the Anthocerotophyta (hornworts), but this will likely have to be confirmed experimentally given the current lack of resources on DNA sequence data (Table 3.1). A large gap in our knowledge of AOX exists within the vascular seedless plants. Research efforts should focus on investigating AOX in horsetails, whiskferns and their close relatives (Table 3.1). Within the Spermatophyta we have no information about AOX in the Cycadophyta, Ginkophyta and Gnetophyta (Table 3.1). Although AOX has been studied most thoroughly in angiosperms, the focus has been on monocots and dicots. Investigations into the presence of AOX in basal angiosperms would be useful.

References

Archibald, J.M. (2006) Endosymbiosis: Double-take on plastid origins. *Current Biology* **16**: R690–R692.

Arnholdt-Schmitt, B., Costa, J.H. and de Melo, D.F. (2006) AOX – a functional marker for efficient cell reprogramming under stress? *Trends in Plant Science* **11**: 281–287.

Baurain, D., Dinant, M., Coosemans, N. and Matagne, R.F. (2003) Regulation of the alternative oxidase *Aox1* gene in *Chlamydomonas reinhardtii*. Role of the nitrogen source on the expression of a reporter gene under the control of the *Aox1* promoter. *Plant Physiology* **131**: 1418–1430.

Becker, B. (2013) Snow ball earth and the split of Streptophyta and Chlorophyta. *Trends in Plant Science* **18**: 180–183.

Becker, B. and Marin, B. (2009) Streptophyte algae and the origin of embryophytes. *Annals of Botany* **103**: 999–1004.

Bendall, D.S. and Bonner, W.D. (1971) Cyanide-insensitive respiration in plant mitochondria. *Plant Physiology* **47**: 236–245.

Church, A.H. (1908) *Types of Floral Mechanism*. Part I. Oxford Press. pp. 70–74.

Cruz-Hernández, A. and Gómez-Lim, M. (1995) Alternative oxidase from mango (*Mangifera indica*, L.) is differentially regulated during fruit ripening. *Planta* **197**: 569–576.

Cvetkovska, M. and Vanlerberghe, G.C. (2012) Alternative oxidase modulates leaf mitochondrial concentrations of superoxide and nitric oxide. *New Phytologist* **195**: 32–39.

Dinant, M., Baurain, D., Coosemans, N. *et al.* (2001) Characterization of two genes encoding the mitochondrial alternative oxidase in *Chlamydomonas reinhardtii*. *Current Genetics* **39**: 101–108.

Edwards, S.W. and Lloyd, D. (1978) Properties of mitochondria isolated from cyanide-sensitive and cyanide-stimulated cultures of *Acanthamoeba castellanii*. *Biochemical Journal* **174**: 203–211.

Elthon, T.E., Nickels, R.L. and McIntosh, L. (1989) Monoclonal antibodies to the alternative oxidase of higher plant mitochondria. *Plant Physiology* **89**: 1311–1317.

Elthon, T.E. and McIntosh, L. (1987) Identification of the alternative terminal oxidase of higher plant mitochondria. *Proceedings of the National Academy of Sciences USA* **84**: 8399–8403.

Eriksen, N.T. and Lewitus, A.J. (1999) Cyanide-resistant respiration in diverse marine phytoplankton. Evidence for the widespread occurrence of the alternative oxidase. *Aquatic Microbial Ecology* **17**: 145–152.

Farquhar, J., Zerkle, A.L. and Bekker, A. (2011) Geological constraints on the origin of oxygenic photosynthesis. *Photosynthesis Research* **107**: 11–36.

Friedman, C.R., Ross, B.N. and Martens, G.D. (2013) An antibody against a conserved C-terminal consensus motif from plant alternative oxidase (AOX) isoforms 1 and 2 label plastids in the explosive dwarf mistletoe (*Arceuthobium americanum*, Santalaceae) fruit exocarp. *Protoplasma* **250**: 317–323.

Gomes, C.M., Le Gall, J., Xavier, A.V. and Teixeira, M. (2001) Could a diiron containing four-helix-bundle protein have been a primitive oxygen reductase? *ChemBioChem* **2**: 593–587.

Gray, M.W., Burger, G. and Lang, B.F. (2001) The origin and early evolution of mitochondria. *Genome Biology* **2**: 1018.1.

Gupta, K.J., Shah, J.K., Brotman, Y. *et al.* (2012) Inhibition of aconitase by nitric oxide leads to induction of the alternative oxidase and to a shift of metabolism towards biosynthesis of amino acids. *Journal of Experimental Botany* **63**: 1773–1784.

Hill, G.C. and Cross, G.A.M. (1973) Cyanide-resistant respiration and a branched cytochrome system in Kinetoplastidae. *Biochimica et Biophysica Acta* **305**: 590–596.

James, W.O. and Beevers, H. (1950) The respiration of *Arum spadix*. A rapid respiration, resistant to cyanide. *New Phytologist* **49**: 353–374.

Jarmuszkiewicz, W., Behrendt, M., Navet, R. and Sluse, F.E. (2002) Uncoupling protein and alternative oxidase of *Dictyostelium discoideum*: occurrence, properties and protein expression during vegetative life and starvation-induced early development. *FEBS Letters* **532**: 459–464.

Johnson-Flanagan, A.M. and Owens, J.N. (1986) Root respiration in white spruce (*Picea glauca* [Moench] Voss) seedlings in relation to morphology and environment. *Plant Physiology* **81**: 21–25.

Lambowitz, A.M. and Slayman, C.W. (1971) Cyanide-resistant respiration in *Neurospora crassa*. *Journal of Bacteriology* **108**: 1087–1096.

Lang, E.G., Mueller, S.J., Hoernstein, S.N. *et al.* (2011) Simultaneous isolation of pure and intact chloroplasts and mitochondria from moss as the basis for sub-cellular proteomics. *Plant Cell Reports* **30**: 205–215.

Laurin-Lemay, S., Brinkmann, H. and Philippe, H. (2012) Origin of land plants revisited in the light of sequence contamination and missing data. *Current Biology* **22**: R593–R594.

Lemieux, C., Otis, C. and Turmel, M. (2007) A clade uniting the green algae *Mesostigma viride* and *Chlorokybus atmophyticus*. *BMC Biology* **5**: 2.

Macedo, E.S., Sircar, D., Cardoso, H.G. *et al.* (2012) Involvement of alternative oxidase (AOX) in adventitious rooting of *Olea europaea* L. microshoots is linked to adaptive phenylpropanoid and lignin metabolism. *Plant Cell Reports* **31**: 1581–1590.

Li, Y.M., Zhu, L., Xu, B.C. *et al.* (2012) Identification of down-regulated genes modulated by an alternative oxidase pathway under cold stress conditions in watermelon plants. *Plant Molecular Biology Reporter* **30**: 214–224.

Mariano, A.B., Valente, C., Mauer, J.B.B. *et al.* (2008) Functional characterization of mitochondria isolated from the ancient gymnosperm *Araucaria angustifolia*. *Plant Science* **175**: 701–705.

Mathy, G., Cardol, P., Dinant, M. *et al.* (2010) Proteomic and functional characterization of a *Chlamydomonas reinhardtii* mutant lacking the mitochondrial alternative oxidase. *Journal of Proteome Research* **9**: 2825–2838.

Maxwell, D.P., Wang, Y. and McIntosh, L. (1999) The alternative oxidase lowers mitochondrial reactive oxygen production in plant cells. *Proceedings of the National Academy of Sciences of the USA* **96**: 8271–8276.

McDonald, A.E. and Vanlerberghe, G.C. (2006) Origins, evolutionary history, and taxonomic distribution of alternative oxidase and plastoquinol terminal oxidase. *Comparative Biochemistry and Physiology Part D Genomics & Proteomics* **1**: 357–364.

McDonald AE, and Vanlerberghe G. 2004. Branched mitochondrial electron transport in the Animalia: presence of alternative oxidase in several animal phyla. *IUBMB Life.* **56**: 333–341.

McDonald, A.E., Amirsadeghi, S. and Vanlerberghe, G.C. (2003) Prokaryotic orthologues of mitochondrial alternative oxidase and plastid terminal oxidase. *Plant Molecular Biology* **53**: 865–876.

Mhadhbi, H., Fotopoulos, V., Mylona, P.V. *et al.* (2013) Alternative oxidase 1 (*Aox1*) gene expression in roots of *Medicago truncatula* is a genotype-specific component of salt stress tolerance. *Journal of Plant Physiology* **170**: 111–114.

Neimanis, K., Staples, J.F., Hüner, N.P. and McDonald, A.E. (2013) Identification, expression, and taxonomic distribution of alternative oxidases in non-angiosperm plants. *Gene* **526**: 275–286.

Petrussa, E., Bertolini, A., Krajňáková, J. *et al.* (2008) Isolation of mitochondria from embryogenic cultures of *Picea abies* (L.) Karst. and *Abies cephalonica* Loud.: characterization of a K+(ATP) channel. *Plant Cell Reports* **27**: 137–146.

Rasmusson, A.G., Fernie, A.R. and van Dongen, J.T. (2009) Alternative oxidase: a defense against metabolic fluctuations? *Physiologia Plantarum* **137**: 371–382.

Reyes-Prieto, A., El-Hafidi, M., Moreno-Sanchez, R. and Gonzalez-Halphen, D. (2002) Characterization of oxidative phosphorylation in the colorless chlorophyte *Polytomella* sp. Its mitochondrial respiratory chain lacks a plant-like alternative oxidase. *Biochimica et Biophysica Acta – Bioenergetics* **1554**: 170–179.

Rhoads, D.M. and McIntosh, L. (1991) Isolation and characterization of a cDNA clone encoding an alternative oxidase protein of *Sauromatum guttatum* (Schott). *Proceedings of the National Academy of Sciences USA* **88**: 2122–2126.

Sircar, D., Cardoso, H.G, Mukherjee, C. *et al.* (2012) Alternative oxidase (AOX) and phenolic metabolism in methy jasmonate-treated hairy root cultures of *Daucus carota* L. *Journal of Plant Physiology* **169**: 657–663.

Storey, B.T. (1976) Respiratory chain of plant mitochondria. XVIII. Point of interaction of the alternative oxidase with the respiratory chain. *Plant Physiology* **58**: 521–525.

Timme, R.E., Bachvaroff, T.R. and Delwiche, C.F. (2012) Broad phylogenomic sampling and the sister lineage of land plants. *PLoS One* **7**: e29696.

Turmel, M., Otis, C. and Lemieux, C. (2007) An unexpectedly large and loosely packed mitochondrial genome in the charophycean green alga *Chlorokybus atmophyticus*. *BMC Genomics* **8**: 137.

Turner, G. and Rowlands, R.T. (1976) Cytochrome abnormalities and cyanide-resistant respiration in extranuclear mutants of *Aspergillus nidulans*. *Journal of Bacteriology* **125**: 389–397.

Vanderplas, L.H.W. and Wagner, M.J. (1980) Influence of ethanol on alternative oxidase in mitochondria from callus-forming potato-tuber disks. *Physiologia Plantarum* **49**: 121–126.

Weger, H.G. and Guy, R.D. (1991) Cytochrome and alternative pathway respiration in white spruce (*Picea glauca*) roots-effects of growth and measurement temperature. *Physiologia Plantarum* **83**: 675–681.

Xu, F., Yuan, S., Zhang, D.W. *et al.* (2012) The role of alternative oxidase in tomato fruit ripening and its regulatory interaction with ethylene. *Journal of Experimental Botany* **63**: 5705–5716.

Zhang, L., Oh, Y., Li, H.Y. *et al.* (2012) Alternative oxidase in resistance to biotic stresses: *Nicotiana attenuata* AOX contributes to resistance to a pathogen and a piercing-sucking insect but not *Manduca sexta* larvae. *Plant Physiology* **160**: 1453–1467.

Zhang, L.T., Zhang, Z.S, *et al.* (2011) Mitochondrial alternative oxidase pathway protects plants against photoinhibition by alleviating inhibition of the repair of photodamaged PSII through preventing formation of reactive oxygen species in *Rumex* K-1 leaves. *Physiologia Plantarum* **143**: 396–407.

Zhu, L., Li, Y.M., Ara, N. *et al.* (2012) Role of a newly cloned alternative oxidase gene (*BjAOX1a*) in turnip mosaic virus (TuMV) resistance in mustard. *Plant Molecular Biology Reporter* **30**: 309–318.

CHAPTER 4

Alternative pathways and phosphate and nitrogen nutrition

Anna M. Rychter and Bożena Szal
Institute of Experimental Plant Biology and Biotechnology, Faculty of Biology, University of Warsaw, Warsaw, Poland

Introduction

A sufficient nutrient supply is essential for plant growth and development. In their natural environment, plants are often exposed to variations in nutrient availability. Nutrient deficiency or changes in soil nutrient composition is often observed in nature and may inhibit growth and have a severe impact on crop yield. Uptake and assimilation of nutrients require a substantial amount of energy in the form of ATP, and thus reduces the plant resources. Therefore, adaptation to variations in nutrient supply requires changes in metabolism. Plant metabolism is generally flexible because of the presence of alternative respiratory pathways that facilitate adaptation and continuation of growth in a changing environment.

Nutrient uptake mainly occurs in the roots, although the major part of nitrate assimilation occurs in the leaves. The main source of energy for ion uptake and assimilation is derived from respiration. In this chapter, we describe the modifications in respiratory metabolism and the participation of alternative respiratory pathways in plant adaptation to the variations in the supply of two essential macronutrients, namely phosphate (P) and nitrogen (N).

Phosphate limitation

Phosphate, which is mainly derived from the soil in the form orthophosphate $H_2PO_4^-$ (Pi), participates in plant metabolism by regulating the activities of various enzymes. It plays an essential role in energy transduction processes, in the form of ATP or pyrophosphate (PPi). In addition, P is a constituent of several metabolically important metabolites such as sugar and organic acid phosphates, phospholipids and phosphorylated proteins. A decrease in Pi in the environment

Alternative Respiratory Pathways in Higher Plants, First Edition.
Edited by Kapuganti Jagadis Gupta, Luis A.J. Mur and Bhagyalakshmi Neelwarne.

and consequently in plant tissues induces adaptations in both morphology (e.g. an increase in root/shoot ratio, mycorrhizal symbiosis) and metabolism (e.g. secretion of phosphatases and organic acids, increase in activity of Pi transporters, decrease in nitrate uptake and modification of carbon metabolism; for review, see Raghothama, 1999; Vance *et al.*, 2003; Plaxton, 2004; Plaxton and Tran, 2011).

Respiratory pathways such as glycolysis, the tricarboxylic acid (TCA) cycle, and the mitochondrial electron transport chain (mtETC), are necessary for the production of energy and intermediates essential for plant growth and development. Plant respiration is controlled by carbohydrate supply and ADP availability (Azcón-Bieto and Osmond, 1983; Lambers, 1985). Low Pi levels in the tissues is followed by a decrease in ATP and ADP concentration and altered adenylate metabolism (Duff *et al.*, 1989a; Rychter *et al.*, 1992; Plaxton and Podestá, 2006). In contrast to declining ATP concentration, the PPi level remains unchanged (Theodorou and Plaxton, 1993; Rychter and Randall, 1994).

In Pi-deficient conditions, plant growth rate is slower, and the uptake of other nutrients, such as nitrate, may be lower (Gniazdowska *et al.*, 1998); however, plants can survive extended periods of low Pi in the environment. To maintain growth, respiratory metabolism undergoes several modifications to adapt to Pi stress conditions. Modifications of respiratory metabolism involve induction of glycolytic non-phosphorylating bypass enzymes and changes in the function of the respiratory chain, including the participation of non-phosphorylating alternative pathways (Theodorou and Plaxton, 1993; Plaxton and Podestá, 2006) (Figure 4.1).

Glycolysis involves several enzymes that are controlled by adenylates and/or Pi, namely hexokinase, phosphofructokinase (PFK), 3-phosphoglycerate (3-PGA) kinase, pyruvate kinase (PK), and phosphorylating NAD^+-dependent glyceraldehyde-3-phosphate dehydrogenase (GAPDH) (Plaxton, 1996; Plaxton and Podestá, 2006). The influence of Pi deficiency on glycolysis has been extensively studied in *Brassica nigra* suspension cells (Duff *et al.*, 1989a; Duff *et al.*, 1989b; Theodorou *et al.*, 1992; Theodorou and Plaxton, 1994; Moraes and Plaxton, 2000). The pioneering work of Plaxton group elegantly demonstrated that during Pi deficiency in *Brassica nigra* suspension cells, enzymes that omit adenylate and Pi-dependent steps are up-regulated, including sucrose synthase (SuSy), UDP–glucose pyrophosphorylase, PPi-dependent phosphofructokinase (PPi-PFK), non-phosphorylating $NADP^+$-GAPDH, phosphoenolpyruvate (PEP) carboxylase (PEPC), PEP phosphatase, and NAD-malic enzyme (ME) (Figure 4.1), allowing glycolysis to continue despite low adenylate and Pi cell concentrations (reviewed by Plaxton, 1996; Plaxton and Podestá, 2006; Plaxton and Tran, 2011).

The first highly controlled step of glycolysis is phosphorylation of fructose-6-phosphate (Fru-6-P) catalysed by ATP-dependent PFK (Plaxton, 1996; Fernie *et al.*, 2004). Fru-6-P is also phosphorylated by PPi-PFK (Botha and Small, 1987; Dennis and Greysone, 1987; Plaxton and Podestá, 2006)

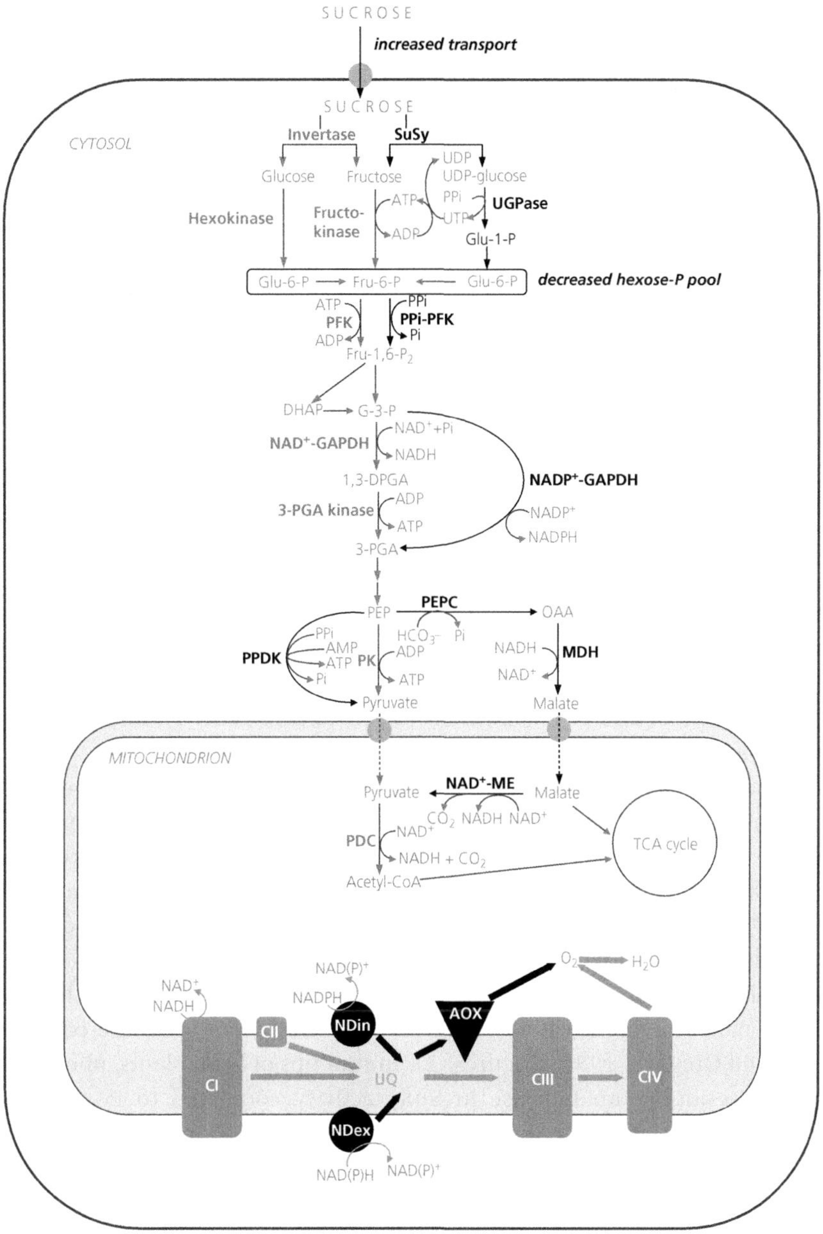

Figure 4.1 Alternative pathways of cytosolic glycolysis and mitochondrial electron transport (indicated in black) engaged in Pi-deficiency. Enzyme abbreviations: SuSy, sucrose synthase; UGPase, UDP-glucose pyrophosphorylase; ATP-PFK, ATP-dependent phosphofructokinase; PPi-PFK, PPi-dependent phosphofructokinase; NAD^+-GAPDH, phosphorylating NAD^+-dependent glyceraldehyde-3-phosphate dehydrogenase; $NADP^+$-GAPDH, non-phosphorylating glyceraldehyde-3-phosphate dehydrogenase; 3-PGA kinase, 3-phosphoglycerate kinase; PEPC, phosphoenolpyruvate carboxylase; PPDK, pyruvate Pi dikinase; PK, pyruvate kinase, MDH, malic dehydrogenase; NAD^+-ME, NAD^+ malic enzyme; PDC, pyruvate dehydrogenase complex; NDin, internal NAD(P)H dehydrogenase; NDex, external NAD(P)H dehydrogenase; AOX, alternative oxidase.

Source: Adapted from Plaxton and Tran (2011).

(Figure 4.1). PPi-PFK is an adaptive enzyme induced during Pi stress and anaerobiosis (Duff *et al.*, 1989a; Mertens, 1991). During Pi deficiency in *Brassica nigra* suspension cells, PPi-PFK activity is increased and acts as a glycolytic bypass to ATP-dependent PFK when ATP level is low, whereas PPi level is unaffected (Duff *et al.*, 1989a; Theodorou and Plaxton, 1993). Induction of PPi-PFK activity during low Pi levels in the cells involves *de novo* synthesis of one of its subunits, resulting in an increased sensitivity to fructose-2,6-bisphosphate (Fru-2,6-P_2) activation (Theodorou *et al.*, 1992). When Pi levels in suspension cells are low, the next steps of glycolysis that involve the activity of phosphorylating (Pi-dependent) NAD^+-GAPDH are bypassed by non-phosphorylating $NADP^+$-GAPDH, giving 3-PGA. The following step of conversion of PEP to pyruvate is catalysed by ADP-dependent PK (Plaxton, 1996), which plays an important role in the control of respiration (Plaxton and Podestá, 2006). As demonstrated in *Brasica nigra* suspension cells subjected to low Pi conditions, the NAD^+-GAPDH phosphorylation step is bypassed through the action of vacuolar PEP phosphatase (Duff *et al.*, 1989a, 1989b) and by the consecutive action of PEPC, NAD^+-malate dehydrogenase (MDH) and $NADP^+$-ME (Figure 4.1) (Theodorou and Plaxton 1993; Nagano *et al.*, 1994). These responses to low Pi and ATP conditions in cell culture allow glycolysis to continue and to supply substrates for mitochondrial respiration.

Alternative glycolytic pathways have also been reported to operate in whole plants grown in Pi-deficient conditions. One of the first metabolic responses in bean and soybean plants to the decreasing Pi concentrations is the increase in sucrose transport and sugar levels in the roots (Fredeen *et al.*, 1989; Cakmak *et al.*, 1994a, 1994b; Rychter and Randall, 1994; Ciereszko *et al.*, 1996). Sucrose is hydrolysed by either invertase (generating glucose and fructose) or SuSy, which yields UDP-glucose and fructose. Thus, to hydrolyse sucrose to Fru-6-P, through the SuSy pathway, requires only half an ATP used in the invertase pathway (Dennis and Greyson, 1987) (Figure 4.1). In root tips of bean plants, phosphate deficiency results in an increase in SuSy activity compared to Pi-sufficient plants, with no significant differences in either acid or neutral invertases (Ciereszko *et al.*, 1998). Despite high sugar levels in Pi-deficient bean plants, the hexose-P pool remains several times lower than that in Pi-sufficient plants (Rychter and Randall, 1994). The low level of hexose phosphates might reflect the depletion of the energy resource pool, as well as a lower phosphorylation rate of hexoses. During prolonged phosphate deficiency, the activities of hexokinases and fructokinase decrease by approximately 30% compared to those in phosphate-sufficient plants (Rychter and Randall, 1994), indicating that the phosphorylation rate may be partially responsible for the occurrence of a low hexose phosphate pool. Prolonged phosphate starvation and low ATP levels in bean plants decreased PFK activity by 50% compared to that in the roots of Pi-sufficient plants, whereas PPi-PFK activity remained unchanged (Rychter and Randall, 1994). Alternative routes for PEP to pyruvate conversion have also been detected in Pi-deficient bean plants, such as an increase in PEPC activity and PEP phosphatase in leaves and roots (Kondracka and Rychter, 1997; Juszczuk and Rychter, 2002). The increase in activity of alternative routes for PEP conversion to pyruvate in bean plants may be a response to the higher demand for pyruvate and the need for Pi recycling.

In phosphate-starved suspension cells of *Brassica nigra* or *Catharanthus roseus*, adenylate-dependent glycolytic enzymes are unaltered or only slightly inhibited (Duff *et al.*, 1989a; Li and Ashihara, 1990). In bean roots, PFK activity is lower, whereas PK remains unchanged (Rychter and Randall, 1994; Juszczuk and Rychter, 2002). It appears that metabolic responses to phosphate starvation may differ between enzymes in cell cultures and whole plants. Restriction of ATP-dependent enzymes does not always result in the induction of ATP-independent enzymes, and thus the estimation of enzyme activity *in vitro* with saturating substrate concentration may not always reflect *in vivo* activity.

In terms of PEP conversion through the action of PEPC, NAD-MDH may also be important for organic acid exudation in the roots when soil inorganic Pi levels are extremely low (Vance *et al.*, 2003). Enhanced activities of these enzymes and citrate synthase were observed together with increased synthesis of malic and citric acid which can be exuded by the roots of Pi-deficient plants (Johnson *et al.*, 1996; Neuman and Romheld, 1999). Previous studies have described a marked transcriptional regulation of genes encoding PEPC isoenzyme *PEPC1* from *Arabidopsis thaliana* (Gregory *et al.*, 2009) and genes related to organic acid metabolism in white lupin (Uhde-Stone *et al.*, 2003). Under Pi-deficient conditions in rice roots, several genes related to glycolysis increased their expression, providing carbon sources for the TCA cycle (Wasaki *et al.*, 2003). A more recent report has described the changes in the expression of genes engaged in Pi uptake and in the glycolytic Pi/ATP-consuming metabolic steps in *Arabidopsis* roots (Lan *et al.*, 2012; Plaxton and Tran, 2011 and references therein).

Adaptive responses of mitochondrial respiration to Pi limitation have been examined in whole plants, cell cultures and isolated mitochondria (Rychter and Mikulska, 1990; Rychter *et al.*, 1992; Hoefnagel *et al.*, 1993; Hoefnagel *et al.*, 1994; Wanke *et al.*, 1998; Parsons *et al.*, 1999; Gonzàlez-Meler *et al.*, 2001; Juszczuk *et al.*, 2001). In bean plants, the decrease in Pi levels slightly affected oxygen uptake in roots, although respiration was mainly resistant to the cyanide (Rychter and Mikulska, 1990). The plant respiratory chain has two pathways of electron transport branching at ubiquinone, UQ, which is the cytochrome pathway coupled to ATP synthesis, and an alternative pathway with alternative oxidase (AOX) not coupled to ATP synthesis, which is responsible for cyanide-resistant respiration. Additionally, external and internal NAD(P)H type II dehydrogenases (NDex and NDin, respectively) transport electrons to UQ, omitting the Complex I phosphorylation site (Figure 4.1). Participation of NDin and NDex and/or the increase of alternative pathway respiration play a role in the maintenance of carbon metabolism during conditions of limited ADP and/or Pi. Therefore, the involvement of alternative pathways in mitochondrial respiration during Pi deficiency may be a similar adaptation to the induction of adenylate and Pi-independent pathways in glycolysis.

Studies of bean root respiration have indicated that concomitantly with declining Pi and ATP tissue concentrations, oxygen uptake becomes insensitive to cyanide and uncoupler addition, thus indicating a dependence of respiration on alternative, non-phosphorylating pathways (Rychter and Mikulska, 1990; Wanke *et al.,* 1998). Moreover, higher cyanide resistance and possible electron flux through alternative pathways were positively correlated with lower cytochrome pathway activity and relative growth rates (Rychter and Mikulska, 1990; Gniazdowska *et al.,* 1998). In mitochondria isolated from phosphate-deficient bean plants, lower cytochrome *c* oxidase (COX) activity, higher expression of AOX protein, and no uncoupler effect were observed, indicating possible AOX involvement (Rychter *et al.*, 1992; Juszczuk *et al.*, 2001).

However, the investigations of regulation of the activity and the *in vivo* participation of AOX indicated that AOX can compete with the cytochrome pathway for the reduced UQ pool; therefore, the use of inhibitors in the estimation of AOX activity (engagement in total respiration) has been questioned (Day *et al.*, 1996). Moreover, it has been reported that an increase in AOX protein alone may not always reflect increased AOX activity (Ribas-Carbo *et al.*, 1995; Lennon *et al.,* 1997). Through the use of inhibitors, maximum electron flux by AOX, termed 'AOX capacity' could be estimated, whereas the actual engagement, 'AOX activity', could be directly determined by the non-invasive technique of isotope discrimination (Ribas-Carbo *et al.*, 1995). Thus, the results indicating the increase in AOX protein level and capacity should be re-examined using the isotope discrimination technique to estimate the actual AOX engagement (AOX activity) in total respiration.

To show the role of AOX in the adaptation to Pi deficiency in tobacco cell cultures, a molecular genetic approach was conducted by the Vanlerberghe group (Parsons *et al.,* 1999). The growth and respiration of wild-type tobacco cells grown on Pi-sufficient medium were compared to transgenic tobacco cells (AS8) that harboured an antisense construct of the tobacco gene, *Aox1*. Pi deficiency had no influence on the respiration rate of wild-type cells, but respiration in the AS8 cells was repressed. AOX protein levels in the wild-type cells grown on Pi-sufficient medium were almost undetectable, but when cells were transferred to Pi-deficient medium, the levels of AOX protein increased significantly, concomitant with a high rate of cyanide-resistant respiration (AOX capacity). No AOX proteins were detected in the AS8 cells and almost no AOX capacity was observed in either Pi-sufficient or Pi-deficient media. Thus, transgenic AS8 cells subjected to Pi-deficient conditions, despite altered metabolism and growth, are unable to induce AOX expression, in contrast to the wild-type tobacco cells (Parsons *et al.,* 1999). In mitochondria isolated from tobacco suspension cells, higher AOX, NDin and NDex capacities were observed in response to low-Pi medium (Sieger *et al.*, 2005). Therefore, in Pi deficiency, oxidation of cytosolic or mitochondrial NADH can bypass all phosphorylation sites (Figure 4.1). In *Arabidopsis* plants cultured in Pi-deficient medium, both ND and AOX capacities increase. ND genes with increased transcript levels in response to Pi deficiency include *Atnda2* (NDin) and *Atndb2* (NDex) (Vijayraghavan and Soole, 2010). Earlier microarray studies (Hammond *et al.*, 2003) indicated increased transcript levels of *Ataox1a* in *Arabidopsis* in response to limited Pi nutrition. Therefore, in *Arabidopsis*, Pi deficiency results in the bypass of adenylate control, which ensures alternative electron flow pathways in the mtETC, consisting of both internal and external dehydrogenases and the synthesis of AOX.

The use of the oxygen isotope technique has allowed investigations of the regulation and alternative pathway activity acting as an electron bypass to the cytochrome path during Pi-limited conditions in plant leaf tissues and tissue cultures (Gonzàlez-Meler *et al.*, 2001). Although Pi deficiency was found to reduce cytochrome pathway activity in both bean and tobacco leaves, alternative pathway and AOX protein levels were shown to increase only in bean. This confirmed the involvement of AOX in the total respiration of bean plants, as suggested by previous reports (Rychter *et al.*, 1992; Juszczuk *et al.*, 2001). In tobacco leaf tissues, alternative pathway activity decrease compared to that in plants grown in Pi-sufficient medium. The response to Pi deficiency of tobacco cell cultures, in which AOX capacity and protein level increase, and tobacco leaves, in which AOX protein level remains unchanged, indicate that the metabolic responses of cell cultures and whole plants are different (Gonzàlez-Meler *et al.*, 2001). In *Gliricidia sepiu*, low phosphate concentrations result in an increase in alternative pathway activity whereas the cytochrome pathway activity remains unchanged compared to that in plants grown in a full nutrient medium. It was concluded that the role of the alternative pathway as a bypass mechanism for the restricted cytochrome pathway is species-dependent (occurring in bean plants but not in tobacco) and the increase in protein levels does not necessarily reflect higher AOX activity (Gonzàlez-Meler *et al.*, 2001). Similarly, in the leaves of MSC16 cucumber mutants, an increase in AOX protein level compared to that in wild plants does not correspond to an increase in alternative pathway activity (Juszczuk *et al.*, 2007).

The increase in alternative pathway activity lowers respiratory chain reactive oxygen species (ROS) formation by modulating the reduction state of respiratory chain components (Millenaar *et al.*, 1998, Maxwell *et al.*, 1999; Møller, 2001). Phosphate deficiency has been reported to cause oxidative stress in bean plants (Juszczuk *et al.*, 2001; Malusá *et al.*, 2002) and tobacco cell cultures (Parsons *et al.*, 1999; Sieger *et al.*, 2005). In low Pi conditions in *Arabidopsis* plants and tobacco cell cultures, an enhanced expression of genes encoding proteins engaged in several aspects of oxidative stress has been observed (Hernández *et al.*, 2007; Vijayraghavan and Soole, 2010).

Increased AOX capacity, protein levels, and transcript levels have been observed in tobacco cells in response to added H_2O_2 (Vanlerberghe and McIntosh, 1996). Moreover, antisense knockout of AOX in tobacco cells and in *Arabidopsis* has been reported to increase mitochondrial ROS formation (Maxwell *et al.*, 1999; Umbach *et al.*, 2005). These data indicate the involvement of AOX in the oxidative defence system, as observed with other stresses (reviewed by Rasmusson *et al.*, 2009; Vanlerberghe *et al.*, 2009; Vanlerberghe, 2013).

During Pi limitation, the induction and the activity of alternative respiratory pathways may be species-dependent and may vary among different plants. Moreover, as previously indicated, numerous studies on plant cell cultures have estimated the metabolic responses to Pi limitation, but the results should be not extended to whole plants because these data may not always correspond to *in vivo* conditions (Gonzàlez-Meler *et al.*, 2001).

The flexibility of respiratory metabolism in plants enables these organisms to survive in Pi-limited conditions for a period of time. A switch to alternative, energy-saving pathways is important for the temporary adaptation to changes in Pi concentrations. Previous studies have shown that during Pi deficiency in bean plants, AOX acts together with glycolytic bypass mechanisms (Rychter *et al.*, 1992; Rychter and Randall, 1994; Juszczuk *et al.*, 2001). Thus, modification of respiratory chain activity (participation of alternative pathway) allows carbon flow during glycolysis despite changes in the level of ATP control. However, the engagement of alternative respiratory pathways, resulting in a decrease in ATP tissue concentration, has negative effects on plant development, growth rate and uptake of other nutrients, such as nitrate (Gniazdowska *et al.*, 1998). Thus, plants also develop other strategies for Pi acquisition, such as increased root development, organic acid exudation and mycorrhizal symbiosis (reviewed by Vance *et al.*, 2003), which functions in the retrieval of other Pi sources in the environment and relieves plants from Pi deficiency.

Nitrogen nutrition and respiratory pathways

Plant productivity is largely determined by N nutrition, and the influence of cellular N status on whole plant metabolism has been extensively studied for several decades. However, compared to phosphate nutrition, studies on N nutrition appear to be more complicated. Firstly, some authors have compared the metabolism of N-deficient plants with that in control plants grown on N-replete medium. Others have described changes in metabolism/gene regulation in response to N supply after a period of N deprivation. In addition to tissue- and organ-specific responses to N status (culture *in vitro* versus whole plants or roots versus shoots), N metabolism is also highly dependent on photoperiod (Matt *et al.*, 2001; Nunes-Nesi *et al.*, 2010). Moreover, even though nitrate is the major source of nitrogen in most plants, ammonium is also taken up by plants. The form in which N is supplied exerts a specific influence on the cellular oxidation-reduction status of a plant. More importantly, because N is a major constituent of various metabolites and its limitation leads to decreased protein levels, calculations of the activity of specific enzymes related to this pathway may often be misleading.

Nitrogen deficit and respiratory metabolism

In nitrogen-limited plants, carbohydrate metabolism is generally altered. A striking negative relationship between N fertilization and starch level has been reported (Scheible *et al.*, 2004). Additionally, in N-limited plants, sucrose and hexose levels are elevated during restricted growth (Paul and Driscoll, 1997; Logan *et al.*, 1999; Okazaki *et al.*, 2008; Schlüter *et al.*, 2012). This observation suggests that under N-limited conditions, lowered plant growth rate may not be mainly due to limited assimilate availability, but rather to sucrose degradation and restricted glycolysis (Rufty *et al.*, 1988; Paul and Stitt, 1993). Glycolysis interacts with N assimilation through the production of intermediates, such as PEP, oxaloacetic acid and pyruvate, all of which may serve as precursors of their respective large families of amino acids. The TCA cycle is a source of C skeletons, mainly 2-oxoglutarate (2-OG), which is needed for the proper action of the chloroplastic glutamine (Gln) synthetase–Gln:2-OG aminotransferase (GS-GOGAT) cycle (see review by Szal and Podgórska, 2012). Under N-limited conditions, the demand for C intermediates decreases and energy costs for N assimilation, protein turnover and phloem loading are restricted. Therefore, in plants grown under N-limited conditions, enzymes for sucrose degradation and most of the enzymes involved in the glycolytic pathway (Table 4.1 and references therein) and the TCA cycle (Lancien *et al.*, 1999; Peng *et al.*, 2007) are repressed at the transcriptional and/or post-transcriptional level. In contrast to these observations, the activities of the TCA cycle-related enzymes have been reported to be up-regulated in N-limited plants (Makino and Osmond, 1991; Noguchi and Terashima, 2006; Watanabe *et al.*, 2010) (Table 4.1). It has been suggested that these enzymes may contribute to the consumption of excess carbohydrates and suppression of the rise in the C/N ratio (Noguchi and Terashima, 2006). An important player in the regulation of glycolysis in the context of N status is Fru-2,6-P_2. The correlation between Fru-2,6-P_2 concentrations and N tissue content has been reported in soybean (Rufty *et al.*, 1989), *Selenastum minutum* (Turpin *et al.*, 1990), *Ricinus* (Geigenberger and Stitt, 1991), tobacco (Paul and Stitt, 1993) and maize (Schlüter *et al.*, 2012). When N is added to N-deprived plants, the activation of Fru-6-P,2-kinase leads to an increase in Fru-2,6-P_2 concentrations and to the activation of PPi-PFK. This may be a result of a decreased concentration of 3-PGA (Geigenberger and Stitt, 1991), which is an inhibitor of Fru6P,2-kinase. The decrease in the concentration of metabolites of the glycolytic pathway is a simple consequence of increased turnover of these components in N-replete conditions (Geigenberger and Stitt, 1991). Furthermore, a decrease in the level of PEP due to the activation of PEPC and PK (Table 4.1 and references therein), leads to the feedback activation of PFK, after adding N to the N-deprived plants (Dennis and Greyson, 1987). A brief overview of the observed changes (Table 4.1) gives the impression that there is no induction/repression of the specific bypass mechanisms of glycolysis under conditions of N. However, it should also be noted that a decrease in N availability results in the accumulation of phosphate and in the strong down-regulation of genes normally involved in the Pi starvation response (Schlüter *et al.*, 2012). Therefore, some effects observed under N starvation are probably secondary responses towards increased phosphate levels.

Table 4.1 Changes in glycolytic pathway and PEPC engagement in response to the N status of plant cells.

Enzyme	Low N versus high N conditions	Species	Level	References
Phosphoglucose isomerase	↓	*Arabidopsis*	Transcript	Wang *et al.*, 2003
	↓	*Arabidopsis*	Transcript	Scheible *et al.*, 2004
Phosphofructokinase	↓	*Lemna minor*	Activity	Humphrey *et al.*, 1977
	↓	tobacco	Activity	Paul and Stitt, 1993
PPi-dependent phosphofructokinase	↓	tobacco	Activity	Paul and Stitt, 1993
	↓	maize	Transcript	Schlüter *et al.*, 2012
Phosphoglycerate mutase	↓	*Arabidopsis*	Transcript	Wang *et al.*, 2003
	↓	*Arabidopsis*	Transcript	Scheible *et al.*, 2004
Pyruvate kinase	↓	maize	Transcript	Schlüter *et al.*, 2012
	↓	tobacco	Activity	Scheible *et al.*, 2000
	↓	*Arabidopsis*	Transcript	Scheible *et al.*, 2004
Phosphoenolpyruvate carboxylase	↓	*Lemna minor*	Activity	Humphrey *et al.*, 1977
	↓	tobacco	Transcript and activity	Scheible *et al.*, 1997
	↓	tobacco	Transcript and activity	Scheible *et al.*, 2000
	↓	*Arabidopsis*	Transcript	Scheible *et al.*, 2004
	↓	*Medicago truncatula*	Transcript	Ruffel *et al.*, 2008

Low N availability affects mtETC functioning. Total respiratory O_2 uptake may increase, decrease, or not change in response to N deprivation (Geigenberger and Stitt, 1991; Scheible *et al.*, 2004; Brück and Guo, 2006; Noguchi and Terashima, 2006; Watanabe *et al.*, 2010), but engagement of individual dehydrogenases and oxidases in respiration is usually modified. An increase in the capacity/protein level of AOX or transcription level under N-limitation stress conditions has been reported in *Catharantus roseus* (Hoefnagel *et al.*, 1993), tobacco suspension cells (Sieger *et al.*, 2005), spinach (Noguchi and Terashima, 2006) and *Arabidopsis* plants (Watanabe *et al.*, 2010). In contrast to the alternative pathway, the capacity of the cytochrome pathway is significantly reduced under low N conditions (Gonzàlez-Meler *et al.*, 1997; Sieger *et al.*, 2005). Under N-limited conditions in *Arabidopsis*, the expression of the type II NADH dehydrogenase gene is also higher compared to that under non-limiting N conditions (Scheible *et al.*, 2004; Watanabe *et al.*, 2010). The non-phosphorylating pathway activities may consume an excess of sugars and to some extent modify the C/N ratio under N stress conditions (Lambers, 1982; Sieger *et al.*, 2005). In contrast to the results obtained by Scheible *et al.* (2004) and Watanabe *et al.* (2010) using seedlings, no activation of type II dehydrogenases was detected for suspension cells under N-starvation stress conditions (Sieger *et al.*, 2005).

Under limiting conditions, the induction of non-phosphorylating pathways may also result from the increased demand for the oxidation of excess reductants. When glucose utilization is restricted, the down-regulation of Rubisco and ATP synthases is more rapid compared to that in light-harvesting complex II (Kilb *et al.*, 1996). The imbalance between light absorption processes and CO_2 assimilation may lead to an over-reduction of chloroplasts and increased chloroplastic ROS production. To prevent such a situation, an elevated export of reductants into the cytosol is induced. The activation of type II dehydrogenases linked to higher AOX activity lowers the cellular reduction state and counteracts the inhibition of photosynthesis (Yoshida *et al.*, 2006).

Respiratory activity under ammonium nutrition

Ammonium, when present in excess, is deleterious to many plant species ('ammonium toxicity') and a first visible symptom of ammonium toxicity is stunted growth. Similar to N deprivation, ammonium toxicity is not related to a depletion of carbon sources because the accumulation of carbohydrates and sugar phosphates has been reported to be a response to ammonium supply (Hachiya *et al.*, 2012). At the cellular level, the assimilation of NO_3^-, compared to the assimilation of NH_4^+, results in differences in the cellular oxidation–reduction state (Figure 4.2). Nitrate reduction, which involves the conversion of nitrate to ammonium, consumes large amounts of reducing equivalents (Noctor and Foyer, 1998). In contrast, ammonium assimilation requires mainly C skeletons (organic acids), resulting in an increased cell reduction state.

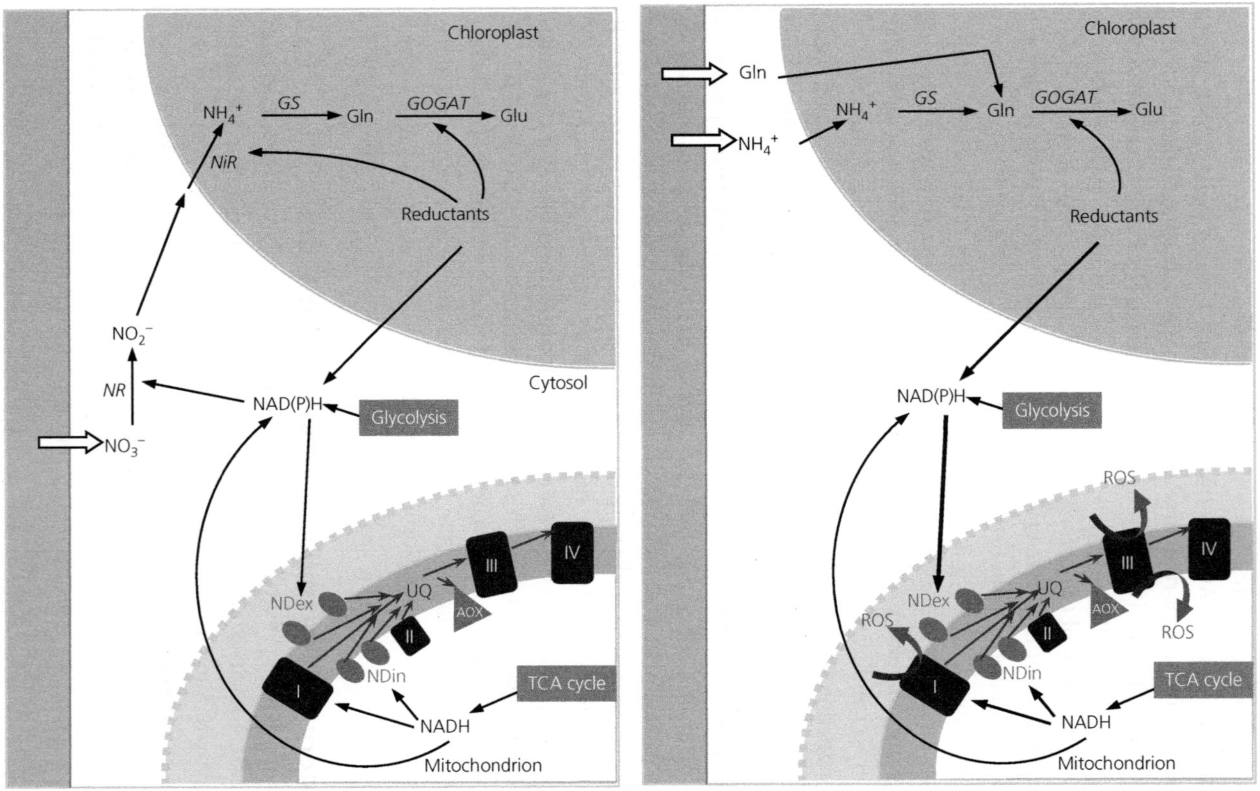

Figure 4.2 The influence of N source on the redox status of individual compartments of leaf cells. Abbreviations: AOX, alternative oxidase; Gln, glutamine; GOGAT, glutamine: 2-oxoglutarate aminotransferase; GS, glutamine synthetase; NDin/ex, internal and external type II dehydrogenases, respectively; NiR, nitrite reductase; NR, nitrate reductase; ROS, reactive oxygen species; I, II, III, Complexes of the mitochondrial electron transport chain.

Source: Modified from Escobar *et al.* (2006).

Analyses of primarily cytosolic-localized catabolic pathways have also revealed modification in respiratory metabolism in ammonium-grown plants, compared to nitrate-grown plants. Nitrate-fed plants develop a higher activity of SuSy, whereas ammonium-fed plants show an enhanced invertase activity (Raab and Terry, 1995). Ammonium treatment also results in higher activity of PPi-PFK (Raab and Terry, 1995). Most probably, the elevated rate of glycolysis is necessary to provide a higher amount of carbon skeletons for the transamination of root-to-shoot-transported glutamine and for incorporation of ammonium (Rufty *et al.*, 1988).

As previously mentioned, an inorganic N source generally leads to an activation of glycolysis, including PEPC (Table 4.1). However, it has also been shown that PEPC activity differs depending on the form of N that was supplied. Several experiments have shown that nitrate-fed plants show higher PEPC activity compared to ammonium-fed plants (e.g. Foyer *et al.*, 1994; Pasqualini *et al.*, 2001). A widely accepted explanation for this observation is the engagement of PEPC and malate metabolism in pH homeostasis under nitrate assimilation, thus preventing cellular sap alkalinization. Confirming this hypothesis, it has been shown that pH is an important factor for the regulation of PEPC activity (Iglesias and Andreo, 1984). However, in many species, PEPC activity has been shown to be up-regulated in ammonium-supplied plants (Britto and Kronzucker, 2005 and references therein). Britto and Kronzucker (2005) have proposed that the accumulation of malate in plants under nitrate nutrition reflects a lesser anaplerotic requirement (lowered requirement for organic acids) compared to that in ammonium conditions. To confirm this, it has been shown that a higher PEPC activity associated with ammonium nutrition is more apparent in roots where primary ammonium assimilation takes place (Britto and Kronzucker, 2005 and references therein).

Ammonium treatment results in an increase in the activity of pyruvate dehydrogenase complex (PDC) in sugar beet leaves (Raab and Terry, 1995) and pea roots (Lasa *et al.*, 2002a) and in the *PDC1* expression in *Arabidopsis* shoots (Hachiya *et al.*, 2012). PDC is regulated by both product inhibition (NADH and acetyl-CoA) and by protein phosphorylation/dephosphorylation (Miernyk and Randall, 1987). The phosphorylation state of PDC is determined by the combined action of PDC kinase and phosphatase, and ammonium ions have been shown to be activators of PDC kinase, leading to PDC inactivation (Schuller and Randall, 1989; Tovar-Méndez *et al.*, 2003). Under ammonium nutrition, the cellular concentration of NH_4^+ increases significantly (Hachiya *et al.*, 2012) and this may potentially result in the inhibition of PDC. On the other hand, PDC activity and the TCA pathway, up to the isocitrate formation step, are required for the production of organic acids for the GS-GOGAT cycle (reviewed by Szal and Podgórska, 2012). On the basis of the results obtained by Raab and Terry (1995) and by our laboratory (B. Szal, unpublished results), we propose that mitochondrial ammonium content under ammonium nutrition does not accumulate to the level that may inhibit PDC. Contrary to this hypothesis, Hachiya *et al.* (2012) have shown that the pyruvate/TCA organic acid ratio increases under ammonium nutrition; this may indicate that, at least partially, inactivation of PDC limits the input of pyruvate into the TCA cycle.

Mitochondrially localized steps of respiration (TCA cycle and mtETC activity) may be differently stimulated according to N source. Weger and Turpin (1989) referred to a different effect of nitrate or ammonium nutrition on mitochondrial metabolism in *Selenstum minutum* cultures. Under nitrate nutrition, when reductants of mitochondrial origin are needed to support cytosolic nitrate reductase (NR) activity (Bloom *et al.*, 2010), the TCA cycle activity is strongly increased, but mtETC activity remains unchanged. Under ammonium assimilation, when intermediates of the TCA cycle (2-OG or citrate) are needed as carbon skeletons and the excess of reductants has to be oxidized simultaneously, both TCA cycle and mtETC activities increase (Weger and Turpin, 1989). On the basis of the respiratory quotient (RQ) parameter (the ratio of CO_2 evolution to O_2 consumption), it was proposed that a similar scenario also occurs in higher plant cells. The decreased RQ ratios in ammonium-supplied barley, wheat, maize and pea plants indicate a higher electron flux in mtETC compared to the TCA cycle activity (de Visser, 1985; Bloom *et al.*, 1992; Cramer and Lewis, 1993).

Stimulation of oxygen uptake under ammonium supply has been observed in several higher plant species (Rigano *et al.*, 1996; Lasa *et al.*, 2002b; Brück and Guo, 2006; Escobar *et al.*, 2006; Podgórska *et al.*, 2013). Ammonium supply results in a higher engagement of alternative pathways in mtETC, most probably in response to an oxidation–reduction imbalance. During nitrate nutrition, a large portion of cytosolic NADH is consumed by the NR, which has a high affinity for the substrate (NR *Km*(NADH) approx. 1.4 uM). When ammonium is used as a nitrogen source and the reaction catalysed by NR is omitted, excess reducing power in the cytosol may occur. Recently, Podgórska *et al.* (2013) have shown experimentally that under long-term ammonium nutrition, the extrachloroplast $NAD(P)H/NAD(P)^+$ ratio increases significantly. This may promote an induction of type II dehydrogenases. Indeed, the genes encoding NDin/NDex are up-regulated during ammonium nutrition (Escobar *et al.*, 2006; Patterson *et al.*, 2010; Hachiya *et al.*, 2012; Podgórska *et al.*, 2013). There is a lack of clarity on which of the mitochondrial oxidases are preferentially induced under ammonium nutrition. An increase in alternative pathway engagement in respiration has been found (e.g. by Barneix *et al.*, 1984; Blacquière and de Visser, 1984; Escobar *et al.*, 2006; Podgórska *et al.*, 2013). An increase in AOX capacity and protein level is largely the result of up-regulation of the *AOX2* gene (Escobar *et al.*, 2006; Podgórska *et al.*, 2013). The expression of *AOX2* increases in response to redox signals (Clifton *et al.*, 2006); therefore, this observation is consistent with the previously mentioned redox imbalance under ammonium nutrition. In contrast, Hachiya *et al.* (2010) showed that the activity of COX, but not AOX, is enhanced in response to ammonium supply. According to Hachiya *et al.* (2010), enhanced COX activity may be related to high-energy demands in ammonium-grown plants, resulting from the high demand for energy needed for ATP-dependent futile cycling of NH_4^+ across the plasma membrane (Britto *et al.*, 2001). It should also be taken into account that this discrepancy may be due to species-dependent differences in sensitivity, or developmental- or organ-specific traits. Lasa *et al.* (2002b) have found that in the roots of ammonium-sensitive spinach plants, the activity of the COX pathway increases and that of the AOX pathway decrease in response to ammonium. In contrast, in the roots of ammonium-tolerant pea plants, the capacity of COX remained unchanged, but AOX was highly induced (Lasa *et al.*, 2002b).

Ammonium supply leads to an increased ROS content in plant tissues (Escobar *et al.*, 2006) and consequently to oxidative stress (Podgórska *et al.*, 2013). Guo *et al.* (2005) hypothesized that higher ROS generation in ammonium grown plants is largely caused by superfluous redox equivalents from the photosynthetic electron transport chain to mitochondria. AOX, which is not controlled by adenylate status, may facilitate the oxidation of excess reductants and prevent overproduction of mitochondrial ROS (Møller, 2001). Therefore, AOX may also be involved in modulation of retrograde signal transduction under ammonium supply, namely, in controlling mitochondrially derived H_2O_2 production (Vanlerberghe *et al.*, 2009). Confirming this hypothesis, an increased mitochondrial H_2O_2 concentration, together with higher AOX protein/capacity, was recently found in *Arabidopsis* leaf tissues under ammonium stress (Podgórska *et al.*, 2013).

Ammonium nutrition also activates some alternative pathways of the TCA cycle. An induction of glutamate dehydrogenase (GDH) and proline dehydrogenase under ammonium supply has been reported in *Arabidopsis* tissues (Fizames *et al.*, 2004; Patterson *et al.*, 2010). GDH activity, which has been shown to depend on the mtETC redox state (Tarasenko *et al.*, 2009), may provide 2-OG that is further incorporated into the TCA cycle (Masclaux-Daubresse *et al.*, 2006). Slightly surprising is the activation of proline dehydrogenase under ammonium supply (Patterson *et al.*, 2010) because its oxidation, in addition to providing of 2-OG, also delivers additional electrons into the mtETC.

Summary

The flexibility of respiratory pathways enables plant growth and development in different nutrient conditions. As discussed earlier, the limited availability of Pi, and consequently restricted ATP synthesis, influence energy metabolism, activating bypasses and omitting ATP/Pi-dependent steps in the glycolytic pathway and mtETC. Low N supply, due to a decreased energy demand required for biosyntheses, ion uptake and transport, results in the general suppression of glycolytic and cytochrome pathways in the cytosol and mitochondria, respectively, but activates mitochondrial AOX. Recently, in our laboratory, Juszczuk and Ostaszewska (2011) have shown that lack of sulfur in the growth medium of beans leads to suppression of mitochondrial complex I and activation of NDin but not NDex. Most nutritional stresses result in an increased reduction state of cells or individual compartments. No doubt, due to the branched structure of mtETC (possessing NDin/NDex and AOX), the mitochondria are important players in the regulation of oxidation–reduction homeostasis (Noctor *et al.*, 2007).

References

Azcón-Bieto, J. and Osmond, C.B. (1983) Relationship between photosynthesis and respiration. The effect of carbohydrate status on the rate of CO_2 production by respiration in darkened and illuminated wheat leaves. *Plant Physiology* **71**: 574–581.

Barneix, A.J., Breteler, H. and van de Geijn, S.C. (1984) Gas and ion exchanges in wheat roots after nitrogen supply. *Physiologia Plantarum* **61**: 357–362.

Blacquière, T. and de Visser, R. (1984) Capacity of cytochrome and alternative path in coupled and uncoupled root respiration of *Pisum* and Plantago. *Physiologia Plantarum* **6**: 427–432.

Bloom, A.J., Burger, M., Asensio, J.S.R. and Cousins, A.B. (2010) Carbon dioxide enrichment inhibits nitrate assimilation in wheat and *Arabidopsis*. *Science* **328**, 899–903.

Bloom, A.J., Sukrapanna, S.S. and Warner, R.L. (1992) Root respiration associated with ammonium and nitrate absorption and assimilation by barley. *Plant Physiology* **99**: 1294–1301.

Botha, F.C. and Small, J.G.C. (1987) Comparison of the activities and some properties of pyrophosphate and ATP dependent fructose-6-phosphate 1-phospotransferases of *Phaseolus vulgaris* seeds. *Plant Physiology* **83**: 772–777.

Britto D.T. and Kronzucker H.J. (2005) Nitrogen acquisition, PEP carboxylase, and cellular pH homeostasis: new views on old paradigms. *Plant, Cell and Environment* **28**, 1396–1409.

Britto, D.T., Siddiqi, M.Y., Glass A.D. and Kronzucker H.J. (2001) Futile transmembrane NH_4^+ cycling: A cellular hypothesis to explain ammonium toxicity in plants. *Proceedings of the National Academy of Sciences of the USA* **98**: 4255–4258.

Brück, H. and Guo, S. (2006) Influence of N form on growth and photosynthesis of *Phaseolus vulgaris* L plants. *Journal of Plant Nutrition and Soil Science* **169**: 849–856.

Cakmak, I., Hengeler, C. and Marschner, H. (1994a) Partitioning of shoot and root dry matter and carbohydrates in bean plants suffering from phosphorus, potassium and magnesium deficiency. *Journal of Experimental Botany* **45**: 1245–1250.

Cakmak, I., Hengeler, C. and Marschner, H. (1994b) Changes in phloem export of sucrose in leaves in response to phosphorus, potassium and magnesium deficiency in bean plants. *Journal of Experimental Botany* **45**: 1251–1257.

Ciereszko, I., Gniazdowska, A., Mikulska, M. and Rychter, A.M. (1996) Assimilate translocation in bean plants (*Phaseolus vulgaris* L.) during phosphate deficiency. *Journal of Plant Physiology* **149**: 343–348.

Ciereszko, I., Zambrzycka, A. and Rychter, A.M. (1998) Sucrose hydrolysis in bean roots (*Phaseolus vulgaris* L.) under phosphate deficiency. *Plant Science* **133**: 139–144.

Clifton, R., Millar, A.H. and Whelan, J. (2006) Alternative oxidases in *Arabidopsis*: A comparative analysis of differential expression in the gene family provides new insights into function of non-phosphorylating bypasses. *Biochimica et Biophysica Acta* **1757**: 730–741.

Cramer, M.D. and Lewis, O.A.M. (1993) The influence of NO_3^- and NH_4^+ nutrition on the gas exchange characteristics of the roots of wheat (*Triticum aestivum*) and maize (*Zea mays*) plants. *Annals of Botany* **72**: 37–46.

Day, D.A., Krab, K., Lambers, H. *et al.* (1996) The cyanide-resistant oxidase: To inhibit or not to inhibit, that is the question. *Plant Physiology* **110**: 1–2.

Dennis, D.T. and Greyson, M. (1987) Fructose-6-phosphate metabolism in plants. *Physiologia Plantarum* **69**: 395–404.

Duff, S.M.G., Moorhead, G.B.G., Lefebvre, D.D. and Plaxton, W.C. (1989a) Phosphate starvation inducible "bypases" of adenylate and phosphate dependent glycolytic enzymes in *Brassica nigra* suspension cells. *Plant Physiology* **90**: 1275–1278.

Duff, S.M.G., Lefebvre, D.D. and Plaxton W.C. (1989b) Purification and characterization of a phosphoenolpruvate phosphatases from *Brasica nigra* suspension cell culture. *Plant Physiology* **90**: 734–741.

Escobar, M.A., Geisler, D.A. and Rasmusson, A.G. (2006) Reorganization of the alternative pathways of the Arabidopsis respiratory chain by nitrogen supply: opposing effects of ammonium and nitrate. *The Plant Journal* **45**: 775–788.

Fernie, A.R., Carrari, F. and Sweetlove, L.J. (2004) Respiratory metabolism: Glycolysis, TCA cycle and mitochondrial electron transport. *Current Opinion in Plant Biology* **7**: 254–261.

Fizames, C., Muños, S., Cazettes, C. *et al.* (2004) The *Arabidopsis* root transcriptome by serial analysis of gene expression. Gene identification using the genome sequence. *Plant Physiology* **134**: 67–80.

Foyer, C.H., Noctor, G., Lelandais, M. *et al.* (1994) Short-term effects of nitrate, nitrite and ammonium assimilation on photosynthesis, carbon partitioning and protein phosphorylation in maize. *Planta* **192**: 211–220.

Fredeen, A.L., Rao, I.M. and Terry, N. (1989) Influence of phosphorus nutrition on growth and carbon partitioning in *Glycine max*. *Plant Physiology* **89**: 225–230.

Geigenberger, P. and Stitt, M. (1991) Regulation of carbon partitioning between sucrose and nitrogen assimilation in cotyledons of germinating *Ricinus communis* L. seedlings. *Planta* **185**: 563–568.

Gniazdowska, A., Mikulska, M. and Rychter, A.M. (1998) Growth, nitrate uptake and respiration rate in bean roots under phosphate deficiency. *Biologia Plantarum* **41**: 217–226.

Gonzàlez-Meler, M.A., Giles, L., Thomas, R.B. and Siedow, J.N. (2001) Metabolic regulation of leaf respiration and alternative pathway activity in response to phosphate supply. *Plant, Cell and Environment* **24**: 205–215.

Gonzàlez-Meler, M.A., Matamala, R. and Peñuelas, J. (1997) Effects of prolonged drought and nitrogen deficiency on the respiratory O_2 uptake of bean and pepper leaves. *Photosynthetica* **34**: 505–512.

Gregory, A.L., Hurley, B.A., Tran, H.T. *et al.* (2009) *In vivo* regulatory phosphorylation of the phosphoenolpyruvate carboxylase At PPC 1 in phosphate-starved *Arabidopsis thaliana*. *Biochemical Journal* **420**: 57–65.

Guo, S., Schinner, K., Sattelmacher, B. and Hansen, U.P. (2005) Different apparent CO_2 compensation points in nitrate- and ammonium-grown *Phaseolus vulgaris* and the relationship to non-photorespiratory CO_2 evolution. *Physiologia Plantarum* **123**: 288–301.

Hachiya, T., Watanabe, C.K., Boom, C. *et al.* (2010) Ammonium-dependent respiratory increase is dependent on the cytochrome pathway in *Arabidopsis thaliana* shoots. *Plant, Cell and Environment* **33**: 1888–1897.

Hachiya, T., Watanabe, C.K., Fujimoto, M. *et al.* (2012) Nitrate alleviates ammonium toxicity without lessening ammonium accumulation, organic acid depletion and inorganic cation depletion in *Arabidopsis thaliana* shoots. *Plant and Cell Physiology* **53**: 577–591.

Hammond, J.P., Bennett, M.J., Bowen, H.C. *et al.* (2003) Changes in gene expression in *Arabidopsis* shoots during phosphate starvation and the potential for developing smart plants. *Plant Physiology* **132**: 578–596.

Hernández, G., Ramirez, M., Valdés-López, O. *et al.* (2007) Phosphorus stress in common bean: Root transcript and metabolic responses. *Plant Physiology* **144**: 752–767.

Hoefnagel, M.H.N., van Iren, F. and Libbenga, K.R. (1993) In suspension cultures of *Catharanthus roseus* the cyanide-resistant pathway is engaged in respiration by excess sugar in combination with phosphate or nitrogen starvation. *Physiologia Plantarum* **87**: 297–304.

Hoefnagel, M.H.N., van Iren, F., Libbenga, K.R. and Van der Plas, L.H.W. (1994) Possible role of adenylates in the engagement of the cyanide-resistant pathway in nutrient-starved *Catharanthus roseus* cells. *Physiologia Plantarum* **90**: 269–278.

Humphrey, T.J., Sarawek, S. and Davies, D.D. (1977) The effect of nitrogen deficiency on the growth and metabolism of *Lemna minor* L. *Planta* **137**: 259–264.

Iglesias, A.A. and Andreo, C.S. (1984) On the molecular mechanism of maize phosphoenolpyruvate carboxylase activation by thiol compounds. *Plant Physiology* **75**: 983–987.

Johnson, J.F., Vance, C.P. and Allan, D.L. (1996) Phosphorus deficiency in *Lupinus albus.* Altered lateral root development and enhanced expression of phosphoenolpyruvate carboxylase. *Plant Physiology* **112**: 31–41.

Juszczuk, I.M. and Ostaszewska, M. (2011) Respiratory activity, energy and redox status in sulphur-deficient bean plants. *Environmental and Experimental Botany* **74**: 245– 254.

Juszczuk, I.M. and Rychter, A.M. (2002) Pyruvate accumulation during phosphate deficiency stress of bean roots. *Plant Physiology and Biochemistry* **40**: 783–788.

Juszczuk, I.M., Flexas, J., Szal, B. *et al.* (2007) Effect of mitochondrial genome rearrangement on respiratory activity, photosynthesis, photorespiration and energy status of MSC16 cucumber (*Cucumis sativus* L.) mutant. *Physiologia Plantarum* **131**: 527–541.

Juszczuk, I.M., Malusá, E. and Rychter, A.M. (2001) Oxidative stress during phosphate deficiency in roots of bean plants (*Phaseolus vulgaris* L). *Journal of Plant Physiology* **158**: 1299–1305.

Kilb, B., Wietoska, K. and Godde, D. (1996) Changes in the expression of photosynthetic genes precede the loss of photosynthetic activities and chlorophyll when glucose is supplied to mature spinach leaves. *Plant Science* **115**: 225–235.

Kondracka, A. and Rychter, A.M. (1997) The role of Pi recycling processes during photosynthesis in phosphate-deficient bean plants. *Journal of Experimental Botany* **48**: 1461–1468.

Lambers, H. (1982) Cyanide-resistant respiration: a non-phosphorylating electron transport pathway acting as an energy overflow. *Physiologia Plantarum* **55**: 478–485.

Lambers, H. (1985) Respiration in intact plants and tissues: Its regulation and dependence on environmental factors, metabolism and invaded organisms. In *Encyclopedia of Plant Physiology,* vol. 18, *Higher Plant Respiration* (C.R. Duce and D.A. Day, eds), pp. 444–448. Springer-Verlag, Berlin.

Lan, P., Li, W. and Schmidt, W. (2012) Complementary proteome and transcriptome profiling in phosphate-deficient *Arabidopsis* roots reveals multiple levels of gene regulation. *Molecular and Cellular Proteomics* **11**: 1156–1166.

Lancien, M., Ferrario-Méry, S., Roux, Y. *et al.* (1999) Simultaneous expression of NAD-dependent isocitrate dehydrogenase and other Krebs cycle genes after nitrate resupply to short-term nitrogen-starved tobacco. *Plant Physiology* **120**: 717–725.

Lasa, B., Frechilla, S., Aparicio-Tejo, P.M. and Lamsfus, C. (2002a) Role of glutamate dehydrogenase and phosphoenolpyruvate carboxylase activity in ammonium nutrition tolerance in roots. *Plant Physiology and Biochemistry* **40**: 969–979.

Lasa, B., Frechilla, S., Aparicio-Tejo, P.M. and Lamsfus, C. (2002b) Alternative pathway respiration is associated with ammonium ion sensitivity in spinach and pea plants. *Plant Growth Regulation* **37**: 49–55.

Lennon, A.M., Neuenschwander, U.H., Ribas-Carbo, M. *et al.* (1997) The effects of salicylic acid and tobacco mosaic virus infection on the alternative oxidase of tobacco. *Plant Physiology* **115**: 783–791.

Li, X.N. and Ashihara, H. (1990) Effects of inorganic phosphate on sugar catabolism by suspension-cultured *Catharanthus roseus*. *Phytochemistry* **29**: 497–500.

Logan, B.A., Demming-Adams, B., Rosenstiel, T.N. and Adams, III, W.W. (1999) Effect of nitrogen limitation on foliar antioxidants in relationship to other metabolic characteristics. *Planta* **209**: 213–220.

Makino, A. and Osmond, B. (1991) Effects of nitrogen nutrition on nitrogen partitioning between chloroplasts and mitochondria in pea and wheat. *Plant Physiology* **96**: 355–362.

Malusá, E., Laurenti, E., Juszczuk, I.M. *et al.* (2002) Free radical production in roots of *Phaseolus vulgaris* subjected to phosphate deficiency stress. *Plant Physiology and Biochemistry* **40**: 963–967.

Masclaux-Daubresse C., Reisdorf-Cren M., Pageau K., Lelandais M., Grandjean O., Kronenberger J., Valadier M.H., Feraud M., Jouglet T., Suzuki A. (2006) Glutamine synthetase/glutamate synthase pathway and glutamate dehydrogenase play distinct roles for sink/source nitrogen cycle in tobacco (*Nicotiana tabacum* L.). *Plant Physiology* **140**: 444–456.

Matt, P., Geiger, M., Walch-Liu, P. *et al.* (2001) Elevated carbon dioxide increases nitrate uptake and nitrate reductase activity when tobacco is growing on nitrate, but increases ammonium uptake and inhibits nitrate reductase activity when tobacco is growing on ammonium nitrate. *Plant, Cell and Environment* **24**: 1119–1137

Maxwell, D.P., Wang, Y. and McIntosh, L. (1999) The alternative oxidase lowers mitochondrial reactive oxygen production in plant cells. *Proceedings of the National Academy of Sciences of the USA* **96**: 8271–8276.

Mertens, E. (1991) Phyrophosphate-dependent phosphofructokinase: An anaerobic glycolytic enzyme? *FEBS Letters* **285**: 1–5.

Miernyk, J.A. and Randall, D.D. (1987) Some kinetic properties of the pea mitochondrial pyruvate dehydrogenase complex. *Plant Physiology* **83**: 306–310.

Millenaar, F.F., Benschop, J.J., Wagner, A.M. and Lambers, H. (1998) The role of alternative oxidase in stabilizing the *in vivo* reduction state of ubiquinone pool and the activation state of alternative oxidase. *Plant Physiology* **118**: 599–607.

Møller, I.M. (2001) Plant mitochondria and oxidative stress: electron transport, NADPH turnover, and metabolism of reactive oxygen species. *Annual Review of Plant Physiology and Plant Molecular Biology* **52**: 561–591.

Moraes, T. and Plaxton, W.C. (2000) Purification and characterization of phosphoenolpyruvate carboxylase from *Brasica napus* (rapeseed) suspension cell cultures. Implications for phosphoenolpyruvate carboxylase regulation during phosphate starvation, and the integration of glycolysis with nitrogen assimilation. *European Journal of Biochemistry* **267**: 4465–4476.

Nagano, M., Hachiya, A. and Asihara, H. (1994) Phosphate starvation and glycolytic bypass catalyzed by phosphoenolopyruvate carboxylase in suspension-cultured *Catharanthus* roseus cells. *Zeitschrift für Naturforshung* **49c**: 742–750.

Neuman, G. and Romheld, V. (1999) Root excretion of carboxylic acids and protons in phosphorus- deficient plants. *Plant and Soil* **211**: 121–130.

Noctor, G. and Foyer, C.H. (1998) A re-evaluation of the ATP: NADPH budget during C3 photosynthesis: A contribution from nitrate assimilation and its associated respiratory activity? *Journal of Experimental Botany* **49**: 1895–1908.

Noctor, G., De Paepe, R. and Foyer, C.H. (2007) Mitochondrial redox biology and homeostasis in plants. *Trends in Plant Science* **12**: 125–134.

Noguchi, K. and Terashima, I. (2006) Responses of spinach leaf mitochondria to low N availability. *Plant, Cell and Environment* **29**: 710–719.

Nunes-Nesi, A., Fernie, A.R. and Stitt, M. (2010) Metabolic and signaling aspects underpinning the regulation of plant carbon nitrogen interactions. *Molecular Plant* **6**: 973–996.

Okazaki, K., Oka, N., Shinano, T. *et al.* (2008) Differences in the metabolite profiles in spinach (*Spinacia oleracea* L.) leaf in different concentrations of nitrate in the culture solution. *Plant and Cell Physiology* **49**: 170–177.

Parsons, H.L., Yip, J.Y.H. and Vanlerberghe, G.C. (1999) Increased respiratory restriction during phosphate-limited growth in transgenic tobacco cells lacking alternative oxidase. *Plant Physiology* **121**: 1309–1320.

Pasqualini, S., Edereli, L., Picconi, C. *et al.* (2001) Metabolic regulation and gene expression of root phosphoenolpyruvate carboxylase by different nitrogen sources. *Plant, Cell and Environment* **24**: 439–447.

Patterson, K., Cakmak, T., Cooper A. *et al.* (2010) Distinct signalling pathways and transcriptome response signatures differentiate ammonium and nitrate-supplied plants. *Plant, Cell and Environment* **33**: 1486–1501.

Paul, M. and Driscoll, S.P. (1997) Sugar repression of photosynthesis during N deficiency. *Plant, Cell and Environment* **20**: 110–116.

Paul, M.J. and Stitt, M. (1993) Effects of nitrogen and phosphorus deficiencies on levels of carbohydrates, respiratory enzymes and metabolites in seedlings of tobacco and their response to sucrose. *Plant, Cell and Environment* **16**: 1047–1057.

Peng, M., Bi, Y.M., Zhu, T. and Rothstein, S. J. (2007) Genome-wide analysis of Arabidopsis responsive transcriptome to nitrogen limitation and its regulation by the ubiquitin ligase gene *NLA*. *Plant Molecular Biology* **65**: 775–797.

Plaxton, W.C. (1996) The organization and regulation of plant glycolysis. *Annual Review of Plant Physiology and Plant Molecular Biology* **47**: 185–214.

Plaxton, W.C. (2004) Plant response to stress: biochemical adaptation to phosphate deficiency. In *Encyclopedia of Plant and Crop Science* (Goodman, R.E., ed.), pp. 976–980. Marcel Dekker, New York.

Plaxton, W.C. and Podestá, F.E. (2006) The functional organization and control of plant respiration. *Critical Reviews in Plant Science* **25**: 159–198.

Plaxton, W.C. and Tran, H.T. (2011) Metabolic adaptations of phosphate-starved plants. *Plant Physiology* **156**: 1006–1015.

Podgórska, A., Gieczewska, K., Łukawska-Kuźma, K. *et al.* (2013) Long-term ammonium nutrition of Arabidopsis increases the extrachloroplastic NAD(P)H/NAD(P)$^+$ ratio and mitochondrial reactive oxygen species level in leaves but does not impair photosynthetic capacity. *Plant, Cell and Environment* DOI: 10.1111/pce.12113.

Raab, Y.K. and Terry, N. (1995) Carbon, nitrogen, and nutrient interactions in *Beta vulgaris* L. as influenced by nitrogen source, NO_3^- versus NH_4^+. *Plant Physiology* **107**: 575–584.

Raghothama, K.G. (1999) Phosphate acquisition. *Annual Review of Plant Physiology and Plant Molecular Biology* **50**: 665–693.

Rasmusson, A.G., Fernie, A.R., van Dongen, J.T. (2009) Alternative oxidase: a defence against metabolic fluctuations? *Physiologia Plantarum* **137**: 371–382.

Rebeille, F., Bligny, R. and Douce, R. (1984) Is the cytosolic Pi concentration a limiting factor for plant respiration? *Plant Physiology* **74**: 355–359.

Ribas-Carbo, M., Berry, J.A., Yakir, D. *et al.* (1995) Electron partitioning between the cytochrome and alternative pathways in plant mitochondria. *Plant Physiology* **109**: 829–837.

Rigano, B.C., Di Martino Rigano, V., Vona, V. *et al.* (1996) Ammonium assimilation by young plants of *Hordeum vulgare* in light and darkness: Effect on respiratory oxygen consumption by roots. *New Phytologist* **132**: 375–382.

Ruffel, S., Freixes, S., Balzergue, S. *et al.* (2008) Systematic signaling of the plant nitrogen status triggers specific transcriptosome responses depending on the nitrogen source in *Medicago truncatula*. *Plant Physiology* **146**: 2020–2035.

Rufty Jr., T.W., Huber, S.C. and Volk, R.J. (1988) Alterations in leaf carbohydrate metabolism in response to nitrogen stress. *Plant Physiology* **88**, 725–730.

Rychter A.M. and Mikulska M. (1990) The relationship between phosphate status and cyanide resistant respiration in bean roots. *Physiologia Plantarum* **79**: 663–667.

Rychter, A.M. and Randall, D.D. (1994) The effect of phosphate deficiency on carbohydrate metabolism in bean roots. *Physiologia Plantarum* **91**: 383–388.

Rychter, A.M., Chauveau, M., Bomsel, J.L. and Lance, C. (1992) The effect of phosphate deficiency on mitochondrial activity and adenylate levels in bean roots. *Physiologia Plantarum* **84**: 80–86.

Scheible, W.R., González-Fontes, A., Lauerer, M. *et al.* (1997) Nitrate acts as a signal to induce organic acid metabolism and repress starch metabolism in tobacco. *The Plant Cell* **9**: 783–798.

Scheible, W.R., Krapp, A. and Stitt, M. (2000) Reciprocal diurnal changes of phosphoenolpyruvate carboxylase expression and cytosolic pyruvate kinase, citrate synthase and NADP-isocitrate dehydrogenase expression regulate organic acid metabolism during nitrate assimilation in tobacco leaves. *Plant, Cell and Environment* **23**: 1155–1167.

Scheible, W.R., Morcuende, R., Czechowski, T. *et al.* (2004) Genome-wide reprogramming of primary and econdary metabolism, protein synthesis, cellular growth process, and regulatory infrastructure of *Arabidopsis* in response to nitrogen. *Plant Physiology* **136**: 2483–2499.

Schlüter, U., Mascher, M., Colmsee, C. *et al.* (2012) Maize source leaf adaptation to nitrogen deficiency affects not only nitrogen and carbon metabolism but also control of phosphate homeostasis. *Plant Physiology* **160**: 1384–1406.

Schuller, K.A. and Randall, D.D. (1989) Regulation of pea mitochondrial pyruvate dehydrogenase complex. Does photorespiratory ammonium influence mitochondrial carbon metabolism? *Plant Physiology* **89**: 1007–1212.

Sieger, S.M., Kristensen, B.K., Robson, C.A. *et al.* (2005) The role of alternative oxidase in modulating carbon use efficiency and growth during macronutrient stress in tobacco cells. *Journal of Experimental Botany* **56**: 1499–1515.

Szal, B. and Podgórska, A. (2012) The role of mitochondria in leaf nitrogen metabolism. *Plant, Cell and Environment* **35**: 1756–1769.

Tarasenko, V.I., Garnik, E.Y., Shmakov, V.N. and Konstantinov, Y.M. (2009) Induction of *Arabidopsis gdh2* gene expression during changes in redox state of the mitochondrial respiratory chain. *Biochemistry (Moscow)* **74**: 47–53.

Theodorou, M.E. and Plaxton, W.C. (1993) Metabolic adaptations of plant respiration to nutritional phosphate deprivation. *Plant Physiology* **101**: 339–334.

Theodorou, M.E. and Plaxton, W.C. (1994) Induction of PPi-dependent phosphofruktokinase by phosphate starvation in seedlings of *Brassica nigra*. *Plant, Cell and Environment* **17**: 287–294.

Theodorou, M.E., Cornel, F.A., Duff, S.M.G. and Plaxton, W.C. (1992) Phosphate starvation inducible synthesis of the α-subunit of pyrophosphate: fructose-6-phosphate phosphotransferase in black mustard cell-suspension cultures. *Journal of Biological Chemistry* **267**: 21901–21905.

Tovar-Méndez, A., Miernyk, J.A. and Randall, D.D. (2003) Regulation of pyruvate dehydrogenase complex activity in plant cells. *European Journal of Biochemistry* **270**: 1043–1049.

Turpin, D.H., Botha, F.C., Smith, R.G. *et al.* (1990) Regulation of carbon partitioning to respiration during dark ammonium assimilation by the green alga *Selenastum minutum*. *Plant Physiology* **93**: 166–175.

Uhde-Stone, C., Gilbert, G., Johnson, J.M. *et al.* (2003) Acclimation of white lupin to phosphorus deficiency involves enhanced expression of genes related to organic acid metabolism. *Plant and Soil* **248**: 99–116.

Umbach, A.L., Fiorani, F. and Siedow, J.N. (2005) Characterization of transformed *Arabidopsis* with altered alternative oxidase levels and analysis of effects on reactive oxygen species in tissue. *Plant Physiology* **139**: 1806–1820.

Vance, C.P., Uhde-Stone, C. and Allan, D.L. (2003) Phosphorus acquisition and use: Critical adaptations by plants for securing a nonrenewable resource. *New Phytologist* **157**: 427–447.

Vanlerberghe, G.C. (2013) Alternative oxidase: A mitochondrial respiratory pathway to maintain metabolic and signaling homeostasis during abiotic and biotic stress in plants. *International Journal of Molecular Sciences* **14**: 6805–6847.

Vanlerberghe, G.C and McIntosh, L. (1996) Signals regulating the expression of the nuclear gene encoding alternative oxidase of plant mitochondria. *Plant Physiology* **111**: 589–595.

Vanlerberghe, G.C., Cvetkovska, M. and Wang, J. (2009) Is the maintenance of homeostatic mitochondrial signaling during stress a physiological role for alternative oxidase? *Physiologia Plantarum* **137**: 392–406.

Vijayraghavan, V. and Soole, K. (2010) Effect of short- and long-term phosphate stress on the non-phosphorylating pathway o mitochondrial electron transport in *Arabidopsis thaliana. Functional Plant Biology* **37**: 455–466.

de Visser, R. (1985) Efficiency of respiration and energy requirements of N assimilation in roots of *Pisum sativum. Physiologia Plantarum* **65**: 209–218.

Wang, R., Okamoto, M., Xing, X. and Crawford, N.M. (2003) Microarray analysis of the nitrate response in *Arabidopsis* roots and shoots reveals over 1,000 rapidly responding genes and new linkages to glucose, trehalose-6-phosphate, iron, and sulfate metabolism. *Plant Physiology* **132**: 556–567.

Wanke, M., Ciereszko, I., Podbielkowska, M. and Rychter, A.M. (1998) Response to phosphate deficiency in bean (*Phaseolus vulgaris* L) roots. Respiratory metabolism, sugar localization and changes in ultrastructure of bean root cells. *Annals of Botany* **82**: 809–819.

Wasaki, J., Yonetani ,R., Kuroda, S. *et al.* (2003) Transcriptomic analysis of metabolic changes by phosphorus stress in rice plant roots. *Plant, Cell and Environment* **26**: 1515–1523.

Watanabe, C.K., Hachiya, T., Takahara, K. *et al.* (2010) Effect of AOX1a deficiency on plant growth, gene expression of respiratory components and metabolic profile under low-nitrogen stress in *Arabidopsis thaliana. Plant and Cell Physiology* **51**: 810–822.

Weger, H.G. and Turpin, D.H. (1989) Mitochondrial respiration can support NO_3^- and NO_2^- reduction during photosynthesis. *Plant Physiology* **89**: 409–415.

Yoshida, K., Terashima, I. and Noguchi, K. (2006) Distinct roles of the cytochrome pathway and alternative oxidase in leaf photosynthesis. *Plant and Cell Physiology* **47**: 22–21.

CHAPTER 5

Structural elucidation of the alternative oxidase reveals insights into the catalytic cycle and regulation of activity

Catherine Elliott, Mary S. Albury, Luke Young, Ben May and Anthony L. Moore
Biochemistry and Molecular Biology, School of Life Sciences, University of Sussex, Falmer, Brighton, East Sussex, UK

Introduction

In addition to the traditional electron transport chain, all plants, some fungi and some protists contain an additional ubiquinol oxidase known as alternative oxidase (AOX; for recent reviews see Millar *et al.*, 2011; Moore *et al.*, 2013). AOX is a monotopic membrane protein, found in the inner-mitochondrial membrane and branching from the traditional electron transport chain at the point of the ubiquinone pool (Storey, 1976; Rich and Moore, 1976; Rich, 1978). Intriguingly, AOX is non-protonmotive (Bendall and Bonner, 1971; Moore *et al.*, 1978), and instead facilitates the four-electron reduction of oxygen to water and oxidation of ubiquinol to ubiquinone (Rich and Moore, 1976; Moore and Siedow, 1991). AOX is insensitive to a number of respiratory inhibitors which are known to affect the other components of the respiratory chain such as cyanide (cytchrome *c* oxidase inhibitor; Keilin and Hartee, 1938; van Buuren *et al.*, 1972) and antimycin A (cytochrome *c* reductase inhibitor; Chance and Williams, 1956; Rieske *et al.*, 1967; for a concise review see Millar *et al.*, 2011). Instead, AOX is sensitive to inhibition by hydroxamic acids such as salicylhydroxamic acid (Schonbaum *et al.*, 1971), and propyl gallate (Siedow and Bickett, 1981). More recently, it has been confirmed that the trypanosomal alternative oxidase (TAO; discussed in more detail later) is sensitive to the antifungal agent ascofuranone (Yabu *et al.*, 2003; Minagawa *et al.*, 1996).

Alternative Respiratory Pathways in Higher Plants, First Edition.
Edited by Kapuganti Jagadis Gupta, Luis A.J. Mur and Bhagyalakshmi Neelwarne.

Table 5.1 A summary of the presence of AOX in several kingdoms; plant species are not listed as the AOX is ubiquitous to all plants.

Kingdom	Examples	Reference(s)
Fungi	*Candida albicans*	Veiga *et al.*, 2003
	Pichia anomala	A.L. Moore and L. Young, unpublished results
	Chalara fraxinea	
Protista	*Trypanosoma bruceii*	Chaudhuri *et al.*, 1995
	Cryptosporidium parvum	Suzuki *et al.*, 2004
		Roberts *et al.*, 2004
	Blastocystis hominis	Stechmann *et al.*, 2008
		Williams *et al.*, 2010
Archaebacteria	*Novosphingobium aromaticivorans*	Finnegan *et al.*, 2003
Animalia	*Crassostrea gigas*	McDonald and Vanlerberghe, 2004
	Meloidogyne hapla	
	Ciona intestinalis	

Function and species spread of alternative oxidase

AOX is ubiquitous in all plants (McDonald *et al.*, 2002; Moore *et al.*, 2013), and is also found in several other species – including several human parasites (summarised in Table 5.1). AOX performs a variety of functions across several groups of organisms (both proven and hypothetical), while the mechanisms of enzymatic activity is the same.

Thermogenic plants

The role of AOX in thermogenic plant tissues is well established – the heat released from the non-protonmotive reduction of oxygen to water is used to volatilise aromatic compounds found in the spathes of thermogenic lilies, in order to attract insect pollinators (Meeuse, 1975; Meeuse and Raskin, 1988). The resulting smell is unsavoury to humans and often likened to rotting flesh, but attracts flies and other carrion insects which become trapped for a short time in the base of the plants before being released, covered in pollen. The largest thermogenic lily, *Amorphophallus titanum,* is referred to colloquially as the 'corpse' flower; one of its smaller relatives is known as the 'dead horse' lily (*Helicodiceros muscivorus*).

Non-thermogenic plants and fungi

The role of AOX is less apparent in non-thermogenic plants, fungi and other species, for which there have been several suggested functions. According to the findings of several groups working with fungal and non-thermogenic plant models, AOX appears not to be constitutively expressed, but rather expressed when the organism experiences stress (such as ageing in potato slices; Hiser and McIntosh, 1990), or disruption of the respiratory chain (such as the effects of chloramphenicol in *Neurospora crassa*; Lambowitz *et al.*, 1989). More specifically in relation to non-thermogenic plants, it was suggested by Bahr and Bonner (1972) and Lambers (1982) that AOX could act as an energy overflow mechanism, deployed when other respiratory chain components cease to function normally. This has been supported by the findings of Moore *et al.* (1988), Millar *et al.* (1993) and Carré *et al.* (2011), showing that AOX can be stimulated by both a highly-reduced Q-pool and the α-keto acid, pyruvate. An increase in mitochondrial pyruvate levels may indicate a decreased Krebs cycle activity, suggesting a negative feedback on the rate of electron transfer. Without electron transfer continuing via AOX, normal function of both glycolysis and the Krebs cycle would be severely hampered, suggesting that pyruvate levels may act as the molecular trigger for a feedforward mechanism, stimulating AOX activity thereby reducing pyruvate levels (reviewed in detail by Finnegan *et al.*, 2004). The nature of expression of AOX in times of stress therefore is temporary, in which respiratory efficiency is sacrificed for survival.

Another theory of the role of AOX was suggested by Purvis and Shewfelt (1993), Wagner and Moore (1997) and Moore *et al.* (2002), whereby AOX acts as a mechanism for the removal of reactive oxygen species (ROS) generated within the organism, should the other respiratory chain components be unable to do so (for example, in the presence of inhibitors). This has been supported by findings suggesting that AOX can reduce observable ROS numbers in plants cells *in vivo*, as reported by Maxwell *et al.* (1999) and Zheng *et al.* (2008). Furthermore, expression of AOX in genetically engineered mice has been shown to reduce ROS produced when the respiratory chain was inactivated in addition to conferring whole-animal resistance to gaseous cyanide (El-Khoury *et al.*, 2013).

Parasites

Members of the *Trypanosoma brucei* subspecies are parasitic kinetoplasts known to cause African trypanosomiasis in humans (*T. b. rhodensiense* and *T. b. gambiense*) and nagana in livestock such as cattle (*T. b. brucei*). According to a recent World Health Organization report (WHO, 2012), approximately 30 000 people across Africa are currently infected and that while the number of reported cases have fallen in recent years, the chemotherapeutic agents available for the treatment of human African trypanosomiasis still require significant improvement. However, the discovery of the apparent down-regulation of mitochondrial cytochromes during the infectious bloodstream form (trypomastigote) of the parasite (Grant and Sargent, 1960, 1961) and the subsequent identification of the presence of a 'plant-like' AOX (Clarkson *et al.*, 1989) has led to the trypanosomal alternative oxidase (TAO) emerging as a key drug target (Minagawa *et al.*, 1996; Nihei *et al.*, 2002; Nakamura *et al.*, 2010; Shiba *et al.*, 2013).

As with AOX, TAO branches from the respiratory chain at the point of the Q pool and catalyses the reduction of oxygen through to water. The function is not for thermogenesis or ROS reduction however; instead TAO maintains the Q pool in an oxidized state in order to facilitate the regeneration of NADH to NAD^+ via the cytosolic and mitochondrial glycerol-3-phosphate dehydrogenases (Clarkson *et al.*, 1989). The supply of NAD^+ ensures that glycolysis – the major source of energy production for the bloodstream trypomastigotes (for a review, see Chaudhuri *et al.*, 2006) – continues to function, thus ensuring survival of the parasite in the host's bloodstream where glucose is readily available. Studies have shown that inhibiting the function of AOX in bloodstream-form of trypanosomes with the inhibitor ascofuranone leads to the death of the parasite (Minagawa *et al.*, 1996; Nihei *et al.*, 2002) and its removal from infected mice. It is hoped that further research on ascofuranone derivatives may prove to be specific, safe and effective replacements for the current drugs used to treat African trypanosomiasis which are (with the exception of eflornithine) relatively ineffective and known to cause significant side effects in patients (pentamidine) and in one case (melarsoprol), the symptoms of arsenic poisoning (compare Fairlamb, 2003; Gadelha *et al.*, 2011).

While AOX is known to be expressed in other parasites related to trypanosomes (such as *Cryptosporidium parvum* and *Blastocystis hominis*, both of which cause disease in humans; Suzuki *et al.*, 2004; Roberts *et al.*, 2004; Williams *et al.*, 2010), it has not been confirmed whether the role of AOX in these organisms is the same, and therefore whether drugs which target AOX – such as ascofuranone – would be effective in treating the diseases caused by these organisms.

The plastid terminal oxidase

A distant relative of AOX, the plastid terminal oxidase (or plastoquinol terminal oxidase; PTOX), also catalyses the reduction of oxygen through to water, although it is localized to the thylakoid membrane of the chloroplast rather than the inner membrane of the mitochondria (Cournac *et al.*, 2000). PTOX is thought to play an essential role in the desaturation of carotenoids (Carol *et al.*, 1999; Josse *et al.*, 2000; Carol and Kuntz, 2001) in addition to providing electron transport when components of the plastid electron transport chain are inhibited (McDonald *et al.*, 2011). Interestingly, in *A. thaliana* variants lacking the PTOX gene, PTOX function can be performed by an AOX targeted to the thylakoid membrane (Fu *et al.*, 2012). Furthermore, when expressed in *E. coli* membranes, PTOX confers the same cyanide-resistant respiration observed in tissues expressing AOX which is inhibited by addition of the AOX-inhibitor octyl-gallate (Josse *et al.*, 2000). Although sequence similarity between the AOX and PTOX families is very low (25%; Josse *et al.*, 2000), similarities in function (as flexible terminal oxidases) and catalytic mechanisms (quinol:oxygen oxidoreductases) suggest that their structures may well be similar.

Structure of the trypanosomal alternative oxidase

The trypanosomal alternative oxidase has recently been determined at 2.85 Å resolution, which revealed that each asymmetric unit contained four monomers that associate to form two homodimers. Each of the monomers contains six long α-helices and four short helices. The six long helices are arranged in an antiparallel fashion with four of the helices forming a four-helix bundle which acts as a scaffold to bind the two iron atoms, comparable to that found in other diiron proteins and in agreement with previous modelling studies. The active-site is located in a hydrophobic environment deep within the molecule and the iron atoms within the diiron centre are ligated by four highly conserved glutamate residues in addition to a hydroxo bridge. Although two histidine residues are also located within the active site, they are too far away from the iron atoms in the oxidized state to act as ligands. Such a primary ligation sphere results in a five-coordinated diiron centre possessing a distorted square pyramidal geometry similar to that observed in the reduced form of the castor acyl-ACP desaturase (Shanklin *et al.*, 2009). Surface representation of the TAO dimer reveals the presence of a large hydrophobic face on one side of the dimer surface which, similar to other monotopic proteins such as the NADH dehydrogenase (Iwata *et al.*, 2012) or prostaglandin H_2 synthase, undoubtedly anchors the protein to the membrane via a series of conserved arginine residues. Close scrutiny of the surface representation reveals there are two hydrophobic cavities, one of which is located in the centre of the hydrophobic face and perpendicular to the membrane surface whereas the second cavity lies parallel to the membrane surface. Although both cavities reach directly into the diiron centre, the cavity perpendicular to the membrane surface is probably the route of ubiquinol entry from within the mitochondrial inner membrane again being comparable to that observed with other monotopic proteins.

Models of the alternative oxidase

Prior to the publication of the crystal structure of TAO (Shiba *et al.*, 2013; Protein Data Bank accession numbers 3VV9, 3VVA and 3W54), there was no definitive structure of AOX. In lieu of this, homology modelling was performed (Andersson and Nordlund, 1999) using a diiron carboxylate, Δ^9-desaturase (1OQ4; Lindqvist *et al.*, 1996) as a template. A representation of this model is shown in Figure 5.1.

In order to identify the potential diiron binding residues of the active site of AOX, sequence alignment and comparison of the AOX and the other diiron carboxylates was undertaken. Six key iron-binding residues were identified (Nordlund and Eklund, 1995), corresponding to known diiron binding motifs

([E] ...[ExxH] ×2). The residues are not sequential, but are spread across the four-helix bundle in both the diiron carboxylates and AOX. When the proteins are fully folded, the residues are brought close enough together to form a scaffold to ligate the diiron centre. The residues are shown in Table 5.2, and are placed in the model in Figure 5.1. By way of confirmation, if any one of the six residues are mutated, respiratory activity via AOX is completely inhibited (Albury *et al.*, 1998; Albury *et al.*, 2002; Shiba *et al.*, 2013; Moore *et al.*, 2013). This has been confirmed in other alternative oxidase isoforms, such as TAO (Ajayi *et al.*, 2002; Nakamura *et al.*, 2005; Kido *et al.*, 2010). Furthermore, it is supported by findings from studies highlighting the necessity of iron for the functionality of the protein in plants (Affourtit and Moore, 2004) and other key organisms (such as Minagawa *et al.*, 1990, using *Hansenula anomala,* now *Pichia anomala*). When inhibitors of ferric iron are present, activity of AOX is inhibited, but when the inhibitor is removed, activity is completely restored (Affourtit and Moore, 2004).

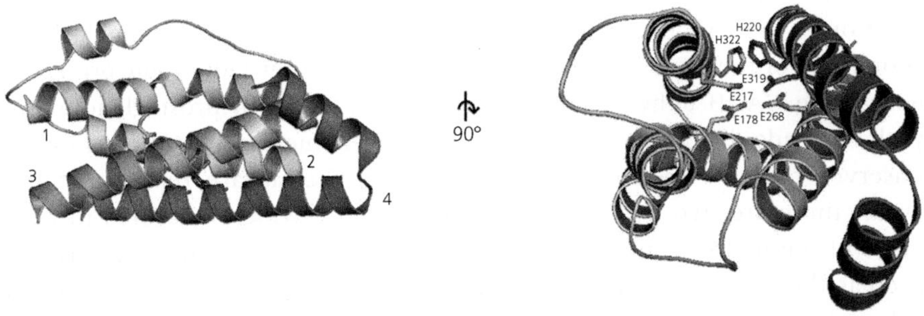

Figure 5.1 A modified version of the 1999 AOX model, indicating iron-binding residues (right, as per Table 5.2) within the four helix bundle (left, numbers indicate helices 1–4). (*See insert for color representation of the figure.*)

BOX 5.1 CREATING AN HOMOLOGY MODEL OF *S. GUTTATUM* ALTERNATIVE OXIDASE

An homology model of the *S. guttatum* alternative oxidase (using sequence P22185) was generated using the SwissMODEL server (Schwede *et al.*, 2003) following the careful alignment of the *S. guttatum* and TAO (Q26710) sequences with ClustalW (Thompson *et al.*, 1994) and the TAO structure 3VV9 as a template. Models were made of chains A and B, which were later combined to simulate the dimeric unit. The monomer was evaluated using ProSA (Weiderstein and Sippl, 2007) with a z-score of −3.93, which is well within the z-score range for other proteins of a similar size in the PDB. The QMEAN score returned by the SwissMODEL server following the homology model generation was a low 0.25. However, taking into account the under-representation of monotopic membrane proteins in the databases against which the approximate free energy for the model are compared, this low score is not particularly informative.

Table 5.2 A list of the six residues proposed to ligate the diiron centre of the alternative oxidase, based on the iron ligation motifs found in other diiron carboxylates. Numbering corresponds to *S.guttatum*

Residue and number	Helix
Glu178	2
Glu217	3
His220	3
Glu268	5
Glu319	6
His322	6

A new model of the alternative oxidase

A new homology model of the AOX has been created using the recently solved TAO crystal structure (Shiba *et al.*, 2013) of the *S. guttatum* alternative oxidase sequence (UniProt accession number P22185). The TAO structure and resulting *S. guttatum* homology model confirms the spatial orientation of the predicted four-helix bundle acting as a scaffold for the six key iron-binding residues. As indicated above, the two key histidine residues (H220 and H322, *S. guttatum* numbering as per Table 5.2), which were predicted to be actively involved in iron-binding in former models (Andersson and Nordlund, 1999; Berthold and Stenmark, 2003), appear to be further away from the diiron centre, suggesting that instead these do not act as Fe ligands but are important for electron and proton transfer (Shiba *et al.*, 2013; Moore *et al.*, 2013; Young *et al.*, 2013). The new model also indicates the presence of two helices (helices 1 and 4) in addition to the central four-helix bundle (comprising helices 2, 3, 5 and 6), which had not been previously predicted (as labelled in Figure 5.2). It is highly likely that these additional helices are involved in membrane association (as demonstrated in Figure 5.3 and Box 5.1), since submersion of these helices into the lipid bilayer would ensure that a hydrophobic tunnel which reaches into the active site was available (compare Hoefnagel *et al.*, 1997). This is also the case with another integral monotopic membrane protein, the external NADH dehydrogenase, the structure of which has recently been determined (Iwata *et al.*, 2012; Protein Data Bank accession numbers 4G9K, 4GAP and 4GAV).

As well as establishing the likely membrane association mechanism, the homology model of *S. guttatum* generated using TAO as a template has provided a strong insight into the region of the protein potentially involved in dimerization. Previously it was suggested that dimerization of AOX either involved an N-terminal domain that included the redox active Cys 122 residue (Siedow and Umbach, 2000), which is highly conserved amongst plant AOX sequences but lacking in almost all protist and fungal AOX sequences (Chaudhuri and Hill, 1996 and Fukai *et al.*, 2002; Sakajo *et al.*, 1991 and Umbach and Siedow, 2000,

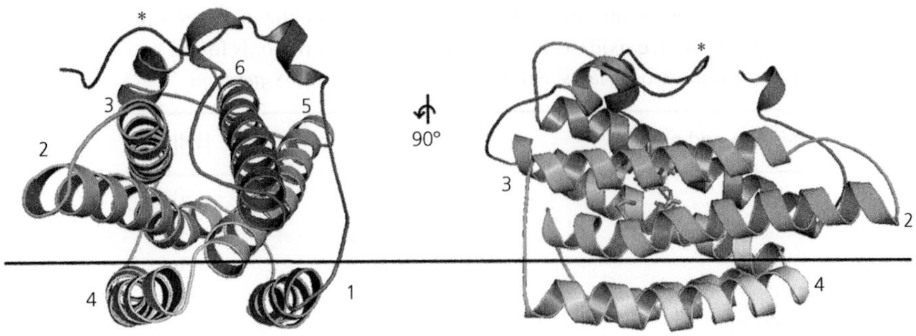

Figure 5.2 The monomeric *S. guttatum* homology model based on TAO (3VV9). On the left, all six helices are labelled and on the right only the nearest are labelled. *, the location of the conserved Cys 122 residue in the unstructured N-terminal region shown in dark blue (see text for details); black line, the approximate placement of the membrane with respect to the protein. The image on the right is the 90° anticlockwise rotation of the image on the left. (*See insert for color representation of the figure.*)

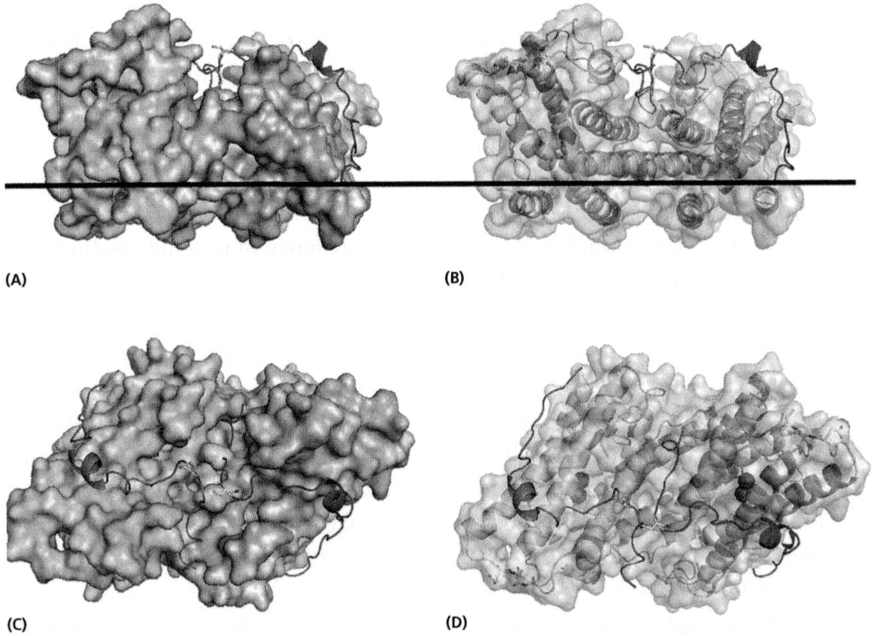

Figure 5.3 The dimeric *S. guttatum* homology model based on TAO (3VV9) is shown here embedded into the inner surface of the inner membrane (see Box 5.1 for further details) with helices 1 and 4, as in Figure 5.2) lying approximately 5 Å below the lipid/solvent interface (solid line). A. Surface representation of the plant AOX showing N-terminal extension and location of Cys 122. B. As A but surface rendered transparent (40%) and showing helices and Fe atoms. C. As A but looped 90°. D. As C but surface rendered transparent (40%) and showing helices and Fe atoms. The yellow and red sticks indicate the position of the QDC motif. (*See insert for color representation of the figure.*)

respectively) or alternatively two hydrophobic regions within the protein (Andersson and Nordlund, 1999). However, the absence of the conserved Cys 122 equivalent in TAO, the crystal structure of which is in fact dimeric (and not monomeric as previously thought; Chaudhuri *et al.*, 2005) would suggest that while the conserved Cys 122, which is highly solvent accessible within the unstructured, non-buried N-terminal region of the protein (as labelled in Figure 5.3), plays an evident role in the regulation of some alternative oxidase isoforms, it is not the primary or universal mechanism of dimerization.

To ascertain other possible sites of dimerization, the dimeric model was studied in detail in conjunction with a carefully constructed multiple alignment containing full AOX sequences from as many species as possible. Residues close enough to form a hydrogen bonding network between the two monomers were identified across helices 2 and 3, with longer range interactions potentially occurring between residues on helix 4 (these residues are shown in Table 5.3) and cross-referenced to ensure that they were relatively well conserved across as many sequences as possible, including conservative substitutions. It is likely that the combination of residues identified using the *S. guttatum* homology model may be more specific to thermogenic plants, for example, but generation of further homology models using AOX sequences from other species could provide key insights into the range of residue combinations found on helices 2, 3 and 4 which make up species-specific dimerization interfaces.

As shown in Figure 5.3, the N-terminal region of the *S. guttatum* AOX is unstructured and therefore potentially has some flexibility with respect to the core of the protein. One of the major differences between the trypanosomal and plant AOX sequences is the length of this N-terminal region, with the TAO sequences consistently shorter than the plant sequences. It has been suggested that this potentially flexible region is a regulatory feature in plants (Ito *et al.*, 2011; Moore *et al.*, 2013), which has been supported by recent recombinant expression of two *S. guttatum* AOX proteins lacking 35 and 70 residues from the

Table 5.3 A list of the residues identified on helices 2 and 3 as potentially involved in formation of a dimer interface, corresponding to those residues shown in Figure 5.4. *S. guttatum* numbering.

Residues on helix 2	Residues on helix 3	Residues on helix 4
T179	I207	R235
A182	R208	V238
M186	L211	Q242
V190	E215	
H193	R218	
L194		
L197		

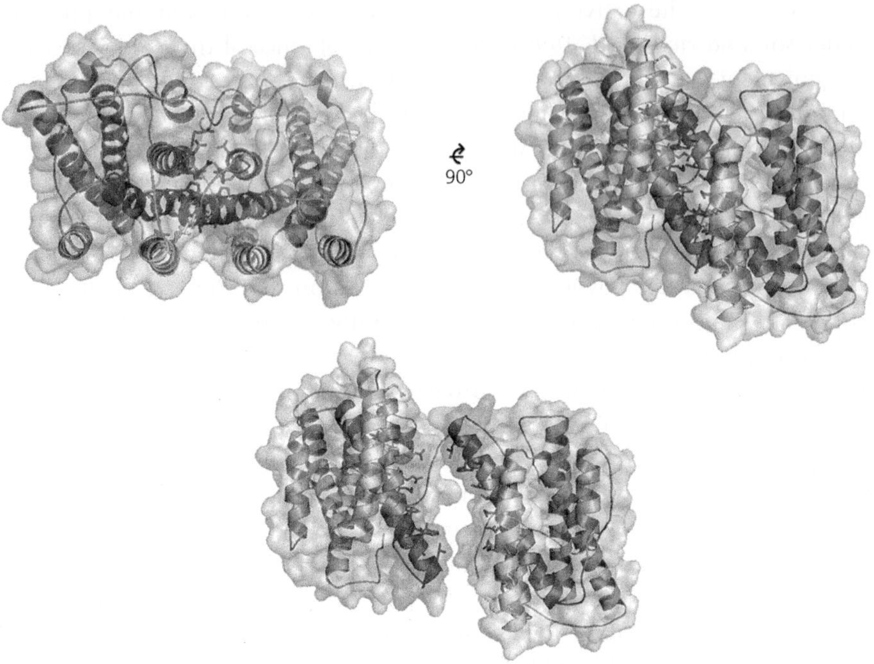

Figure 5.4 A graphical representation of the potential dimer interface, showing conserved residues on helices 2 (red), 3 (teal) and 4 (pale green) as listed in Table 5.3. The top two images represent the whole dimeric model both parallel (top left) and perpendicular (right) to the membrane, whilst the bottom image shows the monomeric models separated artificially to show the extent of tessellation between the two units which overlap rather than lying flush to one another. (*See insert for color representation of the figure.*)

N-terminal, which were very inactive but nevertheless could form dimers (A.L. Moore and M.S. Albury, unpublished data). It is possible that the unstructured N-terminal regions of each monomeric unit are able to for a link via the conserved Cys 122 residue (see Figure 5.3), which is present in this region, thus offering a greater degree of stability for the whole dimeric quaternary structure of the protein. Whether pyruvate is able to interact with the exposed C122 residues remains unknown, although as the residue is solvent-exposed, it is theoretically possible for this to occur.

Modelling the structure of plant alternative oxidase

The oxygen reduction cycle

What does modelling the structure of plant AOX tell us about the oxygen reduction cycle and regulation of activity? Catalytically, AOX is known to reduce oxygen through to water, using ubiquinol as the hydrogen and electron donor (Rich and Moore, 1976; Moore and Siedow, 1991). While the exact catalytic cycle has yet to be completely elucidated, several models to explain the four-electron reduction of oxygen to water have been proposed (Berthold *et al.*, 2000; Affourtit *et al.*, 2002; Maréchal *et al.*, 2009; Faiella *et al.*, 2009; Silverstein 2011; Moore *et al.*, 2013; Young *et al.*, 2013). Although similar in nature, these models differ as to the exact sequence in which oxygen and quinol bind to the enzyme. Furthermore, some of the reaction mechanisms involve one or two protein-based radicals resulting in the generation of high-valent iron intermediates (Affourtit *et al.*, 2002; Maréchal *et al.*, 2009; Silverstein 2011; Moore *et al.*, 2013; Young *et al.*, 2013), whereas others questioned the catalytic necessity of compounds with such strong oxidising potential and hence did not include it in their model (Berthold *et al.*, 2000; Faiella *et al.*, 2009).

Over the past few years it has become apparent that many diiron proteins, including stearoyl-ACP Δ^9-desaturase (Broadwater *et al.*, 1998), MMOH (Gassner and Lippard, 1999) and rubrerythrin (Gomes *et al.*, 2001), are also capable of fully reducing oxygen to water as a side reaction to their main respective catalytic activities. With respect to the oxidase activity of MMOH it has been suggested that this activity is the consequence of reduction of the diferryl MMOH intermediate `Q' (Stahl *et al.*, 2001) and such a mechanism has provided a catalytic precedent for alternative oxidase activity (Affourtit *et al.*, 2002).

Electron paramagnetic resonance (EPR) studies provided the first spectroscopic evidence in favour of the proposal that AOXs contained a diiron carboxylate active-site since EPR signals characteristic of diiron carboxylate proteins were detected both in isolated mitochondria, membrane-bound and purified recombinant alternative oxidases from a variety of organisms (Berthold *et al.*, 2002; Moore *et al.*, 2008). The diiron carboxylate protein family is a functionally diverse group containing a non-haem, diiron centre, and members include methane monooxygenase, ribonucleotide reductase and bacterioferretin. All contain four-helix bundles coordinating the diiron centre within the core of the catalytic units (Berthold and Stenmark, 2003). Most of the members of this family are large, multidomain proteins unassociated with membranes, though some smaller members do exist, such as ferritin and rubrerythrin (compare Berthold and Siedow, 1993; Nordlund and Eklund, 1995). Additionally, both the diiron carboxlyates and AOX lack a spectroscopic absorbance above 340 nm (Berthold and Siedow, 1993). These experimental observations initially led to the inclusion of AOX as a member of the diiron carboxylate family (Siedow *et al.*, 1995; Moore *et al.*, 2008), although the status of the enzyme has recently been reclassified as a family in its own right (E.C. number 1.10.3.11[1]).

The crystal structure of TAO not only confirmed that the redox-active tyrosine (Y275, as per *S. guttatum* AOX numbering) was well within electron transfer range of the diiron centre (<5 Å) and within the primary ligation sphere but

[1]As per ExPASy and BRENDA, May 2013.

was also very close (>4 Å) to the ubiquinol/inhibitor-binding domain (which itself was within 4 Å of the active-site) (Shiba *et al.*, 2013; Moore *et al.*, 2013). Although, as indicated earlier, CAVER visualization software predicted there were two hydrophobic cavities both of which coincided within the active site and could accommodate a ubiquinol molecule suggesting there were two ubiquinol sites (Moore *et al.*, 2013), we are now of the opinion that there is only a substrate-binding domain (Young *et al.*, 2013). The crystal structure also revealed residues within the secondary ligation sphere, which include N216, Y299, W300 and D318. N216 and D318 are situated in the centre of a hydrogen-bond network which connects these residues to the diiron centre involving E178, E217, H220, E268 and H322 and furthermore extends the network to include Y246 and W247.

In light of the above structural information, we have modified our catalytic cycle. The reaction cycle is in essence similar to that proposed previously (Moore *et al.*, 2013), but differs in the nature and positioning of the iron-ligating amino acids within the cycle (Young *et al.*, 2013). The diferrous centre interacts with oxygen to initially establish an superoxo intermediate which – following the extraction of a proton and an electron from tyrosine 275 (thereby generating a tyrosyl radical) – subsequently leads to the formation of a peroxo intermediate. Rearrangement of the peroxo core, followed by proton and electron donation through the proton-coupled electron transfer pathway involving W247 and two ubiquinol molecules, completes the reaction cycle regenerating the diferrous centre and re-reducing the tyrosine and tryptophan residues.

Regulation of the alternative oxidase

It is generally accepted that pyruvate and other α-keto acids can serve as allosteric activators of plant AOX. In addition to activating isolated mitochondria, pyruvate activation is also observed in partially and fully purified preparations from thermogenic tissues (Zhang *et al.*, 1996; Carré *et al.*, 2011) in addition to recombinant alternative oxidases from non-thermogenic tissues (Crichton *et al.*, 2005; Berthold, 1998; Rhoads *et al.*, 1998). The mechanism of α-keto regulation appears to involve both highly conserved Cys residues (Cys 122 and Cys 172), the N-terminal one of which, as indicated earlier, also acts as the site for intermolecular bond formation. Cys172, while highly conserved across all plant species, has a less defined role in α-keto acid activation. Thermogenic tissues tend to have a much more varied response to α-keto acid addition. For instance, both mitochondria and recombinant protein from *S. guttatum* appear constitutively active in the absence of pyruvate (Crichton *et al.*, 2005), whereas *Symplocarpus renifolius* (Onada *et al.*, 2007), some isoforms of *Arum maculatum* (Ito *et al.*, 2011) and *Nelumbo nucifera* (which lack Cys122 gene) (Grant *et al.*, 2009) are sensitive to α-keto acids. Crichton *et al.* (2005) suggested that the sensitivity to α-keto acids depended upon the presence of a QDC or ENV motif located between helices 3 and 4 (see Figure 5.1 and Andersson and Nordlund (1999) for the model) since all tissues possess both Cys residues. Thermogenic tissues that are insensitive to pyruvate possess the QDC motif, while those which are insensitive to α-keto acids (some thermogenic and all non-thermogenic tissues) possess the ENV motif or a variant thereof. Interestingly, of the seven cDNAs encoding AOX in *Arum maculatum* spadices, only *AmAOX1e*, which is only expressed during thermogenesis, possesses the QDC motif and when this gene is expressed in *Schizosaccharomyces pombe* mitochondria, respiration is insensitive to pyruvate addition. The plant AOX model in Figure 5.3D indicates that the QDC/ENV motif is located between helices 5 and 6 close to the surface of the protein but somewhat distant from Cys122. This may not be a particular issue since there is no direct interaction proposed between the regulatory Cys residue and the QDC/ENV motif and furthermore, recent substitution of the ENV motif by QDT in *Arum concinnatum* AOX expressed in HeLa cells diminished catalytic activity, suggesting that functional significance of the N-terminal extension is not particular to this regulatory cysteine (Kakizaki and Ito, 2013). Obviously further clarification of the exact roles of Cys122 and the QDC/ENV motif will have to await elucidation of the crystal structures from thermogenic and non-thermogenic plants.

The limitation of the TAO-based homology model

The homology model of *S. guttatum* AOX based on TAO has provided invaluable insights into the structure, mechanism, membrane association and dimer interface of plant AOX. As discussed in Box 5.1, the evaluation of the model is reasonable and provides a detailed map of the active site of the enzyme. As with any model, however, the limitations need to be considered before firm conclusions are drawn. For example, the TAO and the *S. guttatum* AOX sequence are not particularly closely related (~40% sequence similarity) and there are significant differences in the length of the N-terminal regions of the two sequences, with the TAO sequence being shorter by approximately 21 residues. While this region may play a role in the regulation of plant AOX, it is unlikely to play the same role in the regulation of TAO, as TAO is generally considered to be unregulated and not sensitive to pyruvate stimulation (Chaudhuri *et al.*, 2006). Therefore, the structure of the regulatory region of plant AOX may not be present in the TAO structure and consequently will be absent in the homology model too. As such, this new homology model remains only a very useful guide until the structure of the *S. guttatum* AOX is finally solved.

Summary

AOX is a protein key to the survival of both thermogenic plants and parasitic trypanosomes. It also offers cyanide-insensitive respiration to other plants and fungi when they are wounded, poisoned or otherwise harmed. With ancestral relationships to both diiron carboxylate proteins and the plastid terminal oxidase,

AOX has evolved from a scavenger to a more functionally diverse ubiquinol oxidase with a four-helix bundle at its core. Mutagenesis studies have illustrated the importance of key residues and the most recent, high-quality homology model has answered questions not only about the topography of the active site, but also provided key insights into the dimer interface, substrate access channel and the location of the previously proposed regulatory region at the N-terminus. Understanding the intricate structure–function relationship of this adaptable protein is essential to developing safe new drugs to treat diseases such as African trypanosomiasis and diseases caused by related parasites such as *B. hominis* and *C. parvum*, as well as designing antifungal agents to complement and strengthen existing treatments.

References

Affourtit, C., Albury, M.S., Crichton, P. and Moore, A.L. (2002) Exploring the molecular nature of alternative oxidase regulation and catalysis. *FEBS Letters* **510**: 121–126.

Affourtit, C. and Moore, A.L. (2004) Purification of the plant alternative oxidase from *Arum maculatum*: measurement, stability and metal requirement. *Biochimica et Biophysica Acta* **1608**: 181–189.

Ajayi, W.U., Chaudhuri, M. and Hill, G.C. (2002) Site-directed mutagenesis reveals the essentiality of the conserved residues in the putative diiron active site of the trypanosome alternative oxidase. *Journal of Biological Chemistry* **277**: 8187–8193.

Albury, M.S., Affourtit, C. and Moore, A.L. (1998) A highly conserved glutamate residue (Glu-270) is essential for plant alternative oxidase activity. *Journal of Biological Chemistry* **273**: 30301–30305.

Albury, M.S., Affourtit, C., Crichton, P.G. and Moore, A.L. (2002) Structure of the plant alternative oxidase – site-directed mutagenesis provides new information on the active site and membrane topology. *Journal of Biological Chemistry* **277**: 1190–1194.

Andersson, M.E. and Nordlund, P. (1999) A revised model of the active site of alternative oxidase. *FEBS Letters* **449**: 17–22.

Bahr, J.T. and Bonner, W.D. (1972) Cyanide-insensitive respiration: I. *The steady states of skunk cabbage spadix and bean hypocotyls mitochondria. Journal of Biological Chemistry* **248**: 3441–3445.

Bendall, D.S. and Bonner, W.D. (1971) Cyanide-insensitive respiration in plant mitochondria. *Plant Physiology* **47**: 236–245.

Berthold, D. and Siedow, J.N. (1993) Partial purification of the cyanide-resistant alternative oxidase of skunk cabbage (*Symplocarpus foetidus*) mitochondria. *Plant Physiology* **101**: 113–119.

Berthold, D.A. (1999) Isolation of mutants of the *Arabadopsis thaliana* alternative oxidase (ubiquinol: oxygen oxidoreductase) resistant to salicylhydroxamix acid. *Biochimica et Biophysica Acta* **1364**: 73–83.

Berthold, D., Voevodskaya, N., Stenmark, P. *et al.* (2002) EPR studies of the mitochondrial alternative oxidase. Evidence for a diiron carboxylate centre. *Journal of Biological Chemistry* **277**: 43608–43614.

Berthold, D.A. and Stenmark, P. (2003) Membrane-bound diiron carboxylate proteins. *Annual Review of Plant Biology* **54**: 497–517.

Berthold, D.A., Andersson, M.E. and Nordlund, P. (2000) New insight into the structure and function of the alternative oxidase. *Biochimica et Biophysica Acta* **1460**: 241–254.

Broadwater, J.A., Ai, J., Loehr, T.M. *et al.* (1998) Peroxodiferric intermediate of stearoyl-acyl carrier protein delta (9) destaurase: oxidase reactivity during single turnover and implications for the mechanism of desaturation. *Biochemistry* **37**: 14664–14671.

Carol, P., Stevenson, D., Bisanz, C. *et al.* (1999) Mutations in the arabadopsis gene *IMMUTANS* cause a variegated phenotype by inactivating a chloroplast terminal oxidase associated with phenotype desaturation. *The Plant Cell* **11**: 57–68.

Carol, P. and Kuntz, P. (2001) A plastid terminal oxidase comes to light: implications for carotenoid biosynthesis and chlororespiration. *Trends in Plant Science* **6**: 31–36.

Carré, J.E., Affourtit, C. and Moore, A.L. (2011) Interaction of purified plant alternative oxidase from *Arum maculatum* with pyruvate. *FEBS Letters* **585**: 397–401.

Chance, B. and Williams, G.R. (1956) The respiratory chain and oxidative phosphorylation. *Advances in Enzymology and Related Subjects of Biochemistry* **17**: 65–134.

Chaudhuri, M., Ajayi, W., Temple, S. and Hill, G.C. (1995) Identification and partial purification of a stage specific 33 kDa mitochondrial protein as the alternative oxidase of the *Trypanosoma brucei brucei* bloodstream trypomastigotes. *Journal of European Microbiology* **42**: 467–472.

Chaudhuri, M. and Hill, G.C. (1996) Cloning, sequencing and functional activity of the *Trypanosoma brucei brucei* alternative oxidase. *Molecular and Biochemical Parasitology* **83**: 125–129.

Chaudhuri, M., Ott, R.D., Saha, L. *et al.* (2005) The trypanosome alternative oxidase exists as a monomer in *Trypanosoma brucei* mitochondria. *Parasitology Research* **96**: 178–183.

Chaudhuri, M., Ott, R.D. and Hill, G.C. (2006) Trypanosome alternative oxidase: from molecule to function. *Trends in Parasitology* **22**: 484–491.

Clarkson, A.B., Bienen, E.J., Pollakis, G. and Grady, R.W. (1989) Respiration of bloodstream forms of the parasite *Trypanosoma brucei brucei* is dependent on a plant-like alternative oxidase. *Journal of Biological Chemistry* **264**: 17770–17776.

Cournac, L., Latouche, G., Cerovic, Z. *et al.* (2000) *In vivo* interactions between photosynthesis, mitorespiration and chlororespiration in *Chlamydomonas reingardtii*. *Plant Physiology* **129**: 1921–1928.

Crichton, P.G., Affourtit, C., Albury, M.S. *et al.* (2005) Constitutive activity of *Sauromatum guttatum* alternative oxidase in *Schizosaccharomyces pombe* implicates residues in addition to conserved cysteines in α-keto acid activation. *FEBS Letters* **579**: 331–336.

El-Khoury, R., Dufour, E., Rak, M. *et al.* (2013) Alternative oxidase expression in the mouse enables bypassing cytochrome *c* oxidase blockade and limits mitochondrial ROS overproduction. *PLOS Genetics* **9**: e1003182.

Faiella, M., Andreozzi, C., de Rosales R.T. *et al.* (2009) An artificial di-iron oxo-protein with phenol oxidase activity. *Nature Chemical Biology* **5**: 882–884.

Fairlamb, A.H. (2003) Chemotherapy of human African trypanosomaisis: Current and future prospects. *Trends in Parasitology* **19**: 488–494.

Finnegan, P.M., Umbach, A.L. and Wilce, J.A. (2003) Prokaryotic origins for the mitochondrial alternative oxidase and plastid terminal oxidase nuclear genes. *FEBS Letters* **555**: 425–430.

Finnegan, P.M., Soole, K.L. and Umbach, A.L. (2004) Alternative mitochondrial electron transport proteins in higher plants. In *Plant Mitochondria: From Gene To Function,* (D.A. Day, A.H. Millar and J. Whelan, eds). *Advances in Photosynthesis and Respiration 17*: 163–230. Kluwer, Dordrecht, The Netherlands.

Fu, A., Liu, H., Yu, F. *et al.* (2012) Alternative oxidases (AOX1a and AOX2) can functionally substitude for plastid terminal oxidase in *Arabidopsis* chloroplasts. *The Plant Cell* **24**: 1579–1595.

Fukai, Y., Nihei, C., Yabu, Y. *et al.* (2002) Strain specific difference in amino acid sequences of trypanosome alternative oxidase. *Parasitology International* **51**: 195–199.

Gadelha, C., Holden, J.M., Allison, H.C. and Field, M.C. (2011) Specializations in a successful parasite: What makes the bloodstream-form African trypanosome so deadly? *Molecular and Biochemical Parasitology* **179**: 51–58.

Gassner, G.T. and Lippard, S.J. (1999) Component interactions in the soluble methane monooxygenase system from *Methylococcus capsulatus* (Bath). *Biochemistry* **38**: 12768–12785.

Grant, N., Onda, Y., Kakizaki, Y. *et al.* (2009) Two cys or not two cys? That is the question: alternative oxidase in the thermogenic plant sacred lotus. *Plant Physiology* **150**: 987–995.

Grant, P.T. and Sargent, J.R. (1960) Properties of L-α-glycerophosphate oxidase and its role in the respiration of *Trypanosoma rhodesiense*. *Biochemical Journal* **76**: 229–237.

Grant, P.T. and Sargent, J.R. (1961) Respiratory systems in the trypanosomidae. *Biochemical Journal* **81**: 200–206.

Gomes, C.M., Le Gall, J., Xavier, A.V. and Teixeira, M. (2001) Could a diiron-containing four-helix-bundle protein have been a primitive oxygen reductase? *ChemBioChem* **7**: 583–587.

Hiser, C. and McIntosh, L. (1990) Alternative oxidase of potato is an integral membrane protein synthesized *de novo* during aging of tuber slices. *Plant Physiology* **93**: 312–318.

Hoefnagel, M.H.N., Rich, P.R., Zhang, Q. and Wiskich, J.T. (1997) Substrate kinetics of the plant mitochondrial alternative oxidase and the effects of pyruvate. *Plant Physiology* **115**: 1145–1153.

Ito, K., Ogata, T., Kakizaki, Y. *et al.* (2011) Identification of a gene for pyruvate-insenstive mitochondrial alternative oxidase expressed in the thermogenic appendices in *Arum maculatum*. *Plant Physiology* **157**: 1721–1732.

Iwata, M., Lee, Y., Yamashita, T. *et al.* (2012) The structure of the yeast NADH dehydrogenase (Ndi1) reveals overlapping binding sites for water- and lipid-soluble substrates. *Proceedings of the National Academy of Sciences USA* **109**: 15247–15252.

Josse, E., Simkin, A.J., Gaffé, J. *et al.* (2000) A plastid terminal oxidase associated with carotenoid desaturation during chromoplast differentiation. *Plant Physiology* **123**: 1427–1436.

Kakizaki, Y. and Ito, K. (2013) Engineering plant alternative oxidase function in mammalian cells: substitution of the motif-like sequence ENV for QDT diminishes catalytic activity of *Arum concinnatum* AOX1a expressed in HeLa cells. *Applied Biochemistry and Biotechnology* **170**: 1229–1240. DOI 10.1007/s12010-013-0235-x.

Kido, Y., Sakamoto, K., Nakamura, K. *et al.* (2010) Purification and kinetic characterisation of recombinant alternative oxidase from *Trypanosoma brucei brucei*. *Bioenergetics* **1797**: 443–450.

Keilin, D. and Hartree, E.F. (1938) Letters to the Editor: cytochrome *a* and cytochrome oxidase. *Nature* **141**: 870–871.

Lambers, H. (1982) Cyanide-resistant respiration: a non-phosphorylating electron transport pathway acting as an energy overflow. *Physiologia Plantarum* **55**: 478–485.

Lambowitz, A.M., Sabourin, J.R., Bertrand, H. *et al.* (1989) Immunological identification of the alternative oxidase of *Neurospora crassa* mitochondria. *Molecular and Cellular Biology* **9**: 1362–1364.

Lindqvist, Y., Huang, W., Schneider, G. and Shanklin, J. (1996) Crystal structure of Δ^9 stearoyl-acyl carrier protein desaturase from castor seed and its relationship to other diiron proteins. *EMBO Journal* **15**: 4081–4092.

Maréchal, A., Kido, Y., Kita K. *et al.* (2009) Three redox states of *Trypanosoma brucei* alternative oxidase identified by infrared spectroscopy and electrochemistry. *Journal of Biological Chemistry* **284**: 31827–31833.

Maxwell, D.P., Wang, Y. and McIntosh, L. (1999) The alternative oxidase lowers mitochondrial reactive oxygen production in plant cells. *Proceedings of the National Academy of Sciences* **96**: 8271–8276.

McDonald, A.E., Sieger, S.M. and Vanlerberghe, G.C. (2002) Methods and approaches to study the plant mitochondrial alternative oxidase. *Physiologia Plantarum* **116**: 135–143.

McDonald, A.E. and Vanlerberghe, G.C. (2004) Branched mitochondrial electron transport in the Animalia: presence of alternative oxidase in several animal phyla. *IUBMB Life* **56**: 333–341.

McDonald, A.E., Ivanov, A.G., Bode, R. *et al.* (2011) Flexibility in photosynthetic transport: the physiolological role of plastoquinol terminal oxidase (PTOX). *Biochimica et Biophysica Acta* **1807**: 954–967.

Meeuse, B.J.D. (1975) Thermogenic respiration in Aroids. *Annual Review of Plant Physiology* **26**: 117–126.

Meeuse, B.J.D. and Raskin, I. (1988) Sexual reproduction in the arum lily family, with emphasis on thermogenicity. *Sexual Plant Reproduction* **1**: 3–15.

Millar, A.H., Whelan, J., Soole, K.L. and Day, D.A. (2011) Organization and regulation of mitochondrial respiration in plants. *Annual Reviews of Plant Biology* **62**: 79–104.

Millar, A.H., Wiskich, J.T., Whelan, J. and Day, D.A. (1993) Organic acid activation of the alternative oxidase of plant mitochondria. *FEBS Letters* **329**: 259–262.

Minagawa, N., Sakajo, H., Komiyama, T. and Yoshimoto, A. (1990) Essential role of ferrous iron in cyanide-resistant respiration in *Hansenula anomala*. *FEBS Letters* **267**: 114–116.

Minagawa, N., Yabu, Y., Kita, K. *et al.* (1996) An antibiotic, ascofuranone, specifically inhibits respiration and *in vitro* growth of long slender bloodstream forms of Trypanosoma brucei brucei. *Molecular and Biochemical Parasitology* **81**: 127–136.

Moore, A.L., Bonner, W.D. and Rich, P.R. (1978). The determination of the proton-motive force during cyanide-insensitive respiration in plant mitochondria. *Archives of Biochemistry and Biophysics* **186**: 298–306.

Moore, A.L. and Siedow, J.N. (1991) The regulation and nature of the cyanide-resistant alternative oxidase of plant mitochondria. *Biochimica et Biophysica Acta* **1059**: 121–140.

Moore, A.L., Dry, I.B and Wiskich, J.T. (1988) Measurement of the redox state of the ubiquinone pool in plant mitochondria. *FEBS Letters* **235**: 76–80.

Moore, A.L., Carré, J.E., Affourtit, C. *et al.* (2008) Compelling EPR evidence that the alternative oxidase is a diiron carboxylate protein. *Biochimica et Biophysica Acta – Bioenergetics* **1777**: 327–330.

Moore, A.L., Albury, M.A., Crichton, P. and Affourtit, C. (2002) Function of the alternative oxidase: is it still a scavenger? *Trends in Plant Science* **7**: 478–481.

Moore, A.L., Shiba, T., Young, L. *et al.* (2013) Unravelling the heater – new insights into the structure of the alternative oxidase. *Annual Reviews in Plant Biology* **64**: 637–663.

Nakamura, K., Sakamoto, K., Kido, Y. *et al.* (2005) Mutational analysis of the *Trypanosoma vivax* alternative oxidase: The $E(X)_6Y$ motif is conserved in both mitochondrial alternative oxidase and the plastid terminal oxidase and is indispensable for enzyme activity. *Biochemical and Biophysical Research Communications* **334**: 593–600.

Nakamura, K., Fujioka, S., Fukumoto, S. *et al.* (2010) Trypanosome alternative oxidase, a potential therapeutic target for sleeping sickness, is conserved among *Trypanosoma brucei* subspecies. *Parasitology International* **59**: 560–564.

Nihei, C., Fukai, Y. and Kita, K. (2002) Trypanosome alternative oxidase as a target of chemotherapy. *Biochimica et Biophysica Acta* **1587**: 234–239.

Nordlund, P. and Eklund, H. (1995) Diiron-carboxylate proteins. *Current Opinion in Structural Biology* **5**: 758–766.

Onda, Y., Kato, Y., Abe, Y. *et al.* (2007) Pyruvate-sensitive AOX exists as a non-covalently associated dimer in the homeothermic spadix of the skunk cabbage, *Symplocarpus renifolius*. *FEBS Lett* **581**: 5852–5858.

Purvis, A.C. and Shewfelt, R.L. (1993) Does the alternative pathway ameliorate chilling injury in sensitive plant tissues? *Physiologia Plantarum* **88**: 712–718.

Rhoads, D.M., Umbach, A.L., Sweet, C.R. *et al.* (1998) Regulation of the cyanide-resistant alternative oxidase from plant mitochondria – identification of the cysteine residue involved in α-keto acid stimulation and intersubunit disulphide bond formation. *Journal of Biological Chemistry* **273**: 30750–30756.

Rich, P.R. and Moore, A.L. (1976) The involvement of the protonmotive ubiquinone cycle in the respiratory chain of higher plants and its relation to the branchpoint of the alternate pathway. *FEBS Letters* **65**: 339–344.

Rich, P. R. (1978) Quinol oxidation in *Arum maculatum* mitochondria and its application to the assay, solubilisation and partial purification of the alternative oxidase. *FEBS Letters* **96**: 252–256.

Rieske, J.S., Lipton, S.H., Baum, H. and Silman, H.I. (1967) Factors affecting the binding of antimycin A to complex III of the mitochondrial respiratory chain. *Journal of Biological Chemistry* **242**: 4888–4896.

Roberts, C.W., Roberts, F., Henrique, F.L. *et al.* (2004) Evidence for mitochondrial-derived alternative oxidase in the apicomplexan parasite *Cryptosporidium parvum*: a potential antimicrobial agent target. *International Journal of Parasitology* **34**: 297–308.

Sakajo, S., Minagawa, N. and Yoshimoto, A. (1991) Molecular cloning of cDNA for antimycin A-inducible mRNA and its role in the cyanide-resistant respiration in *Hansenula anomala*. *Biochemica et Biophysica Acta* **1090**: 102–108.

Schonbaum, G.R., Bonner, W., Storey, B.T. and Bahr, J.T. (1971) Specific inhibition of the cyanide-insentive respiratory pathway in plant mitochondria by hydroxamic acids. *Plant Physiology* **47**: 124–128.

Schwede, T., Kopp, J., Guex, N. and Peitsch, M.C. (2003) Swiss-model: an automated protein homology-modelling server. *Nucleic Acids Research* **31**: 3381–3385.

Shanklin, J., Guy, JE., Mishra, G. and Lindqvist, Y. (2009) Desaturases: emerging models for understanding functional diversification of diiron-containing enzymes. *Journal of Biological Chemistry* **284**: 18559–18563.

Shiba, T., Kido, Y., Sakamoto, K. *et al.* (2013) Structure of the trypanosome cyanide-insensitive alternative oxidase, a promising drug target. *Proceedings of the National Academy of Sciences USA* **110**: 4580–4585.

Siedow, J.N. and Bickett, D.M. (1981) Structural features required for inhibition of cyanide-insensitive electron transfer by propyl gallate. *Archives of Biochemistry and Biophysics* **207**: 32–39.

Siedow, J.N. and Umbach, A.L. (2000) The mitochondrial cyanide resistant oxidase: structural conservation amid regulatory diversity. *Biochimica et Biophysica Acta* **1459**: 432–439.

Siedow, J.N., Umbach, A.L. and Moore, A.L. (1995) The active site of the cyanide-resistant oxidase from plant mitochondria contains a binuclear iron centre. *FEBS Letters* **362**: 10–14.

Silverstein, T.P. (2011) Photosynthetic water oxidation versus. mitochondrial oxygen reduction: distinct mechanistic parallels. *Journal of Bioenergetics and Biomembranes* **43**: 437–446.

Stahl, S.S., Francisco, W.A., Merkx, M. *et al.* (2001) Oxygen kinetic isotope effects in soluble methane monooxygenase. *Journal Biological Chemistry* **276**: 4549–4553.

Stechmann, A., Hamblin, K., Pérez-Brocal, V. *et al.* (2008) Organelles in Blastocystis that blur the distinction between mitochondria and hydrogenosomes. *Current Biology* **18**: 580–585.

Storey, B.T. (1976) Respiratory chain of plant mitochondria: XVIII – Point of interaction of the alternative oxidase with the respiratory chain. *Plant Physiology* **58**: 521–525.

Suzuki, T., Hashimoto, T., Yabu, Y. *et al.* (2004) Direct evidence for cyanide-insensitive quinol oxidase (alternative oxidase) in apicomplexan parasite *Cryptosporidium parvum*: phylogenetic and therapeutic implications. *Biochemical and Biophysical Research Communications* **313**: 1044–1052.

Thompson, J.D., Higgins, D.G. and Gibson, T.J. (1994) Clustal W: improving the sensitivity of progressive multiple sequence alignment through sequence weighting, position-specific gap penalties and weight matrix choice. *Nucleic Acids Research* **22**: 4673–4680.

Umbach, A.L. and Siedow, J.N. (2000) The cyanide-resistant alternative oxidase from the fungi *Pichia stipites* and *Neurospora crassa* are monomeric and lack regulatory features of the plant enzyme. *Archives of Biochemistry and Biophysics* **378**: 234–245.

van Buuren, K.J.H., Zuurendonk, P.F., van Gelder, B.F. and Muijsers, A.O. (1972) Biochemical and biophysical studies on cytochrome aa_3: binding of cyanide to cytochrome aa_3. *Biochimica et Biophysica Acta* **256**: 243–257.

Veiga, A., Arrabaça, J.D. and Loureiro-Dias, M.C. (2003) Cyanide-resistant respiration, a very frequent pathway in yeast. *FEMS Yeast Research* **3**: 239–245.

Wagner, A.M. and Moore, A.L. (1997) Structure and function of the plant alternative oxidase: its putative role in the oxygen defence mechanism. *Bioscience Reports* **17**: 319–333.

WHO (2012) World Health Organization Fact Sheet no. 259, Trypanosomiasis, human African (sleeping sickness). http://www.who.int/mediacentre/factsheets/fs259/en/ Published October 2012, accessed September 2014.

Wiederstein, M. and Sippl, M.J. (2007) ProSA-web: interactive web service for the recognition of errors in three-dimensional structures of proteins. *Nucleic Acids Research* **35**: W407–W410.

Williams, B.A.P., Elliott, C., Burri, L. *et al.* (2010) Wide distribution of the mitochondrial alternative oxidase in microsporidia parasites. *Plos Pathogens* **6**: 2 e1000761.

Yabu, Y., Yoshida, A., Suzuki, T. *et al.* (2003) The efficacy of ascofuranone in a consecutive treatment on *Trypanosoma brucei brucei* in mice. *Parasitology International* **52**: 155–164.

Young, L., Shiba, T., Harada, S. *et al.* (2013) The alternative oxidases: simple oxidoreductases proteins with complex functions. *Biochemical Society Transactions* **41**(5): 1305–1311.

CHAPTER 6

The role of alternative respiratory proteins in nitric oxide metabolism by plant mitochondria

Ione Salgado[1] and Halley Caixeta Oliveira[2]

[1] *Departamento de Biologia Vegetal, Instituto de Biologia, Universidade Estadual de Campinas (UNICAMP), São Paulo, Brazil*

[2] *Departamento de Biologia Animal e Vegetal, Centro de Ciências Biológicas, Universidade Estadual de Londrina (UEL), Londrina, Paraná, Brazil*

Introduction

Nitric oxide (NO) is a key signalling molecule that has been reported to be involved in a wide range of processes required for plant growth, development and reproduction, such as seed germination, root growth, leaf expansion and senescence, the establishment of symbiotic interactions, flowering and fruit ripening (Lamattina *et al.*, 2003; Neill *et al.*, 2003; Wendehenne *et al.*, 2004; Baudouin, 2011). NO also mediates adaptive plant responses to various environmental stresses (Wendehenne and Hancock, 2011; Siddiqui *et al.*, 2011). For example, NO is required for plant disease resistance and tolerance to drought, oxygen deficiency and other abiotic stresses (Qiao and Fan, 2008; Baudouin, 2011; Corpas *et al.*, 2011; Oliveira *et al.*, 2013).

The diverse actions of NO in biological systems reflect its physicochemical properties. NO is a gaseous free radical highly mobile in cellular systems, diffusing freely in both aqueous and lipid environments with a relatively long half-life (approximately 5 s) when compared to other radicals (Stamler *et al.*, 1992). NO can be reduced or oxidized to form a nitroxyl ion (NO^-) or nitrosonium ion (NO^+), respectively. NO reacts with O_2 to produce nitrogen dioxide (NO_2), which rapidly reacts with a further NO to generate dinitrogen trioxide (N_2O_3) (Brown, 2007). NO may also react at a diffusion-limited rate with the superoxide anion (O_2^-) to form peroxynitrite ($ONOO^-$) (Radi *et al.*, 2002). Therefore, despite the apparently simple structure of NO, its complex chemical properties in biological systems allow the formation of multiple secondary and tertiary products known collectively

Alternative Respiratory Pathways in Higher Plants, First Edition.
Edited by Kapuganti Jagadis Gupta, Luis A.J. Mur and Bhagyalakshmi Neelwarne.

as reactive nitrogen species (RNS). NO and each of these RNS can interact with the redox centres of various proteins and transiently or permanently alter their functions and/or activities (Stamler *et al.*, 1992; Gow and Ischiropoulos, 2001; Radi *et al.*, 2002).

NO reacts with Fe^{2+} in haeme or the Fe/S centres of proteins to form nitrosyl complexes. Direct binding of NO to the haeme group of guanylate cyclase (GC), which causes its activation and cGMP production, was identified early and represents the major NO-mediated signalling pathway in mammalian cells (Friebe and Koesling, 2009). NO-mediated increases in cGMP levels have been reported in plant cells (Durner *et al.*, 1998), although an NO-sensitive GC has not been identified in plants (Leitner *et al.*, 2009). Moreover, the attachment of NO to the haeme group of symbiotic and non-symbiotic haemoglobins has been suggested to control the steady-state levels and toxicity of NO during nodule formation (Sánchez *et al.*, 2011) and in root hypoxic responses (Igamberdiev *et al.*, 2005).

NO may reversibly attach to the thiol groups of reduced Cys residues, which results in protein *S*-nitrosylation (Stamler *et al.*, 2001). *S*-nitrosylation induced by NO is indirect and possibly occurs through N_2O_3 (Brown, 2007). Proteins may also be *S*-nitrosylated by direct transference of the NO^+ group between different *S*-nitrosothiols (SNO) in a process known as transnitrosylation (Brown, 2007). Many plant proteins that are candidates for *S*-nitrosylation have been identified by proteomic analysis (Lindermayr *et al.*, 2005; Romero-Puertas *et al.*, 2008), and post-translational modification by *S*-nitrosylation is now recognized as a key mechanism for the establishment of plant disease resistance (Spoel and Loake, 2011; Yu *et al.*, 2012).

NO may also indirectly cause the nitration of Tyr residues through the addition of an NO_2^+ group, which can permanently alter the structural and functional activities of proteins (Radi, 2013). Proteomic analysis of different plant species has revealed a large number of nitrated proteins (Lozano-Juste *et al.*, 2011), and it has been suggested that Tyr nitration is a relevant mechanism of protein modification elicited in response to various environmental stresses (Corpas *et al.*, 2008).

The numerous NO-responsive genes and promoters revealed by large-scale transcriptome analysis have demonstrated an additional role for NO in the control of gene expression in plants (Grün *et al.*, 2006; Palmieri *et al.*, 2008). Indeed, NO modulation of transcript levels of genes involved in various signal transduction pathways, such as those involved in disease resistance, stress responses and basic metabolism, has been demonstrated (Grün *et al.*, 2006; Ferrarini *et al.*, 2008; Palmieri *et al.*, 2008; Vitor *et al.*, 2013).

Recent studies have demonstrated that mitochondria may play a central role in various NO-mediated effects in plants because the synthesis and degradation of NO are developed within these organelles. Additionally, NO and its derivatives have multiple effects on mitochondrial bioenergetics, thereby affecting overall cell physiology. Therefore, as has been shown in mammals, the interaction between NO and the mitochondrial respiratory chain may mediate the biological effects of NO in plants.

The general organization of the electron transport pathway in the inner mitochondrial membrane of plant mitochondria is similar to that of mitochondria from other eukaryotes. Electrons enter the respiratory chain through complexes I and II and are transferred through complexes III and IV to O_2. Coupled to this electron flow, protons are pumped to the outside of the inner membrane, which generates an electrochemical gradient that is used for ATP synthesis (Millenaar and Lambers, 2003). In addition to these core proteins, plant mitochondria may also express non-proton-pumping NAD(P)H dehydrogenases on each side of the inner membrane that allow an alternative pathway of electron transport in the respiratory chain that bypasses complex I and directly reduces the ubiquinone pool (Rasmusson *et al.*, 2004). The mitochondrial respiratory chain of plants and many fungi and protists may also express an alternative oxidase (AOX) that can accept electrons directly from reduced ubiquinone and transfer them to O_2, thereby diverting the electron flow from complexes III and IV (Millenaar and Lambers, 2003). Electron transport through AOX also does not involve proton translocation and consequently does not contribute to the generation of the proton motive force and conservation of energy into ATP production, and as a result, energy dissipates as heat (Siedow and Umbach, 1995; Arnholdt-Schmitt *et al.*, 2006). The uncoupling of pathways for electron transport also prevents over-reduction of the electron carriers and is proposed to have physiological significance for redox homeostasis (Millenaar and Lambers, 2003; Rasmusson *et al.*, 2004) and the control of NO levels (de Oliveira *et al.*, 2008; Wulff *et al.*, 2009).

In this review, the role of the mitochondrial respiratory chain in regulating NO homeostasis (see Figure 6.1) and its biological effects in plant cells will be discussed. This review will initially focus on the effects of NO on plant mitochondrial respiration. Then, the mitochondrial mechanisms for NO degradation and synthesis will be discussed. Special emphasis will be placed on the involvement of alternative plant mitochondrial proteins in these processes.

Targets of NO in mitochondria

NO has been determined to regulate the respiration of mammalian cells by causing the reversible inhibition of cytochrome *c* oxidase (COX; complex IV), the terminal enzyme in the mitochondrial respiratory chain (Cleeter *et al.*, 1994). Nanomolar concentrations of NO reversibly inhibit synaptosomal respiration by competing with O_2 at the COX level, and the degree of inhibition depends on the NO:O_2 ratio in the medium, becoming more effective with decreasing O_2 levels (Brown and Cooper, 1994). Inhibition of O_2 consumption has been shown to

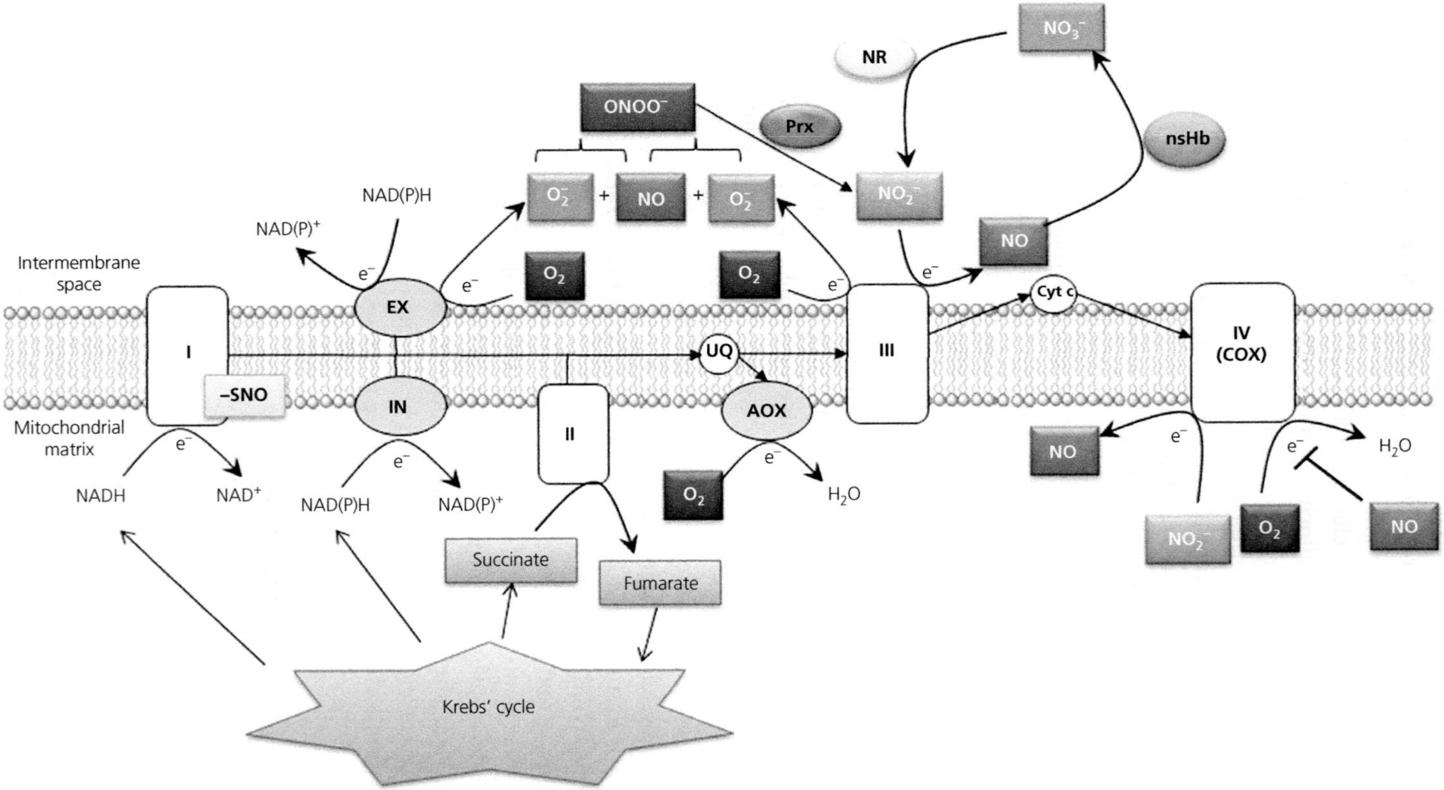

Figure 6.1 Schematic model for the maintenance of NO homeostasis by plant mitochondria. Nitrate (NO_3^-) is reduced to nitrite (NO_2^-) by cytosolic nitrate reductase (NR). NO_2^- is then reduced to NO by cytochrome *c* oxidase (COX) or complex III of mitochondrial respiratory chain. At physiological levels NO causes reversible inhibition of COX and can also lead to *S*-nitrosylation of complex I. The resulting restriction of electron flux through the cytochrome pathway stimulates production of superoxide (O_2^-) by external NAD(P)H dehydrogenases (EX) and complex III. These enzymes then contribute to NO degradation because NO promptly reacts with O_2^- to produce peroxynitrite (ONOO⁻). Conversely, alternative oxidase (AOX) allows mitochondrial electron flow in the presence of NO and decreases electron leakage and NO consumption. ONOO⁻ can be metabolised back to NO_2^- by peroxiredoxins (Prx) and NO can also be metabolised to NO_3^- by cytosolic class 1 non-symbiotic haemoglobins (nsHb), closing the cycle. (*See insert for color representation of the figure.*)

result from the reversible binding of NO to the Fe^{2+}-haeme group in cytochrome a_3 at its O_2-binding site to produce a ferrous-haeme nitrosyl complex (Cleeter *et al.*, 1994). This mechanism of regulating mitochondrial O_2 consumption by NO has been proposed to be physiologically significant through improvement of the oxygenation of tissues distant from blood vessels (Hagen *et al.*, 2003). A role for NO in the improvement of energy metabolism has also been proposed based on the observation that slight inhibition of COX by NO increases the efficiency of oxidative phosphorylation (Clerc *et al.*, 2007).

Studies with isolated mitochondria have demonstrated that plant COX is similarly sensitive to NO (Millar and Day, 1996; Yamasaki *et al.*, 2001). COX from various plant species can be reversibly inhibited by NO through a mechanism similar to that reported for mammals (Millar and Day, 1996; Yamasaki *et al.*, 2001; Zottini *et al.*, 2002; Martí *et al.*, 2013). This reversible and competitive inhibition of COX by NO has been suggested to play a role in controlling oxygen consumption and preventing anoxia during germination of soybean and pea seeds (Borisjuk *et al.*, 2007) and in contributing to hypoxic acclimation of maize roots subjected to low O_2 supply (Mugnai *et al.*, 2012).

In addition to physiological control of COX, it is well known that depending on the concentration and duration of exposure, NO can cause mitochondrial nitrosative stress in mammalian cells, which is observed in various pathophysiological conditions (Cooper and Giulivi, 2007) and may result from inhibition of the mitochondrial respiratory chain at multiple sites. In this case, complex I can be persistently inactivated by *S*-nitrosylation of critical Cys residues (Clementi *et al.*, 1998) and nitration of Tyr groups in the enzymatic complex (Yamamoto *et al.*, 2002). Slow reaction of NO with the bc_1 segment may result in inhibition of complex III, leading to increased O_2^- generation and $ONOO^-$ formation (Poderoso *et al.*, 1996, 1999; Cadenas *et al.*, 2000). Aconitase, complex II and other mitochondrial proteins may also be inhibited by NO when high levels of $ONOO^-$ are formed (Brown, 2007). The resulting persistent inhibition of mitochondrial respiration may cause the mitochondrial permeability transition (MPT), which represents a dramatic increase in the permeability of the inner mitochondrial membrane to small molecules that leads to cell death through necrosis or apoptosis (Brown, 2007).

In *Citrus* suspension cultures, prolonged exposure to NO has been shown to induce apoptosis-like cell death by affecting mitochondrial respiration and inducing MPT (Saviani *et al.*, 2002). Mitochondrial activity in *Arabidopsis thaliana* cells was recently shown to be modulated by the ratio of NO and SNO controlled by differential expression of *S*-nitrosoglutathione reductase (GSNOR) (Frungillo *et al.*, 2013). GSNOR metabolizes GSNO, a reservoir and donor of NO, and has been proposed to play an important role in the modulation of NO-mediated processes (Liu *et al.*, 2001). Increased levels of NO produced by *A. thaliana* cells under nutritional stress have been correlated with down-regulation of complex I, whereas the activity of complex II was not affected (Frungillo *et al.*, 2013). The expression and activation of external NADH dehydrogenase are also sensitive to NO and SNO levels in cell culture (Frungillo *et al.*, 2013). These results suggest a role for GSNOR in modulating mitochondrial respiratory activity and energy conservation in plant cells.

A burst in NO and H_2O_2 has been proposed to play a key role in the induction of cell death in the hypersensitive response (HR) during incompatible plant–pathogen interactions (Delledonne, 2005). There are indications that mitochondria-produced NO plays an active role in programmed cell death during biotic stress responses (Modolo *et al.*, 2005; Amirsadeghi *et al.*, 2007). Although NADPH oxidases have been shown to be required for the accumulation of ROS during HR (Torres *et al.*, 2002; Yun *et al.*, 2011), mitochondria are major sites for H_2O_2 production (Moller, 2001) and therefore may contribute to H_2O_2 generation during HR. Additionally, NO causes *S*-nitrosylation of the mitochondrial glycine decarboxylase complex (GDC), a key enzyme of the photorespiratory C_2 cycle (Palmieri *et al.*, 2010). The inhibition of GDC by mitochondria-generated NO in response to pathogen attack could limit the supply of NADH to the electron transport chain, resulting in an increase in ROS generation, change in the cellular redox status and promotion of cell death (Gupta *et al.*, 2011a).

Mitochondrial NO degradation

The steady-state levels of NO within cells are determined by a balance between the rates of production and consumption of NO, which may undergo auto-oxidation to nitrite in aqueous solutions (Kharitonov *et al.*, 1994). However, this reaction is not sufficiently rapid to explain the extremely short biological half-life of NO, and other reactions, such as those mediated by lipoxygenases, have been proposed to compete for NO in mammalian cells (Coffey *et al.*, 2001). NO dioxygenase activity originally identified in bacterial flavohaemoglobins that promote the haeme-dependent oxidation of NO to NO_3^- has also been observed in mammalian and plant cells (Gardner, 2005). The negative correlation between non-symbiotic haemoglobin expression and NO levels in various systems suggests an NO detoxification function for haemoglobins in plants during normoxia (Perazzolli *et al.*, 2004) and under hypoxic (Dordas *et al.*, 2003) and anoxic (Dordas *et al.*, 2004) stresses.

In non-respiring mitochondria isolated from mammalian sources, NO may be consumed through its reaction with O_2 within the mitochondrial membrane (Shiva *et al.*, 2001). However, the relevance of this mechanism for *in vivo* NO consumption is unclear (Brown, 2007). When isolated mitochondria are energized by the presence of respiratory substrates, NO consumption is increased (Poderoso *et al.*, 1996; Gupta *et al.*, 2005). Given that respiring mitochondria are a source of reactive oxygen species (ROS), the increased consumption of NO under these conditions was demonstrated to result from the reaction of NO with O_2^- to generate $ONOO^-$ (Poderoso *et al.*, 1996), which can in turn be reduced to nitrite by COX (Pearce *et al.*, 2002) or peroxiredoxin (Romero-Puertas *et al.*, 2007). COX can also oxidize NO to NO_2^- at its active site (Sarti *et al.*, 2000). Together with the reduction of NO to NO^- by ubiquinol (Poderoso *et al.*, 1999; Cadenas *et al.*, 2000), these mechanisms have been proposed to contribute to respiration-dependent mitochondrial NO consumption.

Recent studies with mitochondria isolated from potato tubers and *A. thaliana* cells have demonstrated that the reaction of NO with O_2^- is an important mechanism for NO consumption in plant cells (de Oliveira *et al.*, 2008; Wulff *et al.*, 2009). NO degradation by isolated plant mitochondria was shown to be abolished by anoxia and superoxide dismutase, which indicates that NO is consumed by its reaction with O_2^-. The use of various electron donors and inhibitors of mitochondrial electron transport permitted the identification of sites of electron leakage from the respiratory chain involved in NO degradation by plant mitochondria. In isolated potato tuber mitochondria respiring with malate or succinate (electron donors for complex I and II, respectively), inhibitors of complex III antimycin-A (Anti-A) and myxothiazol had different effects on NO degradation. Whereas Anti-A stimulated NO degradation, this process was prevented by myxothiazol (de Oliveira *et al.*, 2008). In mammalian mitochondria (Brand *et al.*, 2004), myxothiazol is known to inhibit complex III at centre *o* (the site of ubiquinol oxidation), whereas Anti-A inhibits centre *i* (the site of ubiquinone reduction). Thus, myxothiazol inhibits the formation of unstable ubisemiquinone, thereby preventing electron leakage from complex III, whereas Anti-A favours the formation of ubisemiquinone, and its auto-oxidation generates O_2^- (Fang and Beattie, 2003). Therefore, the opposing effects of Anti-A and myxothiazol on NO degradation by mitochondria respiring with malate or succinate indicate that electron leakage from complex III contributes to NO degradation in isolated potato mitochondria. These findings are consistent with studies of mitochondria and submitochondrial particles from various plant species that indicate that complex III is an important site for the reduction of O_2 to O_2^- (Moller, 2001).

In mitochondria from animal sources, electron leakage from the ubiquinone cycle of complex III has been suggested to represent the main source of O_2^- for NO degradation (Poderoso *et al.*, 1996, 1999; Chen *et al.*, 2006), although other mitochondrial enzymes, such as complex I (Brand *et al.*, 2004), succinate dehydrogenase and outer mitochondrial membrane cytochrome *b5* reductase (Andreyev *et al.*, 2005) have been proposed as additional sites for electron leakage and O_2^- generation. Complex I inhibitors such as rotenone have been shown to nearly abolish the high rate of O_2^- production by intact mammalian mitochondria during succinate oxidation. In contrast, the low rate of O_2^- production during the oxidation of NAD-linked substrates is increased by rotenone, although not to the same extent as in succinate-energized mitochondria (Brand *et al.*, 2004). These results show that most of the O_2^- generated at complex I by intact mammalian mitochondria is derived from reverse electron transport from succinate, whereas forward electron transport into complex I from NAD-linked substrates produces less O_2^- (Brand *et al.*, 2004). In plants, complexes I and II have also been reported to be sites of O_2^- generation (Braidot *et al.*, 1999). However, in isolated potato tuber mitochondria, electron leakage from complex I and II was shown not to contribute to NO degradation (de Oliveira *et al.*, 2008); in particular the rate of NO degradation was not altered in the presence of the complex I inhibitors rotenone or capsaicin in mitochondria respiring with malate or in the presence of rotenone in succinate-energized mitochondria (de Oliveira *et al.*, 2008).

NO degradation by external NAD(P)H dehydrogenases

In addition to complex III, external NAD(P)H dehydrogenases, which provide an alternative pathway for electron transport in the mitochondrial respiratory chain, have been identified as important contributors to NO degradation in plant mitochondria (de Oliveira *et al.*, 2008; Wulff *et al.*, 2009). As discussed above, this alternative respiration is not coupled to chemical energy production and is thought to contribute to control of the redox balance of the cell (Millenaar and Lambers, 2003; Rasmusson *et al.*, 2004). Recently, a role for these enzymes in the energy dissipation system of thermogenic plants was proposed based on the observation that in addition to AOX, external NAD(P)H dehydrogenases were abundant in the thermogenic appendices of *Arum maculatum* (Kakizaki *et al.*, 2012).

Although alternative NAD(P)H dehydrogenases have a potential role in providing flexibility for the oxidation of cytosolic and matrix NAD(P)H, their possible role in preventing the generation of ROS in response to different environmental stresses remains unclear (Moller, 2001; Rasmusson *et al.*, 2004). Early reports describing potato submitochondrial particles and isolated mitochondria from green pepper fruit suggested that external NAD(P)H dehydrogenases are indeed sites for O_2^- generation (Rich and Bonner, 1978; Purvis *et al.*, 1995). Accordingly, in a study with isolated potato tuber mitochondria, NAD(P)H-respiring mitochondria degraded NO at much higher rates than mitochondria energized with malate or succinate. Although it was stimulated by Anti-A, NAD(P)H-dependent NO consumption was not prevented by myxothiazol, which suggests electron leakage upstream of complex III when NAD(P)H is the respiratory substrate (de Oliveira *et al.*, 2008). Furthermore, increased O_2^- production was positively correlated with higher rates of NO consumption in NAD(P)H-energized mitochondria isolated from potato tubers and *A. thaliana* cells than in mitochondria respiring with complex I and II substrates (de Oliveira *et al.*, 2008; Wulff *et al.*, 2009). Consistent with this observation, a positive correlation between the Ca^{2+}-induced respiratory activity of these alternative enzymes and the rate of NO degradation was also observed (de Oliveira *et al.*, 2008; Wulff *et al.*, 2009). Moreover, constitutive activation of external NADH dehydrogenase was correlated with low levels of NO emission by *A. thaliana* cells (Frungillo *et al.*, 2013). These results revealed a previously unrecognized role for external NAD(P)H dehydrogenases in NO degradation by plant mitochondria (Figure 6.1). NAD(P)H-dependent degradation of NO by mitochondria isolated from potato tubers was shown to accelerate the recovery of O_2 consumption and the re-establishment of the electrical potential across the inner mitochondrial membrane after perturbation by NO (de Oliveira *et al.*, 2008), which suggests that this may be a mechanism by which NO can be consumed in the vicinity of the inner mitochondrial membrane to prevent prolonged inhibition of mitochondrial respiration.

Regardless, O_2^- dependent NO degradation generates $ONOO^-$, which is considered a potent oxidative intermediate that can react with biological molecules involved in cellular signalling, resulting in oxidation, nitrosation and nitration (Radi *et al.*, 2002). However, some reports have suggested that under physiological conditions, the levels of $ONOO^-$ produced in mitochondria would be very low given its extremely short half-life (3–5 ms) and the rates of NO and O_2^- production (Radi *et al.*, 2002; Chen *et al.*, 2006), and it is therefore unlikely that $ONOO^-$ production could affect mitochondrial respiration, as has been demonstrated for rat heart submitochondrial particles (Poderoso *et al.*, 1996) and isolated potato tuber mitochondria (de Oliveira *et al.*, 2008). Additionally, compared with animal cells, plants have been shown to be more resistant to $ONOO^-$ because treatment of soybean cells with $ONOO^-$ at concentrations up to 1 mM did not induce death (Delledonne *et al.*, 2001).

Even in mammalian cells, it has been demonstrated that the primary consequence of O_2^- generation concomitant with NO production is not toxicity associated with the formation of RNS; rather, it represents an important regulatory mechanism that modulates signalling pathways by limiting steady-state levels of NO and preventing formation of H_2O_2 and hydroxyl radicals from O_2^- (Thomas *et al.*, 2006). In potato plants, the interaction between NO and O_2^- was proposed to be important for reducing the oxidative stress induced by the use of herbicides by minimizing the overproduction of ROS within chloroplasts (Beligni and Lamattina, 1999). Therefore, the physiological role of the reaction between NO and O_2^- is being increasingly recognized.

Involvement of AOX in NO signalling and homeostasis

The major functions of the alternative pathway of electron transport enabled by AOX have been proposed to include its contribution to the thermogenic process of some flowers and the prevention of excess ROS production induced by a wide range of environmental stresses (Siedow and Umbach, 1995; Millenar and

Lambers, 2003; Arnholdt-Schmitt *et al.*, 2006). The physiological significance of the AOX pathway in stabilizing the redox state of the ubiquinone pool and continued activity of the Krebs cycle has also been widely discussed (Millenar and Lambers, 2003).

AOX and COX are differentially inhibited; AOX is insensitive to cyanide, azide and carbon monoxide, which are classic inhibitors of COX, and can be specifically inhibited by salicylhydroxamic acid (SHAM) and propyl gallate (Siedow and Umbach, 1995). Moreover, NO has been shown to inhibit COX to a much greater extent than AOX in mitochondria isolated from soybean cotyledons (Millar and Day, 1996), mung bean seedlings (Yamasaki *et al.*, 2001) and pea leaves (Martí *et al.*, 2013). Further studies have demonstrated the involvement of NO in retrograde signalling in plant mitochondria; inhibition of cytochrome-dependent respiration by NO was correlated with increased nuclear AOX gene expression and increased contribution of the alternative pathway to total respiration in carrot and *A. thaliana* cells (Huang *et al.*, 2002; Zottini *et al.*, 2002). It was recently proposed that the effect of NO on AOX expression is indirect; NO produced in *A. thaliana* roots was shown to inhibit aconitase, which was previously identified as a molecular target of NO in the Krebs cycle (Navarre *et al.*, 2000), leading to a marked increase in levels of citrate, which acts as a potent inducer of AOX activity and expression (Gupta *et al.*, 2012). ROS have also been reported to be involved in the induction of AOX genes (Li *et al.*, 2013).

A role for AOX in controlling NO levels in plants has also been proposed (Wulff *et al.*, 2009). In mitochondria isolated from *A. thaliana* cells, when electron flow is only directed toward AOX, NO does not affect respiration and is slowly degraded because of reduced production of O_2^-. Conversely, when AOX is inhibited by propyl gallate and NO binds to COX, electron flow is completely abolished, and NO is rapidly consumed by its reaction with O_2^- (Wulff *et al.*, 2009). Therefore, in addition to having the beneficial effect of enabling electron flow in the presence of NO, AOX diminishes electron leakage from the respiratory chain, thereby increasing the half-life of NO.

Oxidative pathways for NO synthesis

Mitochondria have been suggested to be a site for the synthesis of NO in plant cells. Mechanisms of L-arginine oxidation and nitrite reduction for NO generation by mitochondria in plants have been proposed.

In mammals, a group of enzymes termed nitric oxide synthases (NOS) is well established as the main system for NO synthesis. The NOS enzyme family catalyzes the five-electron oxidation of the guanidine nitrogen of the amino acid L-arginine to L-citrulline with concomitant formation of NO, using O_2 and NADPH as co-substrates (Alderton *et al.*, 2001). The NO generated by this pathway plays important roles in various metabolic functions and may also be involved in various pathophysiological processes (Moncada *et al.*, 1991). The different mammalian NOS isoforms already identified (endothelial NOS, neuronal NOS and inducible NOS) are designated mitochondrial NOS (mtNOS) when they are found attached to or within mitochondria. However, because several research groups have not found NOS activity in mitochondria, the existence of mtNOS in animal cells remains controversial (Brown, 2007).

Evidence for the existence of NOS-like activity in plants has been presented for several tissues and species based on the conversion of L-arginine to L-citrulline and the observation that inhibitors of mammalian NOS inhibit various NO-dependent processes (del Rio *et al.*, 2004). However, no gene with significant homology to mammalian NOS has been identified in higher plants thus far, despite the sequencing of several plant genomes. A NOS-like enzyme with no sequence similarity to any mammalian isoform and encoded by a distinct *AtNOS1* gene has been proposed to be responsible for L-arginine-dependent NO synthesis in *A. thaliana* (Guo *et al.*, 2003). AtNOS1 has been reported to be involved in hormonal signalling (Guo *et al.*, 2003) and to be localized to mitochondria (Guo and Crawford, 2005). However, NOS activity in the isolated protein could not be demonstrated, and the gene was shown to code for a GTPase with no L-arginine-dependent NOS activity (Zemojtel *et al.*, 2006). AtNOS1 was renamed AtNO-Associated 1 (AtNOA1) because of the low NO emission and impaired NO-mediated responses of the *atnoa1* mutant (Crawford *et al.*, 2006). In isolated barley root mitochondria, no NOS activity was detected by chemiluminescence and the apparent NOS activity was low and untypical regarding its response to inhibitors, substrates and cofactors when estimated with diaminofluorescein (Gupta and Kaiser, 2010). Other subcellular compartments, such as peroxisomes, have been proposed to harbor L-arginine-dependent NOS activity in plants (Corpas *et al.*, 2004). However, the presence of a NOS-like enzyme in plant mitochondria or another subcellular compartment remains under question, and molecular evidence is lacking.

Reductive pathways for NO synthesis

Nitrite is an alternative source for the synthesis of NO and, in recent years, nitrite reduction has been associated with various NO-mediated processes in plants.

Although the main reaction catalysed by nitrate reductase (NR) is the NAD(P)H-dependent reduction of nitrate to nitrite, NR may exhibit a secondary nitrite-reducing activity that appears when the O_2 concentration in the medium is low and pH decreases, resulting in the accumulation of nitrite (Rockel *et al.*, 2002). The involvement of NO in the regulation of various processes in plants, including stomatal movement, pathogen defence, floral repression, activation of antioxidant enzymes, osmotic stress, auxin-induced root lateral formation and hypoxic responses, has been suggested to result from the nitrite-reducing activity of NR (Gupta *et al.*, 2011b). In several of these reports, the proposed nitrite-reducing activity of NR is based on the inability of NR-deficient plants to produce NO. However, leaf extracts from mutant *A. thaliana* plants deficient in the two NR structural genes (*NIA1* and *NIA2*) were shown to synthesize NO at the same rate when nitrite was exogenously provided (Modolo *et al.*, 2005). These results suggest that participation of NR in NO synthesis may be related to production of the substrate nitrite (Figure 6.1). Furthermore, the observation that the nitrite-reducing activity detected in NR-deficient leaves could be abolished in the presence of inhibitors of mitochondrial respiration suggests that electrons leaking from the mitochondrial respiratory chain are responsible for this activity (Modolo *et al.*, 2005). This mitochondrial nitrite-reducing activity in *A. thaliana* leaves was identified as the main mechanism for NO synthesis during its incompatible interaction with *Pseudomonas syringae*, revealing that NR contributes to plant defence by providing the substrate nitrite for synthesis of NO (Modolo *et al.*, 2005).

Use of nitrite as an acceptor of electrons leaked from the mitochondrial respiratory chain has also been reported for the algae *Chlorella sorokiniana* (Tischner *et al.*, 2004), tobacco suspension cells (Planchet *et al.*, 2005) and mitochondria isolated from pea seeds (Benamar *et al.*, 2008), *A. thaliana* cells (Wulff *et al.*, 2009) and roots from diverse plant species (Gupta *et al.*, 2005; Stoimenova *et al.*, 2007).

Although mitochondria-dependent nitrite reduction has been detected in *A. thaliana* leaf extracts (Modolo *et al.*, 2005), it has been suggested that this activity only develops in roots and not in the leaves of higher plants (Gupta *et al.*, 2005). The methods employed to detect this activity could explain, at least in part, these different results. NO production in *A. thaliana* leaf extracts was measured by electron paramagnetic resonance (EPR), in which the NO produced is rapidly sequestered by the spin trap, thus allowing the detection of high quantities of the radical without its accumulation in the reaction medium (Modolo *et al.*, 2005). In contrast, when gas-phase chemiluminescent detection is used, substantial quantities of NO can be detected only in the absence of O_2 (Planchet *et al.*, 2005). Recently, Cvetkovska and Vanlerberghe (2012) reported that substantial quantities of NO originating from mitochondrial-dependent nitrite reduction were detected in tobacco leaves under aerobic conditions using a diaminofluorescein probe. These results support the previous observation that mitochondrial nitrite-reducing activity can take place in leaves (Modolo *et al.*, 2005).

NO synthesis is greatly enhanced under hypoxic conditions when nitrite accumulates (Planchet *et al.*, 2005). Indeed, nitrite-reducing activity in plants has been implicated in the control of tissue oxygenation and energy maintenance during hypoxic conditions, such as those occurring during seed germination (Borisjuk *et al.*, 2007; Benamar *et al.*, 2008) and in roots subjected to flooding stress (Oliveira *et al.*, 2013). The use of nitrite as an alternative terminal electron acceptor in the respiratory chain under hypoxic and anoxic conditions is thought to be important for maintaining NADH reoxidation, electron transport and anaerobic ATP synthesis (Stoimenova *et al.*, 2007). Accordingly, when nitrite-reducing activity in mitochondria isolated from *A. thaliana* cells was measured according to the inhibitory effect of nitrite on respiration, the inhibition was inversely correlated with the concentration of O_2 in the reaction medium (Wulff *et al.*, 2009). Additionally, NO produced in the presence of O_2 can be consumed by its reaction with O_2^- (de Oliveira *et al.*, 2008) and by the aerobic activity of NO conversion to nitrate by non-symbiotic haemoglobins (Perazzolli *et al.*, 2004).

Sites of nitrite reduction in the mitochondrial respiratory chain

In mammalian mitochondria, it was initially proposed that electrons that leak from complex III due to the suppression of O_2 reduction by COX could be used for nitrite reduction (Kozlov *et al.*, 1999). However, COX was later implicated as the main enzyme in the mitochondrial respiratory chain responsible for NO synthesis from nitrite (Castello *et al.*, 2006). It was also demonstrated that nitrite must reach the mitochondrial matrix to be reduced to NO by COX, which suggests that electrons that leak to the matrix side of the inner mitochondrial membrane are used for nitrite reduction (Castello *et al.*, 2006). Recently, cytochrome *c* from mammalian mitochondria was also reported to catalyse reduction of nitrite to NO (Basu *et al.*, 2008).

In plant mitochondria (Figure 6.1), COX has also been suggested to be the most plausible site for reduction of nitrite to NO (Gupta and Igamberdiev, 2011), although other sites of electron leakage for nitrite reduction, such as complex III (Stoimenova *et al.*, 2007) and AOX, have been proposed (Planchet *et al.*, 2005). The suggestion that AOX possesses nitrite-reducing activity was based on the inhibitory effect of SHAM on NO emission by tobacco suspension cells (Planchet *et al.*, 2005). However, it was recently proposed that AOX has a negative effect on mitochondrial NO production by decreasing electron leakage from the mitochondrial respiratory chain to nitrite (Cvetkovska and Vanlerberghe, 2012). Moreover, AOX inhibitors have been shown to have no effect on nitrite reduction to NO in *A. thaliana* leaf extracts (Modolo *et al.*, 2005) and alfalfa root nodules (Horchani *et al.*, 2011). These results indicate that more direct evidence is necessary to confirm nitrite-reducing activity of AOX.

Summary

Alternative pathways for electron flow in the mitochondrial respiratory chain may play a major role in the control of NO homeostasis and signal transduction in plant cells. As depicted in Figure 6.1, physiological levels of NO produced under aerobic conditions restrain the cytochrome pathway, which causes over-reduction of the electron carriers. Consequently, increased production of ROS occurs at the level of external NAD(P)H dehydrogenases and complex III. The consumption of both radicals by the spontaneous reaction of NO with O_2^- to generate $ONOO^-$ helps to minimize ROS and NO levels and allows the recovery of O_2 consumption by mitochondria. Insensitive to NO, AOX enables electron flow by the respiratory chain in the presence of NO, thus limiting ROS production, NO degradation and $ONOO^-$ accumulation. Therefore, a role for external NAD(P)H dehydrogenases and AOX in maintaining NO homeostasis in plant cells seems quite possible. Further studies are necessary to demonstrate the physiological relevance of these mechanisms in plant tissues. Steady-state levels of NO are also controlled by the rate of synthesis of this radical. Although various cellular sources of NO have been proposed, the reduction of nitrite by the mitochondrial respiratory chain has emerged as an important mechanism for NO production in plants. This mechanism is enhanced as O_2 concentrations decrease, and COX is the main candidate for this activity, with cytosolic NR providing nitrite. When the production of NO is overwhelmed and high levels of RNS accumulate, the mitochondrial respiratory chain may be affected at multiple sites, which may cause the irreversible inhibition of respiration and lead to cell death. In this scenario, the tolerance of plants to different types of stresses may be related to their ability to control steady-state levels of NO, in which the induction of alternative respiratory pathways could play a central role.

Acknowledgments

We thank the Conselho Nacional de Desenvolvimento Científico e Tecnológico (CNPq, Grant 473090/2011–2) for financial support.

References

Alderton, W.K., Cooper, C.E. and Knowles, R.G. (2001) Nitric oxide synthases: structure, function and inhibition. *Biochemical Journal* **357**: 593–615.

Amirsadeghi, S., Robson, C.A. and Vanlerberghe, G.C. (2007) The role of the mitochondrion in plant responses to biotic stress. *Physiologia Plantarum* **129**: 253–266.

Andreyev, A.Y., Kushnareva, Y.E. and Starkov, A.A. (2005) Mitochondrial metabolism of reactive oxygen species. *Biochemistry* **70**: 200–214.

Arnholdt-Schmitt, B., Costa, J.H. and de Melo, D.F. (2006) AOX – a functional marker for efficient cell reprogramming under stress? *Trends in Plant Science* **11**: 281–287.

Basu, S., Azarova, N.A., Font, M.D. *et al.* (2008) Nitrite reductase activity of cytochrome *c*. *Journal of Biological Chemistry* **283**: 32590–32597.

Baudouin, E. (2011) The language of nitric oxide signaling. *Plant Biology* **13**: 233–242.

Beligni, M.V. and Lamattina, L. (1999) Nitric oxide protects against cellular damage produced by methylviologen herbicides in potato plants. *Nitric Oxide* **3**: 199–208.

Benamar, A., Rolletschek, H., Borisjuk, L. *et al.* (2008) Nitrite-nitric oxide control of mitochondrial respiration at the frontier of anoxia. *Biochimica et Biophysica Acta* **1777**: 1268–1275.

Borisjuk, L., Machere, I D., Benamar, A. *et al.* (2007) Low oxygen sensing and balancing in plant seeds: a role for nitric oxide. *New Phytologist* **176**: 813–823.

Braidot, E., Petrussa, E., Vianello, A. and Macri, F. (1999) Hydrogen peroxide generation by higher plant mitochondria oxidizing complex I or complex II substrates. *FEBS Letters* **451**: 347–350.

Brand, M.D., Affourtit, C., Esteves, T.C. *et al.* (2004) Mitochondrial superoxide: Production, biological effects, and activation of uncoupling proteins. *Free Radical Biology & Medicine* **37**: 755–767.

Brown, G.C. (2007) Nitric oxide and mitochondria. *Frontiers in Bioscience* **12**: 1024–1033.

Brown, G.C. and Cooper, C.E. (1994) Nanomolar concentrations of nitric oxide reversibly inhibit synaptosomal respiration by competing with oxygen at cytochrome oxidase. *FEBS Letters* **356**: 295–298.

Cadenas, E., Poderoso, J.J., Antunes, F. and Boveris, A. (2000) Analysis of the pathways of nitric oxide utilization in mitochondria. *Free Radical Research* **33**: 747–756.

Castello, P., David, P., McClure, T. *et al.* (2006) Mitochondrial cytochrome oxidase produces nitric oxide under hypoxic conditions: implications for oxygen sensing and hypoxia signaling in eukaryotes. *Cell Metabolism* **3**: 277–287.

Chen, Y., Chen, C., Yeh, A. *et al.* (2006) Direct and indirect roles of cytochrome *b* in the mediation of superoxide generation and NO catabolism by mitochondrial succinate-cytochrome *c* reductase. *Journal of Biological Chemistry* **291**: 13159–13168.

Cleeter, M.W., Cooper, JM., Darley-Usmar, V.M. *et al.* (1994) Reversible inhibition of cytochrome *c* oxidase, the terminal enzyme of the mitochondrial respiratory chain, by nitric oxide. Implications for neurodegenerative diseases. *FEBS Letters* **345**: 50–54.

Clementi, E., Brown, G.C., Feelisch, M. and Moncada, S. (1998) Persistent inhibition of cell respiration by nitric oxide: Crucial role of *S*-nitrosylation of mitochondrial complex I and protective action of glutathione. *Proceedings of the National Academy of Science USA* **95**: 7631–7636.

Clerc, P., Rigoulet, M., Leverve, X. and Fontaine, E. (2007) Nitric oxide increases oxidative phosphorylation efficiency. *Journal of Bioenergetics and Biomembranes* **39**: 158–166.

Coffey, M.J., Natarajan, R., Chumley, P.H. *et al.* (2001) Catalytic consumption of nitric oxide by 12/15-lipoxygenase: Inhibition of monocyte soluble guanylate cyclase activation. *Proceedings of the National Academy of Science USA* **98**: 8006–8011.

Cooper, C.E. and Giulivi, C. (2007) Nitric oxide regulation of mitochondrial oxygen consumption II: Molecular mechanism and tissue physiology. *American Journal of Physiology: Cell Physiology* **292**: C1993–C2003.

Corpas, F.J., Barroso, J.B., Carreras, A. *et al.* (2004) Cellular and subcellular localisation of endogenous nitric oxide in young and senescent pea plants. *Plant Physiology* **136**: 2722–2733.

Corpas, F.J., Chaki, M., Fernández-Ocaña, A. *et al.* (2008) Metabolism of reactive nitrogen species in pea plants under abiotic stress conditions. *Plant & Cell Physiology* **49**: 1711–1722.

Corpas, F.J., Leterrier, M., Valderrama, R. *et al.* (2011) Nitric oxide imbalance provokes a nitrosative response in plants under abiotic stress. *Plant Science* **181**: 604–611.

Crawford, N.M., Galli, M., Tischner, R. *et al.* (2006) Response to Zemojtel *et al.*: plant nitric oxide synthase: back to square one. *Trends in Plant Science* **11**: 526–527.

Cvetkovska, M. and Vanlerberghe, G.C. (2012) Alternative oxidase modulates leaf mitochondrial concentrations of superoxide and nitric oxide. *New Phytologist* **195**: 32–39.

Delledonne, M. (2005) NO news is good news for plants. *Current Opinion in Plant Biology* **8**: 390–396.

Delledonne, M., Zeier, J., Marocco, A. and Lamb, C. (2001) Signal interactions between nitric oxide and reactive oxygen intermediates in the plant hypersensitive disease resistance response. *Proceedings of the National Academy of Science USA* **98**: 13454–13459.

Dordas, C., Hasinoff, B.B., Igamberdiev, A.U. *et al.* (2003) Expression of a stress-induced hemoglobin affects NO levels produced by alfalfa root cultures under hypoxic stress. *Plant Journal* **35**: 763–770.

Dordas, C., Hasinoff, B.B., Rivoal, J. and Hill, R.D. (2004) Class-1 hemoglobins, nitrate and NO levels in anoxic maize cell-suspension cultures. *Planta* **219**: 66–72.

Durner, J., Wendehenne, D. and Klessig, D.F. (1998) Defense gene induction in tobacco by nitric oxide, cyclic GMP, and cyclic ADP-ribose. *Proceedings of the National Academy of Science USA* **95**: 10328–10333.

Fang, J. and Beattie, D.S. (2003) External alternative NADH dehydrogenase of *Saccharomyces cerevisiae*: a potential source of superoxide. *Free Radical Biology & Medicine* **34**: 478–488.

Ferrarini, A., De Stefano, M., Baudouin, E. *et al.* (2008) Expression of *Medicago truncatula* genes responsive to nitric oxide in pathogenic and symbiotic conditions. *Molecular Plant-Microbe Interactions* **21**: 781–790.

Friebe, A. and Koesling, D. (2009) The function of NO-sensitive guanylyl cyclase: What we can learn from genetic mouse models. *Nitric Oxide* **21**: 149–156.

Frungillo, L., de Oliveira, J.F., Saviani, E.E. *et al.* (2013) Modulation of mitochondrial activity by S-nitrosoglutathione reductase in *Arabidopsis thaliana* transgenic cell lines. *Biochimica et Biophysica Acta* **1827**: 239–247.

Gardner, P.R. (2005) Nitric oxide dioxygenase function and mechanism of flavohemoglobin, hemoglobin, myoglobin and their associated reductases. *Journal of Inorganic Biochemistry* **99**: 247–266.

Gow, A.J. and Ischiropoulos, H. (2001) Nitric oxide chemistry and cellular signaling. *Journal of Cell Physiology* **187**: 277–282.

Guo, F.Q. and Crawford, N.M. (2005) *Arabidopsis* nitric oxide synthase1 is targeted to mitochondria and protects against oxidative damage and dark-induced senescence. *Plant Cell* **17**: 3436–3450.

Guo, F.Q., Okamoto, M. and Crawford, N.M. (2003) Identification of a plant nitric oxide synthase gene involved in hormonal signalling. *Science* **302**: 100–103.

Gupta, K.J. and Igamberdiev, A.U. (2011) The anoxic plant mitochondria as a nitrite:NO reductase. *Mitochondrion* **11**: 537–543.

Gupta, K.J. and Kaiser, W.M. (2010) Production and scavenging of nitric oxide by barley root mitochondria. *Plant & Cell Physiology* **51**: 576–584.

Gupta, K.J., Fernie, A.R., Kaiser, W.M. and van Dongen, J.T. (2011a) On the origins of nitric oxide. *Trends in Plant Science* **16**: 160–168.

Gupta, K.J., Igamberdiev, A.U., Manjunatha, G. *et al.* (2011b) The emerging roles of nitric oxide (NO) in plant mitochondria. *Plant Science* **181**: 520–526.

Gupta, K.J., Shah, J.K., Brotman, Y. *et al.* (2012) Inhibition of aconitase by nitric oxide leads to induction of the alternative oxidase and to a shift of metabolism towards biosynthesis of amino acids. *Journal of Experimental Botany* **63**: 1773–1784.

Gupta, K.J., Stoimenova, M. and Kaiser, W.M. (2005) In higher plants, only root mitochondria, but not leaf mitochondria reduce nitrite to NO, *in vitro* and *in situ*. *Journal of Experimental Botany* **56**: 2601–2609.

Grün, S., Lindermayr, C., Sell, S. and Durner, J. (2006) Nitric oxide and gene regulation in plants. *Journal of Experimental Botany* **57**: 507–516.

Hagen, T., Taylor, C.T., Lam, F. and Moncada, S. (2003) Redistribution of intracellular oxygen in hypoxia by nitric oxide: Effect on HIF1 alpha. *Science* **302**: 1975–1978.

Horchani, F., Prévot, M., Boscari, A. *et al.* (2011) Both plant and bacterial nitrate reductases contribute to nitric oxide production in *Medicago truncatula* nitrogen-fixing nodules. *Plant Physiology* **155**: 1023–1036.

Huang, X., von Rad, U. and Durner, J. (2002) Nitric oxide induces transcriptional activation of the nitric oxide-tolerant alternative oxidase in *Arabidopsis* suspension cells. *Planta* **215**: 914–923.

Igamberdiev, A.U., Baron, K., Manac'h-Little, N. *et al.* (2005) The haemoglobin/nitric oxide cycle: involvement in flooding stress and effects on hormone signalling. *Annals of Botany* **96**: 557–564.

Kakizaki, Y., Moore, A.L. and Ito, K. (2012) Different molecular bases underlie the mitochondrial respiratory activity in the homoeothermic spadices of *Symplocarpus renifolius* and the transiently thermogenic appendices of *Arum maculatum. Biochemical Journal* **445**: 237–246.

Kharitonov, V.G., Sundquist, A.R. and Sharma, V.S. (1994) Kinetics of nitric oxide autoxidation in aqueous solutions. *Journal of Biological Chemistry* **269**: 5881–5883.

Kozlov, A.V., Staniek, K. and Nohl, H. (1999) Nitrite reductase activity is a novel function of mammalian mitochondria. *FEBS Letters* **454**: 127–130.

Lamattina, L., Garcia-Mata, C., Graziano, M. and Pagnussat, G. (2003) Nitric oxide: The versatility of an extensive signal molecule. *Annual Review of Plant Biology* **54**: 109–136.

Leitner, M., Vandelle, E., Gaupels, F. *et al.* (2009) NO signals in the haze: Nitric oxide signaling in plant defence. *Current Opinion in Plant Biology* **12**: 451–458.

Li, C.R., Liang, D.D., Li, J. *et al.* (2013) Unravelling mitochondrial retrograde regulation in the abiotic stress induction of rice alternative oxidase 1 genes. *Plant Cell Environ* **36**: 775–788.

Lindermayr, C., Saalbach, G. and Durner, J. (2005) Proteomic identification of *S*-nitrosylated proteins in *Arabidopsis. Plant Physiology* **137**: 921–930.

Liu, L., Hausladen, A., Zeng, M. *et al.* (2001) A metabolic enzyme for *S*-nitrosothiol conserved from bacteria to humans. *Nature* **410**: 490–494.

Lozano-Juste, J., Colom-Moreno, R. and León, J. (2011) *In vivo* protein tyrosine nitration in Arabidopsis thaliana. *Journal of Experimental Botany* **62**: 3501–3517.

Martí, M.C., Florez-Sarasa, I., Camejo, D. *et al.* (2013) Response of mitochondrial antioxidant system and respiratory pathways to reactive nitrogen species in pea leaves. *Physiologia Plantarum* **147**: 194–206.

Millar, A.H. and Day, D.A. (1996) Nitric oxide inhibits the cytochrome oxidase but not the alternative oxidase of plant mitochondria. *FEBS Letters* **398**: 155–158.

Millenaar, F.F. and Lambers, H. (2003) The alternative oxidase: *in vivo* regulation and function. *Plant Biology* **5**: 2–15.

Modolo, L.V., Augusto, O., Almeida, I.M. *et al.* (2005) Nitrite as the major source of nitric oxide production by *Arabidopsis thaliana* in response to *Pseudomonas syringae. FEBS Letters* **579**: 3814–3820.

Moller, I.M. (2001) Plant mitochondria and oxidative stress: Electron transport, NADPH turnover, and metabolism of reactive oxygen species. *Annual Review of Plant Physiology and Plant Molecular Biology* **52**: 561–591.

Moncada, S., Palmer, R.M. and Higgs, E.A. (1991) Nitric oxide: Physiology, pathophysiology, and pharmacology. *Pharmacological Reviews* **43**: 109–142.

Mugnai, S., Azzarello, E., Baluska, F. and Mancuso, S. (2012) Local root apex hypoxia induces NO-mediated hypoxic acclimation of the entire root. *Plant & Cell Physiology* **53**: 912–920.

Navarre, D.A., Wendehenne, D., Durner, J. *et al.* (2000) Nitric oxide modulates the activity of tobacco aconitase. *Plant Physiology* **122**: 573–582.

Neill, S.J., Desikan, R. and Hancock, J.T. (2003) Nitric oxide signaling in plants. *New Phytologist* **159**: 11–35.

de Oliveira, H.C., Wulff, A., Saviani, E.E. and Salgado, I. (2008) Nitric oxide degradation by potato tuber mitochondria: evidence for the involvement of external NAD(P)H dehydrogenases. *Biochimica et Biophysica Acta* **1777**: 470–476.

Oliveira, H.C., Salgado, I. and Sodek, L. (2013) Involvement of nitrite in the nitrate-mediated modulation of fermentative metabolism and nitric oxide production of soybean roots during hypoxia. *Planta* **237**: 255–264.

Palmieri, M.C., Lindermayr, C., Bauwe, H. *et al.* (2010) Regulation of glycine decarboxylase by *S*-nitrosylation and glutathionylation. *Plant Physiology* **152**: 1514–1528.

Palmieri, M.C., Sell, S., Huang, X. *et al.* (2008) Nitric oxide-responsive genes and promoters in *Arabidopsis thaliana*: A bioinformatics approach. *Journal of Experimental Botany* **59**: 177–186.

Pearce, L.L., Kanai, A.J., Birder, L.A. *et al.* (2002) The catabolic fate of nitric oxide: The nitric oxide oxidase and peroxynitrite reductase activities of cytochrome oxidase. *Journal of Biological Chemistry* **277**: 13556–13562.

Perazzolli, M., Dominici, P., Romero-Puertas, M.C. *et al.* (2004) *Arabidopsis* nonsymbiotic hemoglobin AHb1 modulates nitric oxide bioactivity. *Plant Cell* **16**: 2785–2794.

Planchet, E., Gupta, K.J., Sonoda, M. and Kaiser, W.M. (2005) Nitric oxide emission from tobacco leaves and cell suspensions: Rate limiting factors and evidence for the involvement of mitochondrial electron transport. *Plant Journal* **41**: 732–743.

Poderoso, J.J., Carreras, M., Lisdero, C. *et al.* (1996) Nitric oxide inhibits electron transfer and increases superoxide radical production in rat heart mitochondria and submitochondrial particles. *Archives of Biochemistry and Biophysics* **328**: 85–92.

Poderoso, J.J., Lisdero, C., Schöpfer, F. *et al.* (1999) The regulation of mitochondrial oxygen uptake by redox reactions involving nitric oxide and ubiquinol. *Journal of Biological Chemistry* **274**: 37709–37716.

Purvis, A.C., Shewfelt, R.L. and Gegogeine, J.W. (1995) Superoxide production by mitochondria isolated from green bell pepper fruit. *Physiologia Plantarum* **94**: 743–749.

Qiao, W. and Fan, L.M. (2008) Nitric oxide signaling in plant responses to abiotic stresses. *Journal of Integrative Plant Biology* **50**: 1238–1246.

Radi, R. (2013) Protein tyrosine nitration: Biochemical mechanisms and structural basis of functional effects. *Accounts of Chemical Research* **46**: 550–559.

Radi, R., Cassina, A. and Hodara, R. (2002) Nitric oxide and peroxynitrite interactions with mitochondria. *Biological Chemistry* **383**: 401–409.

Rasmusson, A.G., Soole, K.L. and Elthon, T.E. (2004) Alternative NAD(P)H dehydrogenases of plant mitochondria. *Annual Review of Plant Biology* **55**: 23–39.

Rich, P.R. and Bonner, W.D. (1978) The sites of superoxide anion generation in higher plant mitochondria. *Archives of Biochemistry and Biophysics* **188**: 206–213.

del Rio, L.A., Corpas, F.J. and Barroso, J.B. (2004) Nitric oxide and nitric oxide synthase activity in plants. *Phytochemistry* **65**: 783–792.

Rockel, P., Strube, F., Rockel, A. *et al.* (2002) Regulation of nitric oxide (NO) production by plant nitrate reductase in vivo and in vitro. *Journal of Experimental Botany* **53**: 103–110.

Romero-Puertas, M.C., Campostrini, N., Mattè, A. *et al.* (2008) Proteomic analysis of S-nitrosylated proteins in *Arabidopsis thaliana* undergoing hypersensitive response. *Proteomics* **8**: 1459–1469.

Romero-Puertas, M.C., Laxa, M., Matte, A. *et al.* (2007) *S*-nitrosylation of peroxiredoxin II E promotes peroxynitrite-mediated tyrosine nitration. *Plant Cell* **19**: 4120–4130.

Sánchez, C., Cabrera, J.J., Gates, A.J. *et al.* (2011) Nitric oxide detoxification in the rhizobia-legume symbiosis. *Biochemical Society Transactions* **39**: 184–188.

Sarti, P., Giuffre, A., Forte, E. *et al.* (2000) Nitric oxide and cytochrome oxidase: Mechanisms of inhibition and NO degradation. *Biochemical and Biophysical Research Communications* **274**: 183–187.

Saviani, E.E., Orsi, C.H., Oliveira, J.F.P. *et al.* (2002) Participation of the mitochondrial permeability transition pore in nitric oxide-induced plant cell death. *FEBS Letters* **510**: 136–140.

Shiva, S., Brookes, P.S., Patel, R.P. *et al.* (2001) Nitric oxide partitioning into mitochondrial membranes and the control of respiration at cytochrome *c* oxidase. *Proceedings of the National Academy of Science USA* **98**: 7212–7217.

Siddiqui, M.H., Al-Whaibi, M.H, and Basalah, M.O. (2011) Role of nitric oxide in tolerance of plants to abiotic stress. *Protoplasma* **248**: 447–455.

Siedow, J.N. and Umbach, A.L. (1995) Plant mitochondrial electron transfer and molecular biology. *The Plant Cell* **7**: 821–831.

Spoel, S.H. and Loake, G.J. (2011) Redox-based protein modifications: the missing link in plant immune signalling. *Current Opinion in Plant Biology* **14**: 358–364.

Stamler, J.S., Lamas, S. and Fang, F.C. (2001) Nitrosylation: The prototypic redox-based signaling mechanism. *Cell* **106**: 675–683.

Stamler, J.S., Singel, D.J. and Loscalzo, J. (1992) Biochemistry of nitric oxide and its redox-activated forms. *Science* **258**: 1898–1902.

Stoimenova, M., Igamberdiev, A.U., Gupta, K.J. and Hill, R.D. (2007) Nitrite-driven anaerobic ATP synthesis in barley and rice root mitochondria. *Planta* **226**: 465–474.

Thomas, D.D., Ridnour, L.A., Espey, M.G. *et al.* (2006) Superoxide fluxes limit nitric oxide-induced signaling. *Journal of Biological Chemistry* **281**: 25984–25993.

Tischner, R., Planchet, E. and Kaiser, W.M. (2004) Mitochondrial electron transport as a source for nitric oxide in the unicellular green alga *Chlorella sorokiniana*. *FEBS Letters* **576**: 151–155.

Torres, M.A., Dangl, J.L. and Jones, J.D.G. (2002) *Arabidopsis* gp91(phox) homologues AtrbohD and AtrbohF are required for accumulation of reactive oxygen intermediates in the plant defense response. *Proceedings of the National Academy of Science USA* **99**: 517–522.

Vitor, S.C., Duarte, G.T., Saviani, E.E. *et al.* (2013) Nitrate reductase is required for the transcriptional modulation and bactericidal activity of nitric oxide during the defense response of *Arabidopsis thaliana* against *Pseudomonas syringae*. *Planta* **238**: 475–486.

Wendehenne, D. and Hancock, J.T. (2011) New frontiers in nitric oxide biology in plant. *Plant Science* **181**: 507–508.

Wendehenne, D., Durner, J. and Klessig, D.F. (2004) Nitric oxide: A new player in plant signaling and defence responses. *Current Opinion in Plant Biology* **7**: 449–455.

Wulff, A., Oliveira, H.C., Saviani, E.E. and Salgado, I. (2009) Nitrite reduction and superoxide-dependent nitric oxide degradation by *Arabidopsis* mitochondria: Influence of external NAD(P)H dehydrogenases and alternative oxidase in the control of nitric oxide levels. *Nitric Oxide* **21**: 132–139.

Yamamoto, T., Maruyama, W., Kato, Y. *et al.* (2002) Selective nitration of mitochondrial complex I by peroxynitrite: involvement in mitochondria dysfunction and cell death of dopaminergic SH-SY5Y cells. *Journal of Neural Transmission* **109**: 1–13.

Yamasaki, H., Shimoji, H., Ohshiro, Y. and Sakihama, Y. (2001) Inhibitory effects of nitric oxide on oxidative phosphorylation in plant mitochondria. *Nitric Oxide* **5**: 261–270.

Yu, M., Yun, B.-W., Spoel, S.H. and Loake, G.J. (2012) A sleigh ride through the SNO: Regulation of plant immune function by protein *S*-nitrosylation. *Current Opinion in Plant Biology* **15**: 424–430.

Yun, B.W., Feechan, A., Yin, M.H. *et al.* (2011) *S*-nitrosylation of NADPH oxidase regulates cell death in plant immunity. *Nature* **478**: 264–268.

Zemojtel, T., Fröchlich, A., Palmieri, M.C. *et al.* (2006) Plant nitric oxide synthase: A never-ending story? *Trends in Plant Science* **11**: 524–525.

Zottini, M., Formentin, E., Scattolin, M. *et al.* (2002) Nitric oxide affects plant mitochondrial functionality *in vivo*. *FEBS Letters* **515**: 75–78.

CHAPTER 7

Control of mitochondrial metabolism through functional and spatial integration of mitochondria

Samir Sharma
Department of Biochemistry, University of Lucknow, Lucknow, India

Introduction

Mitochondria are primarily responsible for respiratory energy transduction and are the major production site for ATP in any cell. Equally important is their ability to provide the carbon skeletons necessary for the myriad of biosynthetic pathways operating in plants, switching between these two modes of action countless times a day. There is a conclusive, fact-based body of evidence to support the extreme importance of these organelles for the cell. The foremost component of this evidence is that it is impossible for the cell to survive more than a few minutes if mitochondrial function is inhibited, as by cyanide. This cannot be said of any other organelle. Secondly, these endosymbiont organelles have withstood selection pressure over billions of years. The coevolution that mitochondria and the eukaryotic cell have undergone has been underpinned by the integration of metabolic activities of the mitochondria and the rest of the cell. Mitochondria are sensitive to the demands of energy and carbon skeleton made by the cell and conduct their activities to support whole-cell metabolism actively. This integration is manifest as the control of metabolic pathways, alterations in mitochondrial organization and morphology, and changes in their position inside the cell.

The demands continuously made by the cell keep varying, starting from the extreme hunger of a dividing cell to the limited demands for the maintenance of state of terminally differentiated cells. These factors determine the proportion in which carbon is distributed between respiratory (oxidative) energy release from reduced substrates and providing carbon skeletons to the cytosol by subtracting them from the TCA cycle. The immediate environment of the mitochondria – the cytosol – has been charged with the task of integrating these demands and modifying mitochondrial metabolism. Control of programmed cell death has also been strongly linked to reactive oxygen species, especially during biotic

Alternative Respiratory Pathways in Higher Plants, First Edition.
Edited by Kapuganti Jagadis Gupta, Luis A.J. Mur, and Bhagyalakshmi Neelwarne.

stress-related defence responses and development. In view of all these, it is imperative that mitochondria be functionally integrated within the metabolic framework of the cell in order to fine tune their metabolism with the requirements of the cell.

There appear to be several layers of regulation, as complex as the processes they control. Expression levels of enzymes and electron transport components, crosstalk through participating metabolites, redox controls, reactive oxygen species (ROS) and reactive nitrogen species (RNS), serve to control and integrate mitochondrial metabolism into that of the whole cell. It is difficult to clearly assign a cause and effect relationship between different events. Some events such as mitochondrial positioning have proven difficult to investigate, while others have been investigated deeply in animals and in yeast. However, the application of knowledge gleaned from those systems to plant mitochondria may not always be appropriate due to the markedly different growth habit of these organisms. Live cell imaging and advances in visualization of functional mitochondria using molecular probes have revealed these organelles to divide (through fission) into smaller mitochondria or fuse together to yield larger mitochondria. Aggregates of glycolytic enzymes have been found to adhere to the outer mitochondrial membrane, ensuring an uninterrupted supply of TCA cycle precursors. Cytoskeleton-mediated mitochondrial positioning within the cell has also been clearly observed. These processes are non-random and directed, and serve to regulate mitochondrial function as well as to integrate mitochondria functionally and spatially within the cell.

Functional and spatial integration: scope of the review

How is integration of mitochondrial metabolism in the broader framework of cellular metabolism different from regulation of its metabolic activities? While the activities of individual enzymes, segments of pathways and of the mitochondrial electron transport system are subject to regulation by individual factors, mitochondrial metabolism overall has to be in tune with the demands of the cell. This requires multiple points of regulation that are not confined to regulating enzyme activities through metabolite concentration or regulating electron flux through the mitochondrial electron transport system through the availability of the reductant molecules. Functional integration is brought about by stoichiometric metabolite exchanges across the mitochondrial membranes, alteration of redox balance and ATP–ADP or ATP–AMP transactions across the mitochondria–cytosol interface. Apart from these molecules, TCA cycle metabolites, as well as those related to photorespiration, experience regular flux. Glycolytic enzymes have been found associated with mitochondria to enable channelling of initial metabolites to mitochondria.

Since function is intimately linked to structure, mitochondria undergo a series of extensive changes such as fission to yield smaller organelles, with dynamic positioning. Fusion, the counterpart of fission, is also frequently observed to yield very large branched mitochondria. Inextricably entwined with mitochondrial fission–fusion events is the active, physical movement of mitochondria to discrete regions inside the cell. Collectively, the mitochondrial fission and fusion events over time are termed mitochondrial dynamics. Mitochondria, especially after fission, are positioned by the elements of, or associated with, the cytoskeleton of the cell. It is now understood with clarity that organization of the cytoskeleton is a precise process, carried out in response to signalling events. These in turn are the consequence of either extracellular cues or alterations in the cell's internal metabolite status. Spatial integration of mitochondria involves changes in mitochondrial morphology and positioning, to optimize metabolic efficiency for the cell at any point of time, stage of development or while reacting to stress.

The present chapter attempts to review and integrate state-of-the-art information regarding individual control mechanisms that influence mitochondrial metabolism to provide a unified view of the functional and spatial integration of mitochondria. Several excellent texts in the present volume carry details of some of the processes outlined here. The intention of the present review is not to be an exhaustive discussion of these processes but to provide a unified model of optimization of mitochondrial metabolism, most appropriate for the ever-changing metabolic status of the cell.

Mitochondria: origins and functions

Respiratory metabolism is central to all aerobic life and represents the main pathway for energy conservation from the oxidation of reduced carbohydrates. The conserved nature of the pathway through different stages of eukaryotic evolution vouches for the essentiality of the pathway. Central to respiratory metabolism are the mitochondria, the principal energy-conserving organelles that are present in virtually every eukaryotic cell, apart from a few exceptions such as mature red blood cells. The mitochondria are largely thought to have evolved through a process of endosymbiosis, with autonomous ancestors transferring large parts of their genomes to the host cell nucleus (Schwartz and Dayhoff, 1978; Gray *et al.*, 1999; Martin, 2010). For a long time they have been recognized as sites of energy conserving oxidative metabolism (Kennedy and Lehninger, 1949), synthesizing most of the ATP produced in plant cells through an oxidative process that derives electrons from reduced substrates. The culmination of this long oxidative process occurs when the respiratory electron transport through the OXPHOS (oxidative phosphorylation) system transfers the electrons derived from reduced substrates to oxygen (Dudkina *et al.*, 2008, 2010). Mitochondria also have a role to play in the expression of the nuclear genome through a pathway called retrograde (RTG) signalling (Butow and Avadhani, 2004) that awaits clearer definition. Although the pathway has not been worked out, qualitative as well as quantitative changes in transcripts have been reported when respiration is compromised or modified (Busi *et al.*, 2011).

Composition, organization and function of mitochondrial respiration in plants

Before discussing the mechanisms, basis and the logic behind various processes that integrate mitochondria into the cell, it would be apt to discuss briefly the structural basis of mitochondrial function. Mitochondria are the seat of aerobic respiration, the last part of a long process of respiration that starts in the cytosol. Mitochondria possess four sets of major, functional components.

1 The tricarboxylic acid cycle (TCA cycle) enzymes perform the oxidative decarboxylation of organic acids, reduce NADP and FAD in the process (Siedow and Day, 2000) and also carry out a step of substrate-level phosphorylation to conserve a small amount of energy. This cycle has essentially evolved to incorporate several freely reversible reactions that allow the freedom of metabolites being moved out of the cycle to provide carbon skeletons fundamental to biosynthesis. As mentioned later, the 'cycle' is quite often not completed at all. The cycle, not discussed in detail here, presents a crucial site of regulation by various internal and external factors.

2 The electron transport chain responsible for the 'classical' process of OXPHOS. OXPHOS complexes take up the baton from the TCA cycle and oxidize the NADPH and $FADH_2$ produced there (Dudkina *et al.*, 2010; Jacoby *et al.*, 2012). This spontaneous process of combination with oxygen at the end of the electron transport chain releases energy to drive H^+ translocation from matrix to the intermembrane space/cytosol, and leads to the formation of a H^+ gradient across the inner mitochondrial membrane. Pulled in by the proton motive force (PMF), the H^+ flow back into the mitochondrial matrix through the inward facing mitochondrial ATP synthase, conserving the energy of the steep H^+ gradient into ATP synthesis. Mitochondria are also the site of synthesis of several essential cofactors like haeme, iron–sulfur clusters and tetrahydrofolate (Meyer *et al.*, 2005; Balk and Pilon, 2011).

3 The third set of components represents major differences from the respiratory set-up found in animal mitochondria (Mackenzie and McIntosh, 1999). One of the most important differences is the existence of the enzyme 'alternative oxidase' (Finnegan *et al.*, 2004) and the integration of a non-energy conserving mode of electron transport component in an otherwise efficient energy-conserving mechanism. This enzyme is virtually ubiquitous in the plant kingdom and has also been reported in some fungi and protista.

Recently, homologues of this enzyme have also been reported from a range of animal systems (McDonald, 2009). Plant mitochondria also possess at least four rotenone-insensitive NADH dehydrogenases in addition to complex I, which are not known to exist in animals (Møller and Rasmusson, 1998). Due to the presence of chloroplasts in green cells, mitochondria also share hosting of a major component of the photorespiratory metabolism, presenting a primary site for transamination and amino acid biosynthesis as well as providing ammonia through photorespiratory deamination for plastids to fix as amino acids (Siedow and Day, 2000).

4 The fourth set comprises carriers, channels and translocators that affect metabolite/ion exchange between the mitochondrial matrix and the cytosol across the mitochondrial membrane (Linka and Weber, 2010; Millar *et al.*, 2011). Apart from these membrane components, every mitochondrion possesses a protein translocation apparatus that carries out the uptake of nuclear-coded proteins targeted to the mitochondria from the cytosol into one of its water compartments (matrix and intermembrane space) or into the lipid matrix of one of the mitochondrial membranes (Wiedemann *et al.*, 2004; Carrie *et al.*, 2009; Schmidt *et al.*, 2010).

Mitochondria play a critical part in the signalling involved in programmed cell death (Kim *et al.*, 2006; Scott and Logan, 2008). They are also extremely important for efficient photosynthesis under stress, being the compartment for an essential segment of the photorespiratory pathway (Maurino and Peterhansel, 2010). The photorespiratory pathway is one of principal mechanisms for avoidance of photoinhibition through consumption of ATP in the chloroplasts, as well as by generating CO_2 internally in the mitochondria. This release of CO_2 is extremely important for survival during periods of water stress when gas exchange is severely limited due to closure of stomata. Photorespiration also functions indirectly as an NADH shuttle between the peroxisomes and mitochondria, thus affecting redox balance. Mitochondrion–nucleus communication, an emerging, albeit extremely challenging area of study, is revealing processes that result in the stoichiometry adjustment of mitochondrial components and coordinate the expression of genes of mitochondrial proteins coded in the mitochondrial and nuclear genome (Ryan and Hoogenraad, 2007).

The sedentary habit and autotrophic mode of life of plants make them radically different from animals. These differences are reflected in the vastly greater ability of plants to adapt to their environment as compared to animals. Expectedly, mitochondria in plants have evolved to be sensitive to this mode of existence. This sensitivity is most significantly conferred by phytochrome action that makes mitochondrial metabolism light responsive to quite an extent. This is brought about especially by the sensitivity of succinate dehydrogenase to phytochrome-mediated inhibition in the day.

Calcium homeostasis: interactions with the cellular calcium pool

Calcium, due to its low background presence in the cytoplasm, high rates of diffusion and because it lends itself freely to reversible chelation by certain proteins, is an extremely effective intracellular messenger (Clapham, 2007). Reversal of this signal requires removal from the cytosol into the storage compartment, via Ca^{2+} pumps. Endoplasmic reticulum (ER), vacuoles and mitochondria serve as internal stores of Ca^{2+} while the apoplast serves as a source of Ca^{2+} in the extracellular milieu. Ca^{2+} is brought into the cytosol along a strong gradient by opening variously gated Ca^{2+} channels from one or more of these locations. This influx is quickly reversed by Ca^{2+} pumps functioning actively to pump Ca^{2+} back into the storage spaces. Mitochondria have been shown to be major participants in this rapid, reversible flux of Ca^{2+} (Giorgi *et al.*, 2009). Recent studies have also revealed the presence of overlapping regions between ER and mitochondria that allow direct physical association of ER proteins with outer mitochondrial membranes (Patergnani *et al.*, 2011). This brings about a modification of calcium signatures in microdomains tenanted by mitochondria at a given point of time (Clapham, 2007; Laude and Simpson, 2009), while at the same time, as a corollary, elevation of mitochondrial calcium up-regulates critical TCA cycle enzymes and alters the status of ATP synthesis. Over-accumulation of calcium in mitochondria results in opening of the mitochondrial permeability transition pore (mPTP) and the release of cytochrome c, bringing about apoptosis. This phenomenon is seen in plants also (Arpagaus *et al.*, 2002; Virolainen *et al.*, 2002). In plants the main store of intracellular Ca^{2+} is the vacuole, while the resting free Ca^{2+} in mitochondria is ~200 nM. However, comparison with animal mitochondria has not been without results. The calcium uniporter has been identified in animals and designated MICU1 (De Stefani *et al.*, 2011; Perocchi *et al.*, 2010). The subunits of this protein that probably oligomerize in the inner mitochondrial membrane have homologues in *Arabidopsis* also, which despite having low homology to their animal counterparts, have retained the oligomerization motif (Stael *et al.*, 2012). In addition to the MICU1 homologues, plant mitochondria also possess homologues of the high affinity mitochondrial calcium transporter (LETM1) present in mammalian cells (Van Aken *et al.*, 2009). Calcium and calmodulin (CaM) have been reported to promote protein import into mitochondria and CaM has also been suggested to occur in plant mitochondria (Kuhn *et al.*, 2009; Bussemer *et al.*, 2009).

Apart from CaM, several proteins with calcium-binding motifs (Chigri *et al.*, 2012) apparently acting as Ca^{2+} sensors, are also found in plants. These proteins are found to reversibly modify the structures of interacting proteins, leading to changes in the functional status of the latter in a Ca^{2+}-dependent manner. Such changes in the TCA cycle-related proteins act as agents of regulation of the TCA cycle, as has been shown in animal mitochondria (Griffiths and Rutter, 2009).

The concentration range in which Ca^{2+} is sensed as well as the kinetics of reversible chelation by sensor proteins, together fine tune cellular responses to concentration changes of the divalent cation. Apart from these two factors, spatiotemporal Ca^{2+} calcium signatures have a lot to do with metabolic control, even though this is an aspect of Ca^{2+} signalling that has just begun to reveal itself.

Plant mitochondria, because of the differences in the Ca^{2+} sensor proteins among other things, exhibit remarkably different calcium dynamics to those observed in the cytosol (Logan and Knight, 2003; Loro *et al.*, 2012). The current status therefore remains that although plant mitochondria do possess the required machinery for calcium sensing, the link between calcium dynamics and physiological regulation is unclear (Schwarzländer and Finkemeier, 2013). The current stream of thought expects a very important, central role for calcium in the regulation of mitochondrial metabolism as well as in metabolic integration of the organelle within the plant cell due to the interaction of mitochondrial and cytosolic pools of calcium. Specific instances of these aspects will present themselves during the course of the next section.

Functional integration of mitochondria in plant cellular metabolism

The respiratory process resident in mitochondria is perhaps the most essential part of plant cellular metabolism. For this reason, optimization of mitochondrial metabolism and maintaining a high degree of coordination between metabolism within and outside this organelle is of utmost importance. There appear to be two broad classes of control mechanisms. The first is a class of interactions in which mitochondria react to metabolite and energy flux across its boundary with the cytosol and with other compartments inside the cell. The other involves large-scale changes in mitochondrial organization through fission and fusion and in mitochondrial positioning inside the cell.

Metabolic regulation

Mitochondrial metabolism has been considered as being central to cellular homeostasis and is established as a process that complements photosynthesis, supplying the cell with ATP as well as with carbon skeletons (Nunes-Nesi *et al.*, 2011; Kramer and Evans, 2011). Having roles as crucial as these, there is an express need for mitochondria to communicate effectively with the cell and be functionally integrated within it. It is understandable therefore that the respiratory machinery of the mitochondria is sensitive to redox changes, a control mechanism that confers a very high degree of flexibility to mitochondrial function. Light represents a factor external to the cell as well as to the plant and controls mitochondrial metabolism via phytochrome action. This can be viewed as a mechanism of control of mitochondrial metabolism parallel to the control of other aspects of metabolism. Last, but not the least, mitochondria exercise strict control over the outward flux of 'deliverables' like carbon skeletons or energy in the form of ATP. This control translates into an almost hardwired form of communication with the cytosol, maintained through a host of transporter molecules residing in the mitochondrial membrane that serve to maintain stoichiometric metabolite exchanges. The resulting change in metabolite concentrations represents another mechanism of metabolic integration in the process of being sensitive to the demands of the cell.

Redox regulation: the synchrony of energy

The respiratory pathway is extremely sensitive to redox changes, altering mitochondrial status, which in turn can trigger 'retrograde signalling' (RTG), mitochondrial language for talking back to the rest of the cell through redox messages (Schwarzländer and Finkemeier, 2013). Reactive oxygen and reactive nitrogen species (ROS/RNS), the redox state of NAD(P)H and antioxidant pools, ATP/ADP ratio, the status of mitochondrial proton motive force and metabolite levels, can all be considered arms of mitochondrial signalling pathways. Of these, perhaps ROS, or more specifically, superoxide producing centres, abound within the mitochondrial electron transport chain. Molecular access and redox potentials make complex I (NADH dehydrogenase) and complex III (cytochrome *b*/*c*1 complex) the most active centres for superoxide production (Turrens, 1997). High matrix NADH/NAD^+ ratio, highly reduced ubiquinone pool and high membrane potential have been known to induce mitochondrial superoxide production in animal systems (Murphy, 2009). Since plants possess a more flexible and adaptive system with a multitude of alternative pathways of electron transport and energy dissipation, the relevance of these studies in plants remains to be ascertained. However, ROS as well as RNS, are relevant as signalling entities, since these reactive species are the hallmark and a compulsory by-product of respiratory metabolism.

Cumulative oxidative damage is said to be one of the main causes of ageing-related deleterious changes (Kregel and Zhang, 2007). The neatest trick employed by mitochondria in their endeavour to employ ROS as signalling entities, is to maintain a delicate balance between production and scavenging of these species, allowing a small amount of irreversible oxidative damage that makes molecules serve as beacons communicating the status of mitochondrial oxidative metabolism to the cell. This can be affected by modulating mitochondrial redox reactions so that energy conservation, provision of carbon skeletons and production of ROS are all balanced. In addition to this, ROS can be kept below dangerous levels by a strong scavenging system, comprising non-enzymatic antioxidants as well as antioxidant enzymes. The latter is the principal ROS modulating system in animals. ROS and other signalling entities responsible for metabolic pace-setting and carbon skeleton diversions are discussed further in the conclusion of this chapter.

The enzyme system working to lessen the concentration of the ROS produced mainly comprises three enzymes working in concert. The first of these, Mn-superoxide dismutase, converts the superoxide produced during mitochondrial electron transport, into hydrogen peroxide. Hydrogen peroxide is converted to water by either by ascorbate peroxidase, using ascorbate, a plant specific antioxidant, as a substrate (Chew *et al.*, 2003). The same is also accomplished through reduced glutathione by glutathione peroxidase (Navrot *et al.*, 2006). It seems obvious that high concentrations of glutathione and ascorbate are required to support these enzyme activities. The efficacy of this antioxidant system is validated by mutant plants with low mitochondrial levels of glutathione being associated with a plethora of growth defects (Zechmann and Müller, 2010). On the contrary, mutants with an extremely low glutathione content in all cell compartments except the mitochondria, exhibit an almost wild-type like phenotype (Zechmann *et al.*, 2008). These results point towards the essentiality of high mitochondrial levels of glutathione (10.4mM) and ascorbate (15mM) for proper plant growth, development and survival, due to their roles in keeping ROS concentration within safe limits. It is also significant to note that the inability of mitochondria to control ROS can trigger cell death (Vianello *et al.*, 2007). Indeed, senescing tissues are characterized by rapidly decreasing concentrations of ascorbate and glutathione, oxidation of the pools for the two antioxidant substances and a concomitant rise of H_2O_2 in these tissues (Jiminez *et al.*, 1998).

ROS concentration in plants is subject to a rigorous and precise system of redox modulation, where alternative electron transport pathways keep electron flow channelled between energy conserving and non-conserving pathways. These aspects have been discussed in great detail in several reviews and research papers (Ernster and Schatz, 1981; Cvetkovska and Vanlerberghe, 2012, 2013; Moore *et al.*, 2013). Research implicates alternative oxidase, the respiratory enzyme unique to plants, in most responses arising due to stress or development. Change in the status of ROS and the functionally related RNS, as well as the oxidative changes brought about by them, have a very large dynamic range. That is, these changes in ROS and RNS and the ensuing oxidative damage, are small for a large change in the redox status of the mitochondria, thus restricting oxidative damage to an easily tolerable level, while keeping these levels changing in synchrony with the redox status of mitochondria. This ensures seamless integration of mitochondrial metabolism in the general metabolic picture of the cell at any given instant.

The fact that aerobic respiration is compartmentalized in mitochondria, has been proposed to be the basis for the greater complexity of eukaryotes as compared to prokaryotes (Martin, 2010). The sessile habit of plants has led to the development of a highly dynamic metabolic system, with mitochondria occupying the veritable functional centre. The split location of mitochondrial genes in both the nuclear and mitochondrial genomes is also said to be driven by the need for redox regulation of gene expression in the mitochondria and is described in the

Colocation for Redox Regulation, or the CoRR hypothesis (Allen, 1993, 2003). Redox changes are said to result in stoichiometric adjustments in the electron transport components through coordinate gene expression between the mitochondrial genes collocated in the nuclear and mitochondrial genomes. Within the mitochondrial compartment, there are several ways of expressing the redox status of the organelle. It can be reflected in the mitochondrial GSH/GSSG ratio (Schwarzländer *et al.*, 2009) and also in the redox status of mitochondrial redoxins, that can act as redox switches capable of detecting shifts towards oxidation in an otherwise highly reducing environment of the mitochondria. Status of protein thiols has not been probed in detail in plants, but is likely to be similar to that in animal mitochondria, where protein thiols are known to play a major role in redox control of gene expression (Requejo *et al.*, 2010). Thiol-based mitochondrial peroxidases that include peroxiredoxins and glutathione peroxidases mentioned earlier in this section, play a major role in modulating the concentration of H_2O_2 (Barranco-Medina *et al.*, 2007), which is one of the major components of the retrograde signalling system employed by the mitochondria due to its membrane permeability and relatively long half-life.

Cellular redox balance is understood to be a common ground for integration of stimuli as diverse as nutrient sensing, pathogen attack, heavy metal exposure and stresses originating from adverse conditions of temperature, light or water availability. This list is still not a comprehensive one. Apart from the factors mentioned therein, the interaction of chloroplast and mitochondrial metabolism is perhaps the most important of all. Chloroplasts harvest light energy and produce reduced equivalents; the mitochondria on the other end, oxidize reduced equivalents, generating energy as well as carbon skeletons. This is an interaction that integrates not just these organelles within cellular metabolism, but integrates the metabolism of the entire plant. With most of the tissues in a plant being heterotrophic, chloroplasts carrying out their reducing activities may be located in the leaf canopy and the mitochondria oxidizing these reduced equivalents in the lowermost root of the tree. It is interesting to note that the primary product of photosynthesis is a reduced carbohydrate while the transported molecule is a non-reducing disaccharide, sucrose.

Rapid cell division or the environmental conditions that constrain it, alter redox balance and usually result in an increased production of ROS, leading to a condition referred to as 'oxidative stress'. It is widely accepted that such changes in ROS/redox status serve as signals that are perceived by different mechanisms to bring about a change in the metabolic state of the cell and nuclear gene expression (Noctor, 2006; Moller and Sweetlove, 2010). Mitochondrial stress response has been defined by monitoring the expression of genes that putatively code mitochondrial proteins. Of more than a thousand such genes, 26 were confirmed to encode proteins that were stress-responsive (Van Aken *et al.*, 2009). Direct oxidative inhibition of complex I and aconitase (Zhang *et al.*, 1990; Verniquet *et al.*, 1991) and inhibition of pyruvate dehydrogenase complex and

2-oxoglutarate dehydrogenase by modification of their lipoic acid residues by 4-hydroxy-2-nonenol, a lipid oxidation product (Millar and Leaver, 2000), has been reported. Among chloroplasts, peroxisomes and mitochondria – three organelles conducting redox metabolism – mitochondria have been found to accumulate oxidatively modified proteins to a several-fold higher concentration than the other two (Bartoli *et al.*, 2004). One of the outcomes of mitochondria being the main targets of such oxidative stress is that the damage reduces mitochondrial ROS production due to the damage caused to its electron transport system (Yao *et al.*, 2002). Thus, ROS production in times of stress is a self-limiting reaction. By sacrificing mitochondrial function to a limited but definite extent, the cell is saved from more extensive ROS-induced damage.

In the final analysis, it would be appropriate to say that the levels of ROS, RNS and antioxidants as well as the redox state of the mitochondria vis-à-vis that of the rest of the cell serve to integrate mitochondrial metabolism into the mainstream quite seamlessly through ROS/RNS-mediated signalling that is amplitude-modulated within safe operational limits by the combined action of enzymatic and non-enzymatic antioxidants (Kocsy *et al.*, 2013).

The calcium connection: crosstalk between the ROS and Ca^{2+} signalling pathways

ROS-mediated signalling has now been suggested by an increasing number of studies to be connected to Ca^{2+} signalling circuits of the cell. The complexity of Ca^{2+} signalling is beyond the scope of the current chapter and can be accessed in excellent reviews by Clapham (2007) and McAinsh and Pittman (2009). Ca^{2+} plays a very important role in integration of mitochondria into cellular metabolism as mitochondria have been shown not only to undergo calcium regulation but also to influence Ca^{2+} signalling pathways in the cytoplasm (Stael *et al.*, 2012). The modification of these signalling pathways is also attributed to the crosstalk between ROS and calcium signalling not only to coordinate mitochondrial metabolism with the metabolic events in the cytosol, but also regulation of nuclear gene expression (Mazars *et al.*, 2010). However, the concept of well-defined Ca^{2+} signatures that characterize Ca^{2+}-mediated signalling in all organisms, are not applied as rigorously to ROS-mediated signalling, but parallels with Ca^{2+} signalling have been suggested (Fedoroff, 2006).

Various families of Ca^{2+} binding proteins are responsible for transducing Ca^{2+} signals and converting them to a change in the metabolic status (Dodd *et al.*, 2010) and annexins are among the several proteins present in cells that serve to transduce Ca^{2+} signals. Annexins are proteins that bind to phospholipid membranes in response to calcium binding and have also been implicated in ROS-mediated signalling pathways. To qualify as an annexin, firstly, the protein has to exhibit Ca^{2+}-dependent binding to negatively charged phospholipid membranes and secondly it must possess a motif comprising approximately 70 amino acid residues, known as the annexin fold (Gerke and Moss, 2002). Isolated plant annexins have been shown to bind membranes including secretory vesicles, plasma membrane and endomembranes. Binding has also been demonstrated for GTP/ATP and interestingly, for F-actin (Mortimer *et al.*, 2008). When viewed with the information that these proteins have been co-localized with regions of high activity, and found to stimulate Ca^{2+}-dependent exocytosis in root cap cells, roles for annexin begin to get into sharper focus (Bassani *et al.*, 2004; Blackbourne and Battey, 1993; Carroll *et al.*, 1998; Clark *et al.*, 2005).

More recently, Laohavisit *et al.*, (2009) have shown that plant annexins are likely to be multifunctional proteins that are capable of peroxidase activity, thereby modulating ROS signatures also. Annexin expression is known to respond to stress conditions such as salinity, drought, nutrient deprivation and cold, which are known to increase cytosolic Ca^{2+} and ROS (Mortimer *et al.*, 2008). Huh *et al.*, (2010) followed the expression and localization of two annexins in *Arabidopsis,* AnnAt1 and AnnAt4, and found the latter localized to ER membranes. This could be related to the formation of voltage-gated channels in the ER, helping modify cytosolic Ca^{2+} signature and perhaps peripherally affecting mitochondrial Ca^{2+} also. Very significantly, in fibroblasts, Annexin A6 (AnxA6) has been shown to regulate mitochondrial morphogenesis, interacting with Drp1, the fission GTPase, in a Ca^{2+}-dependent manner. AnxA6 sequesters Drp1, preventing mitochondrial fission to a great extent. However, when AnxA6 binds Ca^{2+}, it unbinds Drp1 and localizes to the plasma membrane, allowing Drp1 to carry out mitochondrial fission (Chlystun *et al.*, 2013). This shows annexins in a very different light. It has been known for some time that Ca^{2+} binding results in annexin localization to membranes, but that this alteration in localization brings about a change in the mitochondrial dynamics by exercising a kind of negative regulation is indeed significant. The reporting of AnxA6–Drp1 interaction should lead to more studies on protein–protein interactions involving annexins and open new vistas of metabolic control being exercised. It is significant that annexin expression has been found to be induced by auxin (Baucher *et al.*, 2011) and ABA (Lee *et al.*, 2004), also under stress conditions as mentioned earlier. The recent discovery of control of mitochondrial morphogenesis may point towards annexin-mediated control of mitochondrial morphogenesis under stress conditions as well as during development.

Interfacing mitochondrial metabolism with light: phytochrome-mediated regulation of respiratory metabolism

Since light is unequivocally the single most important factor regulating plant growth and development, it is imperative that plants sense light and convert it to signals that coordinate respiratory pathways. Phytochromes are red light/far red light receptors that perceive light and tune plant photomorphogenesis to light quality, quantity and duration (Bae and Choi, 2008; Arsovski *et al.*, 2012; Hughes, 2013). While other plant photoreceptors also sensitize the plant to blue and UV light (Lin, 2002; Möglich *et al.*, 2010), phytochrome is known to make the respiratory process sensitive to light cues. In doing this, phytochrome-regulated aspects of respiratory metabolism translate environmental cues into signals that contribute to the integration of mitochondrial metabolism with that of the rest of the cell. Involvement of phytochrome in the control of TCA cycle activity was suggested by Cedel and Roux (1980a, 1980b), who found plant mitochondrial $NADP^+$ activity to be regulated by mitochondria from oat leaves, pre-illuminated by white or red light prior to extraction. Later, Serlin and Roux (1986) reported the light-induced import of phytochrome into mitochondria. This was further supported by Morohashi *et al.* (1993) who demonstrated the effect of phytochrome on mitochondria manifest as an increase in the total NAD pool in these mitochondria. The main factor underpinning this aspect of integration is the drastic change in photosynthesis with night and day.

Plants undergo a sea change in metabolism as photosynthate levels fluctuate in a day–night rhythm. Mitochondrial metabolism is functionally located on the extreme end of the energy metabolism, with photosynthesis occupying the other extreme. This changes the demands on the TCA cycle and prompts it to change from being principally a source of energy and providing carbon skeletons as a secondary function, to being a process distributing its functions between generating carbon skeletons, redox homeostasis and energy generation (Hanning and Heldt, 1993; Igamberdiev and Gardeström, 2003; Fernie *et al.*, 2004), with citrate being one of the principal exports. Under conditions of illumination the TCA 'cycle' displays exceptional reversibility, with equilibria of various reactions so balanced that the cycle itself is seldom completed. What is accomplished instead is the production of fumarate and glutamate (Tcherkez *et al.*, 2009, 2012). Additionally, light inactivation of succinate dehydrogenase (SDH) has also been reported to occur (Popov *et al.*, 2011). This could be a major reason for the accumulation of fumarate in the day.

Phytochrome is known to link Ca^{2+} flux across the plasma membrane to light. Mitochondrial ATPase activity is modulated as a result of this action and the related Ca^{2+} flux across the mitochondrial membrane (Serlin *et al.*, 1984). Mitochondria experience as well as modify cytosolic Ca^{2+} flux, as do chloroplasts and peroxisomes (Stael *et al.*, 2012). Since Ca^{2+} is one of the most important signalling entities in cells (Dodd *et al.*, 2010; Whalley and Knight, 2012) and phytochrome has been known to modify Ca^{2+} flux in mitochondria, the connection between phytochrome and modification of mitochondrial metabolism assumes importance.

Of all the mitochondrial processes and metabolite levels affected by phytochrome action, the inhibition of SDH is perhaps the most significant. By virtue of being an intrinsic component of the TCA cycle, as well as a major component of the mitochondrial electron transport system (complex II) located in the mitochondrial inner membrane, SDH represents a direct link between the two processes. It is interesting that SDH also represents the point of phytochrome-mediated control of mitochondrial metabolism. Upon deeper study, plant SDH reveals the presence of four plant specific subunits (Millar *et al.*, 2004). Complex II represents a branch in an otherwise linear electron transport chain comprising complexes I, III and IV to function as a parallel source of electrons for the reduction of the ubiquinone pool in the mitochondria. This supplements the electron flow from complex I into the electron transport system. It has also been found that complexes I, III and IV, comprising what is frequently called the respiratory supercomplex, are more strictly conserved as compared to complex II, i.e. SDH (Dudkina *et al.*, 2005). For this apparent reason, the regulation of SDH gene expression may resemble that of the NADH and NADPH dehydrogenases (Escobar *et al.*, 2004) and alternate oxidase (Ribas-Carbo *et al.*, 2008), all exclusive components of the mitochondrial electron transport system in plants. The diurnal variation in SDH activity is, however, opposite to NADPH dehydrogenases and alternative oxidase, which experience phytochrome-mediated inhibition in the night. Light or photoperiod-induced control of SDH activity is not exercised by protein modification or through small molecule binding to the enzyme protein and are manifested as control of gene expression instead. The picture of this control of gene expression is far from clear, but mitochondrial transmembrane potential and Ca^{2+} have been suggested to play a role (Eprintsev *et al.*, 2013; Igamberdiev *et al.*, 2013).

SDH is not the only enzyme affected by phytochrome action. Glycine decarboxylase (GDC) and serine hydroxymethylaminotransferase (SHMT), two enzymes catalysing the decarboxylation as well as amine transfer reactions in the component of photorespiration in mitochondria, also exhibit light modulation of their activities (Morohashi, 1987; McClung *et al.*, 2000). Promoter analysis of the genes for these enzymes reveals the presence of light-dependent promoter elements (Vauclare *et al.*, 1998). The findings are supported by expression profiling carried out using microarray analysis (Tepperman *et al.*, 2004; Thum *et al.*, 2004) where all but the L-protein of the GDC have been shown to be light-controlled. The same data also shows suppression/repression of gene(s) coding for citrate synthase and mitochondrial aconitase. Later data have led to proposing of gene networks regulated by light and metabolite concentrations (Thum *et al.*, 2008).

Metabolite and ion transporters: flow of matter as an integrating factor

Metabolic compartmentation has accompanied the evolution of eukaryotic cells (Lunn, 2007). This has led to pathways being concentrated inside compartments of limited volume, resulting in shorter diffusion distances and greater concentrations of relevant metabolites, enzymes and cofactors, maintenance of the most appropriate pH for a particular pathway, avoidance futile cycles and more. All of these factors have led to more efficiently conducted pathways. Metabolic control is exercised through a variety of means, ranging from metabolite levels and changing concentrations of effector, non-metabolite molecules to covalent modifications and degradation. One of the most important methods of metabolic regulation that does not require alteration of the metabolite or the enzyme is a change in location of the molecules, nearly always the metabolites involved, from one compartment to another (Tegeder and Weber, 2006). This requires the presence of metabolite transporters in the bounding membrane of the concerned compartment. Thus the presence of transporters, specific for particular metabolites or frequently, exchange transporters, overcomes the diffusion barrier presented by the membrane of that compartment and also regulates the pathway by adjusting metabolite concentrations (Linka and Weber, 2010). In this way, metabolite transporters serve to integrate apparently isolated pathways, sequestered in different organelles as well in the cytosol. This in turn brings about fine tuning of the metabolic state of the cell at any given point of time (Schwacke *et al.*, 2003; Schwacke *et al.*, 2004; Lunn, 2007; Linka and Weber, 2010). The database of *Arabidopsis* reported by Schwacke *et al.*, (2003, 2004), has grown to contain 2705 proteins having the required structural characteristics for forming pores that can putatively serve as metabolite transporters.

The outer membrane of the mitochondria does not represent a barrier to metabolites. The inner membrane, like any other lipid bilayer membrane, however, does not allow polar or charged molecules and ions to pass freely. The inner mitochondrial membrane, therefore, presents a highly regulated point of metabolite exchange and incorporates a large number of transporters, translocators, channels and carriers. The integrity of the inner membrane is also the reason for sustenance of the transmembrane H^+ gradient set-up during electron transport. This gradient is responsible for energy transduction as well as for providing the energy input for a large number of energetically unfavourable transport processes (Fernie *et al.*, 2004; Millar *et al.*, 2011). The TCA cycle in mitochondria provides reducing equivalents for other cell compartments and more specifically, in seeds, mitochondria participate in the mobilization of carbon and nitrogen storage compounds during germination (Mackenzie and McIntosh, 1999; Logan, 2006). These metabolic pathways require a regular, rapid and highly specific exchange of molecules. Additionally, in photosynthetically active tissues, photorespiration is the inevitable outcome of the dual activity of Rubisco, beginning with the oxidation of ribulose bisphosphate (RuBP). A significant part of this pathway occurs in the matrix of mitochondria (Tolbert, 1997; Seidow and Day, 2000; Sage *et al.*, 2012) and requires metabolite flux. Plant groups have evolved to concentrate CO_2 either for survival under arid/semi-arid conditions (CAM plants), or to achieve greater photosynthetic efficiency (C4 plants). These plant groups require metabolite transporters for the additional metabolite flux required for the flux of metabolites involved in carbon concentration (Seidow and Day, 2000; Sage *et al.*, 2012). This has led to the evolution of mitochondria to incorporate a large number of transporters in the inner membrane. All of these belong to the mitochondrial carrier family or the MCF (Haferkamp *et al.*, 2002; Picault *et al.*, 2004). These transporters do not possess as great a structural diversity as the metabolites transported by them seem to suggest (Haferkamp, 2007; Haferkamp and Schmitz-Esser, 2012; Picault *et al.*, 2004).

Strangely, the identity of the carrier for pyruvate remains elusive, despite its obvious importance in providing the initial carbon skeletons for the TCA cycle.

Other carriers, no less important, have been characterized. The mitochondrial inner membrane located dicarboxylate/tricarboxylate carrier is one of utmost importance in view of its ability to catalyse the transport of unprotonated dicarboxylates such as oxoglutarate, oxaloacetate, malate, maleate, malonate and succinate against tricarboxylates citrate, isocitrate and aconitate (Picault *et al.*, 2002). Palmieri *et al.* (2008a) have reported that three genes, previously reported as uncoupling proteins (UCPs), actually belong to the dicarboxylate carrier class (DIC 1–3). These carriers are also known to transport phosphate and sulfate. The broad specificity for transported substrates in these transporters endows mitochondria with a lot of flexibility in the efflux and influx of these metabolites. While unprotonated dicarboxylates can be transported into mitochondria against sulfate or phosphate to feed the TCA cycle, the same DIC may function to shuttle malate-oxaloacetate to provide other cell compartments with reducing equivalents (Palmieri *et al.*, 2008b). This feature is extremely handy during C_4 photosynthesis-related carbon flux. Although amino acid flux into mitochondria is mandatory for protein synthesis in organelles as well as to conduct the complicated steps of decarboxylation and deamination of glycine, much is still unknown about mitochondrial amino acid transporters (Picault *et al.*, 2004; Haferkamp, 2007). BAC1 and BAC2, two carriers specific for basic amino acids, have been reported in *Arabidopsis* and are known to prefer arginine because yeast mutants deficient in the ornithine/arginine transporter were relieved of arginine auxotrophy when transformed with the genes for these two carriers (Catoni *et al.*, 2003; Hoyos *et al.*, 2003). Arginine, lysine, histidine and ornithine are the transported metabolites for BAC1, while BAC2 also transports citrulline (Hoyos *et al.*, 2003; Palmieri *et al.*, 2006a), thus constituting an ornithine/citrulline shuttle. BAC1 is a highly expressed protein during seed germination and could be a major conduit for entry of arginine into the mitochondrion to feed mitochondrial protein synthesis (Palmieri *et al.*, 2006b).

One of the most important transporters found in the mitochondrial inner membrane is the ADP/ATP carrier (AAC). This transporter arguably sets the pace for respiration on one hand and communicates the cytosol's demand for ATP to the mitochondrial matrix. It catalyses the stoichiometric exchange of an ADP from the cytosol for a molecule of ATP from the mitochondrial matrix, with the balance of charge being more negative outside due to the negative charge on ATP being one more than that on ADP. This exchange is offset by the H^+ pumping that accompanies mitochondrial electron transport, with these two oppositely electrogenic processes balancing each other.

When the cytosolic demand for ATP is less and cytosolic ADP concentration is low, the AAC cannot affect the exchange and mitochondrial ATP remains in the matrix. The chain of events continues to slow down the matrix facing mitochondrial ATP synthase due to low ADP concentration inside the matrix. This leads to non-dissipation of the H^+ gradient set-up during electron transport leading to the exterior of the mitochondria being maintained positive. The H^+ gradient is the coupling between electron transport and ATP synthesis, and this coupling is jammed as a consequence of low demand for ATP in the cytosol. With the strongly positive outward $\Delta\Psi$ (potential gradient), it becomes increasingly difficult for the electron transport system to pump H^+ into the intermembrane space, making deprotonation of the ubiquinol pool increasingly difficult. The UQH_2/UQ pool gets increasingly reduced and the 'message' of low cytosolic ATP demand gets passed from the potential gradient of H^+ to the redox system of the mitochondrial electron transport system (ΔpH not being significant in mitochondria). A highly reduced electron transport chain finds it difficult to accept electrons from the TCA cycle, leading to high NADH/NAD and $FADH_2$/FAD ratios. With the accumulation of NADH, a potent inhibitor of the TCA cycle, the cycle slows down and TCA cycle metabolite concentrations rise. This may lead to the export of the 'equilibrium' metabolites like citrate and 2-OG from mitochondria into the cytosol through their respective transporters. The AAC can thus be considered one of the foremost regulatory molecules governing not only second-to-second infinitesimal changes in metabolic flux through the cycle, but also the mode in which it is appropriate for the cycle to run, particularly in the context of the needs of the cytosol. In doing so, it acts as the switch that causes the TCA cycle to shift from a pathway providing energy to one providing carbon skeletons, and thus serves as an extremely important component for integrating mitochondrial metabolism with that of the cytosol (Figure 7.1).

The AAC, possibly for this reason, is the most abundant exchange transporter in the mitochondrial membrane (Klingenberg, 2008). The *Arabidopsis* genome codes for three AACs that have been characterized by expression in *E. coli* (Haferkamp *et al.*, 2002) and were found to be high affinity transporters and possess sensitivity to bongkrekik acid and carboxy atractyloside. This makes them very similar to the AACs found in animal and yeast mitochondria (de Marcos Lousa *et al.*, 2002; Gawaz *et al.*, 1990; Heimpel *et al.*, 2001). Of these, AAC1 is the most abundant in the mitochondrial membrane. Plants also possess a carrier unique to their mitochondria. Designated as ADNT1 (Palmieri *et al.*, 2008a), this carrier is remarkably distinct from the AACs found in yeast, humans and *Arabidopsis* in that it primarily carries out an antiport of ATP and AMP, and with a much lower affinity, ADP. This gains importance because in heterotrophic plant tissues (all non-green tissues), AMP is a major metabolite. The ADP/ATP or AMP/ATP exchanger has to be complemented with a carrier for inorganic phosphate, required for ATP synthesis (Rausch and Bucher, 2002). The mitochondrial inorganic phosphate carrier (PIC) carries out electroneutral transport by symporting phosphate with H^+ or by antiporting it against OH^- (Pratt *et al.*, 1991; Stappen and Kramer, 1994).

Control of respiration in plants can be effectively exercised by altering the energy dynamics of the mitochondrial electron transport system. As mentioned

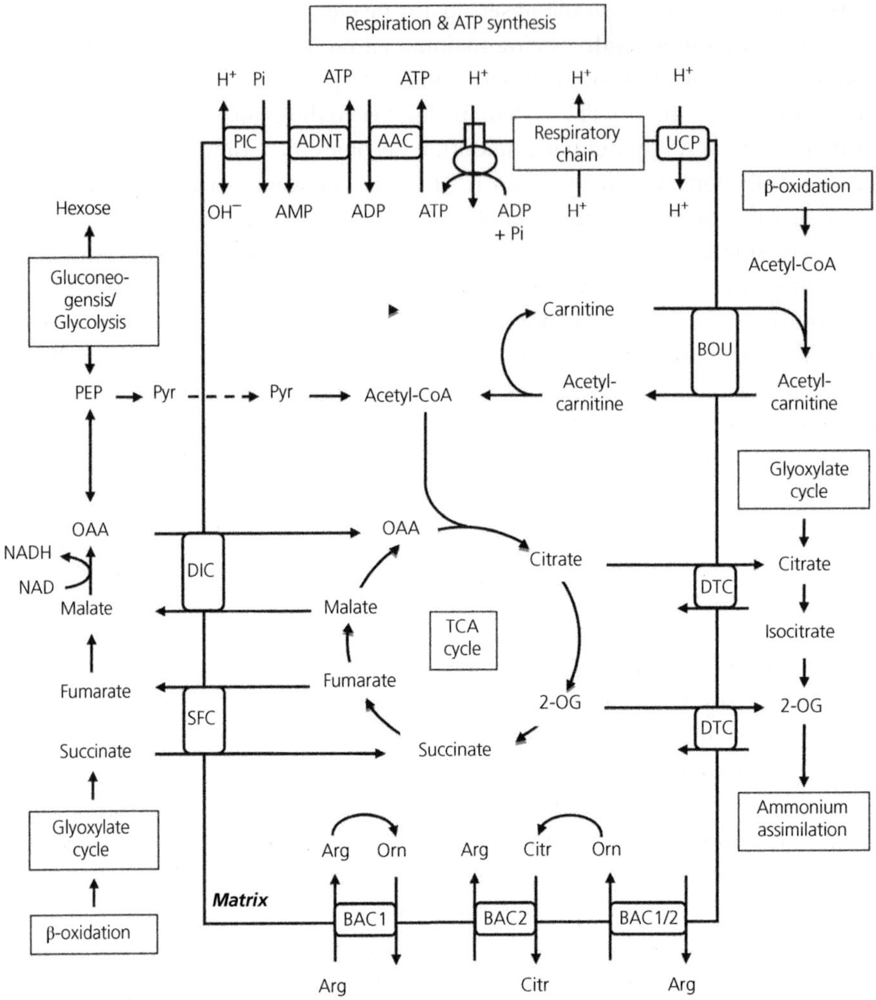

Figure 7.1 Schematic representation of the characterized metabolite transporters across the mitochondrial membrane. 2-OG, 2-oxoglutarate; AAC, ATP/ADP carrier; ADNT, adenine nucleotide carrier; ADP, adenosine diphosphate; AMP, adenosine monophosphate; Arg, arginine; ATP, adenosine triphosphate; BAC, basic amino acid carrier; BOU, carnitine carrier; Citr, citrulline; DTC, dicarboxylate/tricarboxylate carrier; DIC, dicarboxylate carrier; OAA, oxaloacetic acid; Orn, ornithine; PEP, phosphoenolpyruvate; Pi, inorganic phosphate; PIC, phosphate transporter; Pyr, pyruvate; SFC, succinate/fumarate carrier; TCA cycle, tricarboxylic acid cycle; UCP, uncoupling proteins.

Source: Reproduced with permission from Linka and Weber, 2010.

before in this section, the electron transport system can exercise control at several points on the TCA cycle, mainly by controlling the ratio of reduced to non-reduced nucleotides. Thus, any mechanism, however simple, that alters the flow of electrons and/or the H^+ gradient across the mitochondrial inner membrane, can control the rate of respiration. Proteins found in a wide range of plant species, located in the inner mitochondrial membrane, serve to allow an influx of protons into the mitochondrial matrix and thus delink or 'uncouple' ATP synthesis from electron transport. These proteins are consequently referred to as uncoupling proteins or UCPs (Vercesi *et al.*, 2006). Six universally expressed, putative uncoupling protein genes (Borecky *et al.*, 2006) have been identified in *Arabidopsis* (*UCP 1–6*). The primary function of these proteins appears to be to modulate the strength of coupling between mitochondrial electron transport and ATP synthesis (Vercesi *et al.*, 2006; Jarmuszkiewicz *et al.*, 2010). This uncoupling is deemed to play a role in thermogenesis and in response to stress (Chen *et al.*, 2013) possibly due to the ROS stress involved in almost all forms of stress. These and other uncoupling proteins like the alternative oxidase (AOX) in particular, play an extremely important role of interfacing ROS stress and respiration. Their response is manifest as modulation of mitochondrial membrane potential (Jezek *et al.*, 1996) and consequently mitochondrial ROS production (Popov *et al.*, 2011).

Alternative oxidase: large-scale integration

Plant mitochondria are starkly different from their animal counterparts in having two terminal oxidases, one of which performs non-energy conserving oxidation of reduced substrates obtained from the TCA cycle. One of these terminal oxidases is the ubiquitous cytochrome c oxidase (complex IV), while the other is termed alternative oxidase or AOX (Juszczuk and Rychter, 2003). As this discussion progresses, it will become increasingly apparent that the term 'alternative' confers a somewhat secondary status to the AOX, which is possibly due to the fact that traditionally, energy conservation has been accorded primary importance. The AOX however, is an extremely important component of the plant mitochondrion and serves to integrate it not just within the cell but also within the plant as a whole organism. While complex IV deals with molecular oxygen, the sink for electrons flowing through the mitochondrial electron transport system, AOX serves to balance mitochondrial metabolism and fine tune it to the metabolic pace and demands of cellular metabolism as a whole (Moore *et al.*, 2013). This function gains even greater importance in times of biotic and abiotic stresses. Carbon metabolism, electron transport and the H^+ gradient generated ATP production represent a tightly coupled set of reactions that serve extremely well to conserve energy and to ensure metabolic integration of mitochondria, as discussed earlier. AOX reduces the level of this coupling and quite naturally, allows energy to 'leak' from the system. This leak could be used for thermogenesis or to moderate the generation of ROS by the electron transport chain (Møller, 2001). AOX works to couple the oxidation of ubiquinol to the reduction of molecular oxygen to water, short-circuiting the electron flow to avoid their passage through complexes III and IV. This completely avoids the H^+ translocation through these two complexes. Since the H^+ gradient generated is not as steep as in the absence of AOX activity, energy of downhill electron transport is not conserved in ATP synthesis and instead may result in thermogenesis also (Vanlerberghe, 2013).

However, thermogenesis is not a part of life for most plants. Therefore there must be another, compelling reason for the presence of AOX throughout the plant kingdom (McDonald and Vanlerberghe, 2006).

Being sessile, plants have to stay rooted and face all stresses that come their way. The capability to adapt has been central to the progress of plant life, much more than for animal life. To dwell briefly on the subject of stresses, most stresses, perhaps all, effect mitochondrial function in one way or another. Nutrient deficiencies, particularly those of micronutrients, may result in insufficient cofactors/prosthetic groups being produced, temperature stresses on both sides of best growth temperature bring about major changes in membrane fluidity as does drought and salinity stress. Folding and association of membrane complexes may be different from what is most appropriate. Heavy metals and xenobiotics often target electron transport systems and bring about major changes in effective stoichiometry of complexes left active. While this does not mean that heavy metal stresses affect mitochondria alone, they are certainly one of the major targets. When viewed in conjunction with the immense importance of the organelle to cellular metabolism, avoidance of stresses would be central to the effort of the plant to survive, to sustain itself and possibly to evolve. Mitochondrial AOX appears to be an extremely important player in the ability of the organelle to tide over stresses of various natures. This would also be a very strong reason for natural selection to have voted strongly in favour of AOX.

The enzyme is a dimeric diiron carboxylate protein (Maréchal A *et al.*, 2009, 2009) with the N-terminal of one monomer extending into the other and in this way being necessary for dimerization (Umbach and Siedow, 1993). A large hydrophobic region on one side of the dimer and a relatively hydrophilic region on the other side enable this enzyme to bind to the inner mitochondrial membrane in a way that embeds the diiron centre into the membrane in an interfacial fashion (Shiba *et al.*, 2013). Tyr-220, buried deep in the four-helix bundle of the enzyme, just 4.7 Å from the diiron centre, is the most likely candidate for the amino acid radical proposed in the AOX catalytic cycle. There is scant evidence for mitochondrial proteins interacting with AOX, but there is a distinct possibility of the enzyme existing as a multienzyme complex, primarily catalysing reducing equivalents generated by ubiquinone reductases like alternative NAD(P)H dehydrogenases (Kakizaki *et al.*, 2012). Recent research indicates that respiratory supercomplexes are affected to a significant extent by oxygen availability and pH of the mitochondrial matrix, among other intracellular factors (Ramirez-Aguilar *et al.*, 2011). Conserved Cys residues appear to contribute towards making AOX a more active, covalently linked dimer (Umbach and Siedow, 1993) and also sensitize it to regulation by α-keto acids (Umbach *et al.*, 2002).

Due to its ability and propensity to short-circuit mitochondrial electron transport towards a non-energy conserving mode, AOX can be said to play an antioxidant role in plant mitochondria. The enzyme is coded by the nuclear genome as sets of genes categorized as *AOX1* and *AOX2* (Considine *et al.*, 2002), with tissue and development-specific differential expression (Millar *et al.*, 2011). Far from playing this short-circuit, non-energy conserving role arbitrarily, the AOX is subject to intensive regulation (Vidal *et al.*, 2007). Regulation is exercised at the level of gene expression as well as activity of the mature protein. This is mainly accomplished by evaluating redox status of mitochondrial metabolism as a function of the membrane potential developed due to H^+ translocation during electron transport. Insufficient cytochrome pathway activity downstream of UQ reduction by complex I or II, inhibition of ATP synthase and of the TCA cycle or due to uncoupling of the electron transport chain, have all been known to bring about the induction of AOX (Vanlerberghe, 2013). Functional characterization of the *Arabidopsis AOX1a* promoter has identified a repressor element that could bind the transcription factor abscissic acid insensitive 4 (ABI4), which is a molecular component of the chloroplast retrograde signalling pathways as well as the link between AOX expression and the stress hormone ABA. This suggests a correlation between these pathways and also points towards the similar, endosymbiont origin of the two energy transducing organelles. AOX expression leads to lowering of reactive oxygen as well the related reactive nitrogen species concentration that is often seen to rise due to stress-induced changes in membrane dynamics. Levels of AOX expression and activity increase with the accumulation of TCA cycle intermediates and inhibition of the cytochrome pathway downstream of ubiquinone. Exogenous ROS and stresses such as drought, cold and salinity that change membrane dynamics also increase AOX activity. The mitochondrial permeability transition pore (MPTP) has also been implicated in the expression of AOX, such that blocking the pore leads to blocked AOX induction. Opening of the MPTP is known to be promoted by ROS and therefore appears to be an important step in AOX induction (Arpagaus *et al.*, 2002). Relationship between ROS generation and AOX expression was probed by over-expressing Mn-SOD in the matrix (Li *et al.*, 2013), resulting in lowering of superoxide (O_2^-). Plants over-expressing this enzyme showed a lesser accumulation of the AOX under conditions of stress known to increase AOX expression/activity. See Figure 7.2.

AOX, thus represents an optional safety feature of mitochondrial electron transport. A lesion in electron transport chain components downstream of UQH_2, any one or a combination of the myriad stress conditions – biotic and abiotic – that plants are often subject to, in fact anything or event that perturbs respiratory electron transport, leads to at least a portion of the chain becoming excessively reduced. This leads to an increase in ROS and consequently RNS production, leading to over-expression of AOX. The resultant uncoupling decreases the level of reduction of the electron transport chain and helps tide over oxidative stress.

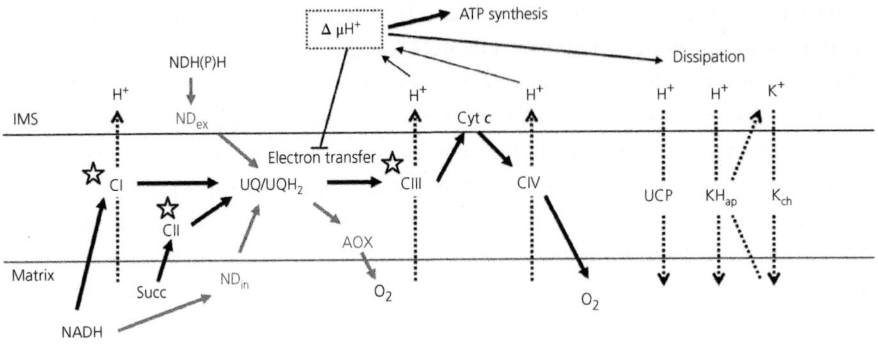

Figure 7.2 Schematic representation of the electron transport pathway in plant mitochondria. ND_{ex} and ND_{in} are the exterior (IM space) facing and interior (matrix) facing NADH dehydrogenases. AOX (alternative oxidase) and UCP (uncoupling protein) delink mitochondrial electrotransport from ATP synthesis and thus dissipate energy. AOX does this by catalysing the reduction of O_2 to H_2O, oxidizing UQH_2 in the process, while the UCP dissipates the H^+ gradient by allowing them passage back into the matrix. ND_{ex}, ND_{in}, AOX and the UCP are molecules exclusive to plant mitochondria.

Source: Reproduced with permission from Atkin and Macherel, 2009.

Microcompartmentation of metabolism: the metabolons

Eukaryotic cells have managed to deal with greatly complicated and diverse metabolic reactions through a process of macrocompartmentation, that involves pathways, or parts of these to be confined to the space limited by the bounding membrane of the organelle. This allows pathways to operate under conditions of pH and ionic strength that best suit them and to optimise metabolite and enzyme concentrations in a temporally coordinated manner. Equally significant is the stoichiometric exchange of critical metabolites across the bounding membrane, that coordinates metabolic reactions in different compartments, and integrates organellar metabolism in the big picture of cellular metabolism. This has been discussed in some detail in the earlier section 'Interfacing mitochondrial metabolism with light: phytochrome-mediated regulation of respiratory metabolism'. Paradoxically, one does run into an apparently logic-defying distribution of pathways in more than one compartment (Lunn, 2007). The distribution of biotin and ascorbate, beginning in the cytosol and terminating in the mitochondrion is one such example (Rebeille *et al.*, 2007). Possibly, this distribution and the duplication found therein, owes its origin to the diverse and scattered origin of various pathways, particularly those resident in organelles of endosymbiont origin (Lunn, 2007; Sweetlove and Fernie, 2013). Additionally, extensive duplication is also thought to have occurred due to horizontal gene transfer.

Macrocompartmentation (organellar localization) of discrete pathways, or significant portions of them, was until recently thought to be the only level of organization of metabolism. Each water compartment, represented by the aqueous matrix of organelles was thought to be distinct from any other compartment, but generally homogenous in composition within itself. This view has been changing, slowly, but very surely and the 'homogenous solution' view is giving way to the vision of a highly structured cytoplasmic matrix having localized, intracellular microenvironments with very few truly 'free' proteins in solution (Srere, 2000; Gierasch and Gershenson, 2009). In the mitochondrial context, there are possibilities that the enzymes of the TCA cycle are bound to the inner mitochondrial membrane via interactions with the respiratory chain complexes (Wang *et al.*, 2010), which are themselves known to be organized into respiratory supercomplexes of different composition (Dudkina *et al.*, 2010). Clustering of enzymes brings with it the advantage of substrate channelling, where metabolites are handed over between enzymes instead of diffusing in bulk water (Srere, 1987). In *Arabidopsis thaliana*, up to 10% of cytosolic isoforms of each glycolytic enzyme are clustered on the mitochondrial outer membrane (Kim *et al.*, 2006; Mustroph *et al.*, 2007). This could represent a mechanism to provide pyruvate close to the mitochondria so that the TCA cycle is fuelled efficiently. Graham *et al.* (2007) investigated the functional significance of the partitioning of glycolytic enzymes to favour mitochondria. It was determined that the partitioning of these enzymes to the surface of mitochondria was dependent on the demand of pyruvate inside mitochondria, as a chemically induced increase or decrease in respiratory rates brought about a similar increase or decrease in the amount of outer membrane associated glycolytic enzymes, supporting the possibility of substrate channelling.

That the cytosolic component of the respiratory pathway senses the demand of metabolites made by the TCA cycle and supports it by substrate channelling at appropriate rates is a major cellular mechanism to support mitochondrial metabolism and integrate it into the broader framework of cellular metabolism. The organizing principle behind microcompartmentation does not appear to be limited to large-scale sequestration within an organelle. It now appears to extend to organization of enzymes into 'metabolons' that act as biochemical force multipliers and bring about metabolic integration on a dynamic as well as a temporal scale. It would not be an exaggeration to propose a vectorial generation as well as transport of small molecule metabolites between pathways to bring about metabolic efficiency as well as to sensitize organelles like mitochondria to the metabolic demands of the cell. See Figure 7.3.

Organization and positioning of mitochondria in the cell

Mitochondria are extremely dynamic organelles in every sense of the word. They can exist as small, oblong organelles one moment, fusing to become one massively reticulate organelle the next and possibly undergo fission to smaller mitochondria once again. They can also undergo positioning changes that appear anything but random. Riding microtubule tracks on motor proteins or moving in association with actin, mitochondria locate themselves in different parts of the cell. The mechanisms are becoming increasingly clearer while the purpose is still

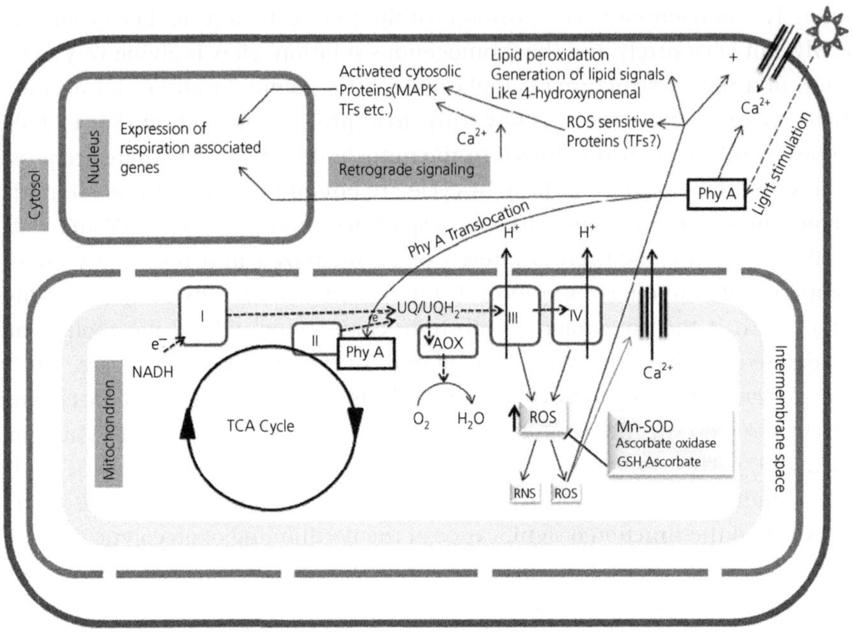

Figure 7.3 Diagrammatic representation of factors controlling mitochondrial metabolism. Electron flow through mitochondrial electron transport system generates ROS mainly through complexes III and IV. AOX activity diverts electron flow and reduces ROS production while scavenging enzymes remove ROS. The small amount of ROS left, serves to signal though oxidation products as well as by altering the Ca^{2+} signature of the cell. This translates into retrograde signalling from mitochondria to nucleus and modifies gene expression to suit the prevailing metabolic situation. The phytochrome relates complex II activity to light cues in light-exposed parts of the plant.

shrouded in mystery, with several competing explanations for what is achieved by repositioning. The explanations for mitochondrial fission and fusion are a lot more coherent and connect the organelles' morphology to redox changes, stress and to changes in metabolic demands brought about by energy-intensive events like the initiation of the cell cycle.

Mitochondrial dynamics: fission and fusion as agents of change

The traditional, textbook illustrations of mitochondria lead us to believe that these organelles are small and individual in existence. These illustrations and photographs came from electron microscopy studies and revealed many a secret (Ernster and Schatz, 1981), stemmed from the static nature of the technique and represented only one stage of the existence of mitochondria inside cells. These furiously dynamic organelles, whose appearance may range from that mentioned earlier, to a large, extensively interconnected, membrane-bound tubular network (Bereiter-Hahn, 1990; Jakobs *et al.*, 2003), are in fact, anything but static. Mitochondria have now come across as highly dynamic organelles that can undergo morphological alterations, brought about largely by fission and fusion (Detmer and Chan, 2007; Rafelski, 2013).

The shape and dimensions of mitochondria are cell type dependent and are strongly affected by environmental conditions (Kuznetsov and Margreiter, 2009). Mitochondria have long been known to divide or undergo fission and yield morphologically distinct, spherical structures in quiescent cells (Skulachev, 2001). In doing so, they differ to a very large extent from another organelle of endosymbiont origin, the chloroplasts. Whereas chloroplasts have retained the bacterial machinery used for division, mitochondria have evolved one that is radically different as far as the molecules involved are concerned (Osteryoung and Nunnar, 2003). On the one hand, in rapidly dividing and metabolically active cells, mitochondria fuse to form extensively interconnected networks (Collins *et al.*, 2002; Westermann, 2010). While the studies on mitochondrial dynamics were initiated on yeast or animal cells, lately, plant mitochondria seem to have caught up. Mitochondrial fusion has been observed in plant species and the time course of fusion has even been followed using the switchable fluorescent protein Kaede (Arimura *et al.*, 2004).

The identity of molecular mediators of mitochondrial fusion in plants is still unclear, but at least a few proteins involved in mitochondrial fission have been identified in *Arabidopsis thaliana.* DRP3A and DRP3B are two dynamin-related proteins known to mediate mitochondrial fission in *Arabidopsis thaliana.* These proteins were later shown to be functionally redundant having incomplete overlaps in their function (Fujimoto *et al.*, 2009). BIGYIN, an orthologue of yeast mitochondrial Fis1 (fission 1) and ELM1 or elongated mitochondria 1 (Arimura and Tsutsumi, 2002; Logan *et al.*, 2004) are other proteins related to the mitochondrial fission–fusion cycle. The primary objective of fission and fusion is probably the optimization of respiratory metabolism to keep it in sync with the energy demands of the cell (Twig *et al.*, 2008). Mitochondria possess the ability to change the level of facilitation of metabolite exchange through dynamic changes and positioning; something we have learnt from animal mitochondria (Nakada *et al.*, 2001; Ono *et al.*, 2001). The maintenance of a good population of healthy mitochondria is also heavily dependent on mitochondrial fusion (Campello and Scorrano, 2010) in the face of the damage caused routinely due to the generation of reactive oxygen. Fusion also brings about a rapid mixing of outer membrane proteins and matrix, along with a slow and rather limited mixing of the components of the inner mitochondrial membrane (Wilkens *et al.*, 2013). Mitochondrial fission thus ensures proper distribution of mitochondria throughout the cell and allows for local demands for ATP to be met or surpassed. It also allows for damaged portions of mitochondria to be removed and disposed of through mitophagy. Mitochondrial fusion, on the other hand, allows exchange or complementation of genetic material as well as mitochondrial functional proteins (Otera *et al.*, 2013).

Recently, Chlystun *et al.* (2013) have reported a role for Ca^{2+}-binding proteins called annexins in mitochondrial morphogenesis. Although the study has been conducted in fibroblasts, there is every chance that the same may occur in plant cells also since the calcium uptake, binding dynamics and the presence of annexins bear similarity with animal systems. Annexin A6 was found to be instrumental in modulation of Ca^{2+} signalling in the cytosol, promoting mitochondrial fragmentation, impairing respiration and causing elevation of mitochondrial membrane potential. Plant annexins form a significant subset of calcium sensors out of a larger set of plant calcium-binding proteins (Reddy *et al.*, 2004) and are also known to mediate an interaction between ROS and calcium flux, linking these two important regulatory forces in plants (Laohavisit *et al.*, 2010). The link between annexin function and mitochondrial dynamics has not been shown in plants yet, but evolutionary adaptation has been shown to diversify annexin molecular structures as well as their interactions and functional roles in membrane and cytoskeletal associations (Clark *et al.*, 2012; Sheahan *et al.*, 2005). See Figure 7.4.

Mitochondrial positioning: optimization of mitochondrial metabolism by spatial organization

Mitochondria are observed to move vectorially within cells, that is, their movements appear to be directed to a particular space within the cell. These movements take place along actin filaments or along microtubules, in association with motor proteins like kinesin (Frederick and Shaw, 2007; Logan, 2010). Live cell imaging shows some of these movements to be extremely complex contortions. The question that arises is 'How important really are mitochondrial movements?' The importance of mitochondrial positioning appears to be related to the level of polarity of a cell. Regions of the cell actively or even potentially engaged in endocytosis or exocytosis have a very high ATP or GTP demand associated with that region and would benefit from a greater number of mitochondria being associated with that space in the cell. This was found to be the case in neurons, a highly polarized cell type, where the site of the synapse is associated with extremely high rates of exocytosis (Verstreken *et al.*, 2005; Hollenbeck and Saxton, 2005). Similarly, the site of bud formation in budding yeast is a region to which mitochondria from the mother cell are targeted (Simon *et al.*, 1995). Cytoskeletal elements, actin microfibrils and microtubules, are known to play an active role in the movement and positioning of plant cell organelles like chloroplasts (Kadota and Wada, 1992; Kandasamy and Meagher, 1999), mitochondria (Van Gestel *et al.*, 2002), nuclei (Chytilova *et al.*, 2000), peroxisomes (Jedd and Chua, 2002; Mathur *et al.*, 2002), endoplasmic reticulum and the Golgi body (Boevink *et al.*, 1998). Although the functional significance of organelle movements in plant cells is not clearly established, its ubiquitous presence throughout the plant kingdom points towards the essentiality of this process. Organelle movements are effected by environmental stimuli like

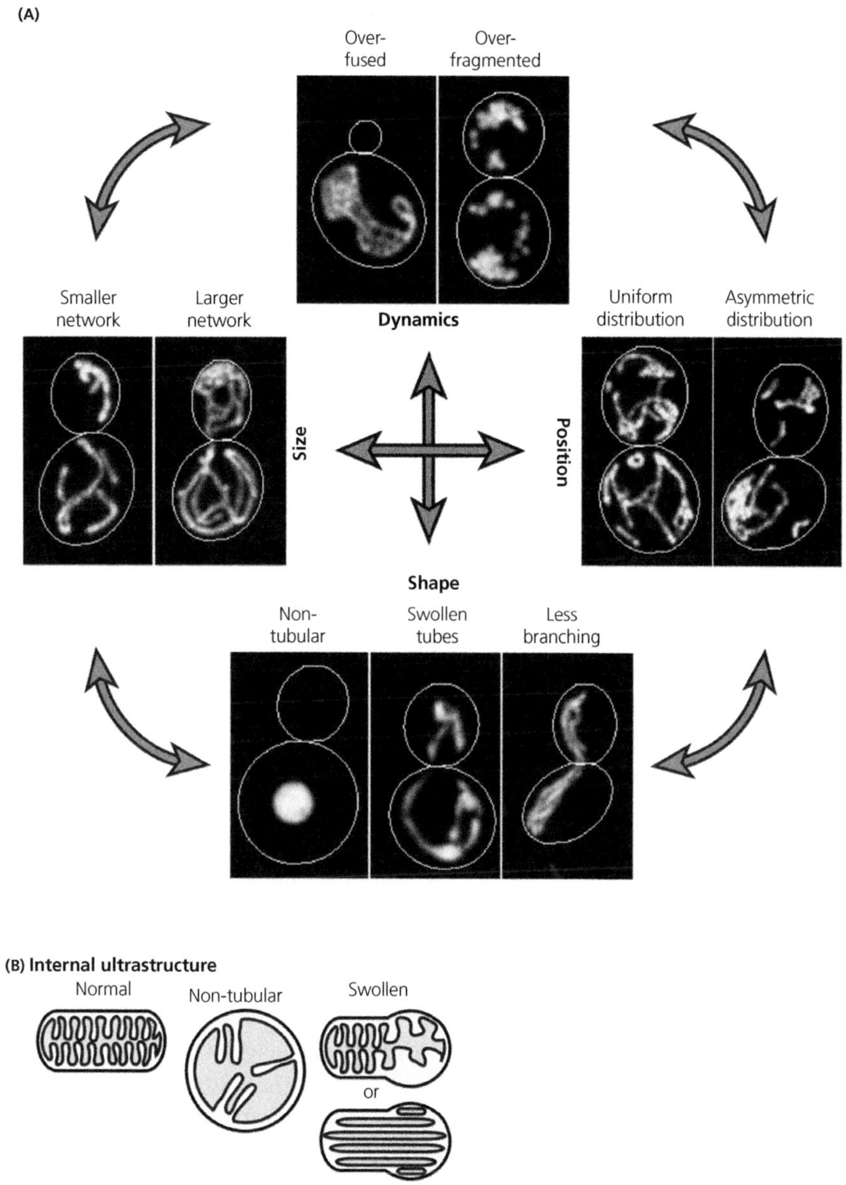

Figure 7.4 (A) Shows variations in morphology and positioning of mitochondria in yeast, tagged by a fluorescent protein. It is apparent that mitochondria undergo remarkable changes from being highly fragmented (top) to existing as organelles fused to different degrees. Along with these changes in mitochondrial dynamics, they also undergo positioning changes (middle right). (B) Diagrammatic representation of the ultrastructural changes in mitochondria accompanying dynamic changes.

Source: Reproduced from Rafelski, 2013.

drought, salinity, light, nutrient deficiency, temperature and physical stresses (Britz, 1979; Nagai, 1993; Wada *et al.*, 2003).

As light is essential for plant life, it is not surprising that organelle positioning is strongly affected by it. Chloroplasts harvest light energy and fix carbon, primarily in the form of triose phosphates. However, excess light can damage the photosynthetic system and lead to photoinhibition. The high-light situation can quickly deteriorate to a life-threatening situation due to excessive production of ROS from the energy-overloaded photosynthetic electron transport system. Chloroplasts have evolved a sophisticated, multi-pronged strategy for avoidance of light-induced damage to the photosynthetic system (photoinhibition), of which chloroplast positioning is an important component. The main agent of positioning events is the protein chloroplast unusual positioning 1 (CHUP1) that promotes actin polymerization along with the front moving end of the chloroplast and leads to positional changes (Wada, 2013). More recently (Kong *et al.*, 2013), it has been confirmed that a chloroplast-specific subset of actin, comprising short actin filaments termed cp-actin, undergo rapid severing and motility under the influence of phototropins, mainly of PHOT1.

For mitochondria, movements supplement the previously discussed dynamic behaviour in keeping mitochondrial metabolism in tune with that of the cell in general. Recently, it has been observed in mitochondria from cotyledons of germinating mung bean (*Vigna radiata*) seeds, that as germination commences, actin is imported into cotyledon mitochondria (Lo *et al.*, 2011). This import is in concert with the conversion of quiescent mitochondria to metabolically active ones. Actin is found localized in the intermembrane space, inner membrane, matrix and contact sites. Interestingly, treatment with latrunculin B, an actin depolymerizing agent, resulted in lowering of membrane potential and release of cytochrome *c*, suggesting a relationship between actin import and control of mitochondrial metabolism as well as programmed cell death. Since cotyledons represent organs that are destined to die after mobilization of reserves has completed, this mechanism could represent at least one of the mechanisms for bringing about the demise of the cotyledons. Mitochondrial actin is also believed to be connected to mtDNA through the mitochondrial protein complex, the mitochore, on one end and to cytosolic actin on the other (Boldogh *et al.*, 2003). This function may have implications in mtDNA inheritance. Two more components of the contact sites, specialized structures found at junctions of the outer membrane and the inner boundary membrane (Harner *et al.*, 2011), include an actin depolymerizing factor, VDACs (voltage dependent anion channels, also called porins), and adenine nucleotide translocator (ANT). These discoveries provide strong circumstantial evidence for dynamic actin–mitochondrion associations.

Like the interaction of actin with mitochondria, strong evidence exists for the interaction between mitochondria and kinesins, the motor proteins found associated with the mitochondria. Yang *et al.* (2011) report a specific interaction between a plant-specific kinesin, Kinesin KP1 and VDAC3, one of the VDACs found in the outer mitochondrial membrane. This kinesin-like protein (KP1 or At KIN14h) from germinating *Arabidopsis thaliana* seeds, was found to localize to mitochondria via its tail domain and interact specifically with the VDAC3 protein present in the outer mitochondrial membrane under conditions of low temperature. Seeds from *kp1* and *vdac3* mutants had increased oxygen consumption, imbalanced energy conserving as well as alternative pathways and low ATP levels, indicating that both proteins were involved in regulating respiration, especially at low temperature. It is possible that the KP1–VDAC3 interaction results in mitochondrial repositioning and brings about fundamental changes in mitochondrial metabolism. The inherent difficulties in studies involving mitochondrial repositioning have been extensively reviewed (Wada, 2013) and a model proposed for a functional cooperation between different motor proteins in mitochondrial repositioning (Cai and Cresti, 2012). Few plant cells are as starkly polarized as neurons in animal cell. In neurons, mitochondria are primarily found to move associated with microtubule tracks (Frederick and Shaw, 2007) and specific motor as well as adapter molecules have been identified. However, most plant cells do not display the extent of polarization or even the physical extension similar to neurons. Studies with elongating cultured tobacco cells also indicate that the primary responsibility of moving mitochondria rests on actin filaments and myosin motor proteins, while their positioning in the cortical cytoplasm is dependent on F-actin as well as on microtubules (Van Gestel *et al.*, 2002).

That mitochondria move and are positioned intentionally is established. The mechanisms and processes responsible for this appear to be different from those in animal cells, with actin turnover and myosin attachment being more rapid forms of transport than kinesin-mediated transport on microtubules, the latter being more important for positioning (Zheng *et al.*, 2009). However, kinesins are more specific for the organelle they transport while myosins do not display such specificity (Romagnoli *et al.*, 2007). Precisely what mitochondria do achieve by movement and positioning is difficult to state with accuracy. That the two processes are essential has been proven by studies that disrupted movement and positioning and severely compromised cellular function. One reason that presents itself as a result of comparisons between plant and animal cells could be the removal of damaged or non-functional mitochondria from the chondriome (Logan, 2010).

Metabolite exchange with physical inter-organelle contact also presents itself as a very important reason for moving mitochondria. Optimal use of oxygen gradients prevailing inside the cell might be another reason, but is proving to be a difficult problem to solve. Identifying subcellular locations for metabolites and enzymes is a daunting task to say the least. It is becoming increasingly apparent, however, that metabolism is spatially organized at a level finer than mere organellar sequestration of pathways and sets of metabolites (Sweetlove and Fernie, 2013). Perhaps the optimal use of this spatial organization of metabolic processes is one of the major reasons for mitochondrial movement and positioning.

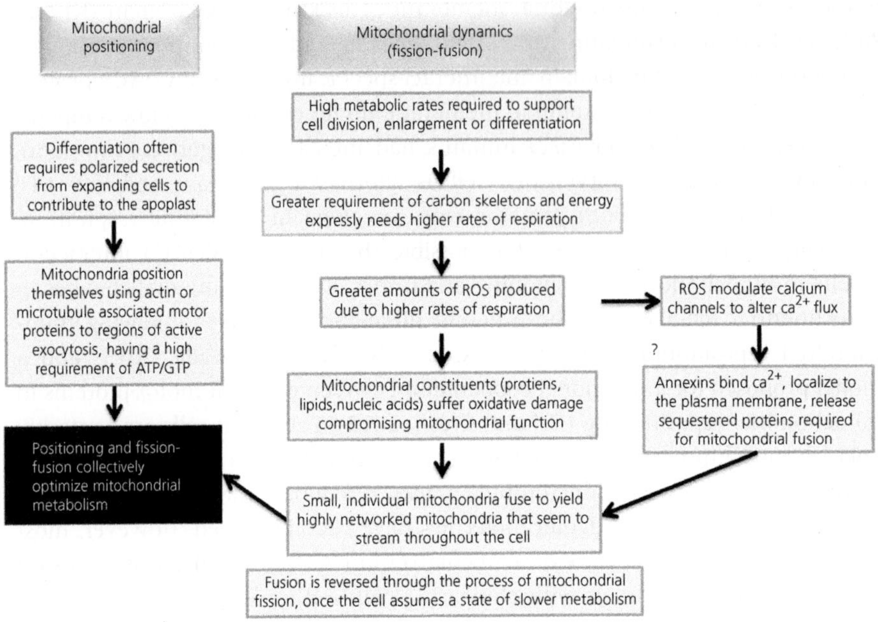

Figure 7.5 Mitochondrial positioning and dynamics optimize mitochondrial function for different levels of metabolic requirement such as that existing between non-dividing, differentiated cells and rapidly dividing or expanding cells. Under conditions of intense respiratory activity characteristic of dividing cells, mitochondrial fusion serves to complement gene and protein function in the face of oxidative damage. Once the massive demand for energy and carbon skeletons is over, mitochondrial fission and repositioning occur. The cues and mechanisms are yet to be defined with authority.

Needless to say, this extremely active area of research thoroughly deserves the attention it is currently getting and will be a goldmine of information in coming years. See Figure 7.5.

Concluding remarks

Mitochondria have several functions that are of utmost importance to the cell. It is therefore of equal importance that mitochondria be functionally integrated into the metabolic framework of the cell to optimize its function and support the cell in its range of endeavours. These functions could range from the relatively inactive life of a terminally differentiated cell to the intense metabolic activity associated with cell division or differentiation. Mitochondrial activity has to respond to cellular demands very rapidly. This requirement of rapid response requires that mitochondria are always kept in a metabolic state fine-tuned to the metabolic status of the cell. In addition to the range of activities found in animal cells, the life of a plant cell is guided as well as fuelled by light, directly or indirectly, presenting mitochondrial metabolism with an additional cue to respond to. The sessile nature of plant habit calls for greater adaptability to the fluctuating conditions of nature and again, mitochondrial metabolism has to rise to the occasion. Plant mitochondria, for these reasons, have additional features not seen in animal mitochondria and the presence of these features translates into complex mechanisms of metabolic integration. External environmental factors, most importantly light, control mitochondrial metabolism as a circuit parallel to other aspects of plant cell metabolism. Control of mitochondrial metabolism by light is a case in point, with phytochrome-mediated control over mitochondrial metabolism being more or less independent of other aspects of phytochrome action. Other control mechanisms stand out in stark contrast to this, with control being exercised by complex, closed but live circuits of signalling. Stoichiometric metabolite exchanges across the mitochondrial inner membrane are a prime example of such control. The decision to export carbon skeletons or ATP, taken qualitatively as well as quantitatively, is perhaps the greatest point of metabolic integration. Redox signalling, including a very significant amount of signalling mediated through respiratory electron transport spin-offs like ROS and RNS, actively links mitochondrial metabolism to that outside mitochondria. Mitochondria modify cellular Ca^{2+}, a well-established signalling element, through a process of reversible storage modifying the Ca^{2+} signature spatiotemporally. Apart from this, ROS–Ca^{2+} crosstalk brings about even tighter integration of mitochondrial metabolism. The control of mitochondrial metabolism through condition-dependent organization of glycolytic metabolons on the mitochondrial membrane, driving substrate channelling to fuel the TCA cycle and ABA-mediated control of AOX represent two additional points by which mitochondria sense cytosolic demands.

Mitochondrial fission and fusion are linked to the status of metabolic activity of the cell, with relatively inactive or quiescent cells favouring distributed mitochondria over the massively networked mitochondria favoured by differentiating or dividing cells. What links the fission–fusion cycle of mitochondria to metabolic demands is still not clear and represents a very active area of research. The same can be said for mitochondrial positioning in response to internal and external cues. It is definite that mitochondrial dynamics and positioning are intimately linked to the metabolic status/demands of the cell, but a lot more needs to be done to unravel the mechanisms and the control logic of these events.

References

Allen, J.F. (1993) Control of gene expression by redox potential and the requirement of chloroplast and mitochondrial genomes. *Journal of Theoretical Biology* **282**: 33583–33592.

Allen, J.F. (2003) The function of genomes in bioenergetics organelles. *Philosophical Transactions of the Royal Society of London B Biological Science* **358**: 19–38.

Arimura, S. and Tsustumi, N. (2002) A dynamin like protein ADL2b, rather than FtsZ, is involved in *Arabidopsis* mitochondrial division. *Proceedings of the National Academy of Sciences USA* **99**: 5727–5731.

Arimura, S., Yamamoto, J., Aida, J.P. *et al.* (2004) Frequent fission and fusion of plant mitochondria with unequal nucleoid distribution. *Proceedings of the National Academy of Sciences USA* **101**: 7805–7808.

Arpagaus, S., Rawyler, A. and Braendle, R. (2002) Occurrence and characteristics of the mitochondrial permeability transition in plants. *Journal of Biological Chemistry* **277**: 1780–1787.

Arsovski, A.A., Galstyan, A., Guseman, J.M. and Nemhauser, J. (2012) Photomorphogenesis. The Arabidopsis Book. e0147. doi: 10.1199/tab.0147.

Atkin, O.K. and Macherel, D. (2009) The crucial role of plant mitochondria in orchestrating drought tolerance. *Annals of Botany* **103**: 581–597.

Bae, G. and Choi, G. (2008) Decoding of light signals by plant phytochromes and their interacting proteins. *Annual Review of Plant Physiology* **59**: 281–311.

Balk, J. and Pilon, M. (2011) Ancient and essential: The assembly of iron-sulfur clusters in plants. *Trends Plant Sci.* **16**: 218–226.

Barranco-Medina, S., Krell, T., Finkemeier, I. *et al.* (2007) Biochemical and molecular characterization of the mitochondrial peroxiredoxin PsPrxII F from *Pisum sativum*. *Plant Physiology and Biochemistry* **45**: 729–739.

Bartoli, C.G., Gómez, F., Martínez, D.E. and Guiamet, J.J. (2004) Mitochondria are the main target for oxidative damage in leaves of wheat (*Triticum aestivum* L.). *Journal of Experimental Botany* **55**: 1663–1669.

Bassani, M., Neumann, P.M. and Gepstein, S. (2004) Differential expression profiles of growth-related genes in the elongation zone of maize primary roots. *Plant Molecular Biology* **56**: 367–380.

Baucher, M., Lowe, Y.O., Vandeputte, O.M. *et al.* (2011) Ntann12 annexin expression is induced by Auxinin tobacco roots. *Journal of Experimental Botany* **62**: 4055–4065.

Bereiter-Hahn, J. (1990) Behavior of mitochondria in the living cell. *International Review of Cytology* **122**: 1–63.

Blackbourne, H.D. and Battey, N.H. (1993) Annexin mediated secretory vesicle aggregation in plants. *Physiologia Plantarum* **89**: 27–32.

Boevink, P., Oparka, K., Santa Crus, S. *et al.* (1998) Stacks on tracks: The plant Golgi apparatus traffics on an actin/ER network. *The Plant Journal* **15**: 441–447.

Boldogh, I., Nowakowski, D.W., Yang, H.C. *et al.* (2003) A protein complex containing Mdm10p, Mdm12p and Mmm1p links mitochondrial membranes and DNA to the cytoskeleton-based segregation machinery. *Molecular Biology of the Cell* **14**: 4618–4627.

Borecky, J., Nogueira, F.T., de Oliviera, K.A. *et al.* (2006) The plant energy-dissipating mitochondrial systems: depicting the genomic structure and the expression profiles of the gene families of uncoupling protein and alternative oxidase in monocots and dicots. *Journal of Experimental Botany* **57**: 849–864.

Britz, S.J. (1979) Chloroplast and nuclear migration. In *Encyclopedia of Plant Physiology, New Series*, vol. VII (W. Haupt and M.E. Fleinleib, eds), pp. 170–205. Springer-Verlag, Heidelberg.

Busi, M.V., Gomez-Lobato, M.E., Rius, S.P. *et al.* (2011) Effect of mitochondrial dysfunction on carbon metabolism and gene expression in flower tissues of *Arabidopsis thaliana*. *Molecular Plant* **4**: 127–143.

Bussemer, J., Chigri, F. and Vothknecht, U.C. (2009) *Arabidopsis* ATPase family gene 1-like protein is a calmodulin-binding AAA^+-ATPase with a dual localization in chloroplasts and mitochondria. *FEBS Journal* **276**: 3870–3880.

Butow, R.A. and Avadhani, N.G. (2004) Mitochondrial signalling: The retrograde response. *Molecular Cell* **14**: 1–15.

Cai, G. and Cresti, M. (2012) Are kinesins required for organelle trafficking in plant cells? *Frontiers in Plant Science* **3**: 170. DOI: 10.3389/fpls.2012.00170.

Campello, S. and Scorrano, L. (2010) Mitochondrial shape changes: orchestrating cell pathophysiology. *EMBO Reports* **11**: 678–684.

Carrie, C., Giraud, E. and Whelan, J. (2009) Protein import in organelles: Dual targeting of proteins into mitochondria and chloroplasts. *FEBS Journal* **276**: 1187–1195.

Carroll, A.D., Moyen, C. Van Kesteren, P. *et al.* (1998) Ca^{2+}, annexins and GTP modulate exocytosis from maize root cap protoplasts. *The Plant Cell* **10**: 1267–1276.

Catoni, E., Desimone, M., Hilpert, M. *et al.* (2003) expression pattern of a nuclear-coded mitochondrial arginine-ornithine translocator gene from Arabidopsis. *BMC Plant Biol.* **3**:1.

Cedel, R.E. and Roux, C.J. (1980a) Further characterization of the *in vitro* binding of phytochrome to a membrane fraction enriched for mitochondria. *Plant Physiology* **66**: 696–703.

Cedel, R.E. and Roux, C.J. (1980b) Modulation of a mitochondrial function by oat phytochrome *in vitro*. *Plant Physiology* **66**: 704–709.

Chen, S., Jin, W., Liu, A.-R. *et al.* (2013) Mitochondrial uncoupling protein silence is compromised in photosynthesis and redox poise. *Acta Physiologiae Plantarum* **35**: 2547–2558.

Chew, O., Whelan, J. and Millar, A.H. (2003) Molecular definition of the ascorbate-glutathione cycle in *Arabidopsis* mitochondria reveals dual targeting of antioxidant defenses in plants. *Journal of Biological Chemistry* **278**: 46869–46877.

Chigri, F., Flosdorff, S., Pilz, S. *et al.* (2012) The *Arabidopsis* calmodulin-like proteins AtCML30 and AtCML3 are targeted to mitochondria and peroxisomes respectively. *Plant Molecular Biology* **78**: 211–222.

Chlystun, M., Campanella, M., Law, A.-L. *et al.* (2013) Regulation of mitochondrial morphogenesis by Annexin A6. *PLoS One* **8**(1): e53774. DOI: 10.1371/ journal.pone.0053774.

Chytilova, E., Macas, J., Sliwinska, E. *et al.* (2000). Nuclear dynamics in *Arabidopsis thaliana*. *Molecular Biology of the Cell* **11**: 2733–2741.

Clapham, D.E. (2007) Calcium signaling. *Cell.* **131**: 1047–1058.

Clark, G.B., Lee, D.W., Dauwalder, M. and Roux, S.J. (2005) Immunolocalization and histochemical evidence for the association of two different *Arabidopsis* annexins with secretion during early seedling growth and development. *Planta* **220**: 621–631.

Clark, G.B., Morgan, R.O., Fernandez, M.-P. and Roux, S.J. (2012) Evolutionary adaptation of plant annexins has diversified their molecular structures, interactions and functional roles. *New Phytologist* **196**: 695–712.

Collins, T.J., Berridge, M.J., Lipp, P. and Bookman, M.D. (2002) Mitochondria are morphologically and functionally heterogenous within cells. *EMBO Journal* **21**: 1616–1627.

Considine, M.J., Holtzapffel, R.C., Day, D.A. *et al.* (2002) Molecular distinction between alternative oxidase from monocots and dicots. *Plant Physiology* **129**: 949–953.

Cvetkovska, M. and Vanlerberghe, G.C. (2012) Alternative oxidase modulates leaf mitochondrial concentrations of superoxide and nitric oxide. *New Phytologist* **195**: 32–39.

Cvetkovska, M. and Vanlerberghe, G.C. (2013) Alternative oxidase impacts the plant response to biotic stress by influencing the mitochondrial generation of reactive oxygen species. *Plant, Cell & Environment* **36**: 721–732.

De Stafani, D., Rafaello, A., Teardo, E. *et al.* (2011) A forty-kilodalton protein of the inner membrane is the mitochondrial calcium uniporter. *Nature* **476**: 336–340.

Detmer, S.A. and Chan, D.C. (2007) Functions and dysfunctions of mitochondrial dynamics. *Nature Reviews: Molecular Cell Biology* **8**: 870–878.

Dodd, A.N., Kudla, J. and Sanders, D. (2010) The language of calcium signaling. *Annual Review of Plant Biology* **61**: 593–620

Dudkina, N.V., Eubel, H., Keegstra, W. *et al.* (2005) Structure of a mitochondria supercomplex formed by respiratory-chain complexes I and III. *Proceddings of the National Academy of Sciences USA* **102**: 3225–3229.

Dudkina, N.V., Heinemeyer, J., Sunderhaus, S. *et al.* (2008) Respiratory chain supercomplexes in the plant mitochondrial membrane. *Trends in Plant Science* **11**: 232–240.

Dudkina, N.V., Kouril, R., Peters, K. *et al.* (2010) Structure and function of mitochondrial supercomplexes. *Biochemica et Biophysica Acta* **1797**: 664–670.

Eprintsev, A.T., Fedorin, D.N. and Igamberdiev, A.U. (2013) Ca^{2+} is involved in phytochrome A-dependent regulation of the succinate dehydrogenase gene *sdh1-2* in Arabidopsis. *Journal of Plant Physiology* **170**: 1349–1352. DOI: 10.1016/j.jplph.2013.04.006.

Ernster, L. and Schatz, G. (1981) Mitochondria: A historical review. *Journal of Cell Biology* **91**: 227s–255s.

Escobar, M.A., Franklin, K.A,, Svensson, A.S. *et al.* (2004) Light regulation of the *Arabidopsis* respiratory chain. Multiple discrete photoreceptor responses contribute to induction of type II NAD(P)H dehydrogenase genes. *Plant Physiology* **136**: 2710–2721.

Fedoroff, N. (2006) Redox regulatory mechanisms in cellular stress responses. *Annals of Botany* **98**: 289–300.

Fernie, A.R., Carrari, F. and Sweetlove, L.J. (2004) Respiratory metabolism: Glycolysis, the TCA cycle and mitochondrial electron transport. *Current Opinion in Plant Biology* **7**: 254–261.

Finnegan, P.M., Soole, K.L. and Umbach, A.L. (2004). Alternative mitochondrial electron transport proteins in higher plants. In *Plant Mitochondria: From Genome to Function* (D.A. Day, A.H. Millar and J. Whelan, eds), pp. 163–230. Kluwer Academic, Dordrecht, The Netherlands.

Frederick, R.L. and Shaw, J.M. (2007) Moving mitochondria: Establishing distribution of an essential organelle. *Traffic* **8**: 1668–1675.

Fujimoto, M., Arimura, S.-I., Mano, S. *et al.* (2009) *Arabidopsis* dynamin-related proteins DRP3A and DRP3B are functionally redundant in mitochondrial fission, but have distinct roles in peroxisomal fission. *The Plant Journal* **58**: 388–400.

Gawaz, M., Douglas, M.G. and Klingenberg, M. (1990) Structure–function studies of denine nucleotide transport in mitochondria. II. Biochemical analysis of distinct AAC1 and AAC2 proteins in yeast. *Journal of Biological Chemistry* **265**: 14202–14208.

Gerke, V. and Moss, S.E. (2002) Annexins: From structure to function. *Physiological Reviews* **82**: 331–371.

Gierasch, L.M. and Gershenson, A. (2009) Post-reductionist protein science, or putting Humpty Dumpty back together again. *Nature Chemical Biology* **5**: 774–777.

Giorgi, C., Stefani, D., Bononi, A. *et al.* (2009) structural and functional link between the mitochondrial network and the endoplasmic reticulum. *Int. J. Biochem. Cell Biol.***41**: 1817–1827.

Graham, J.W.A., Williams, T.C.R., Morgan, M. *et al.* (2007) Glycolytic enzymes associate dynamically with mitochondria in response to respiratory demand and support substrate channeling. *The Plant Cell* **19**: 3723–3738.

Gray, M.W., Burger, G. and Lang, B.F. (1999) Mitochondrial evolution. *Science* **283**: 1476–1481.

Griffiths, E.J. and Rutter, G.A. (2009) Mitochondrial calcium as a key regulator of mitochondrial ATP production in mammalian cells. *Biochimica et Biophysica Acta* **1787**: 1324–1333.

Haferkamp, I. (2007) The diverse family of mitochondrial carrier family in plants. *FEBS Letters* **581**: 2375–2379.

Haferkamp, I. and Schmitz-Esser, S. (2012) he plant mitochondrial carrier family: Functional and evolutionary aspects. *Frontiers in Plant Science* DOI: 10.3389/fpls.2012.00002.

Haferkamp, I., Hackstein, J.H., Voncken, F.G. *et al.* (2002) Functional integration of mitochondrial and hydrogenosomal ADP/ATP carriers in the *Escherichia coli* membrane reveals different biochemical characteristics from plants, mammals and anaerobic chytrids. *European Journal of Biochemistry* **269**: 3172–3181.

Hanning, I. and Heldt, H.W. (1993) On the function of mitochondrial metabolism during photosynthesis in spinach leaves (*Spinacia oleracea* L.). Partitioning between respiration and export of redox equivalents and precursors for nitrate assimilation products. *Plant Physiology* **103**: 1147–1154.

Harner, M., Körner, C., Walther, D. *et al.* (2011) The mitochondrial contact site complex, a determinant of mitochondrial architecture. *Embo Journal* **30**: 4356–4370.

Heimpel, S., Basset, G., Odoy, S. and Klingenberg, M. (2001) Expression of the mitochondrial ADP/ATP carrier in *Escherichia coli*: renaturation, reconstitution and effect of mutations on 10 positive residues. *Journal of Biological Chemistry* **276**: 11499–11506.

Hollenbeck, P.J. and Saxton, W.M. (2005) The axonal transport of mitochondria. *Journal of Cell Science* **118**: 5411–5419.

Hoyos, M.E., Palmieri, L., Wertin, T. *et al.* (2003) Identification of a mitochondrial transporter for basic amino acids in *Arabidopsis thaliana* by functional reconstitution into liposomes and complementation in yeast. *Plant Journal* **33**: 1027–1035.

Hughes, J. (2013) Phytochrome cytoplasmic signaling. *Annual Review of Plant Biology* **64**: 377–402.

Huh, S.M., Noh, E.K., Kim, H.G. *et al.* (2010) *Arabidopsis* annexins AnnAt1 and AnnAt4 interact with each other and regulate drought and salt stress responses. *Plant Cell Physiology* **51**: 1499–1514.

Igamberdiev, A.U. and Gardeström, P. (2003) Regulation of NAD- and NADP-dependent isocitrate dehydrogenases by reduction levels of pyridine nucleotides in mitochondria and cytosol of pea leaves. *Biochimica et Biophysica Acta* **1606**: 117–125.

Igamberdiev, A.U., Eprintsev, A.T., Fedorin, D.N. and Popov, V.N. (2013) Phytochrome-mediated regulation of plant respiration and photorespiration. *Plant, Cell & Environment* **37**: 290–299. DOI: 10.1111/pce.12155.

Jacoby, R.P., Li, L., Huang, S. *et al.* (2012) Mitochondrial composition, function and stress response in plants. *Journal of Integrative Plant Biology* **54**: 887–906.

Jakobs, S., Schauss, A.C. and Hell, S.W. (2003) Photoconversion of matrix targeted GFP enables analysis of continuity and intermixing of the mitochondrial lumen. *FEBS Letters* **554**: 194–200.

Jarmuszkiewicz, W., Woyda-Ploszczyca, A., Antos-Krzeminska, N. and Sluse, F.E. (2010) Mitochondrial uncoupling proteins in unicellular eukaryotes. *Biochimica et Biophysica Acta* **1797**: 792–799.

Jedd, G. and Chua, N.-H. (2002) Visualization of peroxisomes in living plant cells reveals actomyosin-dependent cytoplasmic streaming and peroxisome budding. *Plant Cell Physiology* **43**: 384–392.

Jezek, P., Hanus, J., Semrad, C. and Garlid, K.D. (1996) Photoactivated azido fatty acid irreversibly inhibits anion and proton transport through the mitochondrial uncoupling protein. *Journal of Biological Chemistry* **271**: 6199–6205.

Jiminez, A., Hernández, J.A., Pastori, G. *et al.* (1998) Role of ascorbate glutathione cycle of mitochondria and peroxisomes in the senescence of pea leaves. *Plant Physiology* **118**: 1327–1335.

Juszczuk, I.M. and Rychter, A.M. (2003) Alternative oxidase in higher plants. *Acta Biochimica Polonica* **50**: 1257–1271.

Kadota, A. and Wada, M. (1992). Photoorientation of chloroplasts in protonemal cells of the fern *Adiantum* as analyzed by use of a videotracking system. *Botanical Magazine, Tokyo* **105**: 265–279.

Kakizaki, Y., Moore, A.L. and Ito, K. (2012) Different molecular bases underlie the mitochondrial respiratory activity in the homeothermic spadices of *Symplocarpus renifolius*. *Biochemical Journal* **445**: 237–246.

Kandasamy, M.K. and Meagher, R.B. (1999). Actin-organelle interaction: Association with chloroplast in *Arabidopsis* leaf mesophyll cells. *Cell Motility and the Cytoskeleton* **44**: 110–118.

Kennedy, E.P. and Lehninger, A.L. (1949) Oxidation of fatty acids and tricarboxylic cycle intermediates by isolated rat liver mitochondria. *Journal of Biological Chemistry* **179**: 957–972.

Kim, M., Lim, J.H., Ahn, C.S. *et al.* (2006) Mitochondria-associated hexokinases play a role in the control of programmed cell death in *Nicotiana benthamiana*. *The Plant Cell* **18**: 2341–2355.

Klingenberg, M. (2008) The ADP and ATP transport in mitochondria and its carrier. *Biochimica et Biophysica Acta* **1778**: 1978–2021.

Kocsy, G., Tari, T., Vankova, R. *et al.* (2013) Redox control of plant growth and development. *Plant Science* **211**: 77–91.

Kong, S.-G., Arai, Y., Suetsugu. N. *et al.* (2013) Rapid severing and motility of chloroplast-actin filaments are required for the chloroplast avoidance response in *Arabidopsis*. *The Plant Cell* **25**: 572–590.

Kramer, D.M. and Evans, J.R. (2011) The importance of energy balance in improving photosynthetic productivity. *Plant Physiology* **155**: 70–78.

Kregel, K.C. and Zhang, H.J. (2007) An integrated view of oxidative stress in aging: Basic mechanisms, functional effects, and pathological considerations. *American Journal of Physiology. Regulatory, Integrative and Comparative Physiology* **292**: R18–R36.

Kuhn, S., Bussemer, G., Chigri, F. and Vothknecht, U.C (2009) Calcium depletion and calmodulin inhibition affect the import of nuclear-encoded proteins into plant mitochondria. *Plant Journal* **58**: 694–705.

Kuznetsov, A.V. and Margreiter, R. (2009) Heterogeneity of mitochondria and mitochondrial function within cells as another level of mitochondrial complexity. *International Journal of Molecular Science* **10**: 1911–1929.

Laohavisit, A., Brown, T.A., Cicuta, P. and Davies, J.M. (2010) Annexins: Components of the calcium and reactive oxygen signaling network. *Plant Physiology* **152**: 1824–1829.

Laohavisit, A., Mortimer, J.C., Demidchik, V. *et al.* (2009) *Zea mays* annexins modulate cytosolic free Ca^{2+} and generate a Ca^{2+}-permeable conductance. *The Plant Cell* **21**: 479–493.

Laude, A.J. and Simpson, A.W. (2009) Compartmentalized signaling: Ca^{2+} compartments, microdomains and the many facets of Ca^{2+} signaling. *FEBS Journal* **276**: 1800–1816.

Lee, S., Lee, E.J., Yang, E.J. *et al.* (2004) Proteomic identification of annexins, calcium-dependent membrane binding proteins that mediate osmotic stress and abscissic acid signal transduction in *Arabidopsis*. *The Plant Cell* **16**: 1378–1391.

Lin, C. (2002) Blue light receptors and signal transduction. *Plant Cell Supplement* **2002**: S207–S225.

Li, C.-R., Liang, D.-D., Li, J. *et al.* (2013) Unravelling mitochondrial retrograde regulation in the abiotic stress induction of rice *alternative oxidase 1* genes. *Plant, Cell & Environment* **36**: 775–778.

Linka, N. and Weber, A.P.M. (2010) Intracellular metabolite transporters in plants. *Molecular Plant* **3**: 21–53.

Lo, Y.-S., Cheng, N., Hsiao, L.-J. *et al.* (2011) Actin in mung bean mitochondria and implications for its function. *The Plant Cell* **23**: 3727–3744.

Logan, D.C. (2006) The mitochondrial compartment. *Journal of Experimental Botany* **57**: 1225–1243.

Logan, D.C. (2010) Mitochondrial fusion, division and positioning in plants. *Biochemical Society Transactions* **38**: 789–795.

Logan, D.C. and Knight, M.R. (2003) Mitochondrial and cytosolic calcium dynamics are differently regulated in plants. *Plant Physiology* **133**: 21–24.

Logan, D.C., Scott, I. and Tobin, A.K. (2004) ADL2a, like ADL2b, is involved in the control of higher plant mitochondrial morphology. *Journal of Experimental Botany* **55**: 783–785.

Loro, G., Drago, I., Pozzan, T. *et al.* (2012) Targeting of Cameleons to various subcellular compartments reveals a strict cytoplasmic/mitochondrial Ca^{2+} handling relationship in plant cells. *Plant Journal* **71**: 1–13.

Lunn, J.E. (2007) Compartmentation in plant metabolism. *Journal of Experimental Botany* **58**: 35–47.

Mackenzie, S. and McIntosh, L. (1999) Higher plant mitochondria. *The Plant Cell* **11**: 571–586.

de Marcos Lousa, C., Trezeguet, V., Dianoux, A.C. *et al.* (2002) The human mitochondrial ADP/ATP carriers: Kinetic properties and biogenesis of wild-type and mutant proteins in yeast *S. cerevisiae*. *Biochemistry* **41**: 14412–14420.

Maréchal, A., Kido, Y., Kita, K. *et al.* (2009) Three redox states of *Trypanosoma brucei* alternative oxidase identified by infrared spectroscopy and electrochemistry. *Journal of Biological Chemistry* **284**: 31827–31833.

Martin, W. (2010) Evolutionary origins of metabolic compartmentalization in eukaryotes. *Philosophical Transactions of the Royal Society of London* **365**: 847–855.

Mathur, J., Mathur, N. and Hülskamp, M. (2002) Simultaneous visualization of peroxisomes and cytoskeletal elements reveals actin and not microtubule-based peroxisome motility in plants. *Plant Physiology* **128**: 1031–1045.

Maurino, V.G. and Peterhansel, C. (2010) Photorespiration: Current status and approaches for metabolic engineering. *Current Opinion in Plant Biology* **13**: 249–256.

Mazars, C., Thuleau, P., Lamotte, O. and Bourque, S. (2010) Cross-talk between ROS and calcium in regulation of nuclear activities. *Molecular Plant* **3**: 706–718.

McAinsh, M.R. and Pittman, J.K. (2009) Shaping the calcium signature. *New Phytologist* **181**: 275–294.

McClung, C.R., Hsu, M., Painter, J.E. *et al.* (2000) Integrated temporal regulation of the photorespiratory pathway. Circadian regulation of two *Arabidopsis* genes encoding serine hydroxymethyltransferase. *Plant Physiology* **123**: 381–392.

McDonald, A.E. (2009) Alternative oxidase: What information can protein sequence comparisons give us? *Physiologia Plantarum* **137**: 328–341.

McDonald, A.E. and Vanlerberghe, G.C. (2006) Origins, evolutionary history and taxonomic distribution of alternative oxidase and plastoquinol terminal oxidase. *Comparative Biochemistry and Physiology Part D Genomics & Proteomics* **1**: 357–364.

Meyer, E.H., Geigé, P., Gelhaye, E. *et al.* (2005) AtCCMH, an essential component of the c-type maturation pathway in *Arabidopsis* mitochondria, interacts with apocytochrome c. *Proceedings of the National Academy of Sciences USA* **102**: 16113–16118.

Millar, A.H. and Leaver, C.J. (2000) The cytotoxic lipid peroxidation product, 4-hydroxy-2-nonenal, specifically inhibits decarboxylating dehydrogenases in the matrix of plant mitochondria. *FEBS Letters* **481**: 117–121.

Millar, A.H., Eubel, H., Jänsch, L. *et al.* (2004) Mitochondrial cytochrome *c* oxidase and succinate dehydrogenase complexes contain plant specific subunits. *Plant Molecular Biology* **56**: 77–90.

Millar, A.H., Whelan, J., Soole, K. and David, D.A. (2011) Organization and regulation of mitochondrial respiration in plants. *Annual Review of Plant Biology* **62**: 79–104.

Møller, I.M. (2001) Plant mitochondria and oxidative stress: Electron transport, NADPH turnover, and metabolism of reactive oxygen species. *Annual Review of Plant Physiology and Plant Molecular Biology* **52**: 561–591.

Møller, I.M. and Rasmusson, A.G. (1998) The role of NADP in the mitochondrial matrix. *Trends in Plant Science* **3**: 21–27.

Møller, I.M. and Sweetlove, L.J. (2010) ROS signaling-specificity is required. *Trends in Plant Science* **15**: 370–374.

Moore, A.L., Shiba, T., Young, L. *et al.* (2013) Unraveling the heater: New insights into the structure of the alternative oxidase. *Annual Review of Plant Biology* **64**: 637–663.

Möglich, A., Yang, X., Ayers, R.A. and Moffat, K. (2010) Structure and function of plant photoreceptors. *Annual Review of Plant Biology* **61**: 21–47.

Morohashi, Y. (1987) Phytochrome-mediated development of glycine oxidation by mitochondria in cucumber cotyledons. *Physiologia Plantarum* **70**: 46–50.

Morohashi, Y., Sasanuma, S. and Seto, T. (1993) Phytochrome-mediated control of respiratory activities of mitochondria in cotyledons of dark grown cucumber seedlings. *Physiologia Plantarum* **89**: 295–300.

Mortimer, J.C., Laohavisit, A., Macpherson, N. *et al.* (2008) Annexins: Multifunctional components of growth and adaptation. *Journal of Experimental Botany* **59**: 533–544.

Murphy, M.P. (2009) How mitochondria produce reactive oxygen species. *Biochemical Journal* **417**: 1–13.

Mustroph, A., Sonnewald, U. and Biemelt, S. (2007) Characterization of the ATP-dependent phosphofructokinase gene family from *Arabidopsis thaliana. FEBS Letters* **581**: 2401–2410.

Nagai, R. (1993) Regulation of intracellular movements in plant cells by environmental stimuli. *International Review of Cytology* **145**: 251–310.

Nakada, K., Inoue, K. and Hayashi, J. (2001) Interaction theory of mammalian mitochondria. *Biochemical and Biophysical Research Communications* **288**: 743–746.

Navrot, N., Collin, V., Gualberto, J. *et al.* (2006) Plant glutathione peroxidases are functional peroxiredoxins distributed in several subcellular compartments and regulated during biotic and abiotic stresses. *Plant Physiology* **142**: 1364–1379.

Noctor, G. (2006) Metabolic signaling in defence and stress: the central roles of soluble redox couples. *Plant, Cell & Environment* **29**: 409–425.

Nunes-Nesi, A., Araujo, W.L. and Fernie, A.R. (2011) Targeting mitochondrial metabolism and machinery as a means to enhance photosynthesis. *Plant Physiology* **155**: 101–107.

Ono, T., Isobe, K., Nakada, K. and Hayashi, J.I. (2001) Human cells are protected from mitochondrial dysfunction by complementation of DNA products in fused mitochondria. *Nature Genetics* **28**: 272–275.

Osteryoung, K.W. and Nunnar, J. (2003) The division of endosymbiotic organelles. *Science* **302**: 1698–1704.

Otera, H., Ishihara, N. and Mihara, K. (2013) New insights into the function and regulation of mitochondrial fission. *Biochimica et Biophysica Acta* **1833**: 1256–1268.

Palmieri, L., Arrigoni, R., Blanco, E. *et al.* (2006a) Molecular identification of an *Arabidopsis S*-adenosylmethionine transporter: Analysis of organ distribution, bacterial expression, reconstitution into liposomes, and functional characterization. *Plant Physiology* **142**: 855–865.

Palmieri, L., Todd, C.D., Arrigoni, R. *et al.* (2006b) *Arabidopsis* mitochondria have two basic amino acid transporters with partially overlapping specificities and differential expression in seedling development. *Biochimica et Biophysica Acta* **1757**: 1277–1283.

Palmieri, L., Picault, N., Arrigoni, R. *et al.* (2008a) Molecular identification of three *Arabidopsis thaliana* mitochondrial dicarboxylate carrier isoforms: Organ distribution, bacterial expression, reconstitution into liposomes and functional characterization. *Biochemical Journal* **410**: 621–629.

Palmieri, L., Santoro, A., Carrari, F. *et al.* (2008b) Identification and characterization of ADNT1, a novel mitochondrial adenine nucleotide transporter from *Arabidopsis. Plant Physiology* **148**: 1797–1808.

Patergnani, S., Suski, J.M., Agnoletto, C. *et al.* (2011) Calcium signaling around mitochondria associated membranes (MAMs). *Cell Communication and Signaling* **9**: 19. DOI: 10:1186/1478-811X-9-19.

Perocchi, F., Gohil, V.M., Girgis, H.S. *et al.* (2010) MICU1 encodes a mitochondrial EF hand protein required for Ca^{2+} uptake. *Nature* **467**: 291–296.

Picault, N., Hodges, M., Palmieri, L. and Palmieri, F. (2004) The growing family of mitochondrial carriers in *Arabidopsis. Trends in Plant Science* **9**: 138–146.

Picault, N., Palmieri, L., Pisano, I. *et al.* (2002) identification of a novel transporter for dicarboxylates and tricarboxylates in plant mitochondria: Bacterial expression, reconstitution, functional characterization and tissue distribution. *Journal of Biological Chemistry* **277**: 24204–24211.

Popov, V.N., Eprintsev, A.T. and Maltseva, E.V. (2011) Activation of genes encoding mitochondrial proteins involved in alternative and uncoupled respiration of tomato plants treated with low temperature and reactive oxygen species. *Russian Journal of Plant Physiology* **58**:914–920.

Pratt, R.D., Ferreira, G.C. and Pedersen, P.L. (1991) Mitochondrial phosphate transport. Import of the H+/Pi symporter and role of the presequence. *Journal of Biological Chemistry* **266**: 1276–1280.

Rafelski, S.M. (2013) Mitochondrial network morphology: Building an integrative, geometrical view. *BMC Biology* **11**: 71. http://www.biomedcentral.com/1741-7007/11/71 (accessed October 2014).

Ramirez-Aguilar, S.J., Keuthe, M., Rocha, M. *et al.* (2011) The composition of plant mitochondrial supercomplexes changes with oxygen availability. *Journal of Biological Chemistry* **286**: 43045–43053.

Rausch, C. and Bucher, M. (2002) Molecular mechanism of phosphate transport in plants. *Planta* **216**: 23–37.

Rebeille, F., Alban, C., Bourguignon, J. *et al.* (2007) The role of plant mitochondria in the biosynthesis of coenzymes. *Photosynthesis Research* **92**: 149–162.

Requejo, R., Hurd, T.R., Costa, N.J. and Murphy, M.P. (2010) cysteine residues exposed on protein surfaces are the dominant intramitochondrial thiol and may protect against oxidative damage *FEBS Journal* **277**: 1465–1480.

Reddy, V.S., Day, I.S., Thomas, T. and Reddy, A.S.N. (2004) KIC, a novel Ca^{2+} binding protein with one EF-hand motif, interacts with a microtubule motor protein and regulates trichome morphogenesis. *The Plant Cell* **16**: 185–200.

Ribas-Carbo, M., Giles, L., Flexas, J. *et al.* (2008) Phytochrome-driven changes in respiratory electron transport partitioning in soybean (*Glycine max* L.) cotyledons. *Plant Biol.* **10**: 281–287.

Romagnoli, S., Cai, G., Faleri, C. *et al.* (2007) Microtubule- and actin filament-dependent motors are distributed on pollen tube mitochondria and contribute differently to their movement. *Plant Cell Physiology* **48**: 345–361.

Ryan, M.T. and Hoogenraad, N.J. (2007) Mitochondrial–nuclear communications. *Annual Review of Biochemistry* **76**: 701–722.

Sage, R.F., Sage, T.L. and Kocacinar, F. (2012) Photorespiration and the evolution of C4 photosynthesis. *Annual Review of Plant Biology* **63**: 19–47.

Schmidt, O., Pfanner, N. and Meisinger, C. (2010) Mitochondrial protein import: From proteomics to functional mechanisms. *Nature Reviews: Molecular Cell Biology* **11**: 655–667.

Schwacke, R., Flugge, U.I. and Kunze, R. (2004) Plant membrane proteome databases. *Plant Physiology and Biochemistry* **42**: 1023–1034.

Schwacke, R., Schneider, A., van der Graaff, E. *et al.* (2003) ARAMEMNON, a novel database for *Arabidopsis* integral membrane proteins. *Plant Physiology* **131**: 16–26.

Schwartz, R.M. and Dayhoff, M.O. (1978) Origins of prokaryotes, eukaryotes, mitochondria and chloroplasts. *Science.* **199**: 395–403.

Schwarzländer, M. and Finkemeier, I. (2013) Mitochondrial energy and redox signaling in plants. *Antioxidants & Redox Signaling* **18**: 2122–2144.

Schwarzländer, M., Fricker, M.D. and Sweetlove, L.J. (2009) Monitoring the *in vivo* redox state of plant mitochondria: Effect of respiratory inhibitors, abiotic stress and assessment of recovery from oxidative challenge. *Biochimica et Biophysica Acta* **1787**: 468–475.

Scott, I. and Logan, D.C. (2008) Mitochondrial morphology transition I an early indicator of subsequent cell death in *Arabidopsis*. *New Phytologist* **17**: 90–101.

Serlin, B.S. and Roux, S.J. (1986) Light induced import of chromoprotein, phytochrome, into mitochondria. *Biochimica et Biophysica Acta* **848**: 372–377.

Serlin, B.S., Sopory, S.K and Roux, S.J. (1984) Modulation of oat mitochondrial ATPase activity by Ca^{2+} and phytochrome. *Plant Physiology* **74**: 827–833.

Sheahan, M.B., McCurdy, D.W. and Rose, R.J. (2005) Mitochondria as a connected population: Ensuring continuity of the mitochondrial genome during plant cell dedifferentiation through massive mitochondrial fusion. *Plant Journal* **44**: 744–755.

Shiba, T., Kido, Y., Sakamoto, K. *et al.* (2013) Structure of the trypanosome cyanide-insensitive alternative oxidase. *Proceedings of the National Academy of Sciences USA* **110**: 4580–4585.

Skulachev, V.P. (2001) Mitochondrial filaments and clusters as intracellular power transmitting cables. *Trends in Biochemical Science* **26**: 23–29.

Siedow, J.N. and Day, D.A. (2000) Respiration and photorespiration. In *Biochemistry and Molecular Biology of Plants* (B.B. Buchanan, W. Gruissem and R.L. Jones, eds), pp. 676–728. American Society of Plant Physiologists, Rockville, Maryland.

Simon, V.R., Swayne, T.C. and Pon, L.A. (1995) Actin-dependent mitochondrial motility in mitotic yeast and cell free systems: Identification of a motor activity on the mitochondrial surface. *Journal of Cell Science* **130**: 345–354.

Srere, P.A. (1987) Complexes of sequential metabolic enzymes. *Annual Review of Biochemistry* **56**: 89–124.

Srere, P.A. (2000) Macromolecular interactions: tracing the roots. *Trends in Biochemical Science* **25**: 150–153.

Stael, S., Wurzinger, B., Mair, A. *et al.* (2012) Plant organellar calcium signalling: An emerging field. *Journal of Experimental Botany* **63**: 1525–1542.

Stappen, R. and Kramer, K. (1994) Kinetic mechanism of phosphate/phosphate and phosphate/OH^- antiports catalyzed by reconstituted phosphate carrier from beef heart mitochondria. *Journal of Biological Chemistry* **269**: 11240–11246.

Sweetlove, L.J. and Fernie, A.R. (2013) The spatial organization of metabolism within the plant cell. *Annual Review of Plant Biology* **64**: 723–746.

Tcherkez, G., Mahé, A., Gauthier, P. *et al.* (2009) In folio respiratory fluxomics revealed by ^{13}C isotopic labelling and H/D isotope effects highlight the non-cyclic nature of the tricarboxylic acid 'cycle' in illuminated leaves. *Plant Physiology* **151**: 620–630.

Tcherkez, G., Boex-Fontvieille, E., Mahé. A. and Hodges, M. (2012) Respiratory carbon fluxes in leaves. *Current Opinion in Plant Biology* **15**: 308–314.

Tegeder, M. and Weber, A.P.M. (2006) Metabolite transporters in control of plant primary metabolism. In *Control of Primary Metabolism in Plants* (W.C. Plaxton, M.T. McManus eds), pp. 85–120. Oxford, UK, Blackwell Publishing Ltd.

Tepperman, J.M., Hudson, M.E., Khanna, R. *et al.* (2004) Expression profiling of phyB mutant demonstrates substantial contributions of other phytochromes to red-light-regulated gene expression during seedling de-etiolation. *Plant Journal* **38**: 725–739.

Thum, K.E., Shin, M.J., Palenchar, P.M. *et al.* (2004) Genome-wide investigation of light and carbon signaling interactions in *Arabidopsis*. *Genome Biology* **5**: R10.

Thum, K.E., Shin, M.J., Gutiérrez, R.A. *et al.* (2008) An integrated genetic, genomic and systems approach defines gene networks regulated by the interaction of light and carbon signaling pathways in *Arabidopsis*. *BMC Systems Biology* **2**: 31.

Tolbert, N.E. (1997). The C2 oxidative photosynthetic carbon cycle. *Annual Review of Plant Physiology and Plant Molecular Biology* **48**: 1–25.

Turrens, J.F. (1997) Superoxide production by the mitochondrial respiratory chain. *Bioscience Reports* **17**: 3–8.

Twig, G., Hyde, B. and Shirihai, O.S. (2008) Mitochondrial fusion, fission and autophagy as a quality control axis: The bioenergetic view. *Biochimica et Biophysica Acta* **1777**: 1092–1097.

Umbach, A.L. and Siedow, J.N. (1993) Covalent and noncovalent dimers of the cyanide-resistant alternative oxidase protein in higher plant mitochondria and their relationship to enzyme activity. *Plant Physiology* **103**:845–854.

Umbach, A.L., Gonzàlez-Meler, M.A., Sweet, C.R. and Siedow, J.N. (2002) Activation of the plant mitochondrial alternative oxidase: Insights from site-directed mutagenesis. *Biochimica et Biophysica Acta* **1554**: 118–128.

Van Aken, O., Zhang, B., Uggalla, V. *et al.* (2009) Defining the mitochondrial stress response in *Arabidopsis thaliana*. *Molecular Plant* **2**: 1310–1324.

Van Gestel, K., Kohler, R.H. and Verbelen, J.P. (2002) Plant mitochondria move on F-actin, but their positioning in the cortical cytoplasm depends on both F-actin and microtubules. *Journal of Experimental Botany* **53:** 659–667.

Vanlerberghe, G.C. (2013) Alternative oxidase: A mitochondrial respiratory pathway to maintain metabolic and signaling homeostasis during abiotic and biotic stress in plants. *International Journal of Molecular Science* **14**: 6805–6847.

Vauclare, P., Macherel, D., Douce, R. and Bourguignon, J. (1998) The gene encoding T protein of the glycine decarboxylase complex involved in the mitochondrial step of the photorespiratory pathway in plants exhibits features of light-induced genes. *Plant Molecular Biology* **37**: 309–318.

Vercesi, A.E., Borecky, J., Maia Ide, G. *et al.* (2006) Plant uncoupling mitochondrial proteins. *Annual Review of Plant Biology* **57**: 383–404.

Verniquet, F., Gaillard, J., Neuburger, M. and Douce, R. (1991) Rapid inactivation of plant aconitase by hydrogen peroxide. *Biochemical Journal* **276**: 643–648.

Verstreken, P., Ly, C.V., Venken, K.J. *et al.* (2005) Synaptic mitochondria are critical for mobilization of reserve pool vesicles at *Drosophila* neuromuscular junctions. *Neuron* **47**: 365–378.

Vianello, A., Zancani, M., Peresson, C. *et al.* (2007) Plant mitochondrial pathway leading to programmed cell death. *Physiologia Plantarum* **129**: 242–252.

Vidal, G., Ribas-Carbo, M., Garmier, M. *et al.* (2007) Lack of respiratory chain complex I impairs alternative oxidase engagement and modulates redox signaling during elicitor-induced cell death in tobacco. *Plant Cell* **19**: 640–655.

Virolainen, E., Blokhina, O. and Fagerstedt, K. (2002) Induced high amplitude swelling and cytochrome *c* release from wheat (*Triticum aestivum* L.) mitochondria under anoxic stress. *Annals of Botany* **90**: 509–516.

Wada, M. (2013) Chloroplast movement. *Plant Science* **210**: 177–182.

Wada, M., Kagawa, T. and Sato, Y. (2003) Chloroplast movement. *Annual Review of Plant Biology* **54**: 455–468.

Wang, Q., Yu, L. and Yu, C.A. (2010) Cross-talk between mitochondrial malate dehydrogenase and the cytochrome bc_1 complex. *Journal of Biological Chemistry* **285**: 10408–10414.

Westermann, B. (2010) Mitochondrial fusion and fission in cell life and death. *Nature Reviews: Molecular Cell Biology* **11**: 872–884.

Whalley, H.J. and Knight, M.R. (2012) Calcium signatures are decoded by plants to give specific gene responses. *New Phytologist* **197**: 690–693.

Wiedemann, N., Frazier, A.E. and Pfanner, N. (2004) Protein import machinery of mitochondria. *Journal of Biological Chemistry* **279**: 14473–14476.

Wilkens, V., Kohl, W. and Busch, K. (2013) Restricted diffusion of OXPHOS complexes in dynamic mitochondria delays their exchange between cristae and engenders a transitory mosaic distribution. *Journal of Cell Science* **126**: 103–116.

Yang, X.-Y., Chen, Z.-W., Xu, T. *et al.* (2011) *Arabidopsis* kinesin KP1 specifically interacts with VDAC3, a mitochondrial protein, and regulates respiration during seed germination at low temperature. *The Plant Cell* **23**: 1093–1196.

Yao, N., Tada, Y., Sakamoto, M. *et al.* (2002) Mitochondrial oxidative burst involved in apoptotic response in oats. *Plant Journal* **30**: 567–579.

Zechmann, B. and Müller, M. (2010) Subcellular compartmentation of glutathione in dicotyledonous plants. *Protoplasma* **246**: 15–24.

Zechmann, B., Mausch, F., Sticher, L. and Müller, M. (2008) Subcellular immunocytochemical analysis detects the highest concentrations of glutathione in mitochondria and not in plastids. *Journal of Experimental Botany* **59**: 4017–4027.

Zhang, Y., Marcillat, O., Giulivi, C. *et al.* (1990) The oxidative inactivation of mitochondrial electron transport chain components and ATPase. *Journal of Biological Chemistry* **265**: 16330–16336.

Zheng, M., Beck, M., Müller, J. *et al.* (2009) Actin turnover is required for myosin-dependent mitochondrial movements in *Arabidopsis* root hairs. *PLoS One* **4**(6): e5961. DOI: 10.1371/journal.pone.0005961.

CHAPTER 8

Modes of electron transport chain function during stress: Does alternative oxidase respiration aid in balancing cellular energy metabolism during drought stress and recovery?

Greg C. Vanlerberghe, Jia Wang, Marina Cvetkovska and Keshav Dahal
Department of Biological Sciences and Department of Cell and Systems Biology, University of Toronto Scarborough, Toronto, Ontario, Canada

Introduction

Photosynthesis and respiration comprise the core pathways of primary carbon and energy metabolism in plants, providing the ATP, reducing power [NAD(P)H] and carbon intermediates essential for growth and development. Photosynthesis in the chloroplast harvests light energy and transforms it to usable chemical energy in the form of ATP and NADPH. These are used by the Calvin cycle to produce carbohydrate via the assimilation of atmospheric CO_2 (Stitt *et al.*, 2010; Rochaix, 2011; Foyer *et al.*, 2012). Mitochondrial respiration converts the chemical energy stored in carbohydrate back to ATP and NAD(P)H, thus providing these usable forms of energy for numerous other growth and maintenance processes (Fernie *et al.*, 2004; McDonald and Vanlerberghe, 2006; Plaxton and Podestá, 2006; Millar *et al.*, 2011; Tcherkez *et al.*, 2012).

A defining feature of both chloroplast and mitochondrial metabolism is the presence of specialized membrane systems that are largely responsible for the above energy transformations. These membranes house electron transport chain (ETC) components that allow for step-wise electron transfer reactions. In the case of the thylakoid membrane system of the chloroplast, this step-wise process ultimately transfers electrons from H_2O to $NADP^+$, producing O_2 and NADPH. In the case of the inner mitochondrial membrane, this step-wise process transfers

Alternative Respiratory Pathways in Higher Plants, First Edition.
Edited by Kapuganti Jagadis Gupta, Luis A.J. Mur and Bhagyalakshmi Neelwarne.

electrons from NAD(P)H to O_2, producing H_2O and $NAD(P)^+$. Further, electron transport in each organelle is coupled to proton translocation across the respective membrane. In each case, this generates a proton motive force used by a membrane-localized ATP synthase to generate ATP from ADP and P_i. It should be emphasized that continued electron transport in either membrane system is therefore dependent upon both the availability of a terminal electron acceptor (principally $NADP^+$ and O_2 in the chloroplast and mitochondrion, respectively) and upon the availability of ADP and P_i.

Imbalances in energy metabolism

Both photosynthetic and respiratory metabolism can experience energy imbalances, when there is a mismatch between rates of synthesis and rates of utilization of ATP and/or NAD(P)H. Such imbalances can have broad consequences for plant productivity and performance (De Block and Van Lijsebettens, 2011; Kramer and Evans, 2011). In the chloroplast such an imbalance is perhaps most likely to occur when the use of ATP and NADPH by the Calvin cycle does not keep pace with the harvesting of light energy by the thylakoid membranes. This can result in excess 'excitation energy' that can damage photosynthetic components, such as through the generation of reactive oxygen species (ROS). Similarly, in the mitochondrion, an imbalance could arise when the rate of ATP turnover for growth and maintenance processes is not keeping pace with the metabolism of carbohydrate and oxidation of NAD(P)H. Plants are perhaps most susceptible to imbalances in energy metabolism during periods of abiotic stress such as drought, salinity, nutrient deficiency and temperature extremes (Baena-González and Sheen, 2008; Hüner *et al.*, 2012; Suzuki *et al.*, 2012). First, such stresses can perturb metabolism such as by the disruption of enzymes and membrane processes. Such disruption can differentially impact energy-producing and energy-consuming steps within metabolism. Second, such stresses often dramatically slow growth, a major energy sink in some tissues, while at the same time eliciting cell- and tissue-specific acclimation responses that may be quite energy-intensive.

Both chloroplasts and mitochondria have the potential to experience energy imbalances and for these imbalances to be manifest at the level of their ETC. In this case, individual electron carriers may become overly reduced or oxidized, depending upon the rate of upstream and downstream processes. An important consequence of over-reduction of ETC components is that it can increase the rate of side reactions that result in the generation of excessive ROS (Møller, 2001; Apel and Hirt, 2004; Murphy, 2009; Vass, 2012). Specific components of the ETC are most susceptible to such side reactions. In the chloroplast, single electron leak to O_2 at the acceptor side of photosystem I (PSI) produces superoxide (O_2^-), while in the mitochondrion Complexes I and III are the most likely sites of O_2^- formation. Each organelle contains superoxide dismutase (SOD) isoforms

(FeSOD and CuZnSOD in the chloroplast; MnSOD in the mitochondrion) able to convert the O_2^- to another ROS, H_2O_2. If these ROS are not effectively scavenged, they can give rise to a more damaging ROS, the hydroxyl radical. The chloroplast ETC can also give rise to singlet excited oxygen due to over-reduction at photosystem II (PSII) (Vass, 2012). Since ROS have the potential to damage macromolecules and other cell components, it is important to minimize their generation by preventing over-reduction of the ETC, and also by maintaining effective ROS-scavenging systems throughout the cell (Møller, 2001; Apel and Hirt, 2004; Møller *et al.*, 2007; Foyer and Noctor, 2009).

Some ROS species, particularly H_2O_2, are important signal molecules in the control of diverse cell processes (Apel and Hirt, 2004; Foyer and Noctor, 2009; Miller *et al.*, 2010; Cvetkovska *et al.*, 2013). This may include a role as a retrograde signal from organelle to nucleus, acting to control the expression of nuclear genes encoding organelle proteins. Hence, some minimal level of ROS generation by chloroplast and mitochondrial ETC's is likely important to retain this signalling role. Therefore an over-oxidation of key ETC components may be as detrimental as over-reduction. This has led to the concept of physiological redox poising in which specific ETC components likely have optimal reduction states that support both their metabolic and signalling functions (Foyer and Shigeoka, 2011; Juszczuk *et al.*, 2012; Pfalz *et al.*, 2012; Scheibe and Dietz, 2012; Schwarzländer and Finkemeier, 2013). Examples exist in which genetic manipulation of ETC components has been shown to have a relatively minor impact on overall energy metabolism, but is nonetheless found to dramatically alter gene expression, development and/or growth (Noctor *et al.*, 2004; Giraud *et al.*, 2008; Liu *et al.*, 2009; Yoshida *et al.*, 2011). It is possible that changes in the redox poise of particular ETC components, while not greatly perturbing energy metabolism, is nonetheless acting as a signal in the regulation of these higher level processes. This signalling function may act via changes in ROS generation at the ETC or by some other unknown mechanism deriving from the change in ETC composition.

More recently, the generation of reactive nitrogen species (RNS) has been linked to mitochondria (Modolo *et al.*, 2005; Poyton *et al.*, 2009; Gupta *et al.*, 2010). These include nitric oxide (NO) and peroxynitrite, the product of a reaction between O_2^- and NO. The generation of NO by the plant ETC is poorly understood but likely involves single electron leak from complex III and/or IV to nitrite (Poyton *et al.*, 2009; Cvetkovska and Vanlerberghe, 2012). Like ROS, RNS such as NO have been shown to act as signalling molecules in numerous plant processes, and may act in conjunction with ROS (Baudouin, 2011; Molassiotis and Fotopoulos, 2011; Signorelli *et al.*, 2013).

Drought is an excellent example of a common and widespread abiotic stress that has dramatic impacts on carbon and energy metabolism (Lawlor and Tezara, 2009; Pinheiro and Chaves, 2011), as well as on the production and scavenging of ROS (Cruz de Carvalho, 2008; Miller *et al.*, 2010). Leaves respond to drought by closing their stomata, a means to reduce transpirational water loss. However, stomatal closure also restricts CO_2 diffusion into the leaf, which can result in steep declines in CO_2 assimilation. Under these conditions, a strong imbalance will develop between light energy absorption by the thylakoid membranes and metabolic energy utilization by the stromal Calvin cycle enzymes. Exacerbating this imbalance will be a strong curtailing of growth (an early response to water deficit), a major consumer of metabolic energy.

Chloroplasts and mitochondria appear to have a range of processes to buffer against the development of energy imbalances during stress. The next two sections below provide brief descriptions of some of these strategies, particularly at the ETC level. These sections also discuss current knowledge regarding the strategies that may be most prevalent during drought.

Strategies to combat energy imbalances in the chloroplast electron transport chain

To buffer against energy imbalances, chloroplasts have a number of means by which the electrons derived from water-splitting and resulting in the release of O_2 may be transferred back to O_2. First, the thylakoid membrane includes a protein termed the plastid terminal oxidase (PTOX) that directly catalyses the oxidation of plastoquinol and reduction of O_2 to H_2O (McDonald *et al.*, 2011). In general, the amount and maximum activity of PTOX appear quite low relative to overall rates of photosynthetic electron flow. Nonetheless, a number of studies have shown that the protein is induced under stress conditions, suggesting that it may represent a significant alternate electron sink in some circumstances (Stepien and Johnson, 2009; Ivanov *et al.*, 2012; Laureau *et al.*, 2013). To our knowledge, the significance of PTOX as an electron sink during drought stress has not been critically evaluated, although it has been reported that tobacco PTOX transcript increased under severe drought (Wang and Vanlerberghe, 2013) as did a measure of maximal PTOX activity in isolated thylakoids from *Hibiscus rosa-sinensis* and from *Rosa meillandina* (Muñoz and Quiles, 2013; Paredes and Quiles, 2013).

A second means to transfer electrons in the chloroplast ETC back to O_2 and producing H_2O is via the so-called Mehler reaction, which is in fact a process with ROS as intermediates (Asada, 1999). The Mehler reaction is initiated by the leak of single electrons from PSI to O_2 producing O_2^-. The O_2^- is then converted by SOD to H_2O_2 which is then reduced to H_2O by ascorbate peroxidase, using ascorbate as electron source. The oxidized ascorbate is then converted back to its reduced form by NADPH or ferredoxin. While the Mehler reaction might be considered simply a ROS-scavenging pathway to deal with electron leak at PSI, it is possible that the reaction can be advantageous in terms of balancing energy needs since it not only acts as a sink for PSII-derived electrons, but also allows for the generation of extra ATP relative to NADPH. This is similarly the case with PTOX which, while acting as an electron sink, also supports the generation of ATP. There remains uncertainty whether the Mehler reaction is a significant or minor electron sink during drought (Biehler and Fock, 1996; Badger *et al.*, 2000). A recent study suggests that the capacity of the Mehler reaction (and PTOX) to act as alternate electron sinks may be greater in gymnosperms than angiosperms (Shirao *et al.*, 2013), while most studies of these pathways during drought have examined angiosperms.

A third means to consume O_2 in the chloroplast during photosynthesis is via photorespiration, initiated when Rubisco oxygenates (rather than carboxylating) ribulose bisphosphate. This generates 3-phosphoglycerate and 2-phosphoglycolate, the latter of which is metabolized in the chloroplast and peroxisome (with glycolate and glyoxylate as intermediates) to produce glycine (Foyer *et al.*, 2009; Bauwe *et al.*, 2012). The glycine is then metabolized in the mitochondrion to serine, which is then transferred to the peroxisome and converted to glycerate, with hydroxypyruvate as an intermediate, and with consumption of NADH. Glycerate is then converted to 3-phosphoglycerate in the chloroplast, for use by the Calvin cycle. Conversion of glycine to serine in the mitochondrion involves glycine decarboxylase (GDC), in a reaction that also produces CO_2, NH_3 and NADH (thus balancing the NADH requirement in the peroxisome). Refixation of the CO_2 and NH_3 by the chloroplast requires the consumption of ATP and NADPH, and thus photorespiration acts as a net energy sink. During drought, stomatal closure decreases the ratio of CO_2 to O_2 at Rubisco, favouring the oxygenase reaction. For this reason, it is well accepted that the rate of photorespiration relative to that of CO_2 assimilation increases under drought. However, the absolute rate of photorespiration under drought is more controversial and is likely dependent upon species and drought severity (Biehler and Fock, 1996; Cornic and Fresneau, 2002; Noctor *et al.*, 2002; Guan and Gu, 2009; Abogadallah, 2011). This rate may be slightly increased, unchanged or even declined relative to that seen in well-watered plants, suggesting that the path, while certainly active during drought, may not represent much greater an absolute energy sink than under well-watered conditions, particularly as drought severity increases (Lawlor and Tezara, 2009).

As outlined earlier, there remains uncertainty regarding absolute rates of PTOX, the Mehler reaction and photorespiration as electron sinks during drought. Undoubtedly this is due in part to the technical challenges associated with distinguishing between these O_2-consuming processes in the light. What is clear is that these paths collectively become of increased proportional significance during drought, relative to CO_2 assimilation. In addition to these alternate paths of O_2 consumption, chloroplasts may also utilize other related strategies to combat energy imbalances in their ETC during drought. Four will be briefly highlighted here: cyclic electron transport (CET), down-regulation of linear electron transport (LET), non-photochemical quenching (NPQ), and metabolite shuttles.

Electron flow from H_2O to $NADP^+$ in the thylakoid membrane system is referred to as LET. However, another route(s) of electron transport, referred to as CET, is also possible. While different specific routes of CET have been described, their defining feature is that electrons beyond PSI are cycled back to the plastoquinone pool for transport again through cytochrome (cyt) b_6f (Johnson, 2011). This electron flow generates additional proton motive force for ATP synthesis, but without concomitant generation of NADPH. In this way, CEt alters the stoichiometry between ATP and NADPH synthesis. Changes in the rate of CET could buffer against energy imbalances developing in either or both of these metabolic pools. Nonetheless, this strategy is constrained by the fact that changes in the rate of CET can only have opposing impacts on rates of ATP and NADPH synthesis. By promoting the generation of the pH gradient across the thylakoid membrane, CEt also supports the activation of NPQ (Miyake *et al.*, 2004), another mechanism to combat energy imbalance in the chloroplast (see later). Partitioning of electrons between LET and CET appears to be controlled by the reduction state of the chloroplast pyridine nucleotide pool, with increased NADPH favouring CET, perhaps by promoting the formation of a CET complex (Joliot and Johnson, 2011). Interestingly, a study has shown that the slow growth phenotype of mutant plants defective in CET can be alleviated by mutation of PTOX (Okegawa *et al.*, 2010). This may indicate that the redox poise of the plastoquinone pool, as determined by an interplay of these pathways is critical to plant growth and development. There is strong evidence that CET becomes more prevalent during drought (Golding and Johnson, 2003; Kohzuma *et al.*, 2009), an indication that drought does increase the reduction state of the chloroplast stroma.

Beside the up-regulation of CET during drought, there is strong evidence that LET between PSII and PSI is actively down-regulated during drought (Golding and Johnson, 2003; Kohzuma *et al.*, 2009). The details of this down-regulation are not well understood but likely occur at the level of the cyt b_6f complex, which is usually regarded as the rate-limiting step in photosynthetic electron flow. Generally, there is evidence that regulation of cyt b_6f occurs in response to a low pH of the thylakoid lumen and/or a high stromal NADPH (Hald *et al.*, 2008; Rott *et al.*, 2011). This down-regulation of LET is likely important in preventing over-reduction at PSI.

Another major mechanism available to the chloroplast to achieve energy balance is to directly dissipate excess light energy absorbed at PSII in the form of heat. This heat dissipation is referred to as NPQ and the main mechanisms to increase NPQ occur in response to low lumen pH (de Bianchi *et al.*, 2010; Ruban *et al.*, 2012). Low lumen pH promotes the synthesis of the carotenoid zeaxanthin, as well as the protonation of the PSII-related protein PsbS. While all of the molecular details regarding how these changes lead to increased energy dissipation are still being elucidated, the key factor is that these changes result in a re-organization of the supercomplex consisting of PSII and light-harvesting complex II, resulting in an increased dissipation of the absorbed light energy as heat. Numerous studies have shown that drought stress increases NPQ as a central mechanism of photoprotection (Golding and Johnson, 2003; Lawlor and Tezara, 2009). This increase in NPQ under drought may be supported by increased CET (see earlier).

Chloroplasts have effective metabolite shuttles for the transfer of excess reducing power to the cytosol (Taniguchi and Miyake, 2012). During drought, when Calvin cycle activity is declined, reductant balance in the organelle could be at least partially achieved by the export of reducing power, for consumption by extra-chloroplastic processes, including mitochondrial electron transport. The two metabolite shuttles capable of reductant export are the malate/oxaloacetate (OAA) shuttle, also known as the malate valve, and the triose phosphate/3-phosphoglycerate shuttle. However, the triose phosphate/3-phosphoglycerate shuttle is likely not active under conditions of low Calvin cycle activity because it is dependent upon Calvin cycle intermediates. Hence, the malate valve is likely the key shuttle system that may contribute to reductant balance under drought stress. The components of the malate valve include a malate/OAA exchanger in the inner chloroplast membrane, a NADP-malate dehydrogenase (MDH) in the stroma and a NAD-MDH in the cytosol. Reduction of OAA to malate in the stroma consumes NADPH. Malate is then delivered to the cytosol in exchange for cytosolic OAA, and malate oxidation back to OAA in the cytosol produces NADH. To our knowledge, plants altered in malate valve activity (Hebbelmann *et al.*, 2012) have not yet been used to directly evaluate the role of this pathway during drought stress and little other information appears available regarding the malate valve during drought.

Strategies to combat energy imbalances in the mitochondrial electron transport chain

Plant mitochondria also have several potential mechanisms by which they could balance energy metabolism at the ETC level during drought. Two of these mechanisms, the uncoupling proteins (UCPs) and the alternate dehydrogenases, will only be briefly described here since their potential role during drought stress has not yet been extensively examined. A third mechanism, involving the alternative oxidase (AOX) will be discussed in more detail, discussing its potential role in buffering against energy imbalances, and evaluating the current evidence for its role in combating drought stress.

As is the case in animals, plants contain a family of mitochondrial UCPs that are members of a larger family of anion carriers. UCPs are integral proteins of the inner membrane that can facilitate the conductance of protons down their electrochemical gradient from inner membrane space to matrix (Vercesi *et al.*, 2006). This proton flow across the membrane occurs at the expense of proton translocation through ATP synthase and coupled with ATP generation. Hence, UCPs represent an effective means to uncouple carbon metabolism and electron transport from ATP turnover. The proton conductance activity of UCP can be activated by matrix O_2^- in the presence of fatty acids (Considine *et al.*, 2003; Smith *et al.*, 2004). Specifically, O_2^- catalyzes the generation of the lipid peroxidation product 4-hydroxy-2-nonenal, which then activates the proton conductance. This mode of biochemical control appears well-suited to UCP acting as a means to dampen O_2^- generation by the ETC. Over-reduction of the ETC due to high proton motive force would stimulate O_2^- generation, leading to UCP activation. This in turn would reduce the proton gradient and over-reduction of the ETC, thus lowering the rate of ROS generation.

Recently, Begcy *et al.* (2011) showed that overexpression of an *Arabidopsis* UCP in tobacco reduced leaf amounts of H_2O_2 compared to wild-type plants, particularly under drought (actually watering with mannitol) or high salt conditions. This suggests an ability of UCP to dampen ROS generation, particularly during stress. Significantly, the transgenic plants displayed a pronounced increase in stomatal conductance, which allowed them to maintain higher rates of CO_2 assimilation under stress, and improving their ability to recover from the stresses. These findings suggest that an important link may exist between mitochondrial function (perhaps mitochondrial ROS) and the signal paths controlling stomatal function. In another study, it was shown that knockdown of UCP1 in *Arabidopsis* hampered photorespiration, although this was not specifically examined during drought (Sweetlove *et al.*, 2006). The oxidation of glycine in the mitochondrion, which generates NADH, was restricted as shown by a reduction in the metabolism of ^{13}C-labelled glycine to serine in the *ucp1* mutant. Photosynthesis was also impeded in these plants (Sweetlove *et al.*, 2006), likely since a reduction in photorespiration can feedback and inhibit photosynthesis (Timm *et al.*, 2012). Overall, these results suggest that mitochondrial UCP supports photorespiratory function during drought, either by supporting glycine metabolism and/or by influencing stomatal function.

In addition to complex I, which oxidizes matrix NADH, plants have a series of 'alternate dehydrogenases' embedded on either the inner or outer face of the inner mitochondrial membrane. Unlike complex I, these dehydrogenases are not proton pumping and hence relax the coupling between carbon metabolism, electron transport and ATP turnover (Rasmusson *et al.*, 2004). In *Arabidopsis*, there appear to be seven alternate dehydrogenases (Rasmusson *et al.*, 2008). Three of these, the internal alternate dehydrogenases, are on the matrix side of the membrane and are collectively able to oxidize both NADH and NADPH generated in the matrix. Four others, the external alternate dehydrogenases, are on the external side of the membrane and are collectively able to oxidize NADH and NADPH deriving from the cytosol. The external dehydrogenases appear to require high Ca^{2+} for activity, suggesting that they may become engaged in response to stress. Genetic manipulation of one of the external dehydrogenases altered stem NADPH/NADP+ ratio, which then impacted stem bolting (Liu *et al.*, 2009). The regulation of at least some alternate dehydrogenase genes by light (Escobar *et al.*, 2004) suggests they may function in support of photosynthesis, although more direct evidence for this is still required. To our knowledge, the potential role of the alternate dehydrogenases during drought stress has not been reported. This represents an important area for future study, as these dehydrogenases could facilitate the turnover of excess reductant by relaxing its coupling to ATP synthesis.

Another defining feature of the plant mitochondrial ETC is the presence of two terminal oxidases, the usual energy-conserving cyt oxidase (complex IV) and another termed AOX (Finnegan *et al.*, 2004; Vanlerberghe, 2013). The ETC is essentially bifurcated, such that electrons in the ubiquinone pool are partitioned between the cyt pathway (consisting of complex III, cyt *c* and complex IV) and AOX. AOX directly couples the oxidation of ubiquinol with the reduction of O_2 to H_2O. AOX activity dramatically reduces the energy yield of respiration since it is not proton pumping and since electrons flowing to AOX bypass the proton pumping complexes III and IV. Further, in combination with an alternate dehydrogenase to bypass proton pumping complex I, AOX activity could allow for a completely uncoupled route of electron transport from matrix or cytosolic NAD(P)H to O_2. AOX is an interfacial membrane protein, oriented toward the matrix side of the inner mitochondrial membrane.

The *maximum* possible flux of electrons to AOX is often termed AOX capacity, is typically a reflection of AOX protein abundance, and can be measured in isolated mitochondria or *in vivo* by making use of pathway-specific inhibitors such as the complex IV inhibitor CN and the AOX inhibitor salicylhydroxamic acid (SHAM). The *actual* flux of electrons to AOX *in vivo* is termed AOX activity and is dependent upon the true partitioning of electrons between AOX and complex III. This partitioning of electrons is disrupted by inhibitors, so determination of AOX activity requires a more sophisticated approach. The oxygen isotope discrimination method to measure AOX activity is based on the fact that AOX and cyt oxidase discriminate to different extents against heavy O_2 ($^{18}O^{16}O$) (Guy *et al.*, 1989). In photosynthetic tissues, such measurements must be performed in the dark (due to the opposing gas exchange characteristics of photosynthesis and respiration), thus precluding the determination of AOX activity during photosynthesis.

AOX is encoded by a small gene family. Dicotyledons contain members of two distinct subfamilies, *AOX1 and AOX2*, while monocotyledons contain only *AOX1* genes (Considine *et al.*, 2002). *AOX2* genes show specific developmental and tissue expression, while the expression of *AOX1* genes is highly induced by abiotic and biotic stresses (Clifton *et al.*, 2006; Chai *et al.*, 2010). It has also been established that the stress-inducible AOX1a isoforms in tobacco and *Arabidopsis* are subject to sophisticated biochemical control (Vanlerberghe *et al.*, 1995; Rhoads *et al.*, 1998). It is this biochemical control, rather than simply

AOX protein abundance, that controls AOX activity *in vivo* (Guy and Vanlerberghe, 2005). Through covalent modification and allosteric mechanisms, AOX activity is modulated by upstream respiratory metabolism. Activation of AOX occurs in response to a high reduction state of matrix NAD(P)H, combined with high levels of pyruvate. These are conditions that might be expected to occur when there is an imbalance between the rate of upstream respiratory metabolism and downstream electron transport to O_2. Hence, the biochemical properties that govern AOX activity make it well suited as a mechanism to prevent the energy imbalances that lead to ETC over-reduction. In keeping with this, it was recently shown that transgenic tobacco leaves lacking AOX have increased concentrations of mitochondrial-localized O_2^- and NO, the products that can arise when over-reduced ETC components results in electron leak to O_2 or nitrite (Cvetkovska and Vanlerberghe, 2012). This interpretation is corroborated by experiments with the complex III inhibitor antimycin A. In wild-type plants, antimycin A increased both mitochondrial O_2^- and NO since restriction of electron flow leads to an over-reduction of ETC components. However, in plants over-expressing AOX, O_2^- and NO did not increase in response to antimycin A since these plants are able to maintain high rates of electron flow to O_2, even with the sudden and complete loss of complex III activity (Cvetkovska and Vanlerberghe, 2013).

It was reported that AOX activity may be essential to support mitochondrial glycine oxidation during photorespiration, the pathway which most directly links the mitochondrion to photosynthetic metabolism (Igamberdiev *et al.*, 1997, 2001). However, several studies have examined photosynthesis in *aox1a* mutant *Arabidopsis* plants and, to our knowledge, clear evidence that AOX supports photorespiration has not emerged from these studies (Florez-Sarasa *et al.*, 2011; Yoshida *et al.*, 2011; Gandin *et al.*, 2012). In particular, there is no evidence reported whether glycine metabolism is restricted in *aox1a*. This is unlike the case with the *ucp1 Arabidopsis* mutant, in which glycine metabolism is clearly restricted (Sweetlove *et al.*, 2006, see earlier). Interestingly, this study found that, in the absence of UCP, AOX amount also declined. On the one hand, this response is counter to what one might expect if AOX could step in – at least in the absence of UCP – and support glycine oxidation. On the other hand, it does introduce an uncertainty whether the restriction in glycine metabolism observed in *ucp1* was due to the absence of UCP1 or due to the accompanying decline in AOX. Lack of AOX, with its concomitant increase in NO (Cvetkovska and Vanlerberghe, 2012), could perhaps inactivate GDC, as a mechanism for NO inactivation of GDC has been described (Palmieri *et al.*, 2010).

In recent years, studies have investigated the role of AOX in numerous stress conditions, including drought, and have made use of tools such as oxygen isotope discrimination and plants with manipulated AOX amount (Vanlerberghe, 2013). The next section provides further background regarding plant respiration under drought, as well as providing a comprehensive analysis of studies which have specifically examined the role of AOX during drought stress and recovery.

Plant respiration and alternative oxidase during drought stress

A defining feature of drought stress is that it results in a dramatic decline in the rate of carbon assimilation, by the gradual imposition of a combination of stomatal and biochemical limitations of photosynthesis (Lawlor and Tezara, 2009). Given the declines in carbon assimilation, it might be assumed that another defining feature of drought stress would be a drop in plant carbon status, followed by a decline in respiration rate due to substrate limitation. However, this does not appear to be a typical scenario. First, recent studies suggest that the carbon status of plants during drought stress is relatively robust, particularly compared with the decline in photosynthesis (Muller *et al.*, 2011; Pinheiro and Chaves, 2011). This is likely primarily because overall growth declines relatively more than photosynthesis during drought, thus buffering against a decrease in carbon status (Muller *et al.*, 2011). Second, based on studies to date, there is no clear expectation as to the rate of respiration during drought. In some cases, drought has been reported to have little or no impact on total respiration rate (Ribas-Carbo *et al.*, 2005; Giraud *et al.*, 2008; Gimeno *et al.*, 2010), while other studies have reported decreases (Haupt-Herting *et al.*, 2001; Haupt-Herting and Fock, 2002; Taylor *et al.*, 2005; Vassileva *et al.*, 2009; Galle *et al.*, 2010) or even increases (Bartoli *et al.*, 2005; Feng *et al.*, 2008; Hummel *et al.*, 2010; Begcy *et al.*, 2011). It has also been reported that respiration can decrease in response to mild water deficit but then increase with more severe stress (Wang and Vanlerberghe, 2013).

Despite the variable response of respiration rate to drought, a general conclusion that can be drawn from the literature is that, in most instances, drought causes a substantial increase in the ratio of respiration rate to photosynthetic rate (Flexas *et al.*, 2006; Atkin and Macherel, 2009). In this case, the question of how respiration responds to drought takes on added significance in terms of the overall energy and carbon budget of the plant. Recent studies suggest that enzymes and metabolites in respiratory metabolism stay high or even increase under drought (Vasquez-Robinet *et al.*, 2008; Hummel *et al.*, 2010; Acevedo *et al.*, 2013). Interestingly, Bartoli *et al.* (2004) found that wheat mitochondria suffered relatively more oxidative damage (estimated by protein carbonyl content) in response to drought than did either chloroplasts or peroxisomes. It has also been shown that the expression of MnSOD is drought-inducible in wheat (Wu *et al.*, 1999). Such findings are consistent with the view that respiration remains active during drought and that it may indeed take on additional functional roles and significance in support of acclimation to drought and recovery from drought. The observations also suggest that mitochondrial ROS may be prevalent, perhaps the result of an energy imbalance in this organelle during drought.

As outlined earlier, chloroplast metabolism responds to drought – and the decline in the Calvin cycle as a major energy consumer – with the engagement of multiple mechanisms that likely act in parallel to buffer against energy imbalances. Given the prominent role of respiratory metabolism during drought, interest has turned to whether specific mitochondrial mechanisms able to buffer against energy imbalances are also being engaged during drought. In particular, Table 8.1 provides a summary of studies that have focused on AOX respiration in the response of plants to drought stress. Following are some observations and discussion based on the insights gained from these studies:

1 In several species, including both monocots and dicots, drought has been shown to increase the transcript amount of gene(s) encoding AOX (Table 8.1). Similarly, increases in AOX protein and capacity have often been reported. One possibility is that increased AOX expression during drought is due to changes in abscisic acid (ABA) signalling, although this possibility has not been directly evaluated. Increased ABA is a common response to drought, as this hormone is responsible for important acclimation responses such as stomatal closure (Neill *et al.*, 2008; Kim *et al.*, 2010; Daszkowska-Golec and Szarejko, 2013). In *Arabidopsis*, a molecular link has been made between ABA signalling and the regulation of AOX expression. Functional characterization of the promoter of *Arabidopsis AOX1a* identified a repressor element that was shown to bind the transcription factor abscisic acid insensitive 4 (ABI4) (Giraud *et al.*, 2009). ABI4 is an ABA signalling responsive transcription factor. These results hint that increased ABA during drought could act to de-repress *AOX1a* transcription. Supporting this idea, studies have shown that exogenous ABA treatment of *Arabidopsis* increases *AOX1a* transcript amount (Ghassemian *et al.*, 2008; Giraud *et al.*, 2009; Liu *et al.*, 2010; He *et al.*, 2012; Miura *et al.*, 2013). Interestingly, ABA also increased the transcript amount of several genes encoding alternate dehydrogenases, indicating that the components for a completely non-energy conserving path of mitochondrial electron transport can be induced by ABA (Ghassemian *et al.*, 2008; He *et al.*, 2012).

While most studies examining AOX amount in response to drought have reported increased AOX, there are some exemptions (Table 1). In particular, soybean was shown to dramatically increase AOX activity in response to drought (see below) but without any increase in AOX protein amount (Ribas-Carbo *et al.*, 2005). Also, in some species such as tobacco it was shown that a relatively severe drought stress was required before substantial increases in AOX expression and protein amount were evident (Wang and Vanlerberghe, 2013). The variability between species may relate to their 'non-stress' constitutive level of AOX. For example, soybean is known to have relatively high constitutive amounts of AOX, meaning that an increase in AOX amount in response to drought may not be necessary to allow increased AOX activity. For example, Bartoli *et al.*, (2005) provide evidence that, in wheat, drought stress was associated with an increased conversion of AOX protein from its oxidized

Table 8.1 Studies examining the role of AOX during drought stress.

Plant species	Treatment(s)	Major findings	Reference
Arabidopsis thaliana (WT plants and *aox1a* knockout plants)	A moderate combined stress treatment (increased irradiance and drought) that had no impact on leaf RWC of WT plants but reduced the RWC of knockout plants by approximately 10%.	Compared to WT, mutant had reduced root growth that may have been responsible for its compromised RWC under stress. Compared to WT, mutant leaves under stress accumulated more anthocyanins, displayed some reduction in photosynthetic efficiency, had elevated levels of whole leaf O_2^-, and had generally increased amounts of sugars and decreased amounts of amino and organic acids.	Giraud *et al.*, 2008
Arabidopsis thaliana (WT plants, plants overexpressing *AOX1a* and *aox1a* knockout plants)	Mild osmotic stress (mannitol). Drought.	Under non-stress conditions, growth rate was compromised in over-expressing plants and increased in knockout plants, relative to WT. However, under stress conditions, growth rate was improved in over-expressing plants, while knockout plants were similar to WT. In WT plants, AOX expression was induced by stress, but only in young leaves with predominantly dividing cells, suggesting an important role of AOX in proliferating cells under stress.	Skirycz *et al.*, 2010
Nicotiana tabacum (WT and *aox1a* knockdown plants)	Drought resulting in a progressive decline in leaf RWC. Severe drought combined with increased irradiance. Re-watering.	Mild to moderate drought resulted in a progressive and modest increase in AOX protein amount while severe stress (particularly when combined with increased irradiance) strongly increased AOX. All plant lines displayed similar declines in leaf RWC with increasing stress severity. Under severe stress, knockdown lines exhibited more cellular and oxidative damage than WT, and were found to down-regulate rather than up-regulate the transcript level of several important ROS-scavenging components. Compared to WT, knockdown lines were strongly compromised in their ability to recover from severe stress after re-watering.	Wang and Vanlerberghe, 2013
Nicotiana sylvestris (WT plants and CMSII plants lacking complex I)	Drought resulting in an approximately 15% decline in leaf RWC.	In WT plants, AOX protein increased under drought. Isotope discrimination experiments showed that drought decreased electron flow through the cyt pathway, while electron flow to AOX was maintained.	Galle *et al.*, 2010

(*Continued*)

Table 8.1 (*Continued*)

Plant species	Treatment(s)	Major findings	Reference
Glycine max	Drought resulting in a 3% (mild stress) to 15% (severe stress) decline in leaf RWC.	No change in leaf AOX protein level under drought. Isotope discrimination experiments showed that, in response to severe drought, about 40% of total electron flow occurred through AOX, compared to just 10–12% in well-watered plants or plants experiencing mild drought.	Ribas-Carbo *et al.*, 2005
Triticum aestivum	Drought resulting in a 22% decline in leaf RWC. Some plants treated with 1 mM SHAM to inhibit AOX.	Drought increased the total amount of AOX protein and shifted more of the protein toward its reduced (active) form. SHAM treatment of drought-stressed plants reduced photosynthetic performance, decreasing photochemical quenching and increasing NPQ.	Bartoli *et al.*, 2005
Triticum aestivum	Drought resulting in leaf RWC of approximately 78%.	Drought increased AOX transcript and approximately doubled the AOX capacity of leaves. SHAM treatment of drought-stressed leaves increased H_2O_2 amount.	Feng *et al.*, 2008
Triticum aestivum (several varieties)	Moderate drought stress. Re-watering.	Drought approximately doubled the AOX capacity measured in isolated mitochondria, and remained high three days after re-watering.	Vassileva *et al.*, 2009
Nothofagus solandri and Nothofagus menziesii (beech tree species)	Mild to severe drought. Re-watering.	AOX protein amount increased (relative to a cyt pathway protein) under severe drought and this pattern persisted after re-watering. However, isotope discrimination experiments suggested little change in electron partitioning between AOX and the cyt pathway in response to drought or re-watering.	Sanhueza *et al.*, 2013
Pisum sativum	Drought.	Leaf AOX protein amount increased 2.5-fold by drought.	Taylor *et al.*, 2005
Oryza sativa	Drought resulting in approximately 20% decline in leaf RWC.	Leaf AOX transcript amount increased in response to drought.	Feng *et al.*, 2009
Medicago trunculata	Drought. Re-watering.	AOX transcript levels declined in the leaf and increased in the root in response to drought.	Fillippou *et al.*, 2011

(inactive) to reduced (active) form, indicating a biochemical control of AOX activity due to the prevailing metabolic conditions present during drought. It is worth noting that the majority of drought studies to date have examined AOX amount in leaf, so little is yet known about how root AOX may respond to drought. It is also not known whether changes in AOX amount or activity occur in guard cells, an important ABA target during drought stress. A previous study has reported that respiration rates in pea are several-fold higher in guard cells than mesophyll cells (Vani and Raghavendra, 1994). However, little else is known about respiration in guard cells and, in particular, what role the cyt and AOX pathways may have in terms of stomatal function.

Interestingly, a number of recent studies are suggestive of a link between mitochondrial ROS, ABA signalling and stomatal function. In one case, altered expression of an *Arabidopsis* mitochondrial glutathione peroxidase was shown to disrupt H_2O_2 amount in guard cells and to disrupt ABA-mediated stomatal closure in response to drought (Miao *et al.*, 2006). In another example, mutation of a DEXH box RNA helicase that disrupted complex I resulted in higher levels of mitochondrial O_2^-, which in turn reduced stomatal aperture and improved drought tolerance (He *et al.*, 2012). Similarly, a mitochondrial RNA editing mutant defective in complex I accumulated more H_2O_2 in guard cells after ABA treatment and displayed enhanced drought tolerance (Yuan and Liu, 2012). Finally, the complex I mutant of tobacco (CMSII) is also reported to display reduced stomatal aperture in response to drought, perhaps again through changes in ROS (Djebbar *et al.*, 2012). These studies suggest that ABA control of stomatal aperture may be mediated, at least in part, through changes in mitochondrial ROS amounts. Salicylic acid (SA) can also influence stomatal aperture. Several mutants with increased SA displayed reduced stomatal aperture due to increased ROS amount (Miura *et al.*, 2013), which may have been mitochondrial in origin given the ability of SA to disrupt mitochondrial metabolism (Norman *et al.*, 2004). The study by Miura *et al.* (2013) also found high levels of AOX transcript in guard cells and – through cluster analysis of several microarray datasets – identified AOX as a 'gene of interest' in the regulation of stomatal movement by SA, ROS and drought. Despite the interest of these studies, a unifying model of how AOX and mitochondrial ROS may function in the regulation of stomatal aperture by drought, ABA and/or SA is not yet reported, and will require further study at the guard cell level using plants with modified AOX expression.

2 While increases in AOX transcript, protein and capacity in response to drought suggest that AOX activity may be increased under drought, this can only be directly evaluated using the oxygen isotope discrimination technique. To our knowledge, only three such drought studies involving four plant species (*Glycine max, Nicotiana sylvestris* and two *Nothofagus* tree species) has been reported (Table 8.1). Of these species, soybean showed the most dramatic changes in AOX activity under drought. In well-watered soybean, AOX

activity accounted for 10–12% of total electron flow. Drought stress saw both a decline in absolute cyt pathway activity and an increase in absolute AOX activity such that, during drought, total respiration rate was similar to well-watered plants but with 40% of total electron flow occurring via AOX (Ribas-Carbo *et al.*, 2005). The increase in AOX activity under drought may be facilitated by a high energy charge restricting cyt pathway flow and/or by an abundance of reducing equivalents supplying electrons to the ubiquinone pool via complex I and the alternate dehydrogenases. The fact that the increase in AOX activity was combined with a decrease in cyt pathway activity favours high energy change being responsible for the change in electron partitioning. If high energy charge was not being experienced, but only an abundance of electrons in the ubiquinone pool, one might expect the activity of both AOX and the cyt pathway to increase, but this was not the case. Nonetheless, energy charge was not directly measured in this study, so other possibilities for the decline in cyt pathway activity and the increase in AOX activity are also possible. For example, drought might directly inhibit the cyt pathway by another unknown mechanism. Interestingly, the study with *N. sylvestris* also reported that drought decreased cyt pathway activity (Galle *et al.*, 2010). In this case, AOX activity remained unchanged in response to drought; however, due to the decline in cyt pathway respiration, AOX did represent a higher percentage of total respiration under drought than under well-watered conditions. Finally, a study on two *Nothofagus* species suggested no change in electron partitioning between AOX and the cyt pathway due to drought, although this study was hampered because the end-points for discrimination against $^{18}O_2$ by each pathway could not be determined (Sanhueza *et al.*, 2013). In sum, the available evidence indicates that drought can strongly impact the activity of both cyt and AOX respiration under drought and is suggestive that the ratio of AOX to cyt pathway respiration increases under drought. This is consistent with a need for AOX to dissipate excess energy under drought, more so than under well-watered conditions. In this respect, it is worth noting that energy imbalances during drought would be expected to be greater in the light than dark. Hence, the partitioning of electrons to AOX in the light might be even greater than those estimated in the dark by isotope discrimination. Nonetheless, it is obvious that still too little isotope discrimination data overall is available to conclude that increased AOX is a defining feature of respiratory metabolism under drought.

Given our speculation above that high energy charge may be responsible for the shift in electron partitioning toward AOX during drought, it is worth emphasizing some other studies which suggest that the biochemical impairment of photosynthesis during drought is primarily due to a disabling or down-regulation of the chloroplast ATP synthase (Tezara *et al.*, 1999; Kohzuma *et al.*, 2009; Lawlor and Tezara, 2009). This likely should decrease rather than increase ATP amounts during drought, as some studies have demonstrated (Tezara *et al.*, 2008; Lawlor and Tezara, 2009).

3 There is some evidence that AOX respiration is important to maintain respiratory carbon flow under drought (Table 8.1). This is based primarily upon a study comparing wild-type (WT) *Arabidopsis* with T-DNA mutants lacking AOX1a, and involved a stress that combined mild drought with a shift to higher irradiance (Giraud *et al.*, 2008). A survey of metabolites found that, under stress, mutant plants maintained generally higher levels of carbohydrate and lower levels of amino and organic acids than WT. These differences between lines were not seen under the normal growth condition. The results are consistent with a restriction of respiratory carbon flow through glycolysis and the TCA cycle in the plants lacking AOX. On the other hand, no differences were seen in oxygen uptake by the plants suggesting that, while carbon flow appeared restricted by the lack of AOX, the total rate of electron flow through the ETC to oxygen was normal. It is difficult to reconcile these two findings.

4 In both wheat and *Arabidopsis*, there is some evidence that AOX activity under drought acts in support of photosynthetic metabolism (Table 8.1). In wheat, this is primarily based upon experiments comparing the photosynthetic characteristics of well-watered and drought-stressed plants, in the presence or absence of the AOX inhibitor SHAM (Bartoli *et al.*, 2005). It was found that SHAM had no impact on photosynthesis in well-watered plants. In drought-stressed plants, however, SHAM significantly reduced the efficiency of PSII, while increasing NPQ and decreasing photochemical quenching, compared to drought-stressed plants without SHAM treatment. These effects of SHAM were particularly evident at higher irradiances, consistent with an energy imbalance in the chloroplast in the absence of AOX activity. The authors suggest that the positive impact of AOX may be due to it acting both as a sink for reductant (such as generated by glycine oxidation) and by providing increased respiratory CO_2 release for reassimilation by the Calvin cycle. While it was shown that SHAM itself had no apparent direct effect on photosynthesis in isolated chloroplasts, experiments utilizing SHAM should nonetheless be interpreted with caution due to the potential side-effects of this inhibitor. The impact of AOX on photosynthesis during drought was also investigated in the *Arabidopsis aox1a* mutant subjected to drought combined with a shift to higher irradiance (see above, Giraud *et al.*, 2008). Similar to the studies in wheat, lack of AOX during stress decreased PSII efficiency and increased non-photochemical energy dissipation. This study also reported increased whole leaf levels of O_2^- which was suggested to arise in the chloroplast due to the disrupted photosynthetic metabolism. Consistent with this, there was a strong similarity between the transcriptome changes of the *aox1a* mutant under drought and transcriptome changes previously reported to occur in response to chloroplast-generated ROS (Giraud *et al.*, 2008).

5 Theoretically, AOX activity could negatively impact plant productivity since it reduces the respiratory yield of ATP, an important general requirement for biosynthesis and growth. A study with *Arabidopsis* suggests that AOX amount can influence growth under drought stress (Skirycz *et al.*, 2010) (Table 8.1). This study compared the relative growth rate of WT plants with that of plants either lacking AOX or overexpressing AOX. Under optimal growth conditions all the plants displayed similar relative growth rates. However, under drought stress, plants overexpressing AOX displayed higher relative growth rate than WT. This suggests, paradoxically, that the non-energy conserving nature of AOX can positively impact growth under drought stress. While the specific mechanisms responsible for this growth response still need to be elucidated, one possibility is that higher AOX activity improved energy balance, with positive impacts on metabolism and/or signalling processes.

Interestingly, the study by Giraud *et al.* (2008) also reported a growth phenotype in *Arabidopsis aox1a* mutants. Root growth in vertical agar plates was reduced in the mutant by about 10% compared to WT. This study also found that, after the combined drought/irradiance stress (see earlier), the leaf relative water content (RWC) of the mutant plants had declined by about 10%, while no decline occurred in the WT. It seems possible that the root growth defect could account for the greater leaf water deficit being experienced by the mutant plants. It might also provide an explanation for the decline in photosynthetic performance of the mutant, compared to WT (see earlier). If the mutant plants are experiencing a greater water deficit than the WT, as the RWC measurements suggest, they might also experience a greater stomatal limitation of photosynthesis. Hence, there remains some uncertainly whether lack of AOX in these plants was directly impairing photosynthesis, such as by impairing oxidation of excess chloroplast reductant, or indirectly, by impairing the capacity for water uptake due to reduced root growth. As discussed in the study, another explanation is also possible. The *aox1a* mutant plants display a marked reduced expression of the ABI4 transcription factor that is a negative regulator of *AOX1a* expression, presumably an attempt by the plants to increase AOX1a levels (Giraud *et al.*, 2008, 2009). Given that ABI4 is a central stress responsive transcription factor involved in ABA responses as well as chloroplast retrograde responses (León *et al.*, 2013; Wind *et al.*, 2013), its altered amount in *aox1a* might also contribute to the changes in photosynthetic metabolism during stress.

6 There is some evidence that AOX can protect against oxidative and cellular damage during severe drought stress (Table 8.1). Knockdown of AOX in transgenic tobacco had little impact on the amount of oxidative damage (lipid peroxidation) or cellular damage (electrolyte leakage) during mild to moderate drought. However, in response to severe drought combined with a shift to higher irradiance, the knockdown plants exhibited small but significant increases in both oxidative and cellular damage relative to WT plants with

similar RWC (Wang and Vanlerberghe, 2013). Further, *aox1a Arabidopsis* mutants were unable to survive a stress combination in which drought-stressed plants were subsequently subjected to both increased irradiance and elevated temperature (35 °C) (Giraud *et al.*, 2008). Finally, inhibition of AOX by SHAM during drought stress was shown to increase leaf levels of H_2O_2 in wheat (Feng *et al.*, 2008).

7 There is some evidence that the presence of AOX may be important in the recovery phase from drought stress (Table 8.1). In particular, the tobacco study noted earlier showed that plants lacking AOX were strongly compromised in their ability to recover from severe drought stress when re-watered (Wang and Vanlerberghe, 2013). While all WT plants showed rapid evidence of recovery, the knockdown plants were either significantly delayed in their recovery or did not recover at all during the study period. At present, however, it is difficult to untangle whether the compromised ability of these plants to recover is due to an essential role for AOX during the recovery period itself or whether it is due to the slight increased oxidative and cellular damage experienced by the knockdown plants during the severe drought (see earlier, Wang and Vanlerberghe, 2013). Further, the late stages of severe stress were characterized by a down-regulation of expression of several ROS-scavenging components in the knockdowns, while these were increasing in the WT. This may indicate a re-programming of knockdown plants (perhaps a programmed death or senescence program?), which may have also contributed to their susceptibility during the subsequent recovery period. It would be interesting to examine AOX activity using isotope discrimination in tobacco plants during a recovery period from severe drought to examine whether the pathway is highly engaged under such conditions. The study of Galle *et al.* (2010) reported little impact of re-watering on AOX activity, while cyt pathway activity increased. However, this re-watering followed a much less severe drought treatment than reported by Wang and Vanlerberghe (2013). Finally, in the study with *Nothofagus*, drought increased the ratio of AOX protein to that of a cyt oxidase protein and this increased ratio persisted – or even increased further – following re-watering (Sanhueza *et al.*, 2013). There is evidence that re-watering can actually enhance the oxidative stress being experienced by drought-stressed plants (Mittler and Zilinskas, 1994; Flexas *et al.*, 2006). If this is the case, it could provide some explanation for the high AOX after re-watering.

Conclusions

Drought is a widespread abiotic stress that can have strong negative impacts on plant growth, productivity and survival. There is overwhelming evidence from photosynthesis studies that this stress acts to exacerbate energy imbalances in the chloroplast. Given the connectivity of primary energy metabolism between different cellular compartments and given that mitochondrial components such as AOX may be ideally suited to combat cellular energy imbalances, it is clear that more effort should be directed toward the study of mitochondrial and respiratory metabolism during drought and recovery from drought, and in relation to photosynthetic metabolism (Flexas *et al.*, 2006; Atkin and Macharel 2009; Lawlor and Tezara, 2009). Beside the potential metabolic roles of respiration during drought, the potential signalling roles of the mitochondrion in processes such as stomatal function or cell survival during and following severe stress are also of considerable interest.

Acknowledgements

G.C.V. acknowledges the generous financial support of the Natural Sciences and Engineering Research Council of Canada.

Abbreviations

ABA, abscisic acid; ABI4, abscisic acid insensitive 4; AOX, alternative oxidase; CET, cyclic electron transport; cyt, cytochrome; ETC, electron transport chain; GDC, glycine decarboxylase; MDH, malate dehydrogenase; NO, nitric oxide; LET, linear electron transport; NPQ, non-photochemical quenching; OAA, oxaloacetate; PSI, photosystem I; PSII, photosystem II; PTOX, plastid terminal oxidase; O_2^-, superoxide; RNS, reactive nitrogen species; ROS, reactive oxygen species; RWC, relative water content; SA, salicylic acid; SHAM, salicylhydroxamic acid; SOD, superoxide dismutase; UCP, uncoupling protein; WT, wild-type

References

Abogadallah, G.M. (2011) Differential regulation of photorespiratory gene expression by moderate and severe salt and drought stress in relation to oxidative stress. *Plant Science* **180**: 540–547.

Acevedo, R.M., Maiale, S.J., Pessino, S.C. *et al.* (2013) A succinate dehydrogenase flavoprotein subunit-like transcript is upregulated in *Ilex paraguariensis* leaves in response to water deficit and abscisic acid. *Plant Physiology and Biochemistry* **65**: 48–54.

Apel, K. and Hirt, H. (2004) Reactive oxygen species: metabolism, oxidative stress, and signal transduction. *Annual Review of Plant Biology* **55**: 373–399.

Asada, K. (1999) The water-water cycle in chloroplasts: scavenging of active oxygen and dissipation of excess photons. *Annual Review of Plant and Plant Molecular Biology* **50**: 601–639.

Atkin, O.K. and Macherel, D. (2009) The crucial role of plant mitochondria in orchestrating drought tolerance. *Annals of Botany* **103**: 581–597.

Badger, M.R., von Caemmerer, S., Ruuska, S. and Nakano, H. (2000) Electron flow to oxygen in higher plants and algae: Rates and control of direct photoreduction (Mehler reaction) and rubisco oxygenase. *Philosophical Transactions of the Royal Society of London B* **355**: 1433–1446.

Baena-González, E. and Sheen, J. (2008) Convergent energy and stress signalling. *Trends in Plant Science* **13**: 474–482.

Bartoli, C.G., Gómez, F., Martínez, D.E. and Guiamet, J.J. (2004) Mitochondria are the main target for oxidative damage in leaves of wheat (*Triticum aestivum* L.). *Journal of Experimental Botany* **55**: 1663–1669.

Bartoli, C.G., Gomez, F., Gergoff, G. *et al.* (2005) Upregulation of the mitochondrial alternative oxidase pathway enhances photosynthetic electron transport under drought conditions. *Journal of Experimental Botany* **56**: 1269–1276.

Baudouin, E. (2011) The language of nitric oxide signalling. *Plant Biology* **13**: 233–242.

Bauwe, H., Hagemann, M., Kern, R. and Timm, S. (2012) Photorespiration has a dual origin and manifold links to central metabolism. *Current Opinion in Plant Biology* **15**: 269–275.

Begcy, K., Mariano, E.D., Mattiello, L. *et al.* (2011) An *Arabidopsis* mitochondrial uncoupling protein confers tolerance to drought and salt stress in transgenic tobacco plants. *PLoS ONE* **6**: e23776.

de Bianchi, S., Ballottari, M., Dall'Osto, L. and Bassi, R. (2010) Regulation of plant light harvesting by thermal dissipation of excess energy. *Biochemical Society Transactions* **38**: 651–660.

Biehler, K. and Fock, H. (1996) Evidence for the contribution of the Mehler-peroxidase reaction in dissipating excess electrons in drought-stressed wheat. *Plant Physiology* **112**: 265–272.

Chai, T.-T., Simmonds, D., Day, D.A. *et al.* (2010) Photosynthetic performance and fertility are repressed in GmAOX2b antisense soybean. *Plant Physiology* **152**: 1638–1649.

Clifton, R., Millar, A.H. and Whelan, J. (2006) Alternative oxidases in *Arabidopsis*: A comparative analysis of differential expression in the gene family provides new insights into function of non-phosphorylating bypasses. *Biochimica et Biophysica Acta* **1757**: 730–741.

Considine, M.J., Goodman, M., Echtay, K.S. *et al.* (2003) Superoxide stimulates a proton leak in potato mitochondria that is related to the activity of uncoupling protein. *Journal of Biological Chemistry* **278**: 22298–22302.

Considine, M.J., Holtzapffel, R.C., Day, D.A. *et al.* (2002) Molecular distinction between alternative oxidase from monocots and dicots. *Plant Physiology* **129**: 949–953.

Cornic, G. and Fresneau, C. (2002) Photosynthetic carbon reduction and carbon oxidation cycles are the main electron sinks for photosystem II activity during a mild drought. *Annals of Botany* **89**: 887–894.

Cruz de Carvalho, M.H. (2008) Drought stress and reactive oxygen species. Production, scavenging and signalling. *Plant Signal Behav* **3**, 156–165.

Cvetkovska, M., Alber, N.A. and Vanlerberghe, G.C. (2013) The signalling role of a mitochondrial superoxide burst during stress. *Plant Signaling & Behavior* **8**: e22749.

Cvetkovska, M. and Vanlerberghe, G.C. (2012) Alternative oxidase modulates leaf mitochondrial concentrations of superoxide and nitric oxide. *New Phytologist* **195**: 32–39.

Cvetkovska, M. and Vanlerberghe, G.C. (2013) Alternative oxidase impacts the plant response to biotic stress by influencing the mitochondrial generation of reactive oxygen species. *Plant, Cell & Environment* **36**: 721–732.

Daszkowska-Golec, A. and Szarejko, I. (2013) Open or close the gate – stomata action under the control of phytohormones in drought stress conditions. *Frontiers in Plant Science* **4**: Article 138.

De Block, M. and Van Lijsebettens, M. (2011) Energy efficiency and energy homeostasis as genetic and epigenetic components of plant performance and crop productivity. *Current Opinion in Plant Biology* **14**: 275–282.

Djebbar, R., Rzigui, T., Pétriacq, P. *et al.* (2012) Respiratory complex I deficiency induces drought tolerance by impacting leaf stomatal and hydraulic conductances. *Planta* **235**: 603–614.

Escobar, M.A., Franklin, K.A., Svensson, Å. *et al.* (2004) Light regulation of the *Arabidopsis* respiratory chain. Multiple discrete photoreceptor responses contribute to induction of type II NAD(P)H dehydrogenase genes. *Plant Physiology* **136**: 2710–2721.

Feng, H., Duan, J., Li, H. *et al.* (2008) Alternative respiratory pathway under drought is partially mediated by hydrogen peroxide and contributes to antioxidant protection in wheat leaves. *Plant Production Science* **11**: 59–66.

Feng, H.-Q., Li, H.-Y, and Sun, K. (2009) Enhanced expression of alternative oxidase genes is involved in the tolerance of rice (*Oryza sativa* L.) seedlings to drought stress. *Zeitschrift für Naturforschung* **64c**: 704–710.

Fernie, A.R., Carrari, F. and Sweetlove, L.J. (2004) Respiratory metabolism: Glycolysis, the TCA cycle and mitochondrial electron transport. *Current Opinion in Plant Biology* **7**: 254–261.

Filippou, P., Antoniou, C. and Fotopoulos, V. (2011) Effect of drought and rewatering on the cellular status and antioxidant response of *Medicago truncatula* plants. *Plant Signaling & Behavior* **6**: 270–277.

Finnegan, P.M., Soole, K.L, and Umbach, A.L. (2004) Alternative mitochondrial electron transport proteins in plants. In: Day, D.A., Millar, A.H. and Whelan, J. (eds) *Plant Mitochondria: From Genome to Function*. Kluwer Academic Publishers, Dordrecht, The Netherlands, pp. 163–230.

Flexas, J., Bota, J., Galmés, J. *et al.* (2006) Keeping a positive carbon balance under adverse conditions: responses of photosynthesis and respiration to water stress. *Physiologia Plantarum* **127**: 343–352.

Florez-Sarasa, I., Flexas, J., Rasmusson, A.G. *et al.* (2011) In vivo cytochrome and alternative pathway respiration in leaves of *Arabidopsis thaliana* plants with altered alternative oxidase under different light conditions. *Plant, Cell & Environment* **34**: 1373–1383.

Foyer, C.H., Bloom, A.J., Queval, G. and Noctor, G. (2009) Photorespiratory metabolism: Genes, mutants, energetic, and redox signalling. *Annual Review of Plant Biology* **60**: 455–484.

Foyer, C.H. and Noctor, G. (2009) Redox regulation in photosynthetic organisms: signalling, acclimation, and practical implications. *Antioxidants & Redox Signaling* **11**: 861–905.

Foyer, C.H., Neukermans, J., Queval, G. *et al.* (2012) Photosynthetic control of electron transport and the regulation of gene expression. *Journal of Experimental Botany* **63**: 1637–1661.

Foyer, C.H. and Shigeoka, S. (2011) Understanding oxidative stress and antioxidant functions to enhance photosynthesis. *Plant Physiology* **155**: 93–100.

Galle, A., Florez-Sarasa, I., Thameur, A. *et al.* (2010) Effects of drought stress and subsequent rewatering on photosynthetic and respiratory pathways in *Nicotiana sylvestris* wild type and the mitochondrial complex I-deficient CMSII mutant. *Journal of Experimental Botany* **61**: 765–775.

Gandin, A., Duffes, C., Day, D.A. and Cousins, A.B. (2012) The absence of alternative oxidase AOX1A results in altered response of photosynthetic carbon assimilation in increasing CO_2 in *Arabidopsis thaliana*. *Plant Cell Physiology* **53**: 1627–1637.

Ghassemian, M., Lutes, J., Chang, H.-S. *et al.* (2008) Abscisic acid-induced modulation of metabolic and redox control pathways in *Arabidopsis thaliana*. *Phytochemistry* **69**: 2899–2911.

Gimeno, T.E., Sommerville, K.E., Valladares, F. and Atkin, O.K. (2010) Homeostasis of respiration under drought and its important consequences for foliar carbon balance in a drier climate: Insights from two contrasting *Acacia* species. *Functional Plant Biology* **37**: 323–333.

Giraud, E., Ho, L.H.M., Clifton, R. *et al.* (2008) The absence of ALTERNATIVE OXIDASE1a in *Arabidopsis* results in acute sensitivity to combined light and drought stress. *Plant Physiology* **147**: 595–610.

Giraud, E., Van Aken, O., Ho, L.H.M. and Whelan, J. (2009) The transcription factor ABI4 is a regulator of mitochondrial retrograde expression of *ALTERNATIVE OXIDASE1a*. *Plant Physiology* **150**: 1286–1296.

Golding, A.J. and Johnson, G.N. (2003) Down-regulation of linear and activation of cyclic electron transport during drought. *Planta* **218**: 107–114.

Guan, X. and Gu, S. (2009) Photorespiration and photoprotection of grapevine (*Vitis vinifera* L. cv. Cabernet Sauvignon) under water stress. *Photosynthetica* **47**: 437–444.

Gupta, K.J., Fernie, A.R., Kaiser, W.M. and van Dongen, J.T. (2010) On the origins of nitric oxide. *Trends in Plant Science* **16**: 160–168.

Guy, R.D., Berry, J.A., Fogel, M.L. and Hoering, T.C. (1989) Differential fractionation of oxygen isotopes by cyanide-resistant and cyanide-senstive respiration in plants. *Planta* **177**: 483–491.

Guy, R.D. and Vanlerberghe, G.C. (2005) Partitioning of respiratory electrons in the dark in leaves of transgenic tobacco with modified levels of alternative oxidase. *Physiologia Plantarum* **125**: 171–180.

Hald, S., Nandha, B., Gallois, P. and Johnson, G.N. (2008) Feedback regulation of photosynthetic electron transport by NAD(P)H redox poise. *Biochimica et Biophysica Acta* **1777**: 433–440.

Haupt-Herting, S. and Fock, H.P. (2002) Oxygen exchange in relation to carbon assimilation in water-stressed leaves during photosynthesis. *Annals of Botany* **89**: 851–859.

Haupt-Herting, S., Klug, K. and Fock, H.P. (2001) A new approach to measure gross CO_2 fluxes in leaves. Gross CO_2 assimilation, photorespiration, and mitochondrial respiration in the light in tomato under drought stress. *Plant Physiology* **126**: 388–396.

He, J., Duan, Y., Hua, D. *et al.* (2012) DEXH box RNA helicase-mediated mitochondrial reactive oxygen species production in *Arabidopsis* mediates crosstalk between abscisic acid and auxin signalling. *The Plant Cell* **24**: 1815–1833.

Hebbelmann, I., Selinski, J., Wehmeyer, C. *et al.* (2012) Multiple strategies to prevent oxidative stress in *Arabidopsis* plants lacking the malate valve enzyme NADP-malate dehydrogenase. *Journal of Experimental Botany* **63**: 1445–1459.

Hummel, I., Pantin, F., Sulpice, R. *et al.* (2010) *Arabidopsis* plants acclimate to water deficit at low cost through changes of carbon usage: An integrated perspective using growth, metabolite, enzyme, and gene expression analysis. *Plant Physiology* **154**: 357–372.

Hüner, N.P.A., Bode, R., Dahal, K. *et al.* (2012) Chloroplast redox imbalance governs phenotypic plasticity: The 'grand design of photosynthesis' revisited. *Frontiers in Plant Science* **3**: Article 255.

Igamberdiev, A.U., Bykova, N.V. and Gardeström, P. (1997) Involvement of cyanide-resistant nd rotenone-insensitive pathways of mitochondrial electron transport during oxidation of glycine in higher plants. *FEBS Letters* **412**: 265–269.

Igamberdiev, A.U., Bykova, N.V. and Gardeström, P. (2001) The role of photorespiration in redox and energy balance of photosynthetic plant cells: A study with a barley mutant deficient in glycine decarboxylase. *Physiologia Plantarum* **111**: 427–438.

Ivanov, A.G., Rosso, D., Savitch, L.V. *et al.* (2012) Implications of alternative electron sinks in increased resistance to high light stress in cold-acclimated *Arabidopsis thaliana. Photosynthesis Research* **113**: 191–206.

Johnson, G.N. (2011) Physiology of PSI cyclic electron transport in higher plants. *Biochimica et Biophysica Acta* **1807**: 384–389.

Joliot, P. and Johnson, G.N. (2011) Regulation of cyclic and linear electron flow in higher plants. *PNAS* **108**: 13317–13322.

Juszczuk, I.M., Szal, B. and Rychter, A.M. (2012) Oxidation-reduction and reactive oxygen species homeostasis in mutant plants with respiratory chain complex I dysfunction. *Plant, Cell & Environment* **35**: 296–307.

Kim, T.-H., Böhmer, M., Hu, H. *et al.* (2010) Guard cell signal transduction network: Advances in understanding abscisic acid, CO_2, and Ca^{2+} signalling. *Annual Review of Plant Biology* **61**: 561–591.

Kohzuma, K., Cruz, J.A., Akashi, K. *et al.* (2009) The long-term responses of the photosynthetic proton circuit to drought. *Plant, Cell & Environment* **32**: 209–219.

Kramer, D.M. and Evans, J.R. (2011) The importance of energy balance in improving photosynthetic productivity. *Plant Physiology* **155**: 70–78.

Laureau, C., De Paepe, R., Latouche, G. *et al.* (2013) Plastid terminal oxidase (PTOX) has the potential to act as a safety valve for excess excitation energy in the alpine plant species *Ranunculus glacialis* L. *Plant, Cell & Environment* **36**: 1296–1310.

Lawlor, D.W. and Tezara, W. (2009) Causes of decreased photosynthetic rate and metabolic capacity in water-deficient leaf cells: A critical evaluation of mechanisms and integration of processes. *Annals of Botany* **103**: 561–579.

León, P., Gregorio, J. and Cordoba, E. (2013) ABI4 and its role in chloroplast retrograde communication. *Frontiers in Plant Science* **3**: Article 304.

Liu, Y., He, J., Chen, Z. *et al.* (2010) *ABA overly-insensitive 5* (*ABO5*), encoding a pentatricopeptide repeat protein required for *cis*-splicing of mitochondrial *nad2* intron 3, is involved in the abscisic acid response in *Arabidopsis*. *The Plant Journal* **63**: 749–765.

Liu, Y.-J., Nunes-Nesi, A., Wallström, S.V. *et al.* (2009) A redox-mediated modulation of stem bolting in transgenic *Nicotiana sylvestris* differentially expressing the external mitochondrial NADPH dehydrogenase. *Plant Physiology* **150**: 1248–1259.

McDonald, A.E., Ivanov, A.G., Bode, R. *et al.* (2011) Flexibility in photosynthetic electron transport: The physiological role of plastoquinol terminal oxidase (PTOX). *Biochimica et Biophysica Acta* **1807**: 954–967.

McDonald, A.E. and Vanlerberghe, G.C. (2006) The organization and control of plant mitochondrial metabolism. In: Plaxton, W.C., McManus, M.T. (eds) *Control of Primary Metabolism in Plants (Annual Plant Reviews*, Vol 22). Blackwell Publishing, Oxford, UK, pp. 290–324.

Miao, Y., Lv, D., Wang, P. *et al.* (2006) An *Arabidopsis* glutathione peroxidase functions as both a redox transducer and a scavenger in abscisic acid and drought stress responses. *The Plant Cell* **18**: 2749–2766.

Millar, A.H., Whelan, J., Soole, K.L. and Day, D.A. (2011) Organization and regulation of mitochondrial respiration in plants. *Annual Review of Plant Biology* **62**: 79–104.

Miller, G., Suzuki, N., Ciftci-Yilmaz, S. and Mittler, R. (2010) Reactive oxygen species homeostasis and signalling during drought and salinity stresses. *Plant, Cell & Environment* **33**: 453–467.

Mittler, R. and Zilinskas, B.A. (1994) Regulation of pea cytosolic ascorbate peroxidase and other antioxidant enzymes during the progression of drought stress and following recovery from drought. *The Plant Journal* **5**: 397–405.

Miura, K., Okamoto, H., Okuma, E. *et al.* (2013) *SIZ1* deficiency causes reduced stomatal aperture and enhanced drought tolerance via salicylic acid-induced accumulation of reactive oxygen species in *Arabidopsis*. *The Plant Journal* **73**: 91–104.

Miyake, C., Shinzaki, Y., Miyata, M. and Tomizawa, K. (2004) Enhancement of cyclic electron flow around PSI at high light and its contribution to the induction of non-photochemical quenching of chl fluorescence in intact leaves of tobacco plants. *Plant & Cell Physiology* **45**: 1426–1433.

Modolo, L.V., Augusto, O., Almeida, I.M.G. *et al.* (2005) Nitrite as the major source of nitric oxide production by *Arabidopsis thaliana* in response to *Pseudomonas syringae*. *FEBS Letters* **579**: 3814–3820.

Molassiotis, A. and Fotopoulos, V. (2011) Oxidative and nitrosative signalling in plants. Two branches of the same tree? *Plant Signaling & Behavior* **6**: 210–214.

Møller, I.M. (2001) Plant mitochondria and oxidative stress: electron transport, NADPH turnover, and metabolism of reactive oxygen species. *Annual Review of Plant Physiology and Plant Molecular Biology* **52**: 561–591.

Møller, I.M., Jensen, P.E. and Hansson, A. (2007) Oxidative modifications to cellular components in plants. *Annual Review of Plant Biology* **58**: 459–481.

Muller, B., Pantin, F., Génard, M. *et al.* (2011) Water deficits uncouple growth from photosynthesis, increase C content, and modify the relationship between C and growth in sink organs. *Journal of Experimental Botany* **62**: 1715–1729.

Muñoz, R. and Quiles, M.J. (2013) Water deficit and heat affect the tolerance to high illumination in Hibiscus plants. *International Journal of Molecular Science* **14**: 5432–5444.

Murphy, M.P. (2009) How mitochondria produce reactive oxygen species. *The Biochem Journal* **417**: 1–13.

Neill, S., Barros, R., Bright, J. *et al.* (2008) Nitric oxide, stomatal closure, and abiotic stress. *Journal of Experimental Botany* **59**: 165–176.

Noctor, G., Dutilleul, C., De Paepe, R. and Foyer, C.H. (2004) Use of mitochondrial electron transport mutants to evaluate the effects of redox state on photosynthesis, stress tolerance and the integration of carbon/nitrogen metabolism. *Journal of Experimental Botany* **55**: 49–57.

Noctor, G., Veljovic-Jovanovic, S., Driscoll, S. *et al.* (2002) Drought and oxidative load in the leaves of C_3 plants: A predominant role for photorespiration? *Annals of Botany* **89**: 841–850.

Norman, C., Howell, K.A., Millar, A.H. *et al.* (2004) Salicylic acid is an uncoupler and inhibitor of mitochondrial electron transport. *Plant Physiology* **134**: 492–501.

Okegawa, Y., Kobayashi, Y. and Shikanai, T. (2010) Physiological links among alternative electron transport pathways that reduce and oxidize plastoquinone in *Arabidopsis*. *The Plant Journal* **63**: 458–468.

Palmieri, M.C., Lindermayr, C., Bauwe, H. *et al.* (2010) Regulation of plant glycine decarboxylase by *S*-nitrosylation and glutathionylation. *Plant Physiology* **152**: 1514–1528.

Paredes, M. and Quiles, M.J. (2013) Stimulation of chlororespiration by drought under heat and high light illumination in *Rosa meillandina*. *Journal of Plant Physiology* **170**: 165–171.

Pfalz, J., Liebers, M., Hirth, M. *et al.* (2012) Environmental control of plant nuclear gene expression by chloroplast redox signals. *Frontiers in Plant Science* **3**: Article 257.

Pinheiro, C. and Chaves, M.M. (2011) Photosynthesis and drought: Can we make metabolic connections from available data? *Journal of Experimental Botany* **62**: 869–882.

Plaxton, W.C. and Podestá, F.E. (2006) The functional organization and control of plant respiration. *Critical Reviews in Plant Science* **25**: 159–198.

Poyton, R.O., Ball, K.A, and Castello, P.R. (2009) Mitochondrial generation of free radicals and hypoxic signalling. *Trends in Endocrinology and Metabolism* **20**: 332–340.

Rasmusson, A.G., Geisler, D.A. and Møller, I.M. (2008) The multiplicity of dehydrogenases in the electron transport chain of plant mitochondria. *Mitochondrion* **8**: 47–60.

Rasmusson, A.G., Soole, K.L. and Elthon, T.E. (2004) Alternative NAD(P)H dehydrogenases of plant mitochondria. *Annual Review of Plant Biology* **55**: 23–39.

Rhoads, D.M., Umbach, A.L., Sweet, C.R. *et al.* (1998) Regulation of the cyanide-resistant alternative oxidase of plant mitochondria: Identification of the cysteine residue involved in α-keto acid stimulation and intersubunit disulfide bond formation. *Journal of Biological Chemistry* **273**: 30750–30756.

Ribas-Carbo, M., Taylor, N.L., Giles, L. *et al.* (2005) Effects of water stress on respiration in soybean leaves. *Plant Physiology* **139**: 466–473.

Rochaix, J.-D. (2011) Regulation of photosynthetic electron transport. *Biochimica et Biophysica Acta* **1807**: 375–383.

Rott, M., Martins, N.F., Thiele, W. *et al.* (2011) ATP synthase repression in tobacco restricts photosynthetic electron transport, CO_2 assimilation, and plant growth by overacidification of the thylakoid lumen. *The Plant Cell* **23**: 304–321.

Ruban, A.V., Johnson, M.P. and Duffy, C.D.P. (2012) The photoprotective molecular switch in the photosystem II antenna. *Biochimica et Biophysica Acta* **1817**: 167–181.

Sanhueza, C., Bascuuan-Godoy, L., Corcuera, L.J. and Turnbull, M.H. (2013) The response of leaf respiration to water stress in *Nothofagus* species. *New Zealand Journal of Botany* **51**: 88–103.

Scheibe, R. and Dietz, K.-J. (2012) Reduction-oxidation network for flexible adjustment of cellular metabolism in photoautotrophic cells. *Plant, Cell & Environment* **35**: 202–216.

Schwarzländer, M. and Finkemeier, I. (2013) Mitochondrial energy and redox signalling in plants. *Antioxidants & Redox Signaling* **18**: 2122–2144.

Shirao, M., Kuroki, S., Kaneko, K. *et al.* (2013) Gymnosperms have increased capacity for electron leakage to oxygen (Mehler and PTOX reactions) in photosynthesis compared with angiosperms. *Plant & Cell Physiology* **54**: 1152–1163.

Signorelli, S., Corpas, F.J., Borsani, O. *et al.* (2013) Water stress induces a differential and spatially distributed nitro-oxidative stress response in roots and leaves of *Lotus japonicus*. *Plant Science* **201–202**: 137–146.

Skirycz, A., De Bodt, S., Obata, T. *et al.* (2010) Developmental stage specificity and the role of mitochondrial metabolism in the response of *Arabidopsis* leaves to prolonged mild osmotic stress. *Plant Physiology* **152**: 226–244.

Smith, A.M.O., Ratcliffe, R.G. and Sweetlove, L.J. (2004) Activation and function of mitochondrial uncoupling protein in plants. *Journal of Biological Chemistry* **279**: 51944–51952.

Stepien, P. and Johnson, G.N. (2009) Contrasting responses of photosynthesis to salt stress in the glycophyte *Arabidopsis* and the halophyte *Thellungiella*: Role of the plastid terminal oxidase as an alternative electron sink. *Plant Physiology* **149**: 1154–1165.

Stitt, M., Lunn, J. and Usadel, B. (2010) *Arabidopsis* and primary photosynthetic metabolism – more than the icing on the cake. *The Plant Journal* **61**: 1067–1091.

Suzuki, N., Koussevitzky, S., Mittler, R. and Miller, G. (2012) ROS and redox signalling in the response of plants to abiotic stress. *Plant, Cell & Environment* **35**: 259–270.

Sweetlove, L.J., Lytovchenko, A., Morgan, M. *et al.* (2006) Mitochondrial uncoupling protein is required for efficient photosynthesis. *PNAS* **103**: 19587–19592.

Taniguchi, M. and Miyake, H. (2012) Redox-shuttling between chloroplast and cytosol: Integration of intra-chloroplast and extra-chloroplast metabolism. *Current Opinion in Plant Biology* **15**: 252–260.

Taylor, N.L., Heazlewood, J.L., Day, D.A. and Millar, A.H. (2005) Differential impact of environmental stresses on the pea mitochondrial proteome. *Molecular & Cellular Proteomics* **4**: 1122–1133.

Tcherkez, G., Boex-Fontvieille, E., Mahé, A. and Hodges, M. (2012) Respiratory carbon fluxes in leaves. *Current Opinion in Plant Biology* **15**: 308–314.

Tezara, W., Driscoll, S. and Lawlor, D.W. (2008) Partitioning of photosynthetic electron flow between CO_2 assimilation and O_2 reduction in sunflower plants under water deficit. *Photosynthetica* **46**: 127–134.

Tezara, W., Mitchell, V.J., Driscoll, S.P. and Lawlor, D.W. (1999) Water stress inhibits plant photosynthesis by decreasing coupling factor and ATP. *Nature* **401**: 914–917.

Timm, S., Florian, A., Arrivault, S. *et al.* (2012) Glycine decarboxylase controls photosynthesis and plant growth. *FEBS Letters* **586**: 3692–3697.

Vani, T. and Raghavendra, A.S. (1994) High mitochondrial activity but incomplete engagement of the cyanide-resistant alternative pathway in guard cell protoplasts of pea. *Plant Physiology* **105**: 1263–1268.

Vanlerberghe, G.C. (2013) Alternative oxidase: A mitochondrial respiratory pathway to maintain metabolic and signalling homeostasis during abiotic and biotic stress in plants. *International Journal of Molecular Science* **14**: 6805–6847.

Vanlerberghe, G.C., Day, D.A. *et al.* (1995) Alternative oxidase activity in tobacco leaf mitochondria: Dependence of tricarboxylic acid cycle-mediated redox regulation and pyruvate activation. *Plant Physiology* **109**: 353–361.

Vasquez-Robinet, C., Mane, S.P., Ulanov, A.V. *et al.* (2008) Physiological and molecular adaptations to drought in Andean potato genotypes. *Journal of Experimental Botany* **59**: 2109–2123.

Vass, I. (2012) Molecular mechanisms of photodamage in the photosystem II complex. *Biochimica et Biophysica Acta* **1817**: 209–217.

Vassileva, V., Simova-Stoilova, L., Demirevska, K. and Feller, U. (2009) Variety-specific response of wheat (*Triticum aestivum* L.) leaf mitochondria to drought stress. *Journal of Plant Research* **122**: 445–454.

Vercesi, A.E., Borecký, J., de Godoy Maia, I. *et al.* (2006) Plant uncoupling mitochondrial proteins. *Annual Review of Plant Biology* **57**: 383–404.

Wang, J. and Vanlerberghe, G.C. (2013) A lack of mitochondrial alternative oxidase compromises capacity to recover from severe drought stress. *Physiologia Plantarum* DOI: 10.1111/ppl.12059. [Epub ahead of print]

Wind, J.J., Peviani, A., Snel, B. *et al.* (2013) ABI4: versatile activator and repressor. *Trends in Plant Science* **18**: 125–132.

Wu, G., Wilen, R.W., Robertson, A.J. and Gusta, L.V. (1999) Isolation, chromosomal localization, and differential expression of mitochondrial manganese superoxide dismutase and chloroplastic copper/zinc superoxide dismutase in wheat. *Plant Physiology* **120**: 513–520.

Yoshida, K., Watanabe, C.K., Terashima, I. and Noguchi, K. (2011) Physiological impact of mitochondrial alternative oxidase on photosynthesis and growth in *Arabidopsis thaliana*. *Plant, Cell & Environment* **34**: 1890–1899.

Yuan, H. and Liu, D. (2012) Functional disruption of the pentatricopeptide protein SLG1 affects mitochondrial RNA editing, plant development, and responses to abiotic stresses in *Arabidopsis*. *The Plant Journal* **70**: 432–444.

CHAPTER 9
Regulation of cytochrome and alternative pathways under light and osmotic stress

Padmanabh Dwivedi
Department of Plant Physiology, Institute of Agricultural Sciences, Banaras Hindu University, Varanasi, India

Introduction

The respiratory electron transport pathway of plant mitochondria comprises the cytochrome (Cyt) pathway and an alternative pathway (McDonald *et al.*, 2002). Both the Cyt and the alternative respiratory pathways start at protein complex I when NADH is being oxidized. One H^+ (proton) is transported by the complex I to the inner membrane space, whereas two electrons are transported within the inner membrane by the ubiquinone, which at its reduced state (Q_r) transfers these electrons either to complex III or to another protein known as alternative oxidase (AOX). Ubiquinone is the point at which the reactions can proceed in different ways, and it is called the branch point. The Cyt respiratory pathway is present in all living organisms and proceeds when complex III pulls out a proton from the mitochondrial matrix to the intermembrane space. The electrons are received by cytochrome *c* which spreads up to the outer side of the inner membrane towards protein complex IV, which then pulls out another proton similar to complexes I and III, and transports the electrons back to the inner domain of the mitochondria (see also Chapter 1). As a result, oxygen is consumed with a proton and the two electrons to produce water (Figure 9.1). Electron transfer through the Cyt pathway is coupled with ATP synthesis and is inhibited by cyanide, azide and CO_2.

The alternative respiratory pathway is a feature typical of plants, algae, fungi and to some extent protozoa. Unlike the Cyt pathway, there is no proton gradient formation in the alternative respiratory pathway. This type of respiration is brought about by the protein alternative oxidase (AOX), which is a dimer in its inactive form (oxidized state). Oxygen is consumed and through a reaction with electrons transported to AOX and a proton, water is produced. Thus, both the

Alternative Respiratory Pathways in Higher Plants, First Edition.
Edited by Kapuganti Jagadis Gupta, Luis A.J. Mur and Bhagyalakshmi Neelwarne.

respiratory pathways transfer proton(s) to the intermembrane space, transport a couple of electrons and consume oxygen to produce water; however, the difference between the two is that the Cyt pathway transfers two additional protons from the mitochondrial matrix to the intermembrane space thereby leading to a greater proton gradient (Figure 9.2). Electron flow from ubiquinone is through the alternative pathway, which is non-phosphorylating, cyanide-resistant and can yield only about one third of the ATP compared to that generated by the Cyt pathway. The alternative pathway is inhibited by salicyl-hydroxamic acid (SHAM) and n-propyl gallate. AOX plays an important role in the integration of carbon metabolism and electron transport, besides having a role in specific cellular and developmental processes (Vanlerberghe, 2013).

AOX, for a long time been considered to act as an overflow mechanism (energy overflow) (Lambers, 1985), with the exception or modification that it can compete for and share electrons with Cyt *c* oxidase (Simons and Lambers, 1999). The overflow model associated with AOX appears to strike a balance

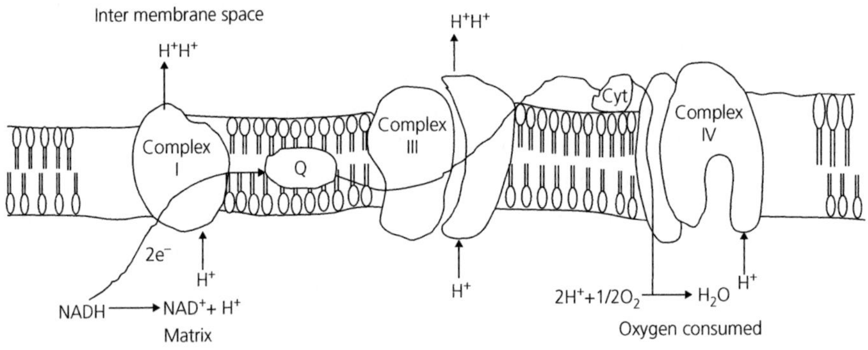

Figure 9.1 Cytochrome respiratory pathway.

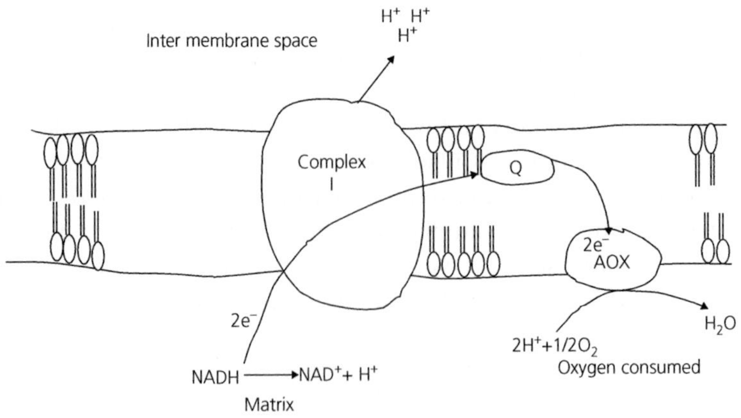

Figure 9.2 Alternative respiratory pathway.

between carbon metabolism and electron transport. This is because a metabolic condition that causes accumulation of either reduced ubiquinone or mitochondrial pyridine nucleotides or pyruvate or citrate has the potential to increase electron flow to the alternative pathway (Moore *et al.*, 2002). The mitochondrial electron transport adjusts its capacity through the alternative pathway, and this has two implications for plant metabolism: if cells need a large amount of carbon sources, without a high ATP demand, then AOX facilitates operation of the TCA cycle. Secondly, AOX prevents over-reduction of respiratory chain components that might result in the production of harmful reactive oxygen species (ROS) (Zhang *et al.*, 2010); thus, AOX plays an important role in the avoidance of cell damage by ROS. This is imperative in *in vitro* studies using tobacco AOX mutants where cells over-expressing AOX contained half as much ROS as control cells (Maxwell *et al.*, 1999), whereas cells with reduced AOX expression due to anti-sensing contained five times more ROS than control cells. Thus, AOX plays a role in preventing the formation of oxygen free radicals. Ubiquinone is a common substrate for both the Cyt and the alternative respiratory pathway. A high reduction state of the ubiquinone pool (Q_r/Q_t) is a feature when the Cyt pathway is inhibited or restricted, and promotes oxygen free radical formation. Respiration via the alternative pathway can help maintain Q_r/Q_t at a low level, probably through stabilizing the reduction state of the mitochondrial ubiquinone pool (Purvis and Shewfelt, 1993).

AOX characteristics: distribution, abundance and activity

AOX is reported in species such as *Arabidopsis thaliana, Oryza sativa, Sauromatum guttatum, Glycine max, Nicotiana tabaccum, Zea mays* and *Pisum sativum*. It is widely accepted that AOX is found throughout the plant kingdom; AOX is reported in the angiosperms, protista, fungi and phytoplanktons (Luz *et al.*, 2002). The transcript level gives an insight of AOX abundance, thereby indicating the manner in which AOX gene expression changes under a given experimental condition, as evident from a study made of *Arabidopsis* in which AOX mRNA was correlated with response to electron transport inhibitors (Saisho *et al.*, 2001). RT-PCR has been employed for AOX transcript measurement in *Arabidopsis* and soybean (Finnegan *et al.*, 1997). A monoclonal antibody raised against a *Sauromatum guttatum* AOX protein facilitated identification and quantification of AOX; it has a highly conserved sequence among plant AOX proteins (Finnegan *et al.*, 1999). Several AOX genes have been isolated and multi-gene families identified from different plant species (Saisho *et al.*, 1997). Sense and antisense constructs of AOX genes have been used to produce transgenic plants which have increased and decreased levels of AOX proteins (Vanlerberghe *et al.*, 1994).

Maximum AOX activity (AOX capacity) of a plant cell or tissue is an estimation of the maximum flux of electrons to AOX, and is measured by addition of a Cyt pathway inhibitor (like CN) followed by addition of an AOX inhibitor (like SHAM). AOX capacity is thus defined as the oxygen uptake resistant to the Cyt pathway inhibitor and sensitive to the AOX inhibitor. An inherent problem of metabolic inhibitors is the possibility of their unspecific and multiple effects on different processes in cells (Moller *et al.*, 1988). As long as the inhibitors are used at a low concentration and for relatively short-term assays, the probability of potential problems is minimal. AOX engagement is a measure of the actual flux of electrons to AOX within a cell, under physiological conditions; but this is more difficult to determine compared to AOX capacity. Generally the ability of an AOX inhibitor to decrease oxygen uptake in the absence of the Cyt pathway inhibitor is examined. But this approach can underestimate AOX engagement in cases where it might have been engaged. Because AOX can compete with the Cyt pathway for electrons, the use of inhibitors for quantifying AOX engagement is discouraged, and the most reliable way of measuring AOX engagement suggested so far is an oxygen isotope discrimination technique (Guy *et al.*, 1989), in which AOX and cytochrome oxidase discriminate to different extents against heavy labelled oxygen.

Structure and regulation of AOX activity

AOX is a mitochondrial inner membrane protein functioning as a component of the plant alternative electron transport chain. AOX, which catalyses four-electron reduction of oxygen to water, branches from the main respiratory chain at the level of ubiquinone. Contrary to electron transfer by the Cyt chain, AOX does not pump H^+ and therefore electron transfer by AOX is not mediated by a transmembrane potential and the drop in free energy between ubiquinol and oxygen is dissipated as heat (Vanlerberghe and McIntosh, 1997). The enzyme, AOX is difficult to purify to homogeneity; however, information obtained from cDNA sequences encoding the AOX protein reveals the AOX structure: AOX from plants is encoded by nuclear genes (Elthon *et al.*, 1989a) and consists of 1–3 proteins between 32 and 39 kDa, depending on species (McIntosh, 1994). It operates as a homodimer with a non-haem diiron centre. *In vitro* studies have shown that AOX activity increases markedly on reduction of the intersubunit disulfide link, thereby producing a non-covalently linked homodimeric protein (Umbach *et al.*, 1994). The reduced enzyme is then activated by pyruvate (α-keto acid), a thiohemiacetal with a protein-derived sulfhydryl moiety (Rhoads *et al.*, 1998; Umbach *et al.*, 2002). Similarly, Berthold *et al.* (2000) proposed the structure of ubiquinone binding sites of AOX.

Gene expression

AOX is encoded by a small family of nuclear genes (*AOX1*, *AOX2*a and *AOX2b*) from a wide variety of non-thermogenic monocots and dicots (Considine *et al.*, 2002). It is proposed that *AOX1* gene expression constitutes the plant's adaptation to stress factors, whereas *AOX2* expression depends on tissue and developmental stage. AOX gene expresses under a variety of biotic and/or abiotic stress conditions indicating thereby that AOX belongs to stress-induced plant proteins. Environmental and developmental conditions involve changes in AOX mRNA, protein, and/or cyanide-resistant respiration. Therefore, AOX gene expression can change in response to an experimental treatment in a developmental specific, tissue-specific and stress-specific manner. In soybean, for instance, expression of three AOX genes in roots and cotyledons differs in the amount of particular gene transcript as well as protein levels. In potato, AOX mRNA and AOX protein accumulate during ageing (Hisher and McIntosh, 1990). Similarly, in bean roots the AOX protein increased under phosphate-deficient conditions (Juszczuk *et al.*, 2001a).

Studies have been made that suggest the role of signal transduction from stressed mitochondria to the nucleus, for transcription of genes. Since partition of electrons between the Cyt chain and AOX is highly regulated and influenced by stress conditions, it implies that signal inducing expression of AOX gene is perceived in the mitochondria and then transmitted to nucleus (McIntosh *et al.*, 1998); both AOX protein concentration and AOX activity increase when plants are subjected to stress conditions such as chilling (Purvis and Shewfelt, 1993) and phosphate deficiency (Juszczuk *et al.*, 2001b). Most of these stress conditions lead to oxidative stress thereby causing an increased production of ROS by the mitochondrial respiratory pathway, and this ROS is considered to be important for the increased AOX protein: addition of 5 mM H_2O_2 to tobacco suspension cells led to increase in AOX1 mRNA levels and AOX capacity (Vanlerberghe and McIntosh, 1996). Similarly, AOX1 gene was induced following treatment with antimycin A or H_2O_2 in tobacco cultured cells (Maxwell *et al.*, 2002). It has been suggested that respiratory-deficient and direct AOX-gene mutants might have a role in analysis of mitochondria-nuclear signalling pathways. Other reports (besides the theory of H_2O_2 and/or other ROS-mediated gene regulation of AOX) indicate that the carbon flux through the TCA cycle can also regulate AOX gene expressions. This notion is supported by the fact that signals affecting AOX1 gene expression are connected with the carbon load and redox status of the mitochondria (Vanlerberghe *et al.*, 2002).

Post-translational control of AOX activity

As there appears to be no direct correlation between AOX protein abundance and its engagement in respiration (McDonald *et al.*, 2002), implies that partitioning of electrons to AOX is determined by post-translational mechanisms. The factors which regulate AOX activity include *in vitro* substrate level, ubiquinone concentration and its redox poise, the redox state of AOX and pyruvate (Siedow and Umbach, 2000; McDonald *et al.*, 2002; Umbach *et al.*, 2002). Voltametric assays and HPLC analysis have been used to study the redox state of ubiquinone in both intact tissues and isolated mitochondria (Ribas-Carbo *et al.*, 1995; Wagner and Wagner, 1995); the result showed that ubiquinone redox poise remained constant over a wide range of AOX engagement in respiration. Some organic acids like glyoxylate, pyruvate, hydroxypyruvate and 2-oxoglutamate activate AOX (Day *et al.*, 1995); these serve as substrates for AOX. In tobacco leaf mitochondria where AOX is oxidized after organelle isolation, AOX activity is very slow until both pyruvate and a reductant are added, thereby suggesting that redox state of AOX guides AOX capacity, whereas pyruvate levels determine how much of that capacity is realized. There is no clear-cut correlation between AOX concentration and its activity *in vivo*.

Cytochrome and alternative respiratory pathways under stress conditions with special reference to light and osmotic stress

The relative contribution of these two pathways to total respiration is flexible and depends on environmental conditions (Gonzalez-Meler *et al.*, 1999). The response of plant respiration to abiotic stress varies with the stress factor as well as the duration of the treatment or exposure to such stress factors (Poorter *et al.*, 1992; Collier and Cummins, 1993; Lambers *et al.*, 1998). Nutrient deficiency, anoxia and low light intensity induced the increased participation of the alternative pathway in plant tissues (Zhou and Solomos, 1998; Millenaar *et al.*, 2000). Operation of the alternative pathway is likely to increase in illuminated plant tissues; AOX level increases upon greening of etiolated leaves (Atkin *et al.*, 1993), and sugars formed during this illumination promote engagement of the alternative pathway (Azcon-Bieto, 1992). The role of mitochondrial oxidative phosphorylation for photosynthetic carbon assimilation is well established; however, the role of the two respiratory pathways in benefiting photosynthetic metabolism has been examined in only a few cases. The importance of both these pathways during photosynthesis was studied in mesophyll protoplasts of pea and barley using the mitochondrial inhibitors oligomycin, antimycin A and SHAM. All three inhibitors decreased the rate of photosynthetic oxygen evolution but had no impact on chloroplast photosynthesis (Kromer *et al.*, 1993; Igamberdiev *et al.*, 1997; Padmasree and Raghavendra, 1999). The sensitivity of photosynthesis to SHAM and antimycin A was indicated as essential for the alternative pathway to photosynthesis. The alternative pathway is also important during interactions between respiration and photosynthesis, as evinced from the sensitivity of light-enhanced dark respiration (LEDR) to SHAM in mesophyll protoplasts of barley (Igamberdiev *et al.*, 1997) and algae *Chlamydomonas reinhardtii* and *Euglena gracilis* (Xue *et al.*, 1996; Ekelund, 2000).

The expression and protein level of AOX is dependent upon irradiance. Under high light (HL) of 3000 μmol m^{-2} s^{-1}, pea mesophyll protoplasts showed decreased rates of $NaHCO_3$-dependent O_2 evolution, whereas the decrease in respiratory uptake was marginal. The AOX pathway showed a significant twofold increase under HL, while the capacity of the Cyt pathway declined by more than 50% when compared to capacities under normal light and darkness. Pyruvate and malate – products of photosynthesis and stimulators of AOX activity – also increased with increased AOX protein under HL (Dinakar *et al.*, 2010).

AOX activity was found to be higher in 'sun' species than in 'shade' species. Noguchi *et al.* (2005) showed that *Alocasia odora* – a shade-loving plant – regulates its respiratory capacity by making changes in the mitochondrial number in leaves when subjected to growth under varied light regimes. It maintained a high AOX capacity whose activity was controlled by keeping AOX protein as inactivated under low light. However, this inactivated, oxidized dimer form was converted to a reduced, active form once the plants were shifted to HL conditions.

There is growing evidence that AOX plays an important role in balancing photosynthesis and respiration metabolism under HL stress. The AOX pathway protects plants from the effects of photoinhibition; the NADPH produced in chloroplasts and transported into mitochondria – via various shuttles such as the malate–oxaloacetate shuttle – is oxidized by mitochondrial AOX. AOX does this without being restricted by a proton gradient across the mitochondrial membrane or the ATP/ADP ratio, as shown in *Rumex* leaves (Zhang *et al.*, 2012). Inhibition of the AOX pathway by SHAM leads to an accumulation of NADPH (reducing equivalents) in chloroplasts causing over-reduction of photosystem I (PSI) acceptor side. As a result, this restriction of photosynthetic electron-flow-generated change of pH of thylakoid and finally non-photochemical quenching (NPQ) was found to be suppressed. This indicated that mitochondrial AOX pathway protects the photosynthetic apparatus against photo-damage by combating over-reduction of PSI acceptor side and also by accelerating induction of NPQ. AOX also imparts protection against photo-oxidation damage by regulating ROS production, which stems from photosynthetic electron transport. ROS production increases with increasing light intensity. Excess ROS causes photo-oxidation damage to photosynthetic apparatus. AOX suppresses ROS production and maintains the photosynthetic electron transport chain in an oxidized state during stress conditions (Zhang *et al.*, 2010). They found increased ROS and reducing equivalents accumulation in *Arabidopsis aox1a* mutant compared to wild-type after HL exposure. Also, enzymes like NADP-MDH, citrate synthase and NADP-ME increased with HL treatment and remained higher in *aox1a* mutant. Thus, increased respiratory rates may lead to ROS production and hence mitochondrial oxidative damage. The overall reduction level of the mitochondrial ubiquinone pool is thought to be the primary determinate of mitochondrial ROS (mtROS) output. ROS formation is prevented via an alternative pathway involving AOX in plants, which is induced under various biotic and abiotic stresses. An increased electron flux through AOX helps to maintain redox levels of the respiratory components relatively oxidized, thereby minimizing ROS generation

Osmotic stress is known to prolong the induction phase, inhibit photosynthetic carbon metabolism and stimulate respiration in protoplasts at 25 °C as well as induce an increased capacity of the alternative pathway (Saradadevi and Raghavendra, 1992). The earlier reports did not give a clear picture of the extent and engagement of these two respiratory pathways under osmotic stress. Osmotic stress was reported to inhibit total respiration (Pheloung and Barlow, 1981). The alternative pathway in mitochondria isolated from mannitol-stressed mung bean was less sensitive to osmotic stress than the Cyt pathway (Schmitt and Dizengremel, 1989). Similarly, leaf discs of *Saxifraga cernua* exposed to a range of sorbitol osmotic potentials from 0.0 to −4.0 MPa did not exhibit any differential response of Cyt and alternative pathways (Collier and Cummins, 1996). However, exposing pea mesophyll protoplasts to osmotic stress (1.0 M sorbitol, hyperosmoticum) led to reduction in the proportion of Cyt pathway from 51 to 32%, and increase in alternative pathway from 25 to 37%, as compared to normal 0.4 M sorbitol (Dwivedi *et al.*, 2003); the extent of engagement of the alternative pathway was less ($\rho = 0.8$) under 0.4 M sorbitol than that under 1.0 M sorbitol ($\rho = 1.0$), reflecting the complete participation of (instead of full engagement of) alternative pathway under hyperosmoticum condition.

José Hélio Costa *et al.* (2007) studied AOX at different levels such as transcript, protein and capacity in response to osmotic stress given to roots of cowpea (*Vigna unguiculata*). Two cultivars used were Vita3 (tolerant) and Vita5 (sensitive) to drought/saline stress. The results demonstrated up- and down-regulation through *VaAox2b* gene in response to osmotic stress. Vita5 cultivar maintained a higher amount of AOX protein, while the sensitive cultivar, Vita5, tended in stress conditions of 100 mM NaCl and PEG to reach that protein level. Similarly, increased AOX transcript as well as its protein concentration have been correlated to salt stress (osmotic effect) in a number of plant species including pea (Marti *et al.*, 2011), *Arabidopsis* (Kreps *et al.*, 2002), poplar (Ottow *et al.*, 2005) and tobacco (Andronis and Roubelakis-Angelakis, 2010). Experiments using isotope discrimination indicated that 14-day salt stress in pea decreased leaf Cyt pathway, whereas the level of AOX pathway respiration was maintained, thus suggesting a key role of AOX in respiratory activity in osmotically stressed pea leaves (Marti *et al.*, 2011).

Arabidopsis plants subjected to salinity stress showed ROS accumulation, increased Na^+ level in shoot and root besides increased transcripts of *Ataox1a, Atndb2* and *Atndb4* genes. Plants over-expressing *Ataox1a* with increased AOX capacity showed lower ROS production, 30–40% better growth rates and lower shoot Na^+ content as compared to controls, under salinity stress. It was demonstrated that more active AOX in root and shoot improved salt tolerance of *Arabidopsis* as evinced by its ability to grow efficiently in the presence of NaCl (Smith *et al.*, 2009).

Other physiological roles of AOX

The respiration in thermogenic inflorescence such as that found in *Arum* lilies takes place through AOX as a result of an increased AOX capacity and a decreased Cyt pathway capacity (Elthon *et al.*, 1989b), probably mediated by salicylic acid. Plants growing at low temperature often show higher rates of respiration compared to those growing at higher temperature when both are measured at same temperature (Collier and Cummins, 1990). This stimulation of respiration by growth at low temperature is considered to be an adaptation of plants growing in cold and arctic regions (McNulty and Cummins, 1987). It is suggested that at low temperature the increased rate of respiration involves a greater participation by the alternative pathway (Purvis and Shewfelt, 1993), due probably to enhanced synthesis of AOX protein (Vanlerberghe and McIntosh, 1992), as low temperature increases the mRNA levels of *aox1a* and *aox1b* genes, as shown for rice (Ito *et al.*, 1997). Chilling stress led to lower Cyt oxidase activity and protein levels in corn seedlings transferred to 14 °C (Prasad *et al.*, 1994) and in mung bean hypocotyls chilled at 0 °C.

Low temperature decreased Cyt pathway capacity by 30% in potato tubers transferred from 10 to 1 °C, but enhanced the capacity of the alternative pathway (Zhou and Solomos, 1998). In maize with a chilling-sensitive genotype, the decrease of root respiration was related to a decline in Cyt pathway activity at 14 °C; however, in chilling-tolerant genotypes, moderate chilling had no effect on root respiration and partitioning of electrons (Luxova and Gasparikova, 1999). Severe chilling stress leads to increased root respiration along with increased alternative pathway capacity and Cyt pathway activity in the tolerant genotype. However, severe chilling (6 °C) for 6 d resulted in an additional increase of the alternative pathway which was accompanied by some loss in Cyt pathway activity (Luxova and Gasparikova, 1999). Low temperature modulates the effect of higher osmoticum stress on photosynthesis and respiration, and results in enhanced participation of the alternative pathway (Dwivedi and Raghavendra, 2004): the protoplasts of pea were exposed to iso-osmoticum (0.4 M) and higher-osmoticum (1.0 M) concentration of sorbitol at 15 °C and 25 °C. At the optimum temperature of 25 °C there was a decline in photosynthesis (<10%) at hyper-osmoticum osmotic effect, whereas respiration increased marginally (by about 15%). Low temperature (15 °C) aggravated the sensitivity of both respiration and photosynthesis to osmotic stress. At 15 °C, the decrease in photosynthesis due to osmotic stress was more than 35%, while the respiration rate was stimulated by 30%. The relative proportion of the Cyt pathway decreased by about 50% at both 15 °C and 25 °C while that of the alternative pathway increased at 15 °C; the engagement of the alternative pathway was higher at 15 °C compared to 25 °C (Dwivedi and Raghavendra, 2004).

High root temperature (38 °C) in *Cucumis sativus* L. cv. 'Sharp I' leads to increased root respiration which is related to the stimulation of alternative respiration. Cyt respiration deteriorated at high root temperature (Du and Tachibana, 1994). Low oxygen suppresses the induction of invertase mRNA and increases the capacity of AOX and fails to prevent a decrease in Cyt capacity (Zhou and Solomos, 1998). A positive relationship between content of carbohydrate and activity of the alternative pathway has been observed in mature leaves of forest and meadow communities of North East Russia (Pystina and Danilov, 2001), growing in natural habitats. Thus, plants appear to adjust to abiotic stress by switching over to the alternative pathway under changing environmental conditions, as evident from the studies mentioned here, which showed that a wide range of such abiotic environmental conditions influence the capacity of the alternative pathway.

It is postulated that under a phosphate-deficient system, the activity of the alternative pathway increases relative to the cytochrome pathway; for example, in *Phaseolus vulgaris* and *Gliricida sepium* leaves, the AOX concentration increases in P-deficient plants. In P-deficient *Phaseolus vulgaris* plants the reduction state of the ubiquinone pool was greater in roots compared to P-enriched plants (Juszczuk *et al.*, 2001b).

AOX has a certain role in floral development: an *Arabidopsis* AOX gene is expressed in tobacco in anti-sense orientation (Kitashiba *et al.*, 1999); one plant showed reduced AOX level in anthers as well as reduced pollen viability. Further, it has been shown that AOX protein is abundantly present in tapetum and meiocytes during microsporogenesis. Fruit ripening is accompanied by a climacteric rise in respiration, induced by endogenous ethylene production. In mango and apple, ripening is associated with increased AOX protein (Cruz-Hernandez and Gomez-Lim, 1995; Duque and Arrabaca, 1999). It is suggested that ethylene can be an important signal for AOX expression, since it could not induce AOX in an *Arabidopsis* mutant lacking in ethylene response (Simons *et al.*, 1999).

AOX respiration has an important role in plant responses to pathogenic attack. For example, AOX has an active role in the resistance response of tobacco to tobacco mosaic virus (Murphy *et al.*, 1999). AOX also has an influence on xylem differentiation, a developmental process which culminates in programmed cell death (Groover and Jones, 1999); differentiating soybean root showed the AOX protein localized to developing xylem tissue, as evinced from an immunohistochemical study (Hilal *et al.*, 1997).

AOX has a certain role in continuation of the citric acid cycle (TCA cycle): TCA cycle operates under conditions of oxidation of NADH to NAD^+, but when

ADP concentration is low, this is difficult because complex I and the Cyt pathway are less active. So, rotenone-insensitive bypass and the alternative pathway become important for the continuation of the TCA cycle, particularly when the Cyt pathway is restricted. This is supported by the observation that addition of an inhibitor of the Cyt pathway leads to an increase in AOX mRNA in *Arabidopsis thaliana* (Saisho *et al.*, 2001).

Future perspectives

Molecular alteration of the mitochondrial electron transport chain components remains an interesting aspect for future investigation. In this context, transgenics with altered alternative pathway capacity will help in a critical analysis of AOX function. Studies on the biochemical analysis of purified and active alternative oxidase enzyme can unravel the intricate properties of AOX. Use of inhibitors to study the engagement of AOX is another area of interest; however, the lack of specificity of inhibitors and problems of their penetration into tissues always remain. Therefore, there is a need to study direct measurements of the *in vivo* partitioning of electrons to AOX in various plant tissues, especially under changed environmental conditions.

References

Andronis, E.A. and Roubelakis-Angelakis, K.A. (2010) Short-term salinity stress in tobacco leads to the onset of animal-like PCD hallmarks *in planta* in contrast to long-term stress. *Planta* **231**: 437–448.

Atkin, O.K., Cummins, W.R. and Collier, D.E. (1993) Light induction of alternative pathway capacity in leaf slices of Belgium endive. *Plant, Cell & Environment* **16**: 231–235.

Azcon-Bieto, J. (1992) Relationship between photosynthesis and respiration in the dark plants. In *Trends in Photosynthesis Research*. (J. Barber, M.G. Guerrero and H. Medrano, eds), pp. 241–253. Intercept, Andover.

Berthold, D.A., Anderson, M.E. and Nordlund, P. (2000) New insight into the structure and function of the alternative oxidase. *Biochimica et Biophysica Acta* **1460**: 241–254.

Collier, D.E. and Cummins, W.R. (1990) The effects of low growth and measurement temperature on the respiratory properties of five temperate species. *Annals of Botany* **65**: 533–538.

Collier, D.E. and Cummins, W.R. (1993) Effect of osmotic stress on the respiratory properties of shoots and roots of *Arnica alpina*. *Canadian Journal of Botany* **71**:1102–1108.

Collier, D.E. and Cummins, W.R. (1996) The rate of development of water deficits affects *Saxifraga cernua* leaf respiration. *Physiologia Plantarum* **96**: 291–297.

Considine, M.J., Holtzapffel, R.C., Day, D.A. *et al.* (2002) Molecular distinction between alternative oxidase from monocots and dicots. *Plant Physiology* **129**: 949–953.

Costa, J.H., Jolivet, Y., Hasenfratz-Sauder, M.-P. *et al.* (2007) Alternative oxidase regulates in roots of *Vigna unguiculata* cultivars differing in drought/salt tolerance. *Journal of Plant Physiology* **164**(4): 718–727.

Cruz-Hernandez, A. and Gomez-Lim, M.A. (1995) Alternative oxidase from mango (*Magnifera indica* L.) is differentially regulated during fruit ripening. *Planta* **197**: 569–576.

Day, D.A., Whelan, J., Miller, A.H. *et al.* (1995) Regulation of alternative oxidase in plants and fungi. *Australian Journal of Plant Physiology* **22**: 497–509.

Dinakar, C., Raghavendra, A.S. and Padmasree, K. (2010) Importance of AOX pathway in optimizing photosynthesis under high light stress: Role of pyruvate and malate in activating AOX. *Physiologia Plantarum* **139**: 13–26.

Du, Y.C. and Tachibana, S. (1994) Effect of supraoptimal root temperature on the growth, root respiration and sugar content of cucumber plants. *Scientia Horticulturae* **58**(4): 289–301.

Dwivedi, P. and Raghavendra, A.S. (2004) Modulation of osmotic stress effects on photosynthesis and respiration by temperature in mesophyll protoplast of pea. *Indian Journal of Experimental Biology* **42**: 1208–1211.

Dwivedi, P., Padmavathi, L., Jagdis Gupta, K. and Raghavendra, A.S. (2003) Increased capacity of alternative pathway of mitochondrial electron transport in mesophyll protoplasts of pea on exposure to osmotic stress. *Indian Journal of Plant Physiology* (Special Issue), 403–406.

Duque, P. and Arrabaca, J.D. (1999) Respiratory metabolism during cold storage of apple fruit. II. Alternative oxidase is inhibited at the climacteric. *Physiologia Plantarum* **107**: 24–31.

Elthon, T.E., Nickels, R.L. and McIntosh, L. (1989a) Monoclonal antibodies to the alternative oxidase of higher plant mitochondria. *Plant Physiology* **89**: 1311–1317.

Elthon, T.E., Nickels, R.L. and McIntosh, L. (1989b) Mitochondrial events during development of thermogenesis in *Sauromatum guttatum* (Schott). *Planta* **180**: 82–89.

Ekelund, N.G. (2000) Interactions between photosynthesis and light enhanced dark respiration (LEDR) in the flagellate *Euglena gracilis* after irradiation with ultraviolet radiation. *Journal of Photochemistry and Photobiology* **55**: 63–69.

Finnegan, P.M., Whelan, J., Miller, A.H. *et al.* (1997) Differential expression of the multigene family encoding the soybean mitochondrial alternative oxidase. *Plant Physiology* **114**: 455–466.

Finnegan, P.M., Wooding, A.R. and Day, D.A. (1999) An alternative oxidase monoclonal antibody recognizes a highly conserved sequence among alternative oxidase subunits. *FEBS Letters* **447**: 21–24.

Gonzalez-Meler, M., Ribas-Carbo, M., Giles, L. and Siedow, J.N. (1999) The effect of growth and measurement temperature on the activity of the alternative respiratory pathway. *Plant Physiology* **120**: 765–772.

Guy, R.D., Berry, J.A., Fogel, M.L. and Hoering, T.C. (1989) Differential fractionation of oxygen isotopes by cyanide-resistant and cyanide-sensitive respiration in plants. *Planta,* **177**: 483–491.

Groover, A. and Jones, A.M. (1999) Tracheary element differentiation uses a novel mechanism coordinating programmed cell death and secondary cell wall synthesis. *Plant Physiology* **119**: 375–384.

Hilal, M., Castagnaro, A., Morano, H. and Massa, E.M. (1997) Specific localization of the respiratory alternative oxidase in meristematic and xylematic tissues from developing soybean roots and hypocotyls. *Plant Physiology* **115**: 1499–1503.

Hisher, C. and McIntosh, L. (1990) Alternative oxidase of potato is an integral membrane protein synthesized *de novo* during aging of tuber slices. *Plant Physiology* **93**: 312–318.

Ito, Y., Saisho, D., Nakazono, M. *et al.* (1997) Trascript levels of tandem-arranged alternative oxidase genes in rice are increased by low temperature. *Gene* **203**: 121–129.

Igamberdiev, A.U., Zhou, G., Malmberg, G. and Gardestrom, P. (1997) Respiration of barley protoplasts before and after illumination. *Physiologia Plantarum* **99**: 15–22.

Juszczuk, I.M., Malusa, E. and Rychter, A.M. (2001a) Oxidative stress during phosphate deficiency in roots of bean plants (*Phaseolus vulagaris* L.). *Journal of Plant Physiology* **158**: 1299–1305.

Juszczuk, I.M., Wagner, A.M. and Rychter, A.M. (2001b) Regulation of alternative oxidase activity during phosphate deficiency in bean roots (*Phaseolus vulgaris*). *Physiologia Plantarum* **113**: 185–192.

Kitashiba, H., Kitazawa, E., Kishitani, S. and Toriyama, K. (1999) Partial male sterility in transgenic tobacco carrying an anti-sense gene for alternative oxidase under the control of a tapetum-specific promoter. *Molecular Breeding* **5**: 209–218.

Kreps, J.A., Wu, Y., Chang, H.-S. *et al.* (2002) Transcriptome changes for *Arabidopsis* in response to salt, osmotic and cold stress. *Plant Physiology* **130**: 2129–2141.

Kromer, S., Malmberg, G. and Gardestrom, P. (1993) Mitochondrial contribution to photosynthetic metabolism. A study with barley (*Hordeum vulgare*) leaf protoplasts at different light intensities and CO2 concentration. *Plant Physiology* **102**: 947–955.

Lambers, H. (1985) Respiration in intact plants and tissues: its regulation and dependence on environmental factors, metabolism and invaded organism. In *Encyclopaedia of Plant Physiology, New Series, Higher Plant Cell Respiration* Vol. 18. (R. Douce and D.A. Day, eds), pp. 418–473. Springer-Verlag, Berlin.

Lambers, H., Chapin, III F.S. and Pons, T.L. (1998) *Plant Physiological Ecology*. Springer, New York.

Luxova, M. and Gasparikova, O. (1999) The effect of low temperature on root respiration in maize. *Biologia* **54**: 453–458.

Luz, B., Barkan, E., Sagi, Y. and Yacobi, Y.Z. (2002) Evaluation of community respiratory mechanisms with oxygen isotopes: a case study in Lake Kinneret. *Limnology and Oceanography* **47**: 33–42.

Marti, M.C., Florez-Sarasa, I., Camejo, D. *et al.* (2011) Response of mitochondrial thioredoxin PsTrxo1, antioxidant enzymes, and respiration to salinity in pea (*Pisum sativum*, L.) leaves. *Journal of Experimantal Botany* **62**: 3863–3874.

Maxwell, D.P., Nickels, R. and McIntosh, L. (2002) Evidence of mitochondrial involvement in the transduction of signals required for the induction of genes associated with pathogen attack and senescence. *Plant Journal* **29**: 269–279.

Maxwell, D.P., Wang, Y. and McIntosh, L. (1999) The alternative oxidase lowers mitochondrial reactive oxygen production in plant cells. *Proceedings of the National Academy of Sciences USA* **96**: 8271–8276.

McDonald, A.E., Sieger, S.M. and Vanlerberghe, G.C. (2002) Methods and approaches to study plant mitochondria alternative oxidase. *Physiologia Plantarum* **116**: 135–143.

McIntosh, L. (1994) Molecular biology of the alternative oxidase. *Plant Physiology* **105**: 781–786.

McIntosh, L., Eichler, G., Gray, G. *et al.* (1998) Biochemical and genetic controls exerted by plant mitochondria. *Biochimica et Biophysica Acta* **1365**: 278–284.

McNulty, A.K. and Cummins, W.R. (1987) The relationship between respiration and temperature in the leaves of the arctic plant *Saxifraga cernua*. *Plant, Cell & Environment* **10**: 319–325.

Millenaar, F.F., Roelofs, R., Gonzalez-Meler, M.A. *et al.* (2000) The alternative oxidase in roots of *Poa annua* after transfer from high-light to low-light condition. *Plant Journal* **23**: 623.

Moller, I.M., Berezi, A., Vander Plas, L.H.W. and Lambers, H. (1988) Measurement of the activity and capacity of the alternative pathway in intact plant tissue: Identification of problems and possible solutions. *Physiologia Plantarum* **72**: 642.

Moore, A.L., Albury, M.S., Crichton, P.G. and Affourtit, C. (2002) Function of the alternative oxidase: Is it still a scavenger? *Trends in Plant Science* **7**: 478–481.

Murphy, A.M., Chivassa, S., Singh, D.P. and Carr, J.P. (1999) Salicyclic acid-induced resistance to viruses and other pathogens: A parity of the ways? *Trends in Plant Science* **4**: 155–160.

Noguchi, K.O., Taylor, N.L., Millar, A.H. *et al.* (2005) Response of mitochondria to light intensity in leaves of sun and shade species. *Plant, Cell & Environment* **28**: 760–771.

Ottow, E.A., Brinker, M., Teichmann, T. *et al.* (2005) *Populus euphratica* displays apoplastic sodium accumulation, osmotic adjustment by decreases in calcium and soluble carbohydrates, and develops leaf succulence under salt stress. *Plant Physiology* **139**: 1762–1772.

Padmasree, K. and Raghavendra, A.S. (1999) Importance of oxidative electron transport over oxidative phosphorylation in optimizing photosynthesis in mesophyll protoplasts of pea (*Pisum sativum* L.). *Physiologia Plantarum* **105**: 546–553.

Pheloung, P. and Barlow, E.W.R. (1981) Respiration and carbohydrate accumulation in the water-stressed wheat apex. *Journal of Experimental Botany* **32**: 921–931.

Poorter, H., Grifford, R.M., Kriedemann, P.E. and Wong, S.C. (1992) A quantitative analysis of dark respiration and carbon content as factors in the growth response of plants to elevated CO_2. *Australian Journal of Botany* **40**: 501–513.

Prasad, T., Anderson, M.D. and Stewart, C.R. (1994) Acclimation, hydrogen peroxide, and abscisic acid protect mitochondria against irreversible chilling injury in maize seedlings. *Plant Physiology* **105**: 619–627.

Purvis, A.C. and Shewfelt, R.L. (1993) Does the alternative pathway ameliorate chilling injury in sensitive plant tissues? *Physiologia Plantarum* **88**: 712–718.

Pystina, N.V. and Danilov, R.A. (2001) Experimental data from three native representatives of natural communities in north-east Russia: Does the activity of the alternative respiratory pathway in leaves depend on the carbohydrate status? *Bulgarian Journal of Plant Physiology* **27**: 104–108.

Rhoads, D.M., Umbach, A.L., Sweet, C.R. *et al.* (1998) Regulation of the cyanide-resistant alternative oxidase of plant mitochondria: Identification of the cysteine residue involved in α-keto acid stimulation and inter subunit disulfide bond formation. *Journal of Biological Chemistry* **273**: 30750–30756.

Ribas-Carbo, M., Aroca, R., Gonzalez-Meler, M.A. *et al.* (1995) The electron partitioning between the cytochrome and alternative respiratory pathways during chilling recovery in two cultivars of maize differing in chilling sensitivity. *Plant Physiology* **122**: 199–204.

Saisho, D., Nakazono, M., Tsutsumi, N. and Hirai, A. (2001) ATP synthesis inhibitors as well as respiratory inhibitors increase steady-state level of alternative oxidase mRNA in *Arabidopsis thaliana. Journal of Plant Physiology* **158**: 241–245.

Saisho, D., Nambara, E., Naito, S. *et al.* (1997) Characterization of the gene family for alternative oxidase from *Arabidopsis thaliana. Plant Molecular Biology* **35**: 585–596.

Saradadevi, K. and Raghavendra, A.S. (1992) Dark respiration protects photosynthesis against photoinhibition in mesophyll protoplasts of pea (*Pisum sativum*). *Plant Physiology* **99**: 1232–1237.

Schmitt, N. and Dizengremel, P. (1989) Effect of osmotic stress on mitochondria isolated from etiolated mungbean and sorghum seedlings. *Plant Physiology and Biochemistry* **27**: 17–26.

Siedow, J.N. and Umbach, A.L. (2000) The mitochondrial cyanide-resistant oxidase: Structural conservation amid regulatory diversity. *Biochimica et Biophysica Acta* **1459**: 432–439.

Simons, B.H. and Lambers, H. (1999) The alternative oxidase – is it a respiratory pathway allowing a plant to cope with stress: In *Plant Responses to Environmental Stresses from Phytohormones to Genomic Reorganization* (H.L. Lerener, ed.), pp. 265–286. Marcel Decker Inc/

Simons, B.H., Millenaar, F.F., Mulder, L. *et al.* (1999) Enhanced expression and activation of the alternative oxidase during infections of *Arabidopsis* with *Pseudomonas syringae* pv tomato. *Plant Physiology* **120**: 529–538.

Smith, C.A., Melino, V.J., Sweetman, C. and Soole, K. (2009) Manipulation of alternative oxidase can influence salt tolerance in *Arabidopsis thaliana. Physiologia Plantarum* **137**(4): 459–472.

Umbach, A.L., Gonzalez-Meler, M.A., Sweet, C.R. and Siedow, J.N. (2002) Activation of plant mitochondrial alternative oxidase: insights from site-directed mutagenesis. *Biochimica et Biophysica Acta* **1554**: 118–128.

Umbach, A.L., Wiskich, J.T and Siedow, J.N. (1994) Regulation of alternative oxidase kinetics by pyruvate and inter molecular disulfide bond redox status in soybean seedling mitochondria. *FEBS Letters* **348**: 181–184.

Vanlerberghe, G.C. (2013) Alternative oxidase: A mitochondrial respiratory pathway to maintain metabolic and signaling homeostasis during abiotic and biotic stress in plants. *Intetnational Journal of Molecular Science* **14**: 6805–6847.

Vanlerberghe, G.C. and McIntosh, L. (1992) Lower growth temperature increases alternative pathway capacity and alternative oxidase protein in tobacco. *Plant Physiology* **100**: 115–119.

Vanlerberghe, G.C. and McIntosh, L. (1996) Signals regulating the expression of the nuclear gene encoding alternative oxidase of plant mitochondria. *Plant Physiology* **111**: 589–595.

Vanlerberghe, G.C. and McIntosh, L. (1997) Alternative oxidase: From gene to function. *Annual Review of Plant Physiology: Plant Molecular Biology* **48**: 703–734.

Vanlerberghe, G.C., Robson, C.A. and Yip, J.Y.H. (2002) Induction of mitochondrial alternative oxidase in response to a cell signal pathway down-regulating the cytochrome pathway prevents programmed cell death. *Plant Physiology* **129**: 1829–1842.

Vanlerberghe, G.C., Vanlerberghe, A.E. and McIntosh, L. (1994) Molecular genetic alteration of plant respiration: silencing and over-expression of alternative oxidase in transgenic tobacco. *Plant Physiology* **106**: 1503–1510.

Wagner, A.M. and Wagner, M.J. (1995) Changes in mitochondrial respiratory chain components of petunia cells during culture in the presence of antimycin A. *Plant Physiology* **115**: 617–622.

Xue, X., Gauthier, D.A., Turpin, D.H. and Weger, H.G. (1996) Interaction between photosynthesis and respiration in the green alga *Chlamydomonas reinhardtii*. Charaterization of light enhanced dark respiration. *Plant Physiology* **112**: 1005–1014.

Zhang, D.W., Xu, F., Zhang, Z.W. *et al.* (2010) Effects of light on cyanide-resistant respiration and alternative oxidase function in *Arabidopsis* seedlings. *Plant, Cell & Environment* **33**: 2121–2131.

Zhang, L.-T., Zhang, Z.-S., Gao, H.Y. *et al.* (2012) The mitochondrial alternative oxidase pathway protects the photosynthetic apparatus against photo damage in *Rumex* k-1 leaves. *BMC Plant Biology* **12**: 40–58.

Zhou, D. and Solomos, T. (1998) Effect of hypoxia on sugar accumulation, respiration, activities of amylase and starch phosphorylase, and induction of alternative oxidase and acid invertase during storage of potato tubers (*Solanum tuberosum* cv. Russet Burbank) at 1 °C. *Physiologia Plantarum* **104**: 255–265.

CHAPTER 10

Alternative respiratory pathway in ripening fruits

Bhagyalakshmi Neelwarne
Plant Cell and Biotechnology Department, CSIR-Central Food Technological Research Institute, Mysore, India

Introduction

Respiration plays a pivotal role in the metabolism of plants by meeting energy needs and providing carbon sources to drive the cellular metabolism and transport processes that are required for well-structured growth and completion of the life cycle. While higher animals have evolved with a marvellous circulatory system to distribute oxygen to each cell and every sub-cellular component, a plant cell needs to ingeniously programme the oxygen sequestering process for its functions irrespective of the bulk of the organ in which it remains buried or fully exposed to hypoxic conditions. Thus plant cells need to survive and perform respiration under a wide array of conditions, and therefore often switch over to alternative respiratory pathways (ARP).

In plants, the vegetative stem apex transforms itself to a floral primordium upon receipt of flowering signals and the resultant flower culminates in fruit formation in fruit-bearing plants. In this entire cycle of flowering to ripened fruit formation, there are two crucial periods that are very short-lived but have very high-speed physiological functions – the flowering stage and the fruit ripening stage. Within these two stages, the rapid respiration that occurs in both flowers and fruits is accompanied by thermogenesis. Although both organs are supported by an alternative respiratory mode in addition to quicker normal respiration, the quantum of respiratory energy (the substrates) in each of these organs varies significantly. Most fruits are storage organs endowed with an enhanced sink capacity, which also displays an altered respiratory metabolism, often instigating alternative oxidases in parallel to normal ATP generation mode. Each fruit has a distinct set of energy source and metabolic profiles that influence the intricately linked ripening-related biochemical changes and respiratory metabolism, which are discussed in the following sections.

Alternative Respiratory Pathways in Higher Plants, First Edition.
Edited by Kapuganti Jagadis Gupta, Luis A.J. Mur and Bhagyalakshmi Neelwarne.

Ethylene triggers normal and alternative respirations during fruit ripening

Fruit ripening is a complex genetically programmed process that brings about dramatic changes in colour, texture, flavour and aroma. Currently it is well-established that this entire set of changes are triggered by ethylene (Alexander and Grierson, 2002; Klee, 2010). There are two main types of ripening mechanisms in fruit: climacteric, where ripening is accompanied by a peak in respiration and a concomitant burst of ethylene, and non-climacteric, where respiration shows no dramatic change and ethylene production remains at a very low level (White, 2002). Ethylene biosynthesis in plant tissues has been studied extensively (Srivastava and Handa, 2005; Argueso *et al.*, 2007); the basal ethylene synthesized in vegetative tissues, including the development of fruit until the onset of ripening (unripe) is constitutively regulated in an auto-inhibitory manner (system-1), whereas the other ethylene biosynthesis operates in an autocatalytic manner (system-2) during the ripening of climacteric fruit and senescence in flower (Barry and Giovannoni, 2007; Yokotani *et al.*, 2009). It is well established that ethylene is synthesized from *S*-adenosyl methionine by the action of two major enzymes – 1-aminocyclopropane-1-carboxylate (ACC) oxidase (ACO) and ACC synthase (ACS) (Wang *et al.*, 2002) – and this has been experimentally confirmed by knocking down the expression of ACO and ACS, which resulted in a strong inhibition of ripening (Hamilton *et al.*, 1990; Oeller *et al.*, 1991). External application of ethylene to climacteric fruit at the mature stage stimulated system-2 ethylene biosynthesis, which in turn orchestrated the ripening process, including the climacteric respiratory peak (Nakatsuka *et al.*, 1998). This ripening-related ethylene is also known to trigger ARP either directly or through nitric oxide (NO) and/or H_2O_2 signalling (Wang *et al.*, 2010). The need for switching on ARP in plants cells may be because rapid respiration is known to invariably result in the over-reduction of the electron transport chain, particularly at the terminal phosphorylating steps of complex III and cytochrome oxidase (COX), which results in the generation of reactive oxygen species (ROS) and cells must handle them by triggering the battalion of ROS-quenching mechanisms (Gandin *et al.*, 2009) (see Chapter 1). ARP on the contrary, has the ability to use excess ubiquinone electron pools, acting as an 'energy overflow' conduit for the cytochrome pathway (Lambers, 1982), thereby avoiding the over-reduction of the electron transport chain. Thus, switching on the ARP is more energy efficient and less stressful; therefore, many plant cells/organs and climacteric fruits are equipped with ARP (catalysed by the enzyme alternative oxidase – AOX) under high respiration rates, where energy is released in the form of heat. The main factors that determine electron partitioning between the COX and AOX, as stated by Gandin *et al.* (2009), are the ratio of reduced ubiquinone to total ubiquinone pools (Wagner *et al.*, 1998), the amount and redox state of AOX proteins (Umbach and Siedow, 1993), the presence of a-keto acids such as pyruvate (Millar *et al.*, 1993; Umbach *et al.*, 1994) and the availability of ADP and Pi (energy status) (Juszczuk *et al.*, 2001). The levels of these specific metabolites can vary with developmental stage and environmental conditions. In general, the protein AOX (isoforms) is thought to primarily help cell adaptation under environmental stresses, such as inhibition of ROS formation, production of heat in thermogenic floral organs and optimization of photosynthesis (Yoshida *et al.*, 2008; Vanlerberghe *et al.*, 2009; Zhang *et al.*, 2010). AOX is encoded by a small nuclear gene family. AOX has a molecular weight between 32 and 39 kD, and is found in almost all plants that can be immunologically detected using antibodies from Voodoo Lilies (*Sauromantum guttatum* S.). The capacity of the protein in respiring mitochondria can be detected by measuring the oxygen consumption when the cytochrome pathway is blocked. In most fruit, ripening is a rapid process wherein high respiratory rates exert demand on the electron transport chain of mitochondria. As a result of these events, the expression of the uncoupling protein –AOX – occurs. The speed at which the fruit ripens depends on the efficiency of dissipation of energy from the proton gradient as heat by AOX. Although several studies have revealed that AOX may play a role in the respiration of climacteric (Duque and Arrabaca, 1999) or post-climacteric senescence processes during fruit ripening (Considine *et al.*, 2001), very limited information is available on the precise extent of involvement of the AOX pathway in fruit ripening. While ARP has been vastly studied in various plant systems, its involvement in the ripening process has been extensively studied in tomato fruit, which undergoes climacteric ripening, although a few other non-climacteric fruits have also been investigated, as discussed later.

ARP in climacteric fruit

Tomato

Tomato has remained a model system for fruit ripening studies for various reasons (Alexander and Grierson, 2002), the major one being its climacteric ripening nature. In tomato, although AOX is known to play a role in fruit development (Kumar *et al.*, 1990; Considine *et al.*, 2001) and certain forms of AOX are specifically induced during climacteric ripening (Xu *et al.*, 2012), the information on the involvement of AOX in fruit development is very limited. In tomato, two types of AOX in four isoforms have been demonstrated, which are differentially expressed. *LeAOX1a* and *LeAOX1b* transcripts were expressed in most tomato tissues, including leaves, root, flowers and fruit. The transcript of *LeAOX2* was detected in carpels and roots, whereas the transcript of *LeAOX1c* was preferentially expressed in roots but not in fruit (Holtzapffel *et al.*, 2003; Fung *et al.*, 2006).

Initial studies on the existence of ARP in tomato fruit, as in other plant species, was demonstrated by the presence of cyaninde (CN)-insensitive respiration. Subsequently the expression of the AOX protein in isolated mitochondria was observed, showing a decreasing trend during post-harvest ripening (Almeida *et al.*, 1999; Costa *et al.*, 1999; Sluse and Jarmuszkiewicz, 2000), whereas AOX protein levels dramatically increased when tomato fruits were ripened on the vine (Holtzapffel *et al.*, 2002). To address such intriguing responses, and to elucidate the role of AOX in climacteric fruit ripening, Xu *et al.*, (2012) explored the role of AOX in ripening tomato fruit through a combination of pharmacological or inhibitor experimental approaches and by transgenic methods. Since CN-insensitive respiration coincided with the climacteric peak, these authors suggested the contribution of ARP during climacteric ripening. In further studies to identify the involvement of AOX genes, the expression patterns of *LeAOX* gene isoforms were followed. There was a significant increase in the expression of *LeAOX1a* during the turning stage of ripening (T stage), which peaked at the pink (P) stage. However, although the other isoforms *LeAOX1b* and *LeAOX2* were also expressed in a similar pattern, their expression levels were relatively low in ripening tomato fruit, indicating that predominantly *LeAOX1a* contributes to the ARP.

Response to ethylene

Although ethylene is not directly involved in ATP generation, its biosynthesis (particularly of the precursor – *S*-adenosyl methionine) is dependent on ATP generation through respiration (Yang and Hoffman, 1984; Genard and Gouble, 2005) and fruit metabolism (Barry and Giovannoni, 2007). As stated by Xu *et al.* (2012), AOX allows carbon flow through glycolysis and the citric acid cycle by way of removing excess sugars and avoiding the over-reduction of the electron transport chain as well (Borecky and Vercesi, 2005). The ARP increases rapidly to accompany the respiratory climacteric, thus assisting in a high rate of carbon turnover, generating a large amount of ATP for system-2 ethylene synthesis and the concomitant series of ethylene-regulated ripening processes. In turn, this increase in ethylene induces CN-insensitive respiration either directly or by its co-product – the CN (Yip and Yang, 1988). CN, probably by acting as stress, activates the AOX genes transcriptionally, as demonstrated in tobacco and maize (Ederli *et al.*, 2006) and causes a rise in respiration and the ripening response in many fruits in a manner very similar to that evoked by ethylene (Solomos and Laties, 1974, 1976; Tucker and Laties, 1984). Therefore, these events go hand-in-hand that also increase ethylene and HCN levels during fruit ripening, which in turn induce AOX expression and trigger CN-insensitive respiration in cyclic manner (Xu *et al.*, 2012). In the *AOX*-silenced tomato fruit, the detectable HCN content was lower than that in the wild-type fruit. Interestingly, mitochondria also houses an enzyme – β-cyanoalanine synthase (β-CAS) – which detoxifies HCN (Millenaar and Lambers, 2003; Ebbs *et al.*, 2010); when the HCN level exceeds the mitochondrial detoxification capacity, AOX is expressed, which promotes CN-insensitive respiration at climacteric ripening. Under this set of conditions, it is not clearly established whether CN acts as a signal molecule or only as a toxic by-product of ethylene metabolism.

The climacteric nature of tomato fruit ripening initiated by ethylene signalling (Alexander and Grierson, 2002) makes it interesting, since it allows validation of the extent of the ARP response to ethylene treatment and ethylene inhibitors such as 1-methylcyclopropene (1-MCP). Treatment of mature tomato fruits with ethylene resulted in climacteric peak as well as ARP expression two days earlier (on 3rd day) than in control fruits (5th day), whereas 1-MCP treatment had an opposite effect on fruit respiration, where the respiratory peak was postponed (11th day). Surprisingly, 1-MCP treatment reduced the transcript levels of *LeAOX1a*, suggesting that its expression is ethylene regulated. No alterations in *LeAOX1b* or *LeAOX2* transcript levels were observed in ethylene treated fruits, although expressions of these genes were repressed by the 1-MCP treatment. Due to the response of *LeAOX1a* to ethylene inhibitor, when this gene was over-expressed (*35S-AOX1a*) or suppressed (*AOX-RNAi*) by transforming tomato plants, there was no change in the pattern of ripening, other than that it countered the inhibitory effect of 1-MCP. In contrast, no significant differences were observed in the expression of *LeAOX1b* and *LeAOX2* between the transgenic and WT plants. Among the *AOX-RNAi* transgenic plants, severe AOX reduction (90% for the *LeAOX1a* transcript and ~50% for the *LeAOX1b* and *LeAOX2* transcripts) was observed. These genes were found to affect only ripening, without causing changes in other features such as flowering. Reduction of AOX by *AOX-RNAi* resulted in the loss of climacteric ripening, with longer ripening time (both on-vine and post-harvest), increased fresh weight, reduced soluble solids and lycopene upon ripening, and hence fruit were paler with a higher loss of firmness when compared with control and *35S-AOX1a* tomatoes. In contrast, the *35S-AOX1a* fruit reached maturity first during on-vine (fruits still attached to the mother plant) or off-vine (harvested) ripening and accumulated more lycopene content at the red stage when compared with control fruit (Xu *et al.*, 2012).

In transgenic tomato fruit, AOX protein level was found altered and barely detectable in *AOX-RNAi* fruit throughout ripening, whereas ethylene production was higher in *35S-AOX1a* fruit than in WT fruit – suggesting that the down-regulation of AOX influences ethylene synthesis. Further characterization of ethylene biosynthesis genes by tracking mRNA abundance revealed that in *AOX-RNAi* fruit, the ACC synthase-4 (*LeACS4*) mRNA was markedly lower at the climacteric (pink) stage than in control fruit, and a 20% suppression was noticed for the *LeACS2* transcript. Such *AOX* repression concomitantly reduced (>40%) the transcript level of ACC oxidase1 (*LeACO1*) – the ethylene catalysing protein – and expressions of its other isoforms, *LeACO4 and LeACS2*, were also greatly repressed, although the mRNA levels of these genes were slightly higher in

AOX1a-over-expressing fruit. These observations show that in tomato fruit the ethylene reduction occurring upon repression of AOX may be attributed to the down-regulation of the key genes involved in ethylene biosynthesis. Inhibition of the ethylene pathway by AOX (*AOX-RNAi*) also reduced the transcript levels of a number of ethylene-regulated genes including polygalacturonase (*LePG*) and the carotenoid synthesis enzyme, phytoene synthase1 (*LePSY1*) (Xu *et al.*, 2012).

Apart from altering ethylene biosynthesis, AOX suppression by *AOX-RNAi* in the transgenic fruit showed that several genes involved in ethylene signal transduction were also simultaneously suppressed, suggesting that AOX might act through the modulation of ethylene signalling flux during ripening. In support of this argument, the transcript levels in mutants *NR*(*LeETR3*) the never-ripe type, *LeETR4* (ethylene repressed), *LeEIL3* (ethylene-insensitive), and *LeERF1* (that codes for ethylene signal transduction factor) were slightly up-regulated in AOX-over-expressing (*35S-AOX1a*) fruit than in control fruit.

When AOX-over-expressing fruits (*35S-AOX1a*) were treated with 1-MCP, the ripening delay was shorter than control fruits, ripening fully in 11 days. Whereas in case of *AOX*-silenced (*AOX-RNAi*) fruits treated with 1-MCP, ripening was nearly blocked, inferring that AOX plays a crucial role in the autocatalysis of ethylene in the ripening of climacteric fruits. These morphological differences in ripening characteristics were consistent with the observed respiration levels and ATP content for ethylene treatment, where total respiration, CN-insensitive respiration and ethylene emission were significantly promoted in control fruit and *35S-AOX1a* fruit, whereas only a slight increase in such parameters occurred (except for ATP content, which was significantly lower) in *AOX-RNAi* fruit, further supporting that AOX plays a key role in ethylene autocatalysis in climacteric fruits, particularly in tomato. Treatment with MCP delayed respiration and ethylene peaks in both control and *AOX1a*-over-expressing fruits, while further suppressing ethylene production in *AOX-RNAi* fruits. This lowered ethylene production in the latter correlated with a significant down-regulation of several key genes involved in ethylene biosynthesis. The content of HCN, a co-product of ethylene biosynthesis, is also indicative of ethylene level, and hence follows a climacteric pattern. The level of HCN remained very low before the initiation of climacteric ripening, was abundant at the climacteric, and then rapidly diminished. Compared with control fruit, the peak HCN content was higher in *35S-AOX1a* fruit but lower in *AOX-RNAi* fruit. Regarding ripening metabolites in these fruits, the lycopene accumulation and the soluble sugar content were substantially reduced in 1-MCP-treated *AOX-RNAi* fruit, whereas these metabolites were higher in *AOX1a*-over-expressing fruits which maintained higher fruit firmness even after 30 days of storage than the 1-MCP-treated control fruit. All in all, no significant change in the pattern of ripening occurred in tomato fruit when *LeAOX1a* was over-expressed although it did offset the inhibitory effect of 1-MCP. In contrast, the reduction of AOX expression (*AOX-RNAi*) affected ethylene perception and delayed ripening, inferring that the AOX pathway is an important component in achieving the respiration peak and that the role of AOX in the tomato respiratory climacteric cannot be substituted by ethylene treatment. Therefore, AOX could be an important target for the regulation of the ripening metabolic network involved in the control of fleshy fruit ripening.

The availability of various types of mutants in tomato with significantly different ripening characteristics such as ripening inhibitor (*RIN*), non-ripening (*NOR*), colourless non-ripening (*CNR*), and never ripe (*NR*) offers an elegant experimental model for elucidating ripening-related morphogenetic networks (Tigchelaar *et al.*, 1978; Wilkinson *et al.*, 1995; Vrebalov *et al.*, 2002; Manning *et al.*, 2006). When the involvement of AOX in these mutants was analysed by tracking the expression profiles, a down-regulation of AOX in *NR* and *CNR* was observed in 1-MCP-treated *AOX-RNAi* fruit, and their transcript levels were found at lower concentrations in *AOX-RNAi* fruit than in control (normal ripening wild type) and *35S-AOX1a* fruit (Xu *et al.*, 2012). The reduction of these transcripts led these authors to suspect that AOX may also play some unexpected roles in fruit ripening since the *NR* gene acts downstream of the ethylene pathway and *CNR* is known to act upstream (Adams-Phillips *et al.*, 2004; Barry and Giovannoni, 2007).These observations were in accordance with the notion that the expression of *NR* is positively regulated by ethylene in tomato fruit (Wilkinson *et al.*, 1995; Nakatsuka *et al.*, 1998), whereas the expression of *RIN* and *NOR* after 1-MCP treatment was similar in WT and transgenic fruit. These observations suggest that AOX plays a partial role in ethylene signal transduction and might be necessary for ethylene autocatalysis even in mutants.

Regulation of AOX by respiratory substrates

Several initial studies demonstrated that upstream respiratory carbon metabolism may also contribute to the regulation of AOX activity *in vivo* in a feed-forward fashion. For instance, intramitochondrial pyruvate, a potent activator of AOX, was demonstrated to stimulate AOX capacity in soybean (Millar *et al.*, 1993), durum wheat (Pastore *et al.*, 2001) and in various other plants (Day and Wiskich, 1995). The rapid stimulation of glycolysis at the climacteric peak is known to increase the flux of pyruvate and its intramitochondrial accumulation leads to metabolic conditions that likely enhance the activity of AOX (Duque *et al.*, 1999) and its importance in climacteric burst and fruit ripening.

Sets of experiments by Xu *et al.*, (2012) demonstrated that AOX-silenced tomato fruit can reach the red stage of ripening even in the absence of ethylene or respiration bursts. This important observation is indicative of the fact that the climacteric is not essential for the ripening process in tomato. Supporting this view, the *AOX-RNAi* fruit treated with 1-MCP (the inhibitor of ethylene perception, and hence ripening) failed to induce ripening providing additional information that the absence of both AOX and ethylene are required to halt tomato ripening completely, and therefore, in tomato both AOX and ethylene contribute to fruit ripening. Since both ethylene-dependent and ethylene-independent regulatory pathways co-exist and orchestrate the ripening process in climacteric fruit (Alba *et al.*, 2000; Pech *et al.*, 2008), more research insights are needed to elucidate the cross-regulation between the AOX pathway, particularly for ripening-associated transcription factors (Xu *et al.*, 2012).

Expression of tomato AOX in other systems

Transgenic petunia lines over-expressing tomato *AOX1a*, showed lowered tomato spotted wilt virus (TSWV) symptoms than that in control plants (Ma *et al.*, 2011). Although it is not clearly established how AOX provides resistance during viral infection, the possibility of strengthening the cellular defence system by reducing oxidative stress by AOX might partially contribute to such mechanisms. This argument is supported by the observation that antisense lines of *AOX1a* magnified ROS generation in a suspension culture of tobacco (Yip and Vanlerberghe, 2001), whereas over-expression resulted in lower ROS abundance with concomitant lower expression of genes encoding ROS scavenging enzymes like SOD and GPX, and in cells lacking AOX transcripts encoding for catalase and pathogenesis-related protein were significantly higher (Maxwell *et al.*, 1999). In general, AOX is considered to prevent the excessive generation of free radicals in the mitochondria (Vanlerberghe *et al.*, 2009) and the various mechanisms involved in accomplishing this are discussed in other chapters of this book.

In chilled tomatoes, *LeAOX1a* and *LeAOX1b* gene transcripts were expressed (Holtzapffel *et al.*, 2003) and their functioning was confirmed by using a yeast expression system, where the *LeAOX1b* protein was shown having altered regulatory properties in comparison to LeAOX1a. The *LeAOX1b*protein was suggested to be a less regulated form of AOX, activated under stress conditions (Holtzapffel *et al.*, 2003).

Imparting stress tolerance by AOX that confer higher fruit storability

Prolonged exposure to stress could convert an epigenetic modification into stable (epi)genetic trait for tolerance or resistance (Boyko and Kovalchuk, 2011). While methylated cytosines are highly prone to spontaneous transition mutations, genomic areas with low levels of methylation may be more inclined to chromosomal rearrangements (Chen and Ni, 2006; Boyko *et al.*, 2007; Boyko and Kovalchuk, 2011). Consequently, the change of methylation pattern in a DNA sequence in response to stress may have a significant impact on the rate and type of genetic changes in that sequence, and may lead to the appearance of new alleles in a population. Since genes involved in stress response (like *AOX1* genes) are highly affected by environmental conditions, it is plausible that different stress-induced epigenetic scenarios around those genes bias the type and frequency of mutations in their sequences, making them rich sources of genomic polymorphisms, which could be exploited for frequent mutants development. Since AOX expression is linked to stress tolerance, be it by retarding the endogenous generation of pro-oxidants or by regulating low temperature stress by liberating heat, or AOX may also provide high temperature tolerance (Wang *et al.*, 2011).

Endogenous ROS production due to the overflow of mitochondrial electrons has been linked to chilling injury in tomato (Purvis *et al.*, 1995). The pre-mRNA levels of COX subunit 2 (COX2), which is involved in electron transport in mitochondria, increased under cold conditions (Kurihara-Yonemoto and Handa, 2001). In addition, cold also renders the functioning of the uncoupling protein due to defective pre-mRNA processing that results in multiple abnormal forms of the uncoupling protein transcripts (Watanabe and Hirai, 2002). In later studies with different systems, it was observed that temperature directly affects gene splicing. For instance, an increase in temperature completely inhibited splicing of the intron for chloroplast NAD(P)H dehydrogenase and *NDHB* genes (Karcher and Bock, 2002). Such post-transcriptional impairment of the RNA processing mechanism is known to affect the respective gene expression, imparting the loss of not only mitochondrial efficacy but also of other organelles such as chloroplast functioning under temperature stress. Fung *et al.* (2006) found a direct relationship between the chilling injury in tomato and the expression of AOX gene family. The accumulation of *LeAOX1* transcripts was highest during cold storage, where *LeAOX1a* mRNA abundance was higher than that of *LeAOX1b* and *LeAOX1c*. Enzymatically, LeAOX1a and 1b proteins were found to vary in their regulatory properties (Holtzapffel *et al.*, 2003), suggesting that closely related AOX isoforms may slightly differ in their biochemical characteristics, and the expressions of distinctly related AOX isoforms may also be regulated by variations in developmental and environmental cues (Considine *et al.*, 2001). Of the environmental cues, the low temperature that leads to chilling injury affected the RNA splicing efficiency of AOX transcripts in tomato fruit, which was partially countered by methyl salicylate (Fung *et al.*, 2006). Here, to find out if altered splicing occurred among the three *LeAOX1* genes at low temperature and if any post-transcriptional regulations also occurred, a RT-PCR method was adopted using gene-specific primers having intron–exon borders. In addition, the expression patterns of several genes involved in RNA metabolism were also followed. The results of this study established that the chilling injury in tomato fruit in terms of decay of fruit to various extents correlated with the expression patterns of the *LeAOX1a* and *LeAOX2* genes and the accumulation of their mRNA. There was reduced decay in fruit that received methyl salicylate (MeSa) treatment, which is indicative of chilling resistance and the treatment also correlated with the expression of *LeAOX1a* and *LeAOX2* genes. Here, it was also found that the splicing of AOX genes was altered by temperature, where the low temperature was found to inhibit LeAOX intron splicing irrespective of transcript abundance. The expressions of genes involved in pre-RNA intron splicing and RNA processing were also significantly altered in cold-stored tomatoes. The correlation of AOX transcript levels with chilling tolerance points towards the existence of a mechanism that ensures the expression of an AOX transcript-specific factor responsible for splicing

(Fung *et al.*, 2006). Such temperature sensitive splicing event occurs not only selectively to the AOX gene but also to other subgroups of genes that coincidentally share mitochondria or plastid organelle evolutionary origins (Fung *et al.*, 2004) (see Introduction). Such regulation, executed by the RNA processing genes (such as splicing factors 9G8-SR, SF2-SR1, fibrillarin and DEAD box RNA helicase) in the control functional proteins, is also indicative of the splicing mechanism acting as a 'master switch' in cold tolerance (Fung *et al.*, 2006) and that there could be a common control even for the set of RNA processing genes that are again responsive to some of the signalling compounds like MeSa.

Chemically induced ARP in tomato

Several compounds alter the ripening process either by blocking ethylene perception or by countering ethylene action downstream. Certain signalling compounds such as salicylic acid, methyl jasmonate and these moieties in such other compounds are extensively studied for characterizing the mechanisms involved in disease resistance, senescence and fruit ripening, including ARP induction and its alteration. Among the other ARP inducers are low temperature, wounding, pathogen attack, elevated carbohydrate status, cell culture stage and elevation of salicylic acid levels (Ding *et al.*, 2002; Ding and Wang, 2003). All these are indicative of the induction of ARP as a response to stress. Tomatoes are sensitive to chilling injury, where AOX genes are up-regulated (Holtzapffel *et al.*, 2003). The application of MeSa vapour enhanced resistance against chilling injury in freshly harvested pink tomatoes, which also increased the transcript levels of AOX. Further analyses of unspliced pre-mRNA transcripts revealed that the intron splicing of *LeAOX1a*, *LeAOX1b* and *LeAOX1c* gene were also affected by cold storage and this alternative splicing event in AOX pre-mRNA molecules occurred, preferentially at low temperature, regardless of mRNA abundance (Fung *et al.*, 2006).

Apple fruit

Apple is one of the earliest studied climacteric fruit, with a large number of varieties that ripen differently and exhibit significant differences in their storability at different temperatures. Duque *et al.* (1999) observed that in apple cv. Reinette du Canada the respiratory pattern in cold-stored (4 °C) was similar to those fruit held at room temperature, although the cold stored apples had a longer shelf life. A deeper insight into the respiratory metabolism at both biochemical and physiological levels indicated that 'the respiratory climacteric does not occur to accommodate extra ATP requirements during sucrose synthesis nor can it be a consequence of an increased supply of respiratory substrate'. Interested with this behaviour, they further studied the respiratory metabolism and demonstrated the presence of a direct link between the increase in respiration and increased AOX capacity during climacteric ripening in apples, where the isolated mitochondria showed an increase in respiratory capacity as well as in the activity of non-phosphorylating alternative pathway at the climacteric (Duque and Arrabaca, 1999). It was further observed by these researchers that alternative oxidation capacity correlated with AOX protein (for which antibodies had been raised against *Sauromatum guttatum* AOX), which was not dependent on the major changes in the oxidative state of the enzyme. Xiaoyong *et al.* (2003), while experimenting with Royal Gala apple fruit, found that exposure of fruit to cold (0 °C for 1 week) and heat (38 °C for 1 h) resulted in the expression of enhanced endogenous ethylene production and alternative oxidase (AOX) protein expression. The presence and the quantity of AOX protein was confirmed using a monoclonal antibody developed for the terminal oxidase of the alternative pathway from *S. guttatum*. Here it was observed that the molecular mass of AOX in Royal Gala apple fruits was approximately 38 kDa, which was similar to those reported in tobacco and tomato and the model plant – *Arabidopsis*. Apples stored in cold showed the suppression of endogenous ethylene levels, prior to climacteric ethylene production, where AOX protein expression was induced. After the cold temperature treatment, the endogenous ethylene peak appearance preceded the maximum AOX expression. As expected, opposite effects occurred in apples held at 38 °C, where both the ethylene and AOX protein expressions were higher than in control. These observations also confirm that the normally occurring climacteric burst of ethylene has no strict coordination between ethylene synthesis and AOX protein level in climacteric fruit.

Mango

Mango (*Mangifera indica*) is a rapidly respiring tropical fruit and its climacteric ripening often results in significant thermogenesis and about a sixfold higher turnover of ethylene. The expression of peroxisomal thiolase (a ripening marker in mango) was investigated by Considine *et al.* (2001) to track the expression profile of AOX and another set of uncoupling proteins (UCP). The latter, by way of bypassing the ATP synthase complex (Almeida *et al.*, 1999; Laloi *et al.*, 1997), may allow the re-entry of protons from the intermembrane space to the mitochondrial matrix. Thus both AOX and UCP non-phosphorylate through different mechanisms, and are known to be regulated differentially in plants (Casolo *et al.*, 2000; Pastore *et al.*, 2000). In mango, genes coding for AOX were differentially expressed during ripening, where the gene expression as well as the final protein versions of the multigenic AOX were abundant, reaching a peak at the climacteric ripe stage. Expression of the single AOX gene peaked at the turning stage and the protein abundance peaked at the ripe stage. However, the accumulation of proteins of the cytochrome chain (COX) peaked at the mature stage of ripening, suggesting that increases in cytochrome chain components played an important role in facilitating the climacteric burst of respiration in mango and that AOX may assist in post-climacteric senescent processes. The primers designed to the putative UCP, resulted in finding a single gene for this protein in mango. Further, Southern analysis with this fragment in mango genomic DNA consistently resulted in confirming a single gene copy (Considine *et al.*, 2001). When gene expressions of mango *AOX* genes were checked, the *MnAOX2* was up-regulated and peaked at the early stage of ripening, declining steadily later at the ripe stage. Contrarily, the *MnAOX1a,* which was very low from the unripe to the turning stage, increased substantially by almost 10-fold in the ripe mango fruit. Similarly, *MnAOX1b* gene expression peaked at the turning stage accounting for an increase of five-fold during ripening, and decreased slightly at the ripe stage. The expression of *MnAOX1c* was not traceable at any stage. The gene expression for the uncoupling protein –*MnUCP1* – also showed a similar expression profile, increasing fivefold from mature to turning stage of ripening, with a small decrease seen at the ripe stage. Upon correction of gene expression amplification efficiency, the *MnUCP1* was found to be the most abundant gene transcript, followed by *MnAOX1a.* Both *MnUCP1* and *MnAOX1a* genes were expressed at 10-fold higher levels, although *MnUCP1*peaked at the turning stage and the *MnAOX1a* peaked at the ripe stage. Southern analyses of the translation products of these genes also correlated with concomitant increase in respective protein levels. Since the gene expression for the AOX and UCP increased in a similar pattern, and their expressions also matched other mictochondrial proteins, it suggests that their expression is not controlled in a reciprocal manner but may be active simultaneously (Considine *et al.*, 2001), similar to vine-ripened tomato fruit (Holtzapffel *et al.*, 2002) and apples (Duque and Arrabaca, 1999). The authors opine that the role of AOX and UCP could be to maintain respiration after a respiratory burst in the presence of high levels of ATP, and thus allow the progression of senescence in ripe mango fruit (Considine *et al.*, 2001).

Banana

A banana bunch as a whole offers an excellent model system to study ripening characteristics because all stages of ripening from mature green to the turning stage can be found on the same bunch. Kumar and Sinha (1992) observed the involvement of AOX in accelerating the thermogenesis in ripening banana (*Musa paradisiaca* var. *Mysore Kadali*) while the fruit were still attached to the bunch. It was observed that the temperature of the youngest (unripe) banana fruit increased from 27·0±0·2°C to 30·8±0·1°C and the total respiration (in nmol O_2min per g dry weight) increased from 1·39·6±5·5 to 167·3±7·0 at the fully ripened stage. Here little change in the capacity for alternative respiration was noted although the actual operation of this pathway increased from 38 to 73% (p=0·38–0·73) during ripening. This trend was also observed at different ripening stages in banana fruit of the central axis, suggesting the contribution of AOX to temperature rise in ripening banana fruit. Ethylene treatment prior to shipment has been a standard practice for banana growers. In such bananas, the oxygen consumption data showed significantly greater respiratory capacity during the green stage, compared to both the yellow (climacteric) and black (over-ripe) stages. The oxygen consumption pattern was in accordance. In such fruit AOX abundance was confirmed by immunoblotting in all three stages of ripening (green, yellow and black according to peel colour), and the greatest amount was recorded in the green stage (Woodward *et al.*, 2009).

AOX in ripening fruit is modulated by yet another set of molecules – NO and H_2O_2 – which act either directly or through ethylene modulation, where the whole chain of events is self-regulatory (Manjunatha *et al.*, 2010, 2012a, 2012b). In stressed vegetative tissues, an increase in NO accumulation activated H_2O_2, which in turn caused ACS activity leading to ethylene-dependent ARP induction in *Arabidopsis* (Wang *et al.*, 2010). Such an increase in ethylene emission correlated with *AOX1a* expression and pyruvate content and enhanced ARP activity. Subsequently the enhanced ARP levels may diminish H_2O_2 generation, thereby avoiding ROS-mediated damage in plant cells (see Salgado and Oliveira, Chapter 6).

A few other fruit where ARP has been recorded during climacteric ripening are listed in Table 10.1.

Table 10.1 Studies that address the involvement of alternative oxidase in the process of fruit ripening.

Fruit name	Botanical name	Context of AOX study	Reference
Climacteric fruit			
Arabidopsis	*Arabidopsis thaliana*		
Banana	*Musa acuminata*	Climacteric ripening	Woodward *et al.*, 2009
Mango	*Mangifera indica*	Regulation by AOX and uncoupling protein	Considine *et al.*, 2001
		Differential regulation of AOX in ripening mango fruit	Bojorquez and Gomez-Lim, 1995
Tomato	*Lycopersicon esculantum* *Solanum lycopersicum* (new synonym)	AOX with altered properties under cold storage	Holtzapffel *et al.*, 2003
		AOX gene expression alterations during ripening and interactions with ethylene	Xu *et al.*, 2012
Avocado	*Persea mexicana*	Confirmation of alternative respiratory pathway	Lange and Kader, 1997
		Ethylene-link with cyanide resistant respiration	Tucker and Laties, 1984
Non-climacteric			
Litchi	*Litchi chinensis*		
Citrus flavedo	*Citrus paradisi*	Cold temperature-induced ARP	Purvis *et al.*, 1988
Raw fruit studies			
Bell pepper	*Capsicum annuum*	AOX regulation by Methyl jasmonate and salicylic acid – countering chilling injury	Fung *et al.*, 2004

ARP in fruits undergoing non-climacteric ripening

Litchi fruit

Litchi is a delicious fruit grown in distinct tropical and subtropical regions and fetches a high commercial value on the international market. Litchi fruit, although non-climacteric, deteriorate rapidly after harvest because of water loss, pericarp browning, rot development (Jing *et al.*, 2013) and reduction of edible portions (Wang *et al.*, 2013). Some of these deteriorations were prevented when exogenous ATP was supplied, which also enhanced antioxidant systems and maintained membrane integrity, prevented loss of fresh weight, and delayed browning and senescence of litchi fruit (Yi *et al.*, 2010). For elucidating the molecular mechanisms underlying these phenomena, Wang *et al.* (2013) isolated full-length sequences of AOX1 (regulated ATP dissipation) as well as other energy pathway genes – *AtpB, UCP1, AAC1* and *SnRK2* from litchi fruit and the transcript abundance of these energy-related genes, respiration intensity and fruit energy status were analysed in developing and post-harvest senescent litchi fruit. The gene transcripts of fruit pericarp were highly expressed and peaked at 70 days after flowering (DAF) in the non-edible portions of litchi and correlated with fruit ADP concentrations. In contrast, the uncoupling mitochondrial protein 1 (UCP1) was predominantly expressed in the plant root, and the ATP synthase beta subunit (AtpB), which was up-regulated significantly before harvest and peaked two days after storage (Wang *et al.*, 2013), indicating that the colour-breaker stage (70 DAF) and two days after storage may be key turning points in litchi fruit energy metabolism. After two days of storage, among the transcript levels of different genes, that of AOX1 increased to much higher levels than of *LcAtpB*. Exogenous ATP significantly down-regulated these gene expressions, while maintaining ATP and energy charge levels, that resulted in delayed senescence.

A few studies recorded ARP in other non-climacteric fruit that are listed in Table 10.1.

Conclusion

Like blooming in flowers, the ripening process is thermogenic. Therefore, the physiological data available for blooming may form a guideline for expediting similar research in fruit ripening. Although the involvement of ARP in different types of fruit during their ripening stages has been reported sporadically for the past three decades, the involvement of AOX at subcellular and molecular levels has been only very recently and rarely reported. The molecular information generated using model systems such as tomato clearly indicates the complexity of AOX regulation in ripening fruit due to the involvement of various signalling molecules and metabolic networks that regulate AOX proteins in very complex but intricately interwoven networks of metabolic events. Nevertheless, simple biochemical data have also been useful as starting points to discover emergent properties. Taken together, the knowledge on the involvement of AOX proteins and their gene regulations appear to generate a higher level of control over the fruit ripening process, with more precision than the hitherto practiced ethylene controls. The available data in model systems such as tomato and *Arabidopsis* may be useful to develop mechanistic models through a systems biology approach for better elucidation of the complete signalling networks that respond to AOX in both climacteric and non-climacteric fruit ripening. Large variations of substrate composition in each variety of fruit makes biochemical and genetic elucidation of AOX interactions, thus supporting the argument that a systems biology approach for each fruit ripening process would result in delivering nutritionally dense fruit commodities with longer-lasting freshness.

References

Adams-Phillips, L., Barry, C. and Giovannoni, J. (2004) Signal transduction systems regulating fruit ripening. *Trends in Plant Science* **9**: 331–338.

Alba, R., Cordonnier-Pratt, M.M. and Pratt, L/H. (2000) Fruit-localized phytochromes regulate lycopene accumulation independently of ethylene production in tomato. *Plant Physiology* **123**: 363–370.

Alexander L. and Grierson D. (2002) Ethylene biosynthesis and action in tomato: a model for climacteric fruit ripening. *Journal of Experimental Botany* **53**: 2039–2055.

Almeida, A.M., Jarmuszkiewicz, W., Khomsi, H. *et al.* (1999) Cyanide-resistant, ATP-synthesis-sustained, and uncoupling-protein-sustained respiration during postharvest ripening of tomato fruit. *Plant Physiology* **119**: 1323–1330.

Argueso, C.T., Hansen, M. and Kieber, J.J. (2007) Regulation of ethylene biosynthesis. *Journal of Plant Growth Regulation* **26**: 92–105.

Barry, C.S. and Giovannoni, J.J. (2007) Ethylene and fruit ripening. *Journal of Plant Growth Regulation* **26**: 143–159.

Bojorquez, G. and Gomez-Lim, M.A. (1995) Peroxisomal thiolase mRNA is induced during mango fruit ripening. *Plant Molecular Biology* **28**: 811–820.

Borecky, J. and Vercesi, A.E. (2005) Plant uncoupling mitochondrial protein and alternative oxidase: energy metabolism and stress. *Bioscience Reports* **25**: 271–286.

Boyko, A., Kathiria, P., Zemp, F.J. *et al.* (2007) Transgenerational changes in the genome stability and methylation in pathogen-infected plants: (virus-induced plant genome instability). *Nucleic Acids Research* **35**: 1714–1725.

Boyko, A. and Kovalchuk, I. (2011) Genetic and epigenetic effects of plant–pathogen interactions: An evolutionary perspective. *Molecular Plant* **4**: 1014–1023.

Casolo, V., Braidot, E., Chiandussi, E. *et al.* (2000) The role of mild uncoupling and non-coupled respiration in the regulation of hydrogen peroxide generation by plant mitochondria. *FEBS Letters* **474**: 53–57.

Chen, Z.J. and Ni, Z. (2006) Mechanisms of genomic rearrangements and gene expression changes in plant polyploids. *Bioessays:* **28**: 240–252.

Considine, M.J., Daley, D.O. and Whelan, J. (2001) The expression of alternative oxidase and uncoupling protein during fruit ripening in mango. *Plant Physiology* **126**: 1619–1629.

Costa, A.D.T., Nantes, I.L., Jezek, P. *et al.* (1999) Plant uncoupling mitochondrial protein activity in mitochondria isolated from tomatoes at different stages of ripening. *Journal of Bioenergetics and Biomembranes* **31**: 527–533.

Day, D.A. and Wiskich, J.T. (1995) Regulation of alternative oxidase activity in higher plants. *Journal of Bioenergetics and Biomembranes* **27**: 379–385.

Ding, C.K., Wang, C., Gross, K. and Smith, D. (2002) Jasmonate and salicylate induce the expression of pathogenesis related protein genes and increase resistance to chilling injury in tomato fruit. *Planta* **214**: 895–901.

Ding, C.K. and Wang, C. (2003) The dual effects of methyl salicylate on ripening and expression of ethylene biosynthetic genes in tomato fruit. *Plant Science* **164**: 589–596.

Duque, P. and Arrabaca, J.D. (1999) Respiratory metabolism during cold storage of apple fruit. II. Alternative oxidase is induced at the climacteric. *Physiologia Plantarum* **107**: 24–31.

Duque, P., Barreiro, M.G. and Arrabaca, J.D. (1999) Respiratory metabolism during cold storage of apple fruit. I. Sucrose metabolism and glycolysis. *Physiologia Plantarum* **107**: 14–23.

Ebbs, S.D., Kosma, D.K., Nielson, E.H. *et al.* (2010) Nitrogen supply and cyanide concentration influence the enrichment of nitrogen from cyanide in wheat (*Triticum aestivum* L.) and sorghum (*Sorghum bicolor* L.). *Plant, Cell and Environment* **33**: 1152–1160.

Ederli, L., Morettini, R., Borgogni, A. *et al.* (2006) Interaction between nitric oxide and ethylene in the induction of alternative oxidase in ozone-treated tobacco plants. *Plant Physiology* **142**: 595–608.

Fung, R.W., Wang, C.Y., Smith, D.L. *et al.* (2004) MeSA and MeJA increase steady-state transcript levels of alternative oxidase and resistance against chilling injury in sweet peppers (*Capsicum annuum* L.). *Plant Science* **166**: 711–719.

Fung, R.W.M., Wang, C.Y., Smith, D.L. *et al.* (2006) Characterization of alternative oxidase (AOX) gene expression in response to methyl salicylate and methyl jasmonate pre-treatment and low temperature in tomatoes. *Journal of Plant Physiology* **163**: 1049–1060.

Gandin, A., Lapointe, L. and Dizengremel, P. (2009) The alternative respiratory pathway allows sink to cope with changes in carbon availability in the sink-limited plant *Erythronium americanum*. *Journal of Experimental Botany* **60**: 4235–4248.

Genard, M. and Gouble, B. (2005) ETHY. A theory of fruit climacteric ethylene emission. *Plant Physiology* **139**: 531–545.

Hamilton, A.J., Lycett, G.W. and Grierson, D. (1990) Antisense gene that inhibits synthesis of the hormone ethylene in transgenic plants. *Nature* **346**: 284–287.

Holtzapffel, R.C., Castelli, J., Finnegan, P.M. *et al.* (2003) A tomato alternative oxidase protein with altered regulatory properties. *Biochimica et Biophysica Acta* **1606**: 153–162.

Holtzapffel, R.C., Finnegan, P.M., Millar, A.H. *et al.* (2002) Mitochondrial protein expression in tomato fruit during on-vine ripening and cold storage. *Functional Plant Biology* **29**: 827–834.

Jing, G., Huang, H., Yang, B. *et al.* (2013) Effect of pyrogallol on the physiology and biochemistry of litchi fruit during storage. *Chemistry Central Journal* **7**: 19. DOI: 10.1186/1752-153X-7-19.

Juszczuk, I.M., Wagner, A.M. and Rychter, A.M. (2001) Regulation of alternative oxidase activity during phosphate deficiency in bean roots (*Phaseolus vulgaris*). *Physiologia Plantarum* **113**: 185–192.

Karcher, D. and Bock, R. (2002) Temperature sensitivity of RNA editing and intron splicing reactions in the plastid *ndhB* transcript. *Current Genetics* **41**: 48–52.

Klee, H.J. (2010) Improving the flavor of fresh fruits: genomics, biochemistry, and biotechnology. *New Phytologist* **187**: 44–56.

Kumar, S., Patil, B.C. and Sinha, S.K. (1990) Cyanide resistant respiration is involved in temperature rise in ripening mangoes. *Biochemical and Biophysical Research Communications* **168**: 818–822.

Kumar, S. and Sinham S. K. (1992). Alternative respiration and heat production in ripening banana fruits (*Musa paradisiaca* van *Mysore Kadali*). *Journal of Experimental Botany* **43**: 1639–1642.

Kurihara-Yonemoto, S. and Handa, H. (2001) Low temperature affects the processing pattern and RNA editing status of the mitochondrial cox2 transcript in wheat. *Current Genetics* **40**: 203–208.

Laloi, M., Klein, M., Riesmeier, J.W. *et al.* (1997) A plant cold-induced uncoupling protein. *Nature* **389**: 135–136.

Lambers, H. (1982) Cyanide-resistant respiration: A nonphosphorylating electron transport pathway acting as an energy overflow. *Physiologia Plantarum* **55**: 478–485.

Lange, D.L. and Kader, A.A. (1997) Changes in alternative pathway and mitochondrial respiration in avocado in response to elevated carbon dioxide levels. *Journal of the American Society of Horticultural Science* **122**: 245–252.

Ma, H., Song, C., Borth, W. *et al.* (2011) Modified expression of alternative oxidase in transgenic tomato and petunia affects the level of tomato spotted wilt virus resistance. *BMC Biotechnology* **11**: 96–102. DOI: 10.1186/1472-6750-11-96.

Manjunatha, G., Lokesh, V. and Bhagyalakshmi N. (2010). Nitric oxide in fruit ripening: Trends and opportunities. *Biotechnology Advances* **28**: 489–499.

Manjunatha, G., Gupta, J.K., Lokesh, V. *et al.* (2012a) Nitric oxide counters ethylene effects on ripening fruits. *Plant Signaling and Behavior* **7**: 476–483.

Manjunatha, G., Lokesh V. and Bhagyalakshmi N. (2012b). Nitric oxide-induced enhancement of banana fruit attributes and keeping quality. *Acta Horticulturae* **934**: 799–806.

Manning, K., Tör, M., Poole, M. *et al.* (2006) A naturally occurring epigenetic mutation in a gene encoding an SBP-box transcription factor inhibits tomato fruit ripening. *Nature Genetics* **38**: 948–952.

Maxwell, D.P., Wang, Y. and McIntosh, L. (1999) The alternative oxidase lowers mitochondrial reactive oxygen production in plant cells. *Proceedings of the National Academy of Sciences USA* **96**: 8271–8276.

Millar, A.H., Wiskich, J.T., Whelan, J. and Day, D.A. (1993) Organic acid activation of the alternative oxidase of plant mitochondria. *FEBS Letters* **329**: 259–262.

Millenaar, F. and Lambers, H. (2003) The alternative oxidase: *in vivo* regulation and function. *Plant Biology* **5**: 2–15.

Nakatsuka, A., Murachi, S., Okunishi, H. *et al.* (1998) Differential expression and internal feedback regulation of 1-aminocyclopropane-1-carboxylate synthase, 1-aminocyclopropane-1-carboxylate oxidase, and ethylene receptor genes in tomato fruit during development and ripening. *Plant Physiology* **118**: 1295–1305.

Oeller, P.W., Lu, M.W., Taylor, L.P. *et al.* (1991) Reversible inhibition of tomato fruit senescence by antisense RNA. *Science* **254**: 437–439.

Pastore, D., Trono, D., Laus, M.N. *et al.* (2001) Alternative oxidase in durum wheat mitochondria. Activation by pyruvate, hydroxypyruvate and glyoxylate and physiological role. *Plant and Cell Physiology* **42**: 1373–1382.

Pech, J.C., Bouzayen, M. and Latche, A. (2008) Climacteric fruit ripening: Ethylene-dependent and independent regulation of ripening pathways in melon fruit. *Plant Science* **175**: 114–120.

Purvis, A.C., Shewfelt, R.L. and Gegogeine, J.W. (1995) Superoxide production by mitochondria isolated from green bell pepper fruit. *Physiologia Plantarum* **94**: 743–749.

Salgado, I., Martínez, M.C., Oliveira, H.C. and Lucas Frungillo, N. (2013) Nitric oxide signaling and homeostasis in plants: A focus on nitrate reductase andS-nitrosoglutathione reductase in stress-related responses. *Brazilian Journal of Botany* **36**: 89–98.

Sluse, F.E. and Jarmuszkiewicz, W. (2000) Activity and functional interaction of alternative oxidase and uncoupling protein in mitochondria from tomato fruit. *Brazilian Journal of Medical and Biological Research* **33**: 259–268.

Solomos, T. and Laties, G.G. (1974) Similarities between the actions of ethylene and cyanide in initiating the climacteric and ripening of avocados. *Plant Physiology* **54**: 506–511.

Solomos, T. and Laties, G.G. (1976) Effects of cyanide and ethylene on the respiration of cyanide-sensitive and cyanide-resistant plant tissues. *Plant Physiology* **58**: 47–50.

Srivastava, A. and Handa, A.K. (2005) Hormonal regulation of tomato fruit development: a molecular perspective. *Journal of Plant Growth Regulation* **24**: 67–82.

Tigchelaar, E., McGlasson, W. and Buescher, R. (1978) Genetic regulation of tomato fruit ripening. *HortScience* **13**: 508–513.

Tucker, M.L. and Laties, G.G. (1984) Interrelationship of gene expression, polysome prevalence, and respiration during ripening of ethylene and/or cyanide-treated avocado fruit. *Plant Physiology* **74**: 307–315.

Umbach, A.L. and Siedow, J.N. (1993) Covalent and noncovalent dimers of the cyanide-resistant alternative oxidase protein in higher plant mitochondria and their relationship to enzyme activity. *Plant Physiology* **103**: 845–854.

Umbach, A.L., Wiskich, J.T. and Siedow, J.N. (1994) Regulation of alternative oxidase kinetics by pyruvate and intermolecular disulphide bond redox status in soybean seedling mitochondria. *FEBS Letters* **348**: 181–184.

Vanlerberghe, G.C., Cvetkovska, M. and Wang, J. (2009) Is the maintenance of homeostatic mitochondrial signaling during stress a physiological role for alternative oxidase? *Physiologia Plantarum* **137**: 392–406.

Vrebalov, J., Ruezinsky, D., Padmanabhan, V. *et al.* (2002) A MADS-box gene necessary for fruit ripening at the tomato ripening-inhibitor (*rin*) locus. *Science* **296**: 343.

Wang, H., Huang, J. and Bi, Y. (2010) Induction of alternative respiratory pathway involves nitric oxide, hydrogen peroxide and ethylene under salt stress. *Plant Signaling & Behavior* **5** (12): 1636–1637.

Wang, H., Qian, Z., Ma, S. *et al.* (2013) Energy status of ripening and postharvest senescent fruit of litchi (*Litchi chinensis* Sonn.). *BMC Plant Biology* **13**: 55.

Wang, J., Rajakulendran, N., Amirsadeghi, S. and Vanlerberghe, G.C. (2011) Impact of mitochondrial alternative oxidase expression on the response of *Nicotiana tabacum* to cold temperature. *Physiologia Plantarum* **142**: 339–351.

Wang, K.L., Li, H., and Ecker, J.R. (2002) Ethylene biosynthesis and signaling networks. *The Plant Cell* **14** (suppl.): S131–S151.

Watanabe, A. and Hirai, A. (2002) Two uncoupling protein genes of rice (*Oryza sativa* L.): Molecular study reveals the defects in the pre-mRNA processing for the heat-generating proteins of the subtropical cereal. *Planta* **215**: 90–100.

White, P.J. (2002) Recent advances in fruit development and ripening: an overview. *Journal of Experimental Botany* **53**: 1995–2000.

Wilkinson, J.Q., Lanahan, M.B., Yen, H.C. *et al.* (1995) An ethylene-inducible component of signal transduction encoded by *Never-ripe*. *Science* **270**: 1807–1809.

Woodward, E.W., Murphy, C.C. and Heckman, N.L. (2009) Alternative oxidase identification and capacity in postharvest ripening stages of banana fruit. *Journal of Undergraduate Chemistry Research* **8**: 52.

Xiaoyong, L., Tian MS., Xiaosong, H. and Yaoren, D. (2003) Relationship between stimulated ethylene production and alternative respiration pathway in Royal Gala' apple fruit. *Tsinghua Science and Technology* **8**: 451–454.

Xu, F., Yuan S., Zhang, D.-W. *et al.* (2012) The role of alternative oxidase in tomato fruit ripening and its regulatory interaction with ethylene. *Journal of Experimental Botany* **63**: 5705–5716.

Yang, S.F. and Hoffman, N.E. (1984) Ethylene biosynthesis and its regulation in higher plants. *Annual Review of Plant Physiology* **35**: 155–189.

Yi, C., Qu, H.X., Jiang, Y.M. *et al.* (2008) ATP-induced changes in energy status and membrane integrity of harvested litchi fruit and its relation to pathogen resistance. *Journal of Phytopathology* **156**: 365–371.

Yip, J.Y.H. and Vanlerberghe, G.C. (2001) Mitochondrial alternative oxidase acts to dampen the generation of active oxygen species during a period of rapid respiration induced to support a high rate of nutrient uptake. *Physiologia Plantarum* **112**: 327–333.

Yip, W.K. and Yang, S.F. (1988) Cyanide metabolism in relation to ethylene production in plant tissues. *Plant Physiology* **88**: 473–476.

Yokotani, N., Nakano, R., Imanishi, S. *et al.* (2009) Ripening-associated ethylene biosynthesis in tomato fruit is autocatalytically and developmentally regulated. *Journal of Experimental Botany* **60**: 3433–3442.

Yoshida, K., Watanabe, C., Kato, Y. *et al.* (2008) Influence of chloroplastic photo-oxidative stress on mitochondrial alternative oxidase capacity and respiratory properties: a case study with *Arabidopsis* yellow variegated 2. *Plant and Cell Physiology* **49**: 592–603.

Zhang, D.W., Xu, F., Zhang, Z.W. *et al.* (2010) Effects of light on cyanide-resistant respiration and alternative oxidase function in *Arabidopsis* seedlings. *Plant, Cell and Environment* **33**: 2121–2131.

CHAPTER 11

Respiratory pathways in bulky tissues and storage organs

Wu-Sheng Liang

Institute of Biotechnology, College of Agriculture and Biotechnology, Zhejiang University, Hangzhou, People's Republic of China

Introduction

Bulky tissues and storage organs have a long storage time, which makes it easy to obtain them for doing research. Furthermore, as they are generally non-green, there is no interference of photosynthesis during respiration rate determination, and it is more convenient to obtain purified mitochondria from such tissues than from green plant tissues such as green leaves. Therefore, they were widely used as material in early research into the mitochondrial electron transport chain. In fact, many of the studies in the review about alternative respiratory pathway (ARP) by Laties in 1982 summarized research results using bulky tissues and storage organs. The equation to estimate the contributions of the ARP and the cytochrome respiratory pathway (CRP) in plant tissues was first introduced with the results from aged potato tuber slices:

$$V_t = \rho \cdot g(i) + V_{cyt} + V_{res},$$

where V_t, V_{cyt} and V_{res} refer to total respiration rate, contribution of CRP, and residual respiration, respectively; $g(i)$ represents ARP capacity, which is also often expressed with Valt; ρ indicates the fraction of ARP engaged in respiration (Theologis and Laties, 1978a).

In studies with bulky plant storage organs, the activities of respiratory pathways are generally determined with prepared slices. Bulky plant storage organs have two categories of response of the freshly prepared slices to cyanide. One group yields cyanide-resistant fresh slices and includes parsnip (*Pastinaca sativa*), carrot (*Daucus carota*) and red sweet potato (*Ipomoea batatas*) (Theologis and Laties, 1978b). The second group yields cyanide-sensitive fresh slices and includes potato (*Solanum tuberosum*), red beet (*Beta vulgaris*) and turnip (*Brassica*

Alternative Respiratory Pathways in Higher Plants, First Edition.
Edited by Kapuganti Jagadis Gupta, Luis A.J. Mur and Bhagyalakshmi Neelwarne.

rapa) (Theologis and Laties, 1978a, 1978b). However, the slices of this group can acquire a certain level of ARP capacity after a process of aging (Laties, 1982).

In this chapter, progresses in research on the respiratory pathways of potato tubers – especially aging potato tuber slices – since the review by Laties in 1982 are summarized. This is intended to be representative of respiratory pathways in bulky tissues and storage organs in general.

The gene encoding potato alternative oxidase and its tissue-specific expression in potato tuber

A cDNA encoding potato alternative oxidase (AOX) has been cloned (Hiser *et al.*, 1996). It was 1254 bp long with a 344-codon open reading frame that encoded a 41 kDa polypeptide. Sequence comparison with AOX proteins from other plants suggested that the encoded polypeptide contained a transit peptide of 59 amino acids. Two conserved cysteine residues (117 and 167) were thought to be responsible for the potential disulfide bond formation (Hiser *et al.*, 1996).

The monoclonal antibody produced against the AOX of *Sauromatum guttatum* reacted with potato AOX. When mitochondrial proteins of potato were probed with the antibody, a closely spaced doublet of proteins was found in the leaf and root mitochondria from the potato variety FL1607. Despite the presence of this doublet in other tissues, only one protein of about 36 kDa was detected in the tuber mitochondria from both potato varieties FL1607 and Russet Burbank (Hiser and Mclntosh, 1990; Hiser *et al.*, 1996). These results indicated that there was a tissue-specific difference between the expression of AOX in tuber and other potato tissues.

Development of alternative respiration pathway capacity of potato tuber slices during the aging process

It is generally assumed that whole potato tubers are cyanide-insensitive because the application of cyanide does not inhibit their respiration (Rychter *et al.*, 1979). However, slices prepared from potato tubers are highly cyanide-sensitive and the ARP capacity (*V*alt) can greatly increase in potato tuber slices during the process of aging (Theologis and Laties, 1978a). During aging, *V*t of the slices notably increased before 12 h, but changed little from 12 to 24 h (Figure 11.1). A determination of the ARP capacity with purified mitochondria from aging slices showed a consistency with the results from whole aging slices (Figure 11.2).

Van Steveninck (1975) has reviewed the 'aging' phenomenon in plant tissues and it is proposed to be adaptive process and there are stimulated

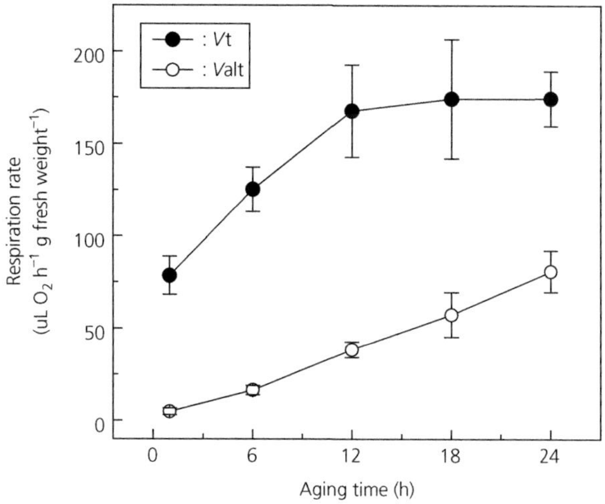

Figure 11.1 Total respiration rate (*V*t) and alternative respiration pathway capacity (*V*alt) of potato tuber slices during aging process of 24 h. Potato tuber slices of 6 mm diameter and 1 mm thickness were prepared and set to age at 27 °C on gauze wetted with distilled water (Liang *et al.*, 1997).

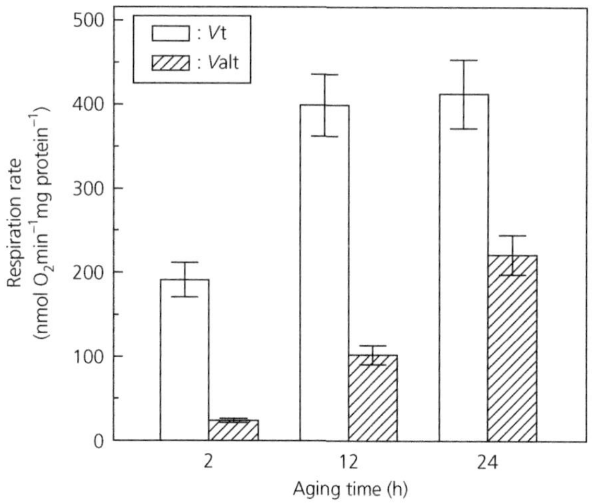

Figure 11.2 Total respiration rate (*V*t) and alternative respiration pathway capacity (*V*alt) of mitochondria purified from aging potato tuber slices (Liang *et al.*, 1997).

synthesis and increased physiological competence during aging of plant tissue slices (Van Steveninck, 1975). Someone may question the scientific usefulness of information based on what may appear to be a highly artificial set of conditions during aging. However, aging potato slices provide one kind of convenient material to study respiratory pathways in bulky tissues and storage organs.

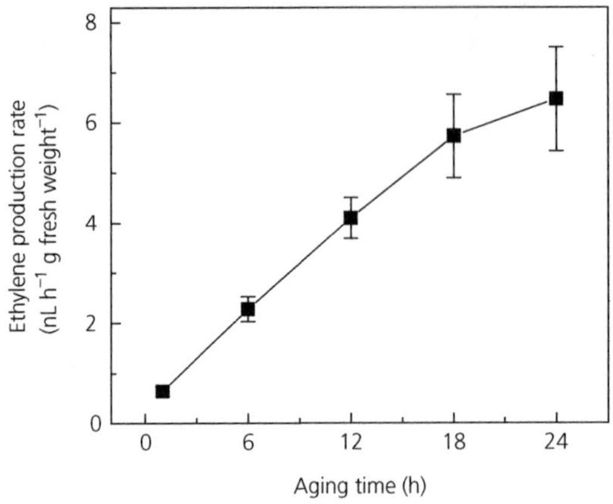

Figure 11.3 Ethylene production rate of potato tuber slices during aging process of 24 h (Liang *et al.*, 1997).

Alternative oxidase in aged potato tuber slices is a protein synthesized *de novo* during the aging process

The mitochondria purified from fresh and aged potato tuber slices were probed with the monoclonal antibody raised against the AOX of *S. guttatum* after one- and two-dimensional gel electrophoresis (Elthon *et al.*, 1989). The relative level of a 36 kDa protein was observed to parallel the rise in ARP capacity, which indicated that the AOX protein in aged potato tuber slices is synthesized *de novo* during the aging process (Hiser and McIntosh, 1990).

The relationship between endogenous ethylene and the development of the alternative respiration pathway capacity of potato tuber slices during the aging process

Aging potato tuber slices were observed to produce ethylene. The ethylene production rate increased with the aging process, and displayed a trend similar to that of *V*alt (Figure 11.3). Application of 1-aminocyclopropane-1-carboxylic acid (ACC), a precursor of ethylene biosynthesis, enhanced both the ethylene production rate and the *V*alt values of the aging potato tuber slices. In contrast, treatment with $CoCl_2$ – which can inhibit the conversion of ACC to ethylene (Wenzel *et al.*, 1995) – resulted in a decrease in both the ethylene production rate and the *V*alt values (Figure 11.4. and Figure 11.5). Western blotting results with a monoclonal antibody against the AOX of *S. guttatum* showed that the expression level of AOX protein was enhanced by ACC, but reduced by $CoCl_2$ (Figure 11.6). These results showed that endogenous ethylene levels were essential to the development of ARP capacity in aging potato tuber slices.

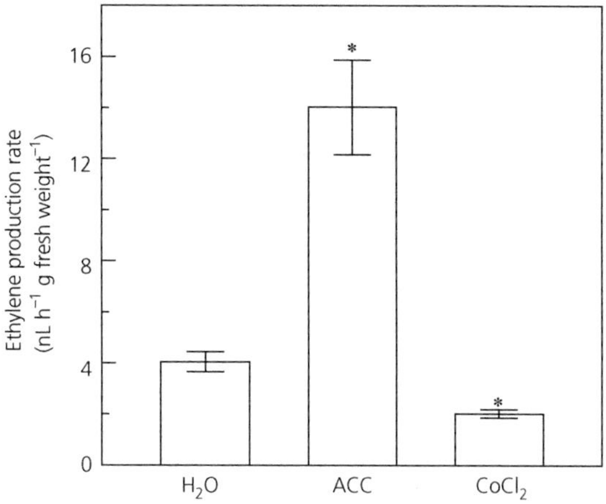

Figure 11.4 Effect of treatment with ACC (1.0 mmol L^{-1}) or $CoCl_2$ (1.0 mmol L^{-1}) on the ethylene production rate of potato tuber slices aged for 12 h. *Means are significantly different from control (H_2O) ($P < 0.05$) (Liang *et al.*, 1997).

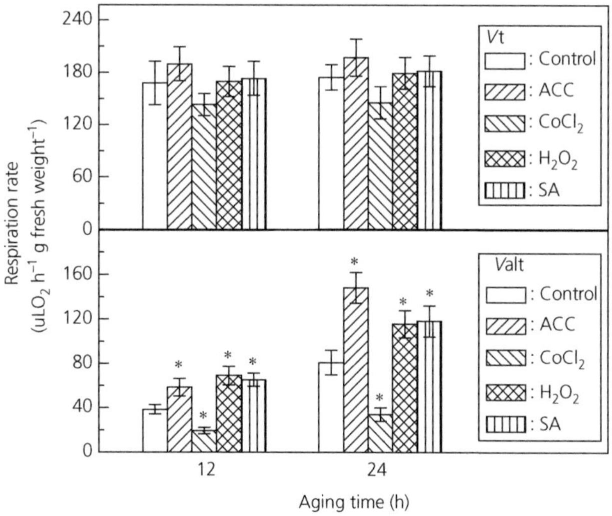

Figure 11.5 Effect of treatment with ACC (1.0 mmol L^{-1}), $CoCl_2$ (1.0 mmol L^{-1}), H_2O_2 (5.0 mmol L^{-1}) and salicylic acid (SA) (0.1 mmol L^{-1}) on the total respiration rate (*V*t) and alternative respiration pathway capacity (*V*alt) of aging potato tuber slices. *Means are significantly different from control (H_2O) ($P < 0.05$) (Liang and Liang, 2002; Liang *et al.*, 1997).

Induction of the ARP by ethylene has also been found in other plant tissues. Ethylene treatment induced ARP in tomato fruit during post-harvest ripening (Xu *et al.*, 2012), tobacco discs (Ederli *et al.*, 2006), and *Arabidopsis*

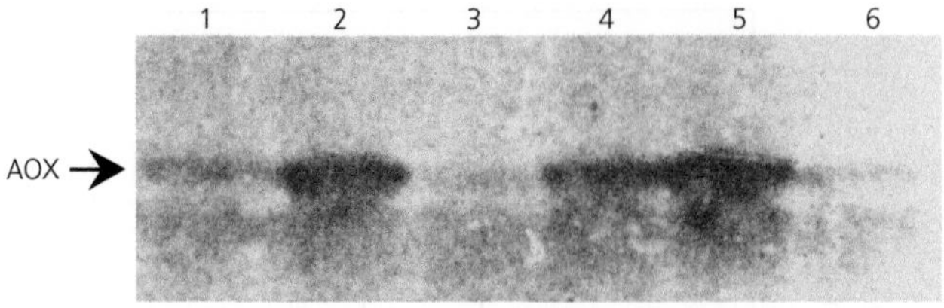

Figure 11.6 Western blotting of AOX in the mitochondria purified from aging potato tuber slices to show the effect of ACC and $CoCl_2$ on AOX expression: 1. Slices aged for 12 h; 2. Slices treated with ACC and aged for 12 h; 3. Slices treated with $CoCl_2$ and aged for 12 h; 4. Slices aged for 24 h; 5. Slices treated with ACC and aged for 24 h; 6. Slices treated with $CoCl_2$ and aged for 24 h. The concentrations of both ACC and $CoCl_2$ were 1.0 mmol L^{-1}. Mitochondrial protein of 200 μg was loaded to each lane. The monoclonal antibody against AOX of *S. guttatum* (gifted by T. E. Elthon) was used as the primary antibody to analyse AOX protein on the nitrocellulose membranes (Liang and Liang, 1999a).

calluses (Wang *et al.*, 2010). In contrast, an ethylene biosynthesis inhibitor, aminoethoxyvinylglycine, inhibited ozone-induced AOX expression in tobacco discs (Ederli *et al.*, 2006), while another ethylene biosynthesis inhibitor, aminooxyacetic acid, inhibited AOX expression induced by salt stress in *Arabidopsis* calluses (Wang *et al.*, 2010).

Alternative respiration pathway capacity can be induced by hydrogen peroxide and salicylic acid in aging potato tuber slices

Hydrogen peroxide is an inducer of AOX expression in *Petunia hybrida* cells (Wagner, 1995), and *Arabidopsis* calluses (Wang *et al.*, 2010). The induction of ARP by H_2O_2 in aging potato tuber slices was observed as well. H_2O_2 (5.0 mmol L^{-1}) had little influence on *V*t of the slices, but showed a significant inducing effect on the *V*alt values of potato tuber slices during 24 h of the aging process (see Figure 11.5) (Liang and Liang, 2002). Western blotting with the monoclonal antibody against AOX showed that H_2O_2 treatment increased the expression of AOX in aging potato tuber slices, which implies that the induction of *V*alt by H_2O_2 was related to AOX expression (Figure 11.7) (Liang and Liang, 2002).

Salicylic acid (SA) is also an inducer of AOX expression in some plant tissues (Chivasa *et al.*, 1997; Lennon *et al.*, 1997). Induction of ARP by SA in aging potato tuber slices was also observed (Wen and Liang, 1994). Western blotting with monoclonal antibody against AOX showed that SA treatment increased the expression of AOX in aging potato tuber slices (Figure 11.7) (Liang and Liang, 2002).

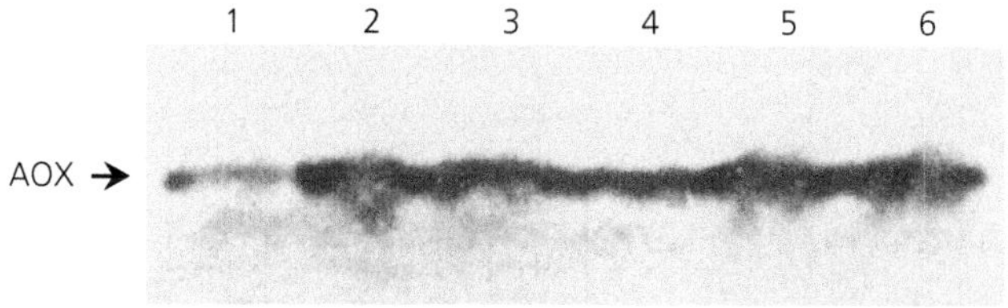

Figure 11.7 Western blotting of AOX in the mitochondria purified from aging potato tuber slices to show induction effect of hydrogen peroxide and salicylic acid on AOX expression. 1–3: aged for 12 h; 4–6: aged for 24 h. 1, 4: control (H_2O); 2, 5: treated with H_2O_2 (5.0 mmol L^{-1}); 3, 6: treated with salicylic acid (0.1 mmol L^{-1}). Mitochondrial protein of 200 µg was loaded to each lane. The monoclonal antibody against AOX of *S. guttatum* (gifted by T. E. Elthon) was used as the primary antibody to analyse AOX protein on the nitrocellulose membranes (Liang and Liang, 2002).

Activation of alternative oxidase by pyruvate in mitochondria of aged potato tuber slices

Plant AOX can exist as a monomer or homodimer. The reduced enzyme is the active form and can be further activated by alpha-keto acids such as pyruvate (Rhoads *et al.*, 1998). When the two subunits are covalently linked by a disulfide bond (i.e. oxidized), the enzyme is then in the form of homodimer, which is essentially inactive and cannot be activated by alpha-keto acids (Umbach and Siedow, 1993; Day *et al.*, 1994; Rhoads *et al.*, 1998).

Hiser *et al.* (1996) observed that both the reduced and the oxidized forms of AOX existed in mitochondria of aged potato tuber slices, and the presence of pyruvate increased the ARP capacity of the mitochondria (Hiser *et al.*, 1996). In order to further test the activating effects of pyruvate on AOX, a series of measurements were carried out with mitochondria purified from potato tuber slices aged for 24 h.

- Measurement (1): the *V*alt values were determined in the absence of pyruvate.
- Measurement (2): pyruvate (5 mmol L^{-1}) was added to the reaction solutions of measurement (1). The determined *V*alt values were quite higher than those in measurement (1).
- Measurement (3): the mitochondria were recovered by centrifugation from the reaction solutions of measurement (2), and were washed with washing buffer. With the washed mitochondria, the determined *V*alt values in the absence of pyruvate decreased to a level similar to the result of measurement (1).
- Measurement (4): pyruvate (5 mmol L^{-1}) was added again to the reaction solutions of measurement (3). The *V*alt values were determined to be much higher than the results of measurement (1) and (3), but at a level similar to that of measurement (2) (Figure 11.8). These results showed that pyruvate could obviously activate the AOX in aging potato tuber slices (Liang *et al.*, 2003).

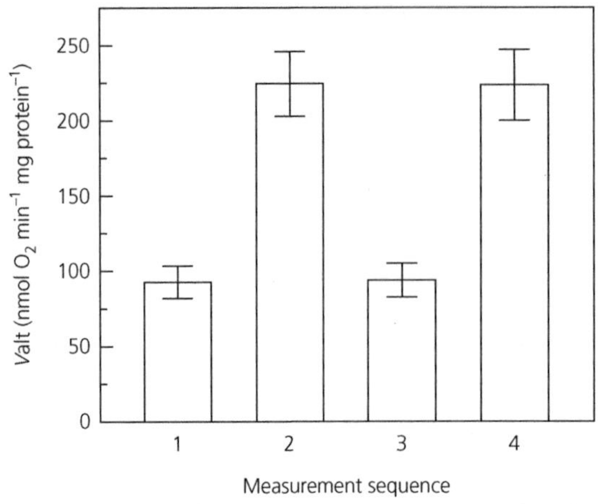

Figure 11.8 Effects of exogenous pyruvate on the *V*alt values of mitochondria purified from potato tuber slices aged for 24h (Liang *et al.*, 2003). 1: The *V*alt values were determined in the absence of pyruvate. 2: Pyruvate (5.0 mmol L^{-1}) was added to the reaction solutions of 1. 3: Mitochondria were recovered by centrifugation from the reaction solutions of 2, and were washed with washing buffer. The *V*alt values were determined with the washed mitochondria in the absence of pyruvate. 4: Pyruvate (5.0 mmol L^{-1}) was added to the reaction solutions of 3.

Oliver *et al.* (2008) generated some transgenic potato plants with strongly reduced expression levels of cytosolic pyruvate kinase by using RNA interference gene silencing under the control of a tuber-specific promoter. The transgenic tubers showed a decrease in the levels of pyruvate. They also showed a strong decrease in the levels of AOX protein and a corresponding decrease in the ARP capacity. External feeding of pyruvate to tuber tissue or isolated mitochondria resulted in activation of ARP (Oliver *et al.*, 2008).

Comparison of the estimated alternative respiration pathway activities of aging potato tuber slices by hydroxamate-inhibition method and oxygen-isotope-fractionation method

The relationship of the contributions of CRP and ARP in plant tissues can be expressed with the equation

$$V_t = \rho \cdot g(i) + V_{cyt} + V_{res}$$

where $g(i)$ can also be expressed as *V*alt, $\rho \cdot g(i)$ (can also be expressed as ρ*V*alt) is the real operating activity of ARP and $\leq$ *V*alt (Theologis and Laties, 1978a).

(Benz)hydroxamates, such as salicylhydroxamic acid (SHAM), are inhibitors of ARP. The ρ*V*alt values of plant tissues and cells are generally determined by the equation

$$\rho V_{alt} = V_t - V(\text{SHAM})$$

Table 11.1 Comparison of the *in vivo* activities of the alternative respiration pathway (ρV_{alt}) in aging potato tuber slices under different treatments determined by the hydroxamate-inhibition method and by the oxygen-isotope-fractionation method.*

Aging time (h)	Treatment	Treatment	
		Hydroxamate-inhibition method	Oxygen-isotope-fractionation method
12	Control	18.8	36.1
	ACC	21.7	40.6
	$CoCl_2$	15.4	30.1
24	Control	21.4	44.5
	ACC	24.9	51.5
	$CoCl_2$	17.8	35.5

*ACC (1.0 mmol L^{-1}) and $CoCl_2$ (1.0 mmol L^{-1}) were applied to do the treatment. The ρV_{alt} values were expressed with nmol O_2 min^{-1} g fresh weight^{-1} as unit (Liang and Liang, 1999b).

where *Vt* refers to total respiration rate determined in the absence of respiratory inhibitors, *V*(SHAM) refers to respiration rate determined in the presence of SHAM (Bingham and Farrar, 1989; McDonnell and Farrar, 1993; Vani and Raghavendra, 1994). This method can be called the hydroxamate-inhibiting method. Because ARP and CRP get electrons from the same ubiquinone pool, the inhibition of ARP may result in the diversion of electrons from ARP to CRP (Wilson, 1988). So it was inferred by some authors that the hydroxamate-inhibiting method could probably cause an underestimation while determining ARP activity (Atkin *et al.*, 1995; Wagner and Krab, 1995).

While questioning the accuracy of the results obtained by the hydroxamate-inhibiting method, the oxygen-isotope-fractionation method was developed to determine ARP activity (Guy *et al.*, 1989). A great difference was observed between the ρ*V*alt values of soybean cotyledons determined by the hydroxamate-inhibiting method and by the oxygen-isotope-fractionation method (Ribas-Carbo *et al.*, 1995).

The ρ*V*alt values of aging potato tuber slices were also compared using the hydroxamate-inhibiting method and the oxygen-isotope-fractionation method. The ρ*V*alt values of aging potato tuber slices determined by the oxygen-isotope-fractionation method were about twice as large as those determined by the hydroxamate-inhibition method. In addition, the ρ*V*alt values measured by the oxygen-isotope-fractionation method in slices treated with ACC and $CoCl_2$ were also twice as much as those determined using the hydroxamate-inhibition method. The results provided more evidence that the hydroxamate-inhibition method results in an underestimation of the *in vivo* ARP activities in plant tissues (Table 11.1) (Liang and Liang, 1999b).

Conclusions

Plant AOX has been confirmed to fulfil some important functions. It helps to produce heat in blooms of the Araceae, such as *S. guttatum,* for a better volatilization of scent compounds to attract insect pollinators (McIntosh, 1994). It may prevent deleterious oxidative stress in plant cells under abiotic stresses (Day *et al.*, 1995; McDonald, 2008), and has been found to induce resistance of plants against viral infection (Chivasa *et al.*, 1997; Chivasa and Carr, 1998; Fu *et al.*, 2010). These functions do not seem to be the role of AOX in bulky tissues and storage organs. The introduction earlier has showed that AOX in the bulky tissues and storage organs can be induced by ACC, hydrogen peroxide and SA, and can be activated by pyruvate, as in other plant tissues. However, the different expression patterns of AOX in tuber and other potato tissues are suggestive of certain specific roles of AOX in bulky tissues and storage organs, which still need to be studied (Hiser and McIntosh, 1990; Hiser *et al.*, 1996).

Acknowledgements

We are grateful to T.E. Elthon (University of Nebraska, Lincoln) for gifting the monoclonal antibody against the AOX of *S. guttatum.* The work in our lab about plant respiration was supported by the National Natural Science Foundation of China (Grant Nos 39670070, 30270127, 30470155, 30870187 and 31370279).

References

Atkin, O.K., Villar, R. and Lambers, H. (1995) Partitioning of electrons between the cytochrome and alternative pathways in intact roots. *Plant Physiology* **108**: 1179–1183.

Bingham, I.J. and Farrar, J.F. (1989) Activity and capacity of respiratory pathways in barley roots deprived of inorganic nutrients. *Plant Physiology and Biochemistry* **27**: 847–854.

Chivasa, S. and Carr, J.P. (1998) Cyanide restores *N* gene-mediated resistance to tobacco mosaic virus in transgenic tobacco expressing salicylic acid hydroxylase. *The Plant Cell* **10**: 1489–1498.

Chivasa, S., Murphy, A.M., Naylor, M. and Carr, J.P. (1997) Salicylic acid interferes with tobacco mosaic virus replication via a novel salicylhydroxamic acid-sensitive mechanism. *The Plant Cell* **9**: 547–557.

Day, D.A., Millar, A.H., Wiskich, J.T. and Whelan, J. (1994) Regulation of alternative oxidase activity by pyruvate in soybean mitochondria. *Plant Physiology* **106**: 1421–1427.

Day, D.A., Whelan, J., Millar, A.H. *et al.* (1995) Regulation of the alternative oxidase in plants and fungi. *Australian Journal of Plant Physiology* **22**: 494–509.

Ederli, L., Morettini, R., Borgogni, A. *et al.* (2006) Interaction between nitric oxide and ethylene in the induction of alternative oxidase in ozone-treated tobacco plants. *Plant Physiology* **142**: 595–608.

Elthon, T.E., Nickels, R.L. and McIntosh, L. (1989) Monoclonal antibodies to the alternative oxidase of higher plant mitochondria. *Plant Physiology* **89**: 1311–1317.

Fu, L.J., Shi, K., Gu, M. *et al.* (2010) Systemic induction and role of mitochondrial alternative oxidase and nitric oxide in a compatible tomato-tobacco mosaic virus interaction. *Molecular Plant-Microbe Interactions* **23**: 39–48.

Guy, R.D., Berry, J.A., Fogel, M.L. and Hoering, T.C. (1989) Differential fractionation of oxygen isotopes by cyanide-resistant and cyanide-sensitive respiration in plants. *Planta* **177**: 483–491.

Hiser, C., Kapranov, P. and McIntosh, L. (1996) Genetic modification of respiratory capacity in potato. *Plant Physiology* **110**: 277–286.

Hiser, C. and McIntosh, L. (1990) Alternative oxidase of potato is an integral membrane protein synthesized de novo during aging of tuber slices. *Plant Physiology* **93**: 312–318.

Laties, G.G. (1982) The cyanide-resistant, alternative path in higher plant respiration. *Annual Review of Plant Physiology* **33**: 519–555.

Lennon, A.M., Neuenschwander, U.H., Ribas-Carbo, M. *et al.* (1997) The effects of salicylic acid and tobacco mosaic virus infection on the alternative oxidase of tobacco. *Plant Physiology* **115**: 783–791.

Liang, W.S. and Liang, H.G. (1999a) Induction of alternative oxidase expression by endogenous ethylene in aging potato tuber slices. *Acta Phytophysiologica Sinica* **25**: 205–208.

Liang, W.S. and Liang, H.G. (1999b) Comparison of the estimation of the operation of the alternative respiration pathway between with oxygen-isotope-fractionation method and with hydroxamate-inhibition method. *Acta Phytophysiologica Sinica* **25**: 287–292.

Liang, W.S. and Liang, H.G. (2002) Comparison of the effects of H_2O_2 and salicylic acid on alternative respiratory pathway in aged potato tuber slices. *Journal of Integrative Plant Biology* **44**: 287–291.

Liang, W.S., Pan, J. and Liang, H.G. (2003) Activation of cyanide-resistant respiration by pyruvate in mitochondria of aged potato tuber slices. *Journal of Plant Physiology and Molecular Biology* **29**: 317–321.

Liang, W.S., Wen, J.Q. and Liang, H.G. (1997) Relationship between endogenous ethylene and the development and operation of the alternative pathway in aged potato tuber slices. *Journal of Integrative Plant Biology* **39**: 527–533.

McDonald, A.E. (2008) Alternative oxidase: An inter-kingdom perspective on the function and regulation of this broadly distributed 'cyanide-resistant' terminal oxidase. *Functional Plant Biology* **35**: 535–552.

McDonnell, E. and Farrar, J.F. (1993) Respiratory characteristics of isolated barley mitochondria and intact barley roots. *Journal of Experimental Botany* **44**: 1485–1490.

McIntosh, L. (1994) Molecular biology of the alternative oxidase. *Plant Physiology* **105**: 781–786.

Oliver, S.N., Lunn, J.E., Urbanczyk-Wochniak, E. *et al.* (2008) Decreased expression of cytosolic pyruvate kinase in potato tubers leads to a decline in pyruvate resulting in an in vivo repression of the alternative oxidase. *Plant Physiology* **148**: 1640–1654.

Rhoads, D.M., Umbach, A., Sweet, C.R. *et al.* (1998) Regulation of the cyanide-resistant alternative oxidase of plant mitochondria. *Identification of the cysteine residue involved in α-keto acid stimulation and intersubunit disulfide bond formation. Journal of Biological Chemistry* **273**: 30750–30756.

Ribas-Carbo, M., Berry, J.A., Yakir, D. *et al.* (1995) Electron partitioning between the cytochrome and alternative pathways in plant mitochondria. *Plant Physiology* **109**: 829–837.

Rychter, A., Janes, H.W. and Frenkel, C. (1979) Effect of ethylene and oxygen on the development of cyanide-resistant respiration in whole potato mitochondria. *Plant Physiology* **63**: 149–151.

Theologis, A. and Laties, G.G. (1978a) Relative contribution of cytochrome-mediated and cyanide-resistant electron transport in fresh and aged potato slices. *Plant Physiology* **62**: 232–237.

Theologis, A. and Laties, G.G. (1978b) Cyanide-resistant respiration in fresh and aged sweet potato slices. *Plant Physiology* **62**: 243–248.

Umbach, A. and Siedow, J.N. (1993) Covalent and noncovalent dimers of the cyanide-resistant alternative oxidase protein in higher plant mitochondria and their relationship to enzyme activity. *Plant Physiology* **103**: 845–854.

Vani, T. and Raghavendra, A.S. (1994) High mitochondrial activity but incomplete engagement of the cyanide-resistant alternative pathway in guard cell protoplasts of pea. *Plant Physiology* **105**: 1263–1268.

Van Steveninck, R.F.M. (1975) The "washing" or "aging" phenomenon in plant tissues. *Annual Review of Plant Physiology* **26**: 237–258.

Wagner, A.M. (1995) A role for active oxygen species as second messengers in the induction of alternative oxidase gene expression in Petunia hybrida cells. *FEBS Letters* **368**: 339–342.

Wagner, A.M. and Krab, K. (1995) The alternative respiration pathway in plants: role and regulation. *Physiologia Plantarum* **95**: 318–325.

Wang, H., Liang, X., Huang, J. *et al.* (2010) Involvement of ethylene and hydrogen peroxide in induction of alternative respiratory pathway in salt-treated Arabidopsis calluses. *Plant & Cell Physiology* **51**: 1754–1765.

Wen, J.Q. and Liang, H.G. (1994) Induction of salicylic acid on alternative pathway in aging potato tuber slices. *Chinese Science Bulletin* **39**: 1644–1647.

Wenzel, A.A., Schlautmann, H., Jones, C.A. *et al.* (1995) Aminoethoxyvinylglycine, cobalt and ascorbic-acid all reduce ozone toxicity in mung beans by inhibition of ethylene biosynthesis. *Physiologia Plantarum* **93**: 286–290.

Wilson, S.B. (1988) The switching of electron flux from the cyanide-insensitive oxidase to the cytochrome pathway in mung bean (Phaseolus aureus L.) mitochondria. *The Biochemical Journal* **249**: 301–303.

Xu, F., Yuan, S., Zhang, D.W. *et al.* (2012) The role of alternative oxidase in tomato fruit ripening and its regulatory interaction with ethylene. *Journal of Experimental Botany* **63**: 5705–5716.

SECTION B
From *AOX* diversity to functional marker development

Birgit Arnholdt-Schmitt
EU Marie Curie Chair, ICAAM - Instituto de Ciências Agrárias e Ambientais Mediterrânicas, Universidade de Évora, Évora, Portugal

Introduction

Alternative oxidase (*AOX*) is proposed as a target gene family for functional marker (FM) development. FMs derived from *AOX* genes are expected to assist breeding for robust plants with individual or multi-stress tolerance that are linked to traits such as yield stability (Arnholdt-Schmitt *et al.*, 2006; Arnholdt-Schmitt, 2009; Polidoros *et al.*, 2009) or post-harvest behaviour (Afuape *et al.*, 2013). The utility of general markers across species will depend on the general importance of the trait they mark. For example, if the aim is to find a marker for better rooting under nutrient stress, then identifying a marker that can be used across species is a possibility, but a marker for robust potato yield or starch content and quality would be species-specific. If a gene group has a crucial upstream role in adaptive metabolism, the gene group may be a good target for marker development for diverse traits from various species. The *AOX* gene family is such a gene group which was recently strengthened when postulated to be a general marker for phenotype plasticity (Cardoso and Arnholdt-Schmitt, 2013). Phenotypic plasticity is essential to the emergence of diversity in nature and is known to provide general advantage during evolution. As with evolution, human-driven breeding is a dynamic process. Breeding challenges permanent crop adjustment and improvement according to environmental contexts and human needs. In the future, breeders might benefit from better knowledge of the molecular basis of phenotypic plasticity and FMs from *AOX* genes might assist elite plant selection for crop improvement.

A marker for phenotypic plasticity can also be useful for selection of 'easy-to-propagate' genotypes with efficient adventitious rooting, as proposed for olive propagation (Santos Macedo *et al.*, 2009) or for improving the capacity for micro-propagation through somatic embryogenesis (Frederico *et al.*, 2009a, 2009b; Afuape *et al.*, 2013). The idea of using *AOX* sequences as markers for adventitious organogenesis and somatic embryogenesis came through understanding that these morphogenic processes are based on stress-induced plant responses

Alternative Respiratory Pathways in Higher Plants, First Edition.
Edited by Kapuganti Jagadis Gupta, Luis A.J. Mur and Bhagyalakshmi Neelwarne.

(Zavattieri *et al.*, 2010). The existence of genetic variability for a desired trait and for the related genes is a prerequisite for FM development. Thus, the existence of polymorphisms in *AOX* gene sequences (alleles, haplotypes) is an essential basis for association studies to find links to breeding goals. However, for breeding, it is not important to understand *why* and *how* a gene sequence or polymorphism closely associates to the desired trait. The aim is merely to use the DNA sequence information as a technical help in the selection of superior genotypes. This is different from biological research, where the aim is 'to understand' a given phenomenon. The candidate gene approach for DNA marker development is chosen to increase the probability or efficiency for identifying a sequence that marks a trait. Nonetheless, association between a FM and a trait needs not necessarily be based on (known) biological causality (Brenner *et al.*, 2013). Abe *et al.* (2002) were the first to indicate the relevance of *AOX* polymorphisms for abiotic stress tolerance by identifying a single nucleotide polymorphism (SNP) in the rice *AOX1a* gene, which mapped to a region of a quantitative trait locus (QTL) for low temperature tolerance of anthers at the booting stage. Also, variability for important functional sites in AOX proteins and their metabolic regulation has been identified among various organisms (see overview in Albury *et al.*, 2009; as well as Cardoso *et al.* in Chapter 12.1, McDonald *et al.* in Chapter 12.4 and Elliot *et al.* in Chapter 5 of this book). These results are encouraging for developing FMs. Since 2008, when a symposium that focused exclusively on AOX was held, (the 'First International AOX Symposium'; www.aox2008.uevora.pt), several reports identifying polymorphic patterning in *AOX* gene bodies from various cultivated plant species have been presented (Cardoso *et al.*, 2009; Costa *et al.*, 2009a, 2009b; Ferreira *et al.*, 2009; Frederico *et al.*, 2009b; Santos Macedo *et al.*, 2009). Arnholdt-Schmitt (2009) stressed that all polymorphisms in all parts of the gene body (UTRs, exons and introns) in addition to regulative sequences (promoter, enhancer) have to be taken seriously and should be seen as potentially important for functionality unless shown otherwise.

The possibiliy of *AOX* for marker-assisted plant breeding was advanced by several studies exploiting this gene family for its involvement in breeding traits and/or the occurrence of polymorphic *AOX* gene sequences (Cardoso *et al.*, 2009, 2011; Costa *et al.*, 2009b; Ferreira *et al.*, 2009; Santos Macedo *et al.*, 2009, 2012; Sircar *et al.*, 2012). However, the strategy still poses certain challenges. Bridging genomics and phenotyping is the most critical bottleneck for future molecular breeding and many groups are searching for efficient technical solutions. A molecular-physiological approach that identifies relevant polymorphic sequences in *AOX* genes by simple and rapid screening procedures would be cost- and time-effective by narrowing the pool of material that needs to be screened in field trials for final plant selection. Thus, a sophisticated definition of appropriate 'deep traits' and target cells for final whole plant behaviour and the development of efficient technologies as screening tools for the phenotypic effects of polymorphisms are of utmost importance.

Research lines that are important to advance FM development from *AOX* genes are highlighted here, and are divided into different chapters, each provided by the indicated authors. The first section describes the characterization of *AOX* gene diversity, highlighting examples from *Arabidopsis* ecotypes and emphasizing the importance of developing appropriate bioinformatics tools for *AOX* gene diversity discovery. It indicates the future challenge of *AOX* gene classification. The second section describes theoretical and methodological approaches required to discover functionality due to *AOX* gene diversity. It focuses on current strategies for phenotyping, while also highlighting the recent knowledge on epigenetics and genome organization that help identifying *AOX* gene variability relevant for pre-breeding, and, finally, gives a short review on gene technology applied to *AOX* genes. The third section gives an overview of selected projects for *AOX* marker application. It describes the development of calorespirometry as a novel technology for plant pre-breeding, present ideas linking *AOX* to the quality of herbal tea, and provide short overviews of recent projects or knowledge related to plant-symbiotic arbuscular mycorrhizal fungi (AMF), *AOX* genes in parasitic nematodes, and bacterial *AOX* genes. A general conclusion provides insights for future research on *AOX* gene-based FM development for breeding.

12

Exploring *AOX* gene diversity

CHAPTER 12.1
Natural *AOX* gene diversity

Hélia G. Cardoso, Amaia Nogales, António Miguel Frederico, Jan T. Svensson*, Elisete Santos Macedo, Vera Valadas and Birgit Arnholdt-Schmitt
EU Marie Curie Chair, ICAAM - Instituto de Ciências Agrárias e Ambientais Mediterrânicas, Universidade de Évora, Évora, Portugal
**Current address: Nordic Genetic Resource Center, Alnarp, Sweden*

Variability at family pattern and plant genome organization

In angiosperms *AOX* has been described as a small multigene family composed of three to five genes distributed in two different subfamilies, *AOX1* and *AOX2* (Whelan *et al.*, 1996; Saisho *et al.*, 1997; Vanlerberghe, 2013). Nevertheless, while *AOX1* occurs in both monocots and eudicots, *AOX2* occurs only in eudicots. Neimanis *et al.* (2013) recently suggested that the lack of *AOX2* in monocots is due to a secondary gene loss event during evolution. This theory has been strengthened by recent work demonstrating the presence of members from both *AOX* subfamilies in gymnosperms (Frederico *et al.*, 2009a; Neimanis *et al.*, 2013).

Variability in the number of *AOXs* belonging to each subfamily has been reported across plant species. In eudicots this varies from a single *AOX1* and two *AOX2* in *Vigna ungiculata* (Costa *et al.*, 2004) and *Glycine max* (Thirkettle-Watts *et al.*, 2003) or four *AOX1* and a single *AOX2* in *Arabidopsis thaliana* (Clifton *et al.*, 2006). Nowadays, the increase of data coming from different genome sequencing projects allows a better understanding of the *AOX* multigene family across higher plants. Data from different web-based databases provided information about *AOX* homologoues from 22 plant species (Table 12.1), including monocots and eudicots. A fast screening of this data already raises some interesting questions, including the number of genes per species: *Arabidopsis lyrata* has six *AOX* (four *AOX1* and two *AOX2*), instead of the five previously described as the maximum of *AOX* gene members detected in one species (Clifton *et al.*, 2006). The expansion of the *AOX2* subfamily with more than two members is also visible. Table 12.1 shows *Medicago truncatula* with three *AOX2* gene members, *Malus domestica* with four and several species that only harbour *AOX2* genes and no *AOX1*. Indeed, in all species in which a single member was identified, it belonged to the *AOX2* subfamily. The biological significance of this has yet to be explored. It is widely

---

Alternative Respiratory Pathways in Higher Plants, First Edition.
Edited by Kapuganti Jagadis Gupta, Luis A.J. Mur and Bhagyalakshmi Neelwarne.

Table 12.1 Diversity of *AOX* in terms of gene location, orientation size and exon-intron pattern across higher plants

	Species	AOX subfamily	Gene_id	Chromosome location	Gene orientation	Gene size	Protein lenght	Exon-intron (box-line) gene structure
Eudicots	*Arabidopsis lyrata*	AOX1	AL1G33660	1	–	1319	315	
			AL3G24680	3	+	1237	324	
			AL3G24690	3	+	1518	354	
			AL5G06730	5	–	1290	330	
		AOX2	AL0G08000	nd	–	6095	334	◆
			AL8G30860	8	–	7486	355	◆
	Arabidopsis thaliana	AOX1	AT1G32350	1	–	1333	318	
			AT3G22360	3	+	1229	325	
			AT3G22370	3	+	1527	354	
			AT3G27620	3	–	1307	329	
		AOX2	AT5G64210	5	–	1700	353	◆
	Brassica rapa	AOX1	Bra010153	A6	–	3045	319	
			Bra001865	A3	+	2110	346	
			Bra031351	A5	–	2078	360	
			Bra023835	A1	–	1804	324	
		AOX2	Bra037768	A9	–	1814	295	
	Carica papaya	AOX2	CP00042G00490	Supercontig_42	+	7793	340	
	Fragaria vesca	AOX1	FV5G29310	LG5	+	1323	361	Δ
		AOX2	FV5G21950	LG5	–	2514	341	
	Glycine max	AOX1	GM04G14800	4	+	3196	321	
		AOX2	GM08G07690	8	+	2053	326	
			GM08G07700	8	+	2803	333	
	Lotus japonicus	AOX1	LJ2G020780	2	+	2471	314	
		AOX2	LJ4G005280	4	–	2764	334	
			LJ4G005290	4	–	2349	314	
	Malus domestica	AOX2	MD00G028680	MDC 001236.510	+	1699	293	
			MD00G081720	MDC 003465.658	–	1858	346	

			MD13G026910	13	+	1548	372
			MD16G016620	16	–	1857	346
	Manihot	AOX2	ME10292G00060	scaff_10293	–	2150	350
	esculenta						
	Medicago	AOX1	MT5G026620	5	–	2287	330
	truncatula		MT5G070680	5	–	1690	275
		AOX2	MT5G070870	5	+	1959	324
			MT5G070880	5	+	1834	323
	Populus	AOX1	PT03G09340	3	–	1310	329
	trichocarpa		PT12G01430	12	+	2149	352
			PT12G01440	12	+	2091	350
			PT15G01960	15	–	2140	351
Eudicots	*Ricinus*	AOX2	RC30063G00030	chr 30063	+	3301	352
	communis						
	Solanum	AOX1	Solyc08g005550.2	8	+	2198	366
	lycopersicum		Solyc08g075540.2	8	–	1402	358
			Solyc08g075550.2	8	–	2711	318
		AOX2	Solyc01g105220.2	1	–	2419	348
	Solanum	AOX1	PGSC0003DMG400007613	8	–	2518	321
	tuberosum		PGSC0003DMG400007614	8	–	1524	356
			PGSC0003DMG400018484	8	–	2333	279
		AOX2	PGSC0003DMG400012558	1	–	3382	353
	Theobroma	AOX1	TC03G031300	3	+	2729	326
	cacao	AOX2	TC02G011670	2	+	2925	342
	Vitis vinifera	AOX1	VV02G09030	2	+	1252	322
			VV02G09050	2	–	1245	320
		AOX2	VV00G00110	12	–	11869	327
	Brachypodium	AOX1	BD3G52505	3	–	1308	343
	distachyon		BD5G20540	5	–	1907	333
			BD5G20547	5	–	1165	324
			BD5G20557	5	–	1186	330

(*continued*)

Table 12.1 *(continued)*

	Species	AOX subfamily	Gene_id	Chromosome location	Gene orientation	Gene size	Protein lenght	Exon-intron (box-line) gene structure
Monocots	*Hordeum vulgare*	AOX1	CAJW010038523	nd	–	1815	281	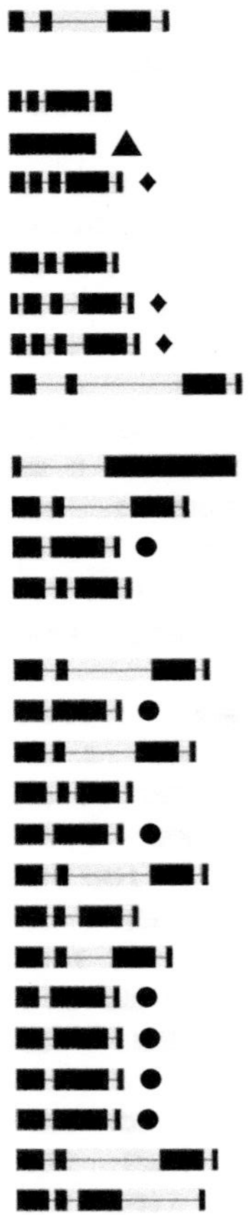
			CAJW011587016	nd	–	1163	270	
			CAJW010099492	2HL	+	975	324	▲
	Musa acuminata	AOX1	GSMUA_Achr5G03810_001	5	–	1277	324	◆
			GSMUA_Achr6G01170_001	6	–	1225	328	
			GSMUA_Achr6G01300_001	6	+	1397	317	◆
			GSMUA_Achr1G27800_001	1	–	1458	327	◆
	Oryza brachyantha	AOX1	OB02G22630	2	+	2617	316	
			OB02G36280	2	–	4950	806	
			OB04G30980	4	–	2009	331	
			OB04G30990	4	–	1219	331	●
	Oryza glaberrima	AOX1	ORGLA02G0249500	2	–	1344	345	
			ORGLA04G0206000	4	–	2216	331	
			ORGLA04G0206100	4	–	1220	335	●
Monocots	*Oryza sativa*	AOX1	BGIOSGA008063	2	+	2061	339	
			BGIOSGA005788	2	–	1344	345	
			BGIOSGA014421	4	–	1227	335	●
			BGIOSGA014422	4	–	2191	332	
	Sorghum bicolor	AOX1	SB04G030820	4	+	1398	346	
			SB06G027410	6	–	1779	331	
			SB06G027420	6	–	1179	314	●
			SB06G027430	6	–	1210	332	●
	Zea mays	AOX1	ZM02G05480	2	+	1215	332	●
			ZM02G05490	2	+	1178	329	●
			ZM02G05500	2	+	2287	329	
			ZM05G37570	5	–	2136	347	

Data retrieved from public web-based databases, freely available (Plaza: http://bioinformatics.psb.ugent.be/plaza/versions/plaza_v2_5/; e!EnsemblPlants: http://plants.ensembl.org/Multi/Search/New?db=core; IPK Barley Blast Server: http://webblast.ipk-gatersleben.de/barley/). Gene draw was performed in FancyGene 1.4 (Rambaldi and Ciccarelli, 2009).

Symbols:: +/–, sense/antisense orientation; nd, not determined; ◆ gain of intron in exon 1; ● loss of intron 2; Δ loss of intron 3; ▲ loss of all introns.

known that *AOX1* gene members are induced as a response to stress during plant adaptation (Considine *et al.*, 2002; Ferreira *et al.*, 2009; Polidoros *et al.*, 2009; Santos Macedo *et al.*, 2009; Vanlerberghe, 2013) while *AOX2* are mostly considered to be constitutive or developmentally expressed, depending on tissues developmental stages (Considine *et al.*, 2002; Frederico *et al.*, 2009b; Polidoros *et al.*, 2009; Vanlerberghe, 2013), being required for 'housekeeping' functions in respiratory metabolism.

The genomic distribution of *AOX* members is also highly variable across species. The most common is the distribution in at least two different chromosomes (see Table 12.1), usually occurring as a single gene with sense or antisense orientation. However, in some genomes *AOX* appears as tandem duplications; for example, *AOX1b* and *AOX1a* in *A. thaliana*, and the three *AOX1* genes from *Sorghum bicolor*, *Zea mays* and *Brachypodium distachyon*. Duplications in a non-tandem distribution were suggested for *AOX1b* and *AOX1a* of *Vitis vinifera*, both being located in chromosome 2.

Gene structure variability

The most common gene structure of *AOX* comprises four exons interrupted by three introns (Table 12.1). Genes sharing this structure usually present exon size conservation for the last three exons (129, 489 and 57 bp, respectively; Campos *et al.*, 2009). This characteristic is responsible for a similar protein size encoded by *AOX* members across plant species. Size variability of *AOX* encoded by genes with a four exon structure is mainly associated with exon 1 variability, although exon size variability can also be observed in the last three exons of *AOX* members. Loss or gain of introns during evolution is responsible for modification in the structure of *AOX* and consequently changes in exons size (see review Polidoros *et al.*, 2009). Examples of this are the loss of intron 2 in *A. thaliana AOX1d* and *AOX1b* and intron 3 in *S. tuberosum AOX1a*, (Considine *et al.*, 2002; Polidoros *et al.*, 2009). Screening the available web databases showed that far more plant species undergo intron losses, mostly in *AOX1* members belonging to monocot plant species. Intron gain is a less frequent event, an example being seen in *Oryza brachyantha* (OB02G36280) with six exons (Table 12.1). Another interesting case is *H. vulgare* with an *AOX1* without introns; the physiological effect of this gene structure has yet to be explored.

Variability at sequence level

AOX encode highly conserved sites in both *AOX*1 and *AOX*2 subfamily members across organisms from diverse kingdoms. These sites are involved in the coordination of the di-iron centre of the enzyme (Siedow *et al.*, 1995;

McDonald, 2008), in *AOX* activity (Moore and Albury, 2008), in anchoring the enzyme to the inner mitochondrial membrane (Crichton *et al.*, 2010), and in the catalytic cycle in respect to its interactions with oxygen (Moore *et al.*, 2008). Conserved sites, also located in a hydrophobic region, are thought to play a role in ubiquinol binding (Albury *et al.*, 2009). Holtzapffel *et al.* (2003) were the first to report variations in the protein functional sites across species (including angiosperms and gymnosperms). In both angiosperms and gymnosperms, the conserved CysI in the N-terminal region of the protein appeared in some plant species as SerI. This substitution consequently changes the enzyme regulation, which is regulated by succinate instead of pyruvate (Holtzapffel *et al.*, 2003; Grant *et al.*, 2009). While the substitution of CysII by SerII was reported by Costa *et al.* (2009a), the physiological consequences of this change were not reported. Variability in conserved sites between *AOX*1 and *AOX*2 which could be useful to discriminate members from both subfamilies was highlighted by Costa *et al.* (2009a) and Frederico *et al.* (2009a, 2009b).

Single nucleotide polymorphism (SNP), insertion and deletion (InDel) events, and transposable elements (TE) are the major driving forces that have shaped genomes (Wessler *et al.*, 1995; Zhang and Gerstein, 2003) and are responsible for phenotype variability in important agronomical traits (Wessler, 1988; Zerjal *et al.*, 2009; Cardoso and Arnholdt-Schmitt, 2013). All forms of polymorphisms mainly occur in the non-coding parts, which could reflect the strict functional requirements of the coding regions, indicating that evolution has worked differently on protein-coding and intron sequences (Wang *et al.*, 2005). Thus, phenotypic variations resulting from differences at the genomic level may determine the capacity of plants to adapt to different environments, where highly robust phenotypes display better stability with higher yields (Cardoso and Arnholdt-Schmitt, 2013). The identification of polymorphisms on *AOX* genes which could be linked to differences at phenotype level is of interest for FM development.

Polymorphisms in protein coding sequences

The genetic code is degenerate, so most amino acids are represented by more than one triplet of nucleotide bases, known as codons, which are considered synonymous. Thus, many SNPs are 'silent' as they result in synonymous codon substitutions. Nevertheless, some synonymous SNPs can yield a protein with a different final structure and function (Kimchi-Sarfaty *et al.*, 2007; Komar, 2007). In other cases, SNPs can conduct to a substitution of the codon (non-synonymous, nsSNP). These nsSNPs may inactivate the functional sites of enzymes, alter splice sites and thereby form defective gene products, destabilize proteins or reduce protein solubility;and may have functional effects on transcriptional regulation by affecting transcription factor binding sites in promoter or intronic enhancer regions, or alternative splicing regulation by disrupting exonic splicing enhancers or silencers (Doss and Sethumadhavan, 2009). nsSNPs can either change protein sequence (missense), or can lead to the insertion of a premature stop codon (nonsense). The predominant consequence of nonsense mutations is not the synthesis of truncated proteins, but the recognition of nonsense transcripts and their efficient degradation by a phenomenon called nonsense-mediated RNA decay (Conti and Izaurralde, 2005). This mechanism guarantees that only full-length proteins are produced (Byers, 2002). In plants there are several examples showing that the nonsense mutations in specific genes are related to phenotype variations (Olsson *et al.*, 2004; Aung *et al.*, 2006; Sattler *et al.*, 2009) and some are used for FM development applicable in plant breeding programmes (Cardoso and Arnholdt-Schmitt, 2013).

SNPs located in the open reading frame (ORF) of *AOX* genes were reported for several species (see Table 12.2). A comparison performed on a partial region of *AOX2* from three Portuguese cultivars of *Olea europaea* revealed nsSNPs related with amino acid substitutions near or within structural elements that were proposed to influence *AOX* regulatory behaviour (Siedow *et al.*, 1995; Andersson and Nordlund, 1999; Crichton *et al.*, 2005) either in the possible membrane-binding domains centre (Andersson and Nordlund, 1999; Berthold *et al.*, 2000), or in a region of the fourth helical regions, previously assumed to be involved in the formation of a hydroxo-bridged binuclear iron centre. It may be assumed that such changes could have a negative effect on a plant's fitness. For example, an olive cultivar-specific nsSNP known as a bad rooter was identified in a position near a highly conserved region across species, the di-iron binding site (RADE_H region). The effects of substitutions in neighbouring residues of the di-iron binding sites have been demonstrated by site-directed mutagenesis in several organisms, like *Trypanosoma vivax* and *Schizosaccharomyces pombe* (Albury *et al.*, 1996; Albury *et al.*, 1998; Affourtit *et al.*,1999; Nakamura *et al.*, 2005), and also in different plant species (see Table 12.2). These mutations were always related to a reduction of enzyme activity or its total inactivation. However, nsSNP may be linked to increased fitness. A nsSNP in *AOX1a* was reported in *O. sativa* (SNP_{297}G/T; see Table 12.2) which was linked to a quantitative trait locus (QTL) for low temperature tolerance (Abe *et al.*, 2002). Polymorphisms in regulatory regions in *OeAOX2* for FM development of relevance to adventitious rooting in olive have been reported (Arnholdt-Schmitt *et al.*, 2006; Santos Macedo *et al.*, 2009, 2012). Two nsSNPs at positions nearby the exon–intron and the intron–exon boundaries, identified in a bad rooting cultivar, appears interesting for FM development if validated through extended association studies and/or mapping studies. SNPs located at introns and exons were related with alternative splicing (Kawase *et al.*, 2007; Seli *et al.*, 2008) with a strongest correlation for those closest to the intron–exon boundaries of the splicing events (Hull *et al.*, 2007). The effects of polymorphisms on splicing may represent an important mechanism by which SNPs influence splicing decisions and induce either exon skipping or intron retention (Aoufouchi *et al.*, 1996; Valentine, 1998). SNPs that affect splicing can have dramatic effects on gene function and consequently the phenotype, usually because the splice mutation results in a shift in the amino acid reading frame. InDel events that lead to a nonsense mutation in *AOX* genes have so far only been reported for *O. europaea AOX2* (Santos Macedo *et al.*, 2009).

Polymorphisms located in intronic sequences

Polymorphisms on introns are expected to play a critical role in gene regulation due to their involvement in a variety of regulatory processes such as RNA stability (Shabalina and Spiridonov, 2004; Haddrill *et al.*, 2005), post-transcriptional gene regulation (Carlini *et al.*, 2001; Shabalina and Spiridonov, 2004), nucleosome formation and chromatin organization (Mattick and Gagen, 2001; Shabalina and Spiridonov, 2004; Vinogradov, 2005), and protein functional domain separation (Duester *et al.*, 1986). Polymorphisms in introns may also influence the binding of transcription factors (Xie *et al.*, 2005), the process of alternative splicing (Noh *et al.*, 2006; Ner-Gaon *et al.*, 2007; Baek *et al.*, 2008), the activity of intron-located promoters, the coding of intronic regulatory elements, such as micro or small nucleolar RNAs (Li *et al.*, 2008; Louro *et al.*, 2007; Nakaya *et al.*, 2007), and nonsense-mediated mRNA decay (Jaillon *et al.*, 2008). Introns can affect either the level or the site of gene expression through intron-mediated enhancement of gene expression and intron-dependent spatial expression, respectively (Morello *et al.*, 2011). Costa *et al.* (2009a) compared primary transcripts (including exons and introns) from different members of the *AOX* multigene family (*AOX1* and *AOX2*) including dicots and monocots. They observed that for *AOX1* gene members the profile length was similar among all species, ranging from 1450 bp in *A. thaliana* to 2809 bp in *G. max*. However, *AOX2* presented a distinct profile with lengths ranging from 1960 bp in *A. thaliana* to 3097 bp in *G. max*, and with different cultivars of *V. vinifera* ranging from 7279 to 12329 bp. New data show that *AOX1* primary transcript has in fact a wide range of sizes (975 bp in *Hordeum vulgare* to 4950 bp in *O. brachyantha*). Nevertheless, longer primary transcripts, with more than 5kb, are always associated with *AOX2* members, and in all cases seem to be related to the presence of TE (Costa *et al.*, 2009b; Macko-Podgorni *et al.*, 2013; see Table 12.2). The role of TE is still uncertain, but it has been suggested that they are involved in gene regulation and contribute to the adaptive fitness of their host (Arnholdt-Schmitt, 2004). TE insertion within a gene or regulatory region can potentially induce alternative splicing and/or change expression patterns, which can result in a relatively rapid change in the function of the gene (Xu and Ramakrishna, 2008). Similarly, insertion of TE in stress-inducible *AOX* could modify gene regulation and plant behaviour relative to adaptive traits (Costa *et al.*, 2009b).

Table 12.2 Natural polymorphisms identified in the different regions of a gene across selected plant species

Polymorphism	Species	No. of genotypes	Gene	ORF	Introns	3′ UTR	Reference
SNPs	*D. carota*	53	*DcAOX2a*	1/0nsSNPs	9	ns	Cardoso *et al.*, 2009
		40	*DcAOX2b*	ns	68	ns	Cardoso *et al.*, 2011
	H. perforatum	6	*HpAOX1b*	-	20	ns	Ferreira *et al.*, 2009
	O. europaea	3	*OeAOX2*	16/7snSNPs	9	6	Santos Macedo *et al.*, 2009
	O. sativa	7	*OsAOX1a*	Lys71 /Asn	ns	ns	Abe *et al.*, 2002
	P. pinea	1	*PpAOX1a*	11/8nsSNPs	3	ns	Frederico *et al.*, 2009b
		1	*PpAOX1b*	11/6nsSNPs	5	ns	
	L. esculentum	1	*LeAOX1b*	CysI/Ser	ns	ns	Holtzapffel *et al.*, 2003
	**A. thaliana*	1	*AtAOX1a*	CysI/Ala	ns	ns	Rhoads *et al.*, 1998
				CysI/Lys			Umbach *et al.*, 2002
				CysI/Arg			Rhoads *et al.*, 1998
				CysI/Gln			
				CysI/Leu			
				CysI/Asp			
				CysI/Ser			
				CysII/Ala			
				CysII/Ser			Djajanegara *et al.*, 1999
	**G. max*	1	*GmAOX2b*	CysII/Ser	ns	ns	
	**N. tabacum*	1	*NtAOX1a*	CysI/Ala	ns	ns	Vanlerberghe *et al.*, 1998
				CysII/Ala	ns	ns	
	**S. guttatum*	1	*SgAOX1*	CysII/Ala	ns	ns	Albury *et al.*, 2002
				Glu217/Ala			
				Tyr253/Phe			
				Tyr275/Phe			

(*continued*)

Table 12.2 *(continued)*

Polymorphism	Species	No. of genotypes	Gene	ORF	Introns	3′ UTR	Reference
InDels	*D. carota*	53	*DcAOX2a*	ns	1event <5 bp 1 event >50bp	ns	Cardoso *et al.*, 2009
		40	*DcAOX2b*	ns	3 events <5 bp 6 events 5-50bp 3 events >50 bp	ns	Cardoso *et al.*, 2011
	H. perforatum	6	*HpAOX1b*	ns	15 events <5 bp 2 events 5-50 bp	ns	Ferreira *et al.*, 2009
	O. europaea	3	*OeAOX2*	1(nm)	1 event <5 bp	3	Santos Macedo *et al.*, 2009
TEs	*D. carota*	53	*DcAOX2a*	ns	MITE	ns	Macko-Podgorni *et al.*, 2013
	V. vinifera	2	*VvAOX2*	ns	LTR retrotransposon	ns	Costa *et al.*, 2009b
	AL0G08000	1	*AlAOX2*	ns	DNA helitron like	ns	not published (*in silico*)
	AL8G30860	1	*AlAOX2*	ns	LINE	ns	not published (*in silico*)
	CP00042G00490	1	*CpAOX2*	ns	LTR retrotransposon	ns	not published (*in silico*)

*Polymorphisms are not provided from natural variability but by site-direct mutagenesis.
nm, associated with a nonsense mutation; ns, not studied.

SNP, Single Nucleotide Polymorphism; InDels, Insertions and Deletions; ns, nonsynonymous.

Intron length polymorphism (ILP) in *AOXs* has been attributed to the insertion of different types of TE. In *V. vinifera AOX2,* a large InDel was attributed to an insertion of a Ty1/copia-LTR (long terminal repeat) retrotransposon with 5028 bp in intron 2 (Costa *et al.*, 2009b). Other types of TE have also been identified in *AOX* introns. Cardoso *et al.* (2009) reported an ILP in intron 3 of *Daucus carota AOX2a* characterized by the existence of an allele 286 bp longer. Recently, Macko-Podgorni *et al.* (2013) discovered that the ILP previously reported in *D. carota AOX2a* was due to the insertion of a *stowaway* element. *Stowaway miniature inverted-repeat transposable elements* (MITEs) are usually the most abundant group of DNA transposons and were initially described in *Z. mays* (Bureau and Wessler, 1994). They are short (<500 bp), AT-rich, relatively hypomethylated and frequently occur in genic regions (Mao *et al.*, 2000; Takata *et al.*, 2007). Some *stowaways* may also provide polyadenylation sites and *cis*-acting regulatory regions to adjacent genes (Macko-Podgorni *et al.*, 2013). Therefore, it is expected that those polymorphisms can have influence on *AOX* regulation that might affect phenotypes. *A. lyrata* and *Carica papaya* also showed the presence of TE (see Table 12.2). An analysis of this sequences using online TE prediction tools such as Censor (http://www.girinst.org/censor/index.php, Kohany *et al.*, 2006) and Plantrepeats (http://plantrepeats.plantbiology.msu.edu/search.html, Ouyang and Buell, 2004) revealed the presence of a DNA helitron-like TE in *A. lyrata* (AL0G08000) intron 5, a non-LRT retrotransposon (LINE) in AL0G08060 intron 3 and possibly an LTR retrotransposon in *C. papaya* (CP00042G00490). These programs confirmed the insertion of an LTR retrotransposon in *V. vinifera.*

Beside TE, the presence of SNPs and InDels is high in introns when compared to exon sequences (Cardoso *et al.*, 2009; Ferreira *et al.*, 2009). Introns have been suggested to provide more genetic flexibility to *AOX* regulation. The high number of polymorphisms in *AOX* intronic regions make them useful for the study and identification of different plant genotypes within species such as *Hypericum perforatum* (Ferreira *et al.*, 2009) and *D. carota* (Cardoso *et al.*, 2009, 2011). An exception to this observation is *O. europaea,* where a relatively low variability was identified in *AOX2* intron 3 (Santos Macedo *et al.*, 2009).

Usually, the first intron of the *AOX* shows the highest number of polymorphic sites, including SNPs and InDels (Cardoso *et al.*, 2011). It is often observed that introns that are most proximal to the 5′ end of a gene are the ones that exert a more pronounced effect on expression (Breviario *et al.*, 2008; Rose, 2008). In *D. carota AOX2a,* the higher number of polymorphic sites is located in intron 3 (Cardoso *et al.*, 2009), and in *H. perforatum,* where both *AOX1b* introns evaluated demonstrated high diversity (Ferreira *et al.*, 2009).

The involvement of introns in the regulation of gene expression can also be due to the coding of regulatory elements such as miRNAs, which inhibit translation of target genes by binding to their mRNAs. Recently, miRNAs have emerged as important players in plant stress responses (Chiou *et al.*, 2006) and development (Achard *et al.*, 2004; Mallory *et al.*, 2004; Wang *et al.*, 2004). Pre-microRNAs have been predicted in intronic regions of *AOX*, such as in *D. carota AOX2a* (Cardoso *et al.*, 2009) and *AOX2b* (Cardoso *et al.*, 2011), *H. perforatum AOX1b* (Ferreira *et al.*, 2009) and *O. europaea AOX2* (Santos Macedo *et al.*, 2009). Pre-miRNAs are the sites for miRNA synthesis and thus important for the regulation of target genes. However, due to the occurrence of InDels or SNPs, it was not possible to predict pre-miRNA sites in some genotypes.

Polymorphisms in untranslated regions (UTRs)

At DNA level, polymorphims at 5′-UTRs could be related with changes in gene regulation by interfering in a mechanism named intron-mediated enhancement (IME). This mechanism is due to the presence of introns in the 5′-UTRs capable of enhancing gene expression. At mRNA level, 5′- and 3′-UTRs play, in eukaryotes, crucial roles in post-transcriptional regulation of gene expression through the modulation of nucleocytoplasmic mRNA transport, translation efficiency, subcellular localization and messenger stability (Grillo *et al.*, 2010; Cenik *et al.*, 2011). Such regulation is mostly mediated by *cis*-acting elements located in those mRNAs regions (Mignone *et al.*, 2002) that govern spatial and temporal mRNA expression (Kuersten and Goodwin, 2003), or by interaction of miRNAs with their specific targets located at 3′-UTRs (Rana, 2007; Flynt and Lai, 2008).

Alternative splicing can produce alternative 5′-UTRs with direct consequences at protein level. Additionally, diversity within the 5′-UTR of a gene enables expression variation, which is dependent on regulatory element effects located at the alternative 5′-UTR (Barrett *et al.*, 2013). Unpublished data shows the occurrence of a shift on the start codon of *O. europaea AOX2* due to an InDel located on the 5′-UTR, which consequently changes the N-terminal region of the protein. Nevertheless, no additional reports show the existence of polymorphisms in the 5′-UTR from *AOX* gene members.

There have also been several studies that investigated variability at the 3′-UTR in *AOX* gene members. Variations within gene sequences at genome level and 3′-UTR microheterogeneity are currently considered as important factors that might cause diseases and differential regulation (Goto *et al.*, 2001; Lambert *et al.*, 2003; Novelli *et al.*, 2007). Polymorphisms, including SNPs and InDels, were identified in *O. europaea AOX2* (Table 12.2). 3′-UTR variation is not restricted to nucleotide polymorphisms but also encompasses length polymorphisms. Variability in the length of 3′-UTR was revealed as a result of both local microheterogeneity and alternative polyadenylation (APA). Microheterogeneity, probably caused by polymerase slippage, could be considered to be due to length variation. The discovery of *AOX* 3′-UTR microheterogeneity in several phylogenetically distinct species strengthens the possibility that this phenomenon is widespread in *AOX* members. Polidoros *et al.* (2005) reported on different 3′-UTR lengths in *AOX1a* from *Z. mays*, and Santos Macedo *et al.* (2009) reported a variation in *O. europaea* ranging between 76 and 301 bp in *OeAOX2*. 3′-UTR microheterogeneity in *O. europaea AOX1a* and *H. perforatum AOX1b* was also identified (Santos Macedo *et al.*, 2009, 2013; Cardoso *et al.*, data not published). Differential gene regulation by 3′-UTR is mostly due to the use of APA sites, which could have an influence at transcriptional or post-transcriptional level. Selection of the incorrect poly(A) site could affect stability, translatability and nuclear-to-cytoplasmic export (Zhao *et al.*, 1999). Additionally, APA sites could also present a role at the control of messenger RNA (mRNA) metabolism and function by regulating the exclusion or inclusion of sequences which control the mRNA metabolism post-transcription (e.g. a miRNA binding site) (Xing and Li, 2011). Several different reports show the existence of APA in several species and demonstrate its functionality in a wide range of biological processes (see review in Xing and Li, 2011).

Post-transcriptional regulation can also be a result of polymorphisms at the 3′-UTR level. In plants, miRNA sites exist anywhere along the target mRNA (Zhang *et al.*, 2006a), including the 3′-UTR (Rhoades *et al.*, 2002). In *Z. mays AOX1a* 3′-UTR, a putative miR163 target site was identified (Polidoros *et al.*, 2009). The *Z. mays AOX1a* is transcribed with variable 3′-UTR lengths, which can be grouped into two major classes (short and long). The miR163 target site is only present in the longer class. A similar result was reported in *AOX2* from *O. europaea*, in which five putative miRNA targets sites were identified, but two of those were absent in the shorter 3′-UTRs of the three classes found (Santos Macedo *et al.*, 2009). Although the functional significance of these sites is still unknown, its differential presence in several 3′-UTRs of *AOX* across species may suggest an important role of 3′-UTR length variations, which may have significant effects on the global regulation of the *AOX* gene.

Conclusions and implications for future studies on FM development

AOX has recently been proposed as a key enzyme coordinating phenotypic changes related to adaptation to environmental changes (plant plasticity), and was for that reason considered as a target for FM development (Arnholdt-Schmitt *et al.*, 2006; Cardoso and Arnholdt-Schmitt, 2013). However, since *AOX* is a gene family and there are no clear rules to name the different gene members (see McDonald *et al.*, Chapter 12.4), the development of FM focused on a specific *AOX* gene in one plant species could not be directly transferred to a different plant species. In addition, the development of FM should also consider the occurrence of natural gene diversity at different levels:

(a) number of *AOX* gene members

(b) their random distribution within the two subfamilies with plant species showing a single gene and others presenting gene members in both subfamilies

(c) no conservation of gene structure within genes from one single species or among species (genes can show several and longer introns and in others introns can be absent).

The existence of allelic variation in different *AOX* genes due to the occurrence of polymorphisms within their genomic sequence is here reported for different plant species. As well as the polymorphisms identified at the protein coding region and 3′-UTR, the main variability was seen at intronic regions and in some cases related with the occurrence/absence of regulatory elements, such as miRNAs and TEs. Association studies should be the next step in FM development focused on *AOX* in order to show the implication of these polymorphisms in plant plasticity upon individual and multiple abiotic stress conditions.

CHAPTER 12.2

AOX gene diversity in *Arabidopsis* ecotypes

José Hélio Costa[1] and Jan T. Svensson[2,*]

[1] *Department of Biochemistry and Molecular Biology, Federal University of Ceara, Fortaleza, Ceara, Brazil*

[2] *EU Marie Curie Chair, ICAAM - Instituto de Ciências Agrárias e Ambientais Mediterrânicas, Universidade de Évora, Évora, Portugal*

**Current address: Nordic Genetic Resource Center, Alnarp, Sweden*

Arabidopsis is a small flowering plant that has been chosen as model organism in plant biology and genetics, and was the first plant genome sequenced in 2000 (The *Arabidopsis* Genome Initiative, 2000). Development of next generation sequencing technologies (NGS) with lower sequencing cost led to launching of the *Arabidopsis* 1001 project, with the goal to discover whole-genome sequence variations in 1001 ecotypes from different geographic locations (Cao *et al.*, 2011). The data represent a powerful tool for studies of genetic heterogeneity in an entire species and can also be used to develop FMs for plant breeding.

AOX in *Arabidopsis* have been extensively studied revealing that this protein is encoded by a multigene family with five members: *AOX1a, AOX1b, AOX1c, AOX1d* and *AOX2* (Saisho *et al.*, 1997, 2001a; Clifton *et al.*, 2006) and that each member presents differential organ and developmental regulation (Clifton *et al.*, 2006). However, no study has been performed focusing on polymorphism in *AOX* sequences in individuals from *Arabidopsis*. In this work, by taking advantage of genome data available from 102 *Arabidopsis* ecotypes, it was possible to identify SNPs and InDels in *AOX* genes. These can be explored for the development of FMs and for studies of gene diversity.

AOX gene polymorphisms

The analyses were conducted using two datasets of *Arabidopsis* ecotypes: 80 ecotypes native in Eurasia specifically from six larger geographic regions – the Iberian Peninsula with North Africa, Southern Italy, Eastern Europe, the Caucasus, Southern Russia, and Central Asia. In addition, two ecotypes from two much smaller regions, Swabia in the southwest of Germany, and South Tyrol in the north of Italy (Cao *et al.*, 2011) and 20 ecotypes from selected

Alternative Respiratory Pathways in Higher Plants, First Edition.
Edited by Kapuganti Jagadis Gupta, Luis A.J. Mur and Bhagyalakshmi Neelwarne.

accessions with maximal genetic diversity spanning different regions of the world (Clark *et al.*, 2007). This analysis is a good example of how to use genomic data for studies of specific genes.

The highest diversity in *AOX* genes were found in ecotypes from Eurasia, perhaps reflecting the large diversity of ecotypes which were originated from varied climates and elevations that cover the high mountains of central Asia to the European Atlantic coast, and from North Africa to the Arctic Circle (Cao *et al.*, 2011). SNPs were detected in exons, introns and UTRs. InDels were observed in non-coding regions of all *AOX* genes, but mainly in *AOX1d* (data not shown). Polymorphisms in the coding region of *AOX* genes appear to be the major target for development of FMs since several nsSNPs and codon deletions were responsible for amino acid (AA) substitutions[a] or AA deletions[a] (Table 12.3). The number and location of SNPs varied among the different *AOX* genes in the 102 ecotypes. The *AOX* genes of the Columbia (Col-0) accession were chosen as reference genes (Tables 12.3 and 12.4).

AOX1a and *AOX1b* showed the lowest number of nsSNPs and the majority of AA changes were observed in the mitochondrial targeting peptide (MTP) (Table 12.3). In mature proteins (MPs) the AA changes occurred only in two positions of both *AOX*1a and *AOX*1b but involved AA with different physicochemical properties (M74T, G95R for *AOX*1a and T121I, R135P for *AOX*1b) (Table 12.4). These changes in MPs were found in a low number of ecotypes (5 for *AOX*1a and 23 for *AOX*1b) from different regions, both at similar latitudes (*AOX*1a; M74T, G95R, and *AOX*1b; R135P) and varied latitudes (*AOX*1b; T121I) (Tables 12.3 and 12.4).

AOX1c had the highest number of nsSNPs (17) in the coding region detected in a high number of ecotypes, while for *AOX1d*, with the highest number of SNPs in the coding region (40), only a few of them were nsSNPs (11) (Table 12.3). In contrast to *AOX*1a and *AOX*1b, the majority of AA changes in both *AOX*1c and *AOX*1d occurred in the MPs instead of MTPs (Table 12.3). In *AOX*1c, 12 AA changes were found in the MP, eight of them involved AA with different physicochemical properties (Table 12.4). In addition, a codon deletion found in one accession (Del-10) resulted in an AA deletion at position 277 (V277Del). Except for the positions 130 and 135, the majority of ecotypes *AOX*1c were similar to the reference Col-0. Four ecotypes (Agu-1, Leo-1, Mer-6, Tuescha-9) showed five AA changes with different physicochemical properties (Table 12.4). In general, changes in the *AOX*1c sequence involved ecotypes from different regions. For *AOX*1d, seven AA changes were found in the MP but only two of them involved AA with different physicochemical properties occurring in ecotypes from different (T148M) or from similar latitudes (K304E) (Table 12.4). An AA change at position 148 can affect *AOX* activity since that position is close to glutamate (E) 147, an AA that is involved in the co-ordination of the di-iron centre of *AOX* (Andersson and Nordlund, 1999; Berthold *et al.*, 2000).

In *AOX1d,* codon deletions at two positions (S47Del, E170Del) were also found. In the first case, S47Del was found in three ecotypes from different regions (North Africa, Spain and Southern Italy). In the second case, the codon deletions E170Del as well as a F169L were both found in the Sha ecotype (latitude 37.29; longitude 71.30 – derived from Cao *et al.,* 2011). Curiously, analysis in another Sha accession (latitude 38.35; longitude 68.48 – derived from Clark *et al.,* 2007) does not reveal this codon deletion. Apart from SNPs and deletions in the coding region, *AOX1d* also showed large insertions/deletions in intron 1 and 2 as well as in the 3′-UTR (data not shown). Introns and 3′-UTRs are known to play crucial roles in gene expression regulation, which involves several regulatory processes (see also Cardoso *et al.,* Chapter 12.1). Thus, these polymorphisms found in non-coding regions of *Arabidopsis AOX1d* also appear as potential candidates to develop FMs.

AOX2 showed the least number of variations among the five *Arabidopsis AOX* genes considering the number of SNPs/mutant ecotypes (Table 12.3). Different from *AOX*1 proteins, no AA change was found in the MTP of *AOX*2. Among the five AA changes in the MP, only one of them involved AA with different physicochemical properties: G92D and A218T. Change at position 218 can affect *AOX*2 activity since as stated above earlier, this position is neighbouring glutamate (E) 221. The majority of mutant ecotypes were originated from similar latitudes (Table 12.4).

Implications of polymorphisms in AOX function

AOX1c and *AOX1d* were the most variable *AOX* genes in the analysed *Arabidopsis* population. *AOX1c* has been linked to plant development since it is co-regulated with components involved with cell division and growth (Ho *et al.,* 2007) and *AOX1d* has been linked to stress conditions. *AOX1d* and *AOX1a* are among the

Table 12.3 Polymorphisms in the coding region of *AOX* genes of the *Arabidopsis* population

Locus ID / Name	sSNPs	nsSNPs	InDels (codon)	MTP AA changes	MP AA changes	No. of ecotypes (MTP/MP)
At3g22370/ *AOX*1a	13	8		6	2	47 / 5
AT3G22360/ *AOX*1b	5	5		3	2	62 / 23
AT3G27620/ *AOX*1c	17	17	1	5	12	21 / 98
AT5G64210/ *AOX*1d	29	11	2	4	7	43 / 52
AT1G32350/ *AOX*2	5	5		-	5	0 / 12

Columbia (Col-0) accession was used as reference.

Abbreviations: AA, amino acid; sSNPs, synonymous; nsSNPs, non-synonymous; InDels, insertion/deletions; MTP, mitochondrial targeting peptide; MP, mature protein.

most stress-responsive genes (at the gene-expression level) encoding mitochondrial proteins (Clifton *et al.*, 2006). Thus, this higher heterogeneity found in both genes may be part of a phenotypic plasticity that evolved in *Arabidopsis* ecotypes related to development and/or adaptation to different environmental constraints. With regard to ecotype origin, AA changes occurring in ecotypes from similar latitudes may indicate *AOX* polymorphisms involved in adaptation to a specific climatic condition. In other cases where AA changes were more extensive, occurring in ecotypes from varied latitudes and longitudes, this could indicate a polymorphism related to adaptation to challenges found elsewhere or as part of a common conserved phenotype. All *AOX* genes in this *Arabidopsis* population present SNPs in the coding region that lead to amino acid changes with different physicochemical properties, which may have implications in *AOX* structure and activity (Table 12.4). With the recent crystallization of trypanosome *AOX* structure (Shiba *et al.*, 2013), homology modelling can be applied to investigate the role of nsSNPs in the flexibility of the *AOX* structure/function of the different rendered proteins. All *AOX*1 proteins revealed several AA changes in the MTP (Table 12.3). This finding raises the question of whether these AA changes will have implications in the *AOX* localization within mitochondria. Experimental assays transiently expressing MTPs with green fluorescent protein (GFP) could be used to study the effect of polymorphisms leading to altered MTPs.

The majority of the mutant ecotypes (against the reference Col-0) had AA changes at a single position and in a single *AOX* protein. Some ecotypes showed AA changes in two *AOX* proteins and one ecotype (Don-0) revealed changes in three *AOX* proteins, both cases at a single position per protein. *AOX*1c was the only protein for which some ecotypes revealed several AA changes (up to five AA with different physicochemical properties) (Table 12.4). From these findings, the next step would be to select specific ecotypes and validate the functional efficiency of different polymorphisms under developmental and/or stress conditions.

Conclusions

The analysis of *AOX* gene variability can be extended to the 500 ecotypes that are available and this would present an atlas of polymorphisms for *AOX*s, useful for development of FM. In order to move towards a holistic view, the analysis can be extended to other 'omics' type of data such as, transcriptome data, methylome, and miRNA data to find *AOXs* with differential expression by comparing different ecotypes in relation to sequence polymorphisms, methylation patterns and miRNA binding sites. Although some data sets are of low resolution, transcriptome, for example, in one tissue type leaf (a few samples of buds) and normal growth conditions revealed a lack of expression of *AOX1a* in one ecotype compared to other (Schmidt *et al.*, 2013). Noceda *et al.* (Chapter 13.2) demonstrates an example of analysis of methylome data for *AOX*.

Table 12.4 Distribution of ecotypes and amino acid changes (with different physicochemical properties) in the mature *AOX* proteins

Protein	AA changes (mature protein)	Ecotypes	Latitude	Longitude
*AOX*1a	M74T,	Don-0^{3}, ICE226^{2}, ICE228^{2}, ICE7^{2}	36.83 to 46.63	−6.36 to 23.50
	G95R	Ped-0	40.74	−3.90
*AOX*1b	T121I	Del-10, Don-0^{3}, Ey15-2, HKT2.4, ICE181^{2}, ICE21, ICE36, ICE61^{2}, ICE7^{2}, ICE73^{2}, Nie1-2, Rue3-1-31, TueSB30-3, Vash-1^{2}, Bor-4, Got-7, Ler-1, Tamm-2	36.83 to 59.58	−6.36 to 60.48
	R135P	Bak-2, Bak-7^{2}, Dog-4^{2}, Nemrut-1^{2}, ICE29^{2}	38.30 to 41.79	23.65 to 43.48
*AOX*1c	E60K	Agu-1$^{(5)}$., Tuescha9$^{2(5)}$., Leo-1$^{(5)}$., Mer-6$^{(5)}$., ICE1$^{(3)}$., Vie-0$^{(2)}$., Bak-7$^{2(3)}$., ICE181$^{2(3)}$., ICE212$^{(3)}$., ICE213$^{(3)}$., ICE226$^{2(3)}$., ICE228$^{2(3)}$.	38.92 to 48.53	−6.34 to 43.48
	G65R	ICE1$^{(3)}$., Vie-0$^{(2)}$., Don-0^{3}	36.83 to 44.46	−6.36 to 25.74
	Q71H	Agu-1$^{(5)}$., Tuescha9$^{2(5)}$., Leo-1$^{(5)}$., Mer-6$^{(5)}$.	38.92 to 48.53	−6.34 to 9.05
	G72Y/C	Agu-1$^{(5)}$., Tuescha9$^{2(5)}$., Leo-1$^{(5)}$., Mer-6$^{(5)}$.	38.92 to 48.53	−6.34 to 9.05
	G75E	Dog-4^{2}, ICE127, ICE130, ICE138, ICE60, ICE61^{2}, ICE71, ICE73^{2}, Koch-1, TueV13	38.30 to 54.09	9.05 to 82.57
	I130F	Agu-1$^{(5)}$., Tuescha9$^{2(5)}$., Leo-1$^{(5)}$., Mer-6$^{(5)}$. and others 58 ecotypes.		
	S220A/P	Agu-1$^{(5)}$., Tuescha9$^{2(5)}$., Leo-1$^{(5)}$., Mer-6$^{(5)}$., ICE1$^{(3)}$., Bak-7$^{4(3)}$., ICE181$^{2(3)}$., ICE212$^{(3)}$., ICE213$^{(3)}$., ICE226$^{2(3)}$., ICE228$^{2(3)}$.	38.92 to 48.53	−6.34 to 25.74
	H329N	Bak-7$^{4(3)}$., ICE181$^{2(3)}$., ICE212$^{(3)}$., ICE213$^{(3)}$., ICE226$^{2(3)}$., ICE228$^{2(3)}$., Nemrut-1^{2}, Vash-1^{2}	38.64 to 46.63	10.82 to 46.37
*AOX*1d	T148M	Bur-0, Cvi-0, Est-1, Lov-5, Ts-1	16 to 62.48	−24 to 25,3
	K304E	Fei-0^{2}, ICE29^{2}	40.92 to 41.43	−8 to 23.65
*AOX*2	G92D	Fei-0^{2}, Tuescha9^{2}, TueWa1-2	40.92 to 48.53	−8 to 9.05
	A218T	ICE92	38.76	16.24

superscript numbers are the number of *AOX* proteins with changes.

superscript bracket indicates the numbers of AA changes in the same *AOX* protein.

[a]Note: AA polymorphisms were described following the nomenclature recommendations for sequence variation (den Dunnen and Antonarakis, 2001). For an amino acid substitution, for example, when an alanine (A) in *AOX* of the reference ecotype was changed for threonine (T) in *AOX* of a mutant ecotype at position 8, this change was written as A8T. Similarly, if the alanine was deleted at position 8, this change was written as A8Del.

CHAPTER 12.3

Artificial intelligence for the detection of *AOX* functional markers

Paulo Quaresma[1], Teresa Gonçalves[1], Salvador Abreu[1], José Hélio Costa[2], Kaveh Mashayekhi[3], Birgit Arnholdt-Schmitt[4] and Jan T. Svensson[4,*]

[1]*Department of Computer Science, University of Évora, Évora, Portugal*

[2]*Department of Biochemistry and Molecular Biology, Federal University of Ceara, Fortaleza, Ceara, Brazil*

[3]*BioTalentum Ltd, Budapest, Hungary*

[4]*EU Marie Curie Chair, ICAAM - Instituto de Ciências Agrárias e Ambientais Mediterrânicas, Universidade de Évora, Évora, Portugal*

**Current address: Nordic Genetic Resource Center, Alnarp, Sweden*

Functional marker (FM) development requires a solid bioinformatics analysis pipeline in order to minimize the need for human curation of data, variant calling and for predicting candidate functional polymorphisms. Here we describe the development of novel tools which reduce the need for human curation of data and improve the accuracy of variant detection. There are several important decisive steps for the identification of FMs: selection of candidate gene; appropriate phenotyping method(s) (including selection of tissue type and developmental stage); selection of germplasm to genotype and phenotype; and the appropriate bioinformatics algorithm and analysis method. The *in silico* aspect of a FM project is crucial for identification of 'true' polymorphisms. Here a short review of current methodologies and development of new tools are presented.

Short review of current methodologies and improved tools

Several steps of bioinformatics/statistical analyses are needed to identify polymorphisms; (i) alignment of sequence reads to a reference gene, (ii) variant detection, and (iii) association of polymorphic alleles with phenotypic variation. Nowadays, the input data for a FM discovery project is sequence reads using either Sanger Sequencing technology (Sanger *et al.*, 1977) or Next Generation

Alternative Respiratory Pathways in Higher Plants, First Edition.
Edited by Kapuganti Jagadis Gupta, Luis A.J. Mur and Bhagyalakshmi Neelwarne.

Sequencing (NGS) technology (for review see Ansorge, 2009). There are many different algorithms for alignment of sequence reads to a reference gene and detection of variants, and these tools can broadly be distinguished based on the nature of input data. Many tools are available for detection of polymorphisms from Sanger sequencing data, for example: PolyPhred (Nickerson *et al.*, 1997) initially used for detection of heterozygous SNPs using peak and base call information, which later was extended for detection of homozygous and heterozygous SNPs and InDels (Bhangale *et al.*, 2006); PolyBayes uses a Bayesian discrimination analysis considering base quality values and depth of coverage to calculate the polymorphic site probability (Marth *et al.*, 1999); NovoSNP uses a cumulative scoring scheme based on the sum of three subgroups (difference, peak shift and feature) for calling of SNPs and InDels (Weckx *et al.*, 2005) and VarDetect is a rule-based polymorphism tool which takes into account common artefacts (Ngamphiw *et al.*, 2008). Both NovoSNP and VarDetect include a graphical user interface (GUI) whereas PolyPhred and PolyBayes are integrated into the Consed viewer (Gordon *et al.*, 1998). The authors of NovoSNP and VarDetect conducted benchmark tests which revealed a large difference in the number of false positive (FP) and false negatives (FN) between the different tools. For example, using stringent settings for SNP calling on *SCN1A*, the rate of false positives were: NovoSNP (15.4% FP), PolyPhred (86.2% FP) and PolyBayes (51.5% FP). The authors of VarDetect compared SNP calling for 77 exonic contigs from 15 human genes for VarDetect, PolyPhred, NovoSNP and a commercial tool Mutation Surveyor. VarDetect had the best F-score (a measure of tests accuracy), here calculated as (2 × precision × recall) / (precision + recall) at 63.25%, followed by PolyPhred (62.94%), Mutation Surveyor (31.84%) and NovoSNP (6.56%). These results illustrate that there is room for algorithm optimization particularly for FM, where for instance the impact of FP is rather high, compromising the all downstream analysis.

Additional methods that can be included in a FM analysis pipeline are predictive tools of deleterious effects (for review see Wu and Jiang, 2013) and for unphased data there is a need for haplotype reconstruction (for review see Browning and Browning, 2011). The haplotype is the nucleotide sequence along a single chromosome whereas the genotype is the mixed haplotype information for a chromosome pair. In order to obtain the haplotypes a reconstruction (phasing) of the haplotypes from the genotype data is needed and this requires specific computational methods (Browning and Browning, 2011).

Development of AOX centric tools

An analysis pipeline for FM discovery using artificial intelligence (AI) was developed. Training data consisted of *AOX* data from both clone (phased) and amplicon (unphased) Sanger sequencing data. Further development will allow input of NGS data in order to provide a bioinformatics pipeline that accepts all types of sequencing data. The developed *AOX* analyses pipeline includes three subdisciplines of AI: (1) natural language processing (NLP) techniques for sequence alignment (matching) of reads to a reference sequence followed by variant discovery, (2) machine learning (ML) and NLP for optimization of variant discovery and prediction of deleterious effects, and (3) constraints-based modelling (CBM) for haplotype phasing.

Natural language processing for alignment to reference sequence and variant detection

NLP is a sub-area of AI aiming to create computational models and procedures able to analyse and represent sentences written in a Natural Language. A typical architecture has the following modules: lexical analysis; syntactical analysis; semantic analysis; and semantic-pragmatic interpretation (Figure 12.1). As there is an obvious analogy between sequence of characters in natural languages and sequences of nucleotides in genes, researchers have been applying NLP techniques to this domain aiming to interpret the sequences of nucleotides (chapter 11 of Baldi and Brunak, 2001). Interestingly, classical sequence alignment algorithms are based on (variations of) the Needleman–Wunsch algorithm (Needleman and Wunsch, 1970), which is equivalent to the Levenshtein algorithm (Levenshtein, 1966) used in the NLP spell-checking task.

In the context of *AOX* gene research, it starts with the first 'lexical' phase analysis (Figure 12.1). Phase 1 in the development focuses on phased Sanger type sequence data of *D. carota AOX*1. (Dc*AOX*1). A new alignment tool was developed – GLocal-UsEr (GLUE) Align *AOX* tool – incorporating many of the

A) "Mary read the book." $\rightarrow_{lexical}$ "Mary", "read", "the", "book"

B) "Mary", "read", "the", "book" $\rightarrow_{syntactical}$ s(np(n("Mary")), vp(v("read"), np(d("the"), n("book"))))

C) s(np(n("Mary")), vp(v("read"), np(d("the"), n("book")))) $\rightarrow_{semantical}$ [X,Y: Mary(X), book(Y), read(X,Y)]

D) [X,Y: Mary(X), book(Y), read(X,Y)] $\rightarrow_{semantic\text{-}pragmatic}$ [X,Y, Z: person(X), thing(Y), event(Z), name(X, Mary), book(Y), action(Z, read), subject(Z,X), object(Z,Y)]

Figure 12.1 Simplified description of natural language architecture. (A) Lexical module, the sequence of characters is analysed and the basic units (words) are identified; (B) syntactical module, a parse tree of the sequence of units is created, associating structural and functional tags to the words (s=sentence; np = noun phrase; vp = verbal phrase; n = noun; d = determiner; v = verb); (C) semantic module, the information conveyed by the sentence is represented, i.e. its meaning is represented (X and Y are referents/entities having some properties); and (D) semantic-pragmatic interpretation, the semantic representation is interpreted taking into account the pre-existing knowledge, usually represented by an ontology (note the use of an external ontology to infer that 'Mary' is a person, 'book' is a 'thing', and 'read' is an event having as action 'read'.).

characteristics of flexible spell-checking programs adapted to this specific domain: it implements a hybrid algorithm based on the global alignment algorithm Needleman–Wunsch (Needleman and Wunsch, 1970), and the local alignment algorithm Smith–Waterman (Smith and Waterman, 1981) supporting weighted scores, parameterized by the users, for each kind of substitution depending on the quality values associated with each nucleotide read. Preliminary evaluation of the alignment tool with 708 reads of the Dc*AOX*1 gene showed excellent results, being able to automatically detect contaminants obtaining a global alignment with only 4% mismatches. Simulation examples with the ART program showed that this tool can handle data originated from different sequencing platforms and produce robust alignments with fewer mismatches compared to the ones obtained directly from ART (Huang *et al.*, 2012). Initial benchmark tests using commercial and freeware tools indicate an improvement in alignment of Dc*AOX*1 to the reference gene, the main difference relating to divergent non-contaminant reads which are not discarded but included in the built contig. Based on conducted experiments, separation of data mainly occurs due to highly divergent introns which leads to loss of information if the sequence read contains both exon and intron data.

From the results of the alignment process, a variant detection program was developed for identification of SNPs and InDels (of any length). GLUE Detect allows the definition of filters, which can depend on a combination of allele frequency, quality average, quality standard deviation, and clustering similarity (haploblocks). This analysis can be seen as a preliminary step of syntactical analysis, aiming to reduce the 'noise'/errors of the input which will be given to the grammar inference process. Additionally syntactical analysis, by using the output in the variant call format (VCF), makes it possible to infer the grammar for the gene language from the consensus sequence and its identified variations (Figure 12.1). Benchmark tests are now in progress to compare GLUE to existing tools both for alignment and build contig quality and for variant detection. The critical step of enabling alignment to reference of sequences with highly divergent introns has been succeeded by allowing the user to define a sliding window and the tolerable percentage of SNPs in that window compared to the reference gene. This step allows for compartmentalization of single sequence reads to highly divergent (due to another evolutionary event) and to a segment that will be included in the variant detection.

Towards a complete analyses pipeline

The initial steps in the GLUE analyses pipeline are (i) alignment to reference gene and (ii) variant detection, which are in the final testing stages, and the development of the other steps are in process, that is, (iii) application and fine tuning of ML methods to increase the accuracy of SNP calls, (iv) NLP and/or ML for prediction of deleterious polymorphisms, and (v) constraints-based modelling to reconstruct haplotypes (Figure 12.2). In addition, for future applications GLUE will be adapted for use with unphased Sanger data and NGS type data.

Application of ML methods aim to improve the variant calling method by using human curated *AOX* sequences from the existing repository as the training set for the development of ML classifiers. Development of a ML method for improvement of the positive predictive values of SNP calls from PolyBayes showed 84.5% accuracy, a five to ten-fold improvement (Matukumalli *et al.*, 2006). ML classifiers will be developed through an iterative process using, for example, sequence depth, haplogroups, quality, agreement forward and reverse read, quality of major/minor alleles and their frequencies. Models will be developed with support vector machines (SVM) and decision trees. SVM (Christianni and Shawe-Taylor, 2000) can process noisy and large data sets effectively, whereas decision trees (Quinlan, 1993) produce models that are easy for biologists to interpret and fine tune.

Many tools have been developed in the medical research community for prediction of deleterious polymorphism, mainly non-synonymous SNPs (Bromberg and Rost, 2007; Jiang *et al.*, 2007; Kumar *et al.*, 2009; Wu and Jiang, 2013). For ML processed variants the aim is development of a ML tool for classification of putative deleterious effects of variants, in effect producing a rank of deleterious variants. Input data is ML processed variant calls together with three main groups of classifiers: sequence, structure and annotation. Support vector machines (Cristianini and Shawe-Taylor, 2000) and random forests

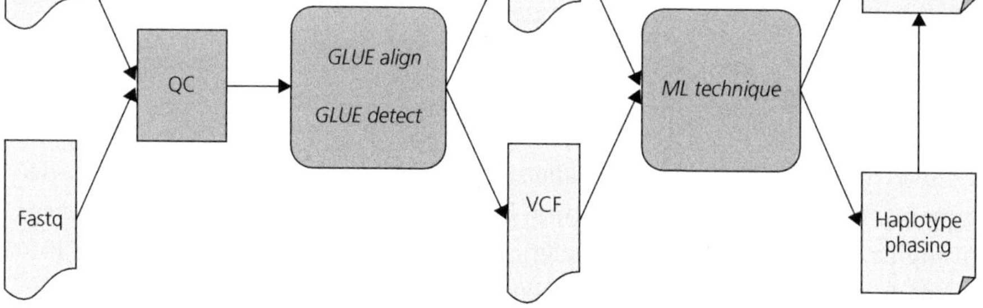

Figure 12.2 GLUE analysis pipeline. (1) Input Sanger sequence data with quality values (phd) or from next generation sequencing (fastq); (2) Removal of low quality segments and vector/adapter trimming (QC); (3) matching of sequences to reference gene and primary variant detection using NLP lexical and semantical modules (GLUE Align and Detect); (4) output alignment to reference gene (SAM); (5) output variant calls (VCF); (6) machine learning module to improve variant calling and for prediction of variants causing a functional effect (ML); (7) output alleles together with a predictive score and (8) for amplicon sequenced data – a module for phasing (haplotype phasing).

(a learning method for classification that operates by constructing a multitude of decision trees at training time and outputting the class that is the mode of the classes output by individual trees) will be used to develop models. In parallel, is the development of an extension of the use of NLP techniques to interpretation of nucleotides sequences, through the implementation of ontology-based approaches to semantic interpretation of gene parse trees. Further development will to focus on research on the third and fourth language analysis phases (Figure 12.1C, D): an important question is how to combine the *AOX* parse trees with functional roles and with ontological information. In order to understand the language of genes it is important to address the phases: C, semantic – what is the `meaning´ of each subset of the parse tree?; D, pragmatic interpretation – how can the semantic of each sequence be interpreted considering an external ontology representing the domain knowledge?

Initial tests have focused on phased sequence data, whereas for unphased data a method for phasing has to be developed. Unphased data has to be phased, for a chromosome with n variants the genotype can be seen as $\{0, 1\}n$ where 0 and 1 represent the two possible haplotypes therefore in a region with n sites there are $2n - 1$ possible haplotypes. The objective of haplotype phasing is to recover the two haplotypes from the $2n - 1$ possible haplotypes. Constraint Programming is well suited for haplotype phasing. It is a set of general-purpose modelling techniques in which a problem is formulated as a set of variables, each with a specified domain, over which a set of relations must hold true. Related approaches include Boolean Satisfiability checking, known as SAT (Graça *et al.*, 2010). In practice, the sheer number and length of the possible haplotypes which match the collected genotypes lead to situations where the solution space is very large. Tools which may exploit parallel computational resources are needed to effectively handle problems such as haplotype inference and phasing at this scale.

Conclusions

A novel tool (GLUE Align) for alignment of sequence reads to a reference sequence was developed using natural language processing. Alignment proceeds through a hybrid local and global alignment method and the output is piped into a variant detection module (GLUE Detect).

CHAPTER 12.4

Evolution of *AOX* genes across kingdoms and the challenge of classification

Allison E. McDonald[1], José Hélio Costa[2], Tânia Nobre[3], Dirce Fernandes de Melo[2] and Birgit Arnholdt-Schmitt[3]

[1] *Department of Biology, Wilfrid Laurier University, Waterloo, Ontario, Canada*

[2] *Department of Biochemistry and Molecular Biology, Federal University of Ceara, Fortaleza, Ceara, Brazil*

[3] *EU Marie Curie Chair, EU Marie Curie Chair, ICAAM - Instituto de Ciências Agrárias e Ambientais Mediterrânicas, Universidade de Évora, Évora, Portugal*

Alternative oxidase (*AOX*), discovered in plants, is responsible for the phenomenon of cyanide-resistant respiration (Bendall and Bonner, 1971). *AOX* is a terminal quinol oxidase found in the mitochondrial electron transport chain that introduces a branch-point at the level of ubiquinol (McDonald, 2008). *AOX* is of research interest for studying the phenomenon of retrograde regulation between the mitochondrion and the nucleus and due to its role in the acclimation of plants to a variety of environmental stresses (McDonald, 2008; Giraud *et al.*, 2009). Recently, *AOX* became of central interest as a gene candidate for functional marker development that helps breeding programmes focused on improving plant stress responses (Arnholdt-Schmitt *et al.*, 2006; Clifton *et al.*, 2006; Arnholdt-Schmitt, 2009; Polidoros *et al.*, 2009).

It is hypothesized that *AOX* arose in prokaryotes and entered the eukaryotic lineage via the primary endosymbiotic event that led to the origin of mitochondria (Finnegan *et al.*, 2003; McDonald *et al.*, 2003; Atteia *et al.*, 2004). This hypothesis is supported by the limited distribution of *AOX* in the proteobacteria and its widespread distribution in many eukaryotic lineages including a wide array of protists, fungi, plants and animals (McDonald, 2008). *AOX* has been most well-studied in the plant kingdom and in particular in angiosperms (i.e. flowering plants).

Recent research on *AOX* has focused on several key questions:

1 What is the physiological role(s) of *AOX*? Under what conditions is it expressed?
2 What is the evolutionary history of *AOX*?

Alternative Respiratory Pathways in Higher Plants, First Edition.
Edited by Kapuganti Jagadis Gupta, Luis A.J. Mur and Bhagyalakshmi Neelwarne.

3 How does *AOX* work (i.e. what is the catalytic cycle of *AOX*) and how is it post-translationally regulated?

The first step in addressing these questions is to devise a robust and definitive classification scheme for *AOX* genes. An effective classification scheme is necessary in order to identify and annotate novel *AOX* sequences, to facilitate comparative studies across species, to attempt expression and association analyses in order to identify *AOX* gene functionality, to investigate the post-translational regulation of *AOX* proteins and, ultimately, to determine the physiological role of *AOX*. In this chapter we define and describe the steps that will be necessary to devise an effective classification scheme for *AOX* genes.

Determining which organisms harbour *AOX* genes

The initial efforts in determining the *AOX* genes present in a particular species focused on gene cloning from genomic DNA or cDNAs. The first discovery and cloning of an *AOX* gene occurred in the thermogenic plant *Sauromatum guttatum* (Schott) (Rhoads and McIntosh, 1991). Almost simultaneously an *AOX* cDNA was also isolated in the fungi *Hansenula anomala* (Sakajo *et al.*, 1991). For many years it was thought that *AOX* was encoded by a single gene, but this hypothesis was overturned with the discovery of several genes in plants such as soybean (Whelan *et al.*, 1996), *Arabidopsis* (Saisho *et al.*, 1997), rice (Ito *et al.*, 1997), mango (Considine *et al.*, 2001) and wheat (Takumi *et al.*, 2002). More than one *AOX* gene was also found in the fungus *Candida albicans* (Huh and Kang, 1999, 2001) and in the green alga *Chlamydomonas reinhardtii* (Dinant *et al.*, 2001). The advantages of the genomic era led to the discovery of *AOX* genes in protists and animals (McDonald, 2008). In the past several years the number of *AOX* DNA/cDNA sequences has grown exponentially in public databases and represents a powerful tool to identify new *AOX* genes in different kingdoms.

AOX belongs to the di-iron carboxylate protein superfamily which includes members that are soluble and members that are membrane-bound. *AOX* and plastoquinol terminal oxidase (PTOX) are the membrane-bound members of this superfamily. Previous work has demonstrated that *AOX* proteins are located in a different clade than PTOX proteins based on protein phylogenies. The first tool needed is a reliable way of differentiating *AOX* from other members of the superfamily. The most common way to do this is *in silico*, by comparisons of nucleotide or protein sequences. This can determine whether a particular sequence is an *AOX* or not. Ultimately, this can establish the presence or absence of an *AOX* gene, or whether multiple *AOX* genes are present in a particular species. Once a putative *AOX* sequence has been found, it should then be classified.

Classifying *AOX* genes

The naming of *AOX* genes originally occurred in the order of their discovery in a species (e.g. *AOX1*, *AOX2*, etc.). As more sequences became available and experiments were performed, an analysis of 18 full length and 30 partial *AOX* sequences provided an initial classification that divided plant *AOX* in two subfamilies (*AOX1*and *AOX2*) while in non-plant species the *AOX* was named as *AOX0* (Considine *et al.*, 2002). Because the initial nomenclature had been somewhat arbitrary, adjustments in gene nomenclature were needed (e.g. soybean *AOX3* was renamed soybean *AOX2*b). In this initial classification, only genes of the fungi *C. albicans* and the green algae *Chlamydomonas reinhardtii*, in addition to plant *AOX*s, were included in the analyses.

A second analysis of *AOX* genes (using 47 plant and fungal *AOX*s) supported the existence of the *AOX1* and *AOX2* subfamilies and indicated the potential existence of a third subfamily *AOX3* (Borecky *et al.*, 2006). Unrooted phylogenies provided evidence for four *AOX* groups: *AOX1* (which contained one class of mostly monocot *AOX*s and a second class of mostly eudicot *AOX*s), *AOX2* (which contained only *AOX*s from eudicots as no *AOX2*s from monocots), and *AOX3* from eudicots (Borecky *et al.*, 2006). Although in practice the *AOX3* designation has not been utilized by the scientific community, three different clades within the *AOX1* subfamily later emerged after the Considine *et al.* (2002) classification. Altogether, evidence points to the need to review and update the limited *AOX* classification scheme. In particular, there is a large quantity of sequences now available, including *AOX* sequences of protists and animals not included in the previous classification, that need to be considered.

Three approaches can potentially be used to categorize genes. Genes could be categorized based on their DNA/ORF/protein sequences by comparing different genes to each other, looking at their intron/exon structure, or based on their expression profiles; or perhaps all of these pieces of information need to be considered simultaneously. An examination of *AOX* gene structure could yield clues about temporal or spatial expression patterns or lead to insights about new functional hypotheses. Computer programs can be used to generate phylogenies of *AOX* proteins or gene family members, and hidden Markov models give us some predictive power. These techniques essentially allow researchers to group genes by similarity, creating categories that include or exclude specific *AOX* members. If the data set has been selected appropriately, this can be a very powerful approach.

With regard to *AOX* expression profiles in plants, the differences in *AOX1* and *AOX2* protein sequence are followed by differences in their mRNA expression. It was generally suggested that *AOX1* is often induced by stress stimuli, while *AOX2* is usually constitutively or developmentally expressed (Considine *et al.*, 2002).

However, this assumption has been challenged since *AOX2b* in *V. unguiculata* (Costa *et al.*, 2010) and *AOX2* in *Arabidopsis* (Clifton *et al.*, 2005) have been shown to be stress-responsive. In addition, an examination of orthologous genes between soybean and *Arabidopsis* did not show similar expression profiles (Thirkettle-Watts *et al.*, 2003). For example, in soybean, *AOX2a* and *AOX2b* are the predominantly expressed genes in a variety of organs at different growth stages, whereas in *Arabidopsis, AOX1a* and *AOX1c* display the highest expression levels (Thirkettle-Watts *et al.*, 2003). It is worth noting that *Arabidopsis* and soybean have very different patterns when it comes to their complement of *AOX* gene family members; while *Arabidopsis* expanded the *AOX1* subtype, soybean expanded the *AOX2* genes. When using expression profiles of *AOX* genes as aids to *AOX* classification, extra care needs to be taken to ensure that proper comparisons are performed.

The intron/exon structure of the majority of identified plant *AOX* presents four exons interrupted by three introns (Considine *et al.*, 2002; Cardoso *et al.*, Chapter 12.1). Variations in this structure are found in *AOX1b* genes of some members of the *Poales,* such as rice sequences that lack the second intron (Considine *et al.*, 2002). When examining algae, however, different patterns are found as *AOX*s from green algae have at least eight exons and seven introns (Dinant *et al.*, 2001). Genome data analyses reveal that other algal species have *AOX* genes lacking introns (e.g. *Ostreococcus tauri*). In fungi, different patterns of intron/exon structure are also found, varying from *AOX* genes without introns to *AOX* genes with five exons and four introns. Analyses of *AOX* from several protists do not reveal any introns. There is therefore a lack of a general pattern across kingdoms with respect to intron/exon structure in *AOX* genes, which indicates that this kind of analysis might be better suited to classifying genes within each kingdom. Taking into consideration this intra- and inter-kingdom variation, intron/exon structure is likely more useful as a validation tool of an *AOX* classification scheme based primarily on protein phylogenies.

Given the earlier considerations, the most reliable parameter for use in the development of a general *AOX* classification scheme across kingdoms is phylogenetic comparisons using DNA/ORF/protein sequences. However, this is not straightforward as comparisons of *AOX* proteins from species belonging to different kingdoms (Table 12.5) reveal low sequence similarity between *AOX* genes (i.e. identities lower than 50% are generally observed).

Recognizing a reliable pattern that could be applied to obtain a robust and definitive classification scheme covering *AOX*s present in different kingdoms is thus still a challenge. Given the circumstances, we believe that an effective classification scheme can only be developed by identifying *AOX* genes within a single species and then expanding the system in incremental steps to include first closely related species, then species within the same kingdom, and finally species in other kingdoms and domains of life.

Table 12.5 Percent identity between *AOX* proteins from a species of plant (*Arabidopsis*), alga (*C. reinhardtii*), bacterium (*Novosphingobium aromaticivorans*), fungus (*Aspergillus. niger*) and animal (*Crassostrea gigas*)

	Plant *AOX* 1a	Plant *AOX* 2	Algae *AOX*	Bacteria *AOX*	Fungi *AOX*	Animal *AOX*
Plant *AOX* 1a	100	61.19	33.62	53.28	33.05	34.34
Plant *AOX* 2		100	32.86	49.78	31.34	34.94
Algae *AOX*			100	41.48	41.03	40.66
Bacteria *AOX*				100	41.92	40.17
Fungi *AOX*					100	41.57
Animal *AOX*						100

Using sequence data to answer questions about *AOX*

Currently *AOX* genes have been identified in a large number of different species and in some cases the number and category of *AOX* genes is known. If these sequences themselves are used as data, as in the past, they might reveal particular patterns that can provide large amounts of information. For example, previous analyses of *AOX* sequence data have revealed information about *AOX* function (e.g. key residues important for catalysis; Albury *et al.*, 2010; Crichton *et al.*, 2010), *AOX* structure (e.g. dimerization; Day and Wiskich, 1995), or regulation (e.g. CysI redox control of enzyme activity; Umbach *et al.*, 2006). Furthermore, analysis of plant *AOX* promoter regions is starting to receive a lot of attention (Giraud *et al.*, 2009; Ng *et al.*, 2013) and this will give us more information about *AOX* gene expression. Effectively, summarizing the existence of different *AOX* gene categories also provides the opportunity to study them in more detail by, for example, using tools such as crystallization to reveal structural information. A key initial question would be to ask which characteristics all *AOX*s have in common. Are there amino acid residues or 3D structures that all *AOX*s share? The other approach is to contrast what we find in different *AOX*s and to determine if there are characteristics that only one taxonomic group possesses and why this might be so.

An examination of *AOX* in plants is a good place to start based on the work already completed and the volume of sequence data available when compared to other taxonomic groups. The vast majority of publications on *AOX* focus on angiosperm plants and recent work has investigated *AOX* in non-angiosperm plants (Frederico *et al.*, 2009b; Neimanis *et al.*, 2013). Comparing *AOX*s in a single plant species can be difficult if the genome is unavailable. The analysis is even more challenging when comparing *AOX* genes in different plant species (as not all of the *AOX* genes have been identified in most species). It is difficult to know where to start when attempting to make comparisons between kingdoms (e.g. plants and fungi) or between different domains of life (e.g. eubacteria and eukaryotes).

Now that research has given us a better understanding of the structure of *AOX* due to crystallization of the trypanosome protein (Shiba *et al.*, 2013), more complex questions can be investigated. For example, does *AOX* exhibit any protein–protein interactions and how is it interacting with ubiquinol and oxygen as substrates in order to produce water? What is the catalytic cycle of the enzyme and is this always the same regardless of species?

These points raise the question of whether each *AOX* gene (and its associated protein product) may have a specific physiological role, and therefore whether the evolutionary divergence of *AOX* subfamilies and classes across plant species might have implications for physiological function (Considine *et al.*, 2002; Borecky *et al.*, 2006). That is, can a link between phylogenetic relationships and gene expression and functionality be made?

Conclusions – Addressing the challenges

Answering these earlier questions requires as initial steps: (i) identification of all currently available *AOX* sequences; (ii) a robust classification scheme for *AOX* subfamilies and classes that provides a consistent means of annotation; and (iii) an analysis of the taxonomic distribution of each subfamily and class in order to detect major trends and investigate the evolutionary history of the enzyme.

A logical starting point would be generating a classification scheme for *AOX* proteins using a large data set of deduced amino acid sequences from angiosperms. This new classification will be based on the phylogenetic analyses of protein sequences, the analysis of specific amino acid sites found to differ between *AOX* subfamilies and classes, and the known evolutionary history of angiosperms. Such a system will have an impact on the identification of trends in *AOX* functionality across species. Improved knowledge of the *AOX* family composition in different angiosperm species will provide a better chance to gain insight into *AOX* regulation and physiological function. In this context, a major challenge will be to understand why some plants that are more phylogenetically divergent have similar multigene families, while more closely related species have such large differences in *AOX* family composition.

13

Towards exploitation of *AOX* gene diversity in plant breeding

CHAPTER 13.1

Functional marker development from *AOX* genes requires deep phenotyping and individualized diagnosis

Amaia Nogales[1], Carlos Noceda[1,*], Carla Ragonezi[1], Hélia G. Cardoso[1], Maria Doroteia Campos[1], Antonio Miguel Frederico[1], Debabrata Sircar[2], Sarma Rajeev Kumar[3], Alexios Polidoros[4], Augusto Peixe[5] and Birgit Arnholdt-Schmitt[1]

[1] *EU Marie Curie Chair, ICAAM - Instituto de Ciências Agrárias e Ambientais Mediterrânicas, Universidade de Évora, Évora, Portugal*

[2] *Biotechnology Department, Indian Institute of Technology Roorkee, Uttarakhand, India*

[3] *Plant Genetic Engineering Laboratory, Department of Biotechnology, Bharathiar University, Coimbatore, India*

[4] *Department of Genetics and Plant Breeding, School of Agriculture, Aristotle University of Thessaloniki, Thessaloniki, Greece*

[5] *Melhoramento e Biotecnologia Vegetal, ICAAM - Instituto de Ciências Agrárias e Ambientais Mediterrânicas, Universidade de Évora, Évora, Portugal*

**Current address: Prometeo Project (SENESCYT), CIBE (ESPOL), Guayaquil, Ecuador*

Marker assisted selection (MAS) is commonly used in plant breeding programmes to select traits with agronomic interest (e.g. productivity, disease resistance, stress tolerance, quality) using molecular markers closely associated to a trait. Functional markers (FM) can be used to detect the presence of allelic or copy number variation for genes underlying a trait, thus increasing the efficiency and precision of plant breeding programmes. For this reason, FM development has become an area of considerable research interest during the past decade (Andersen and Lübberstedt, 2003; Neale and Savolainen, 2004; Arnholdt-Schmitt, 2005; Luebberstedt and Varshney, 2013).

Development of FMs can be laborious and time-consuming, and depending on the nature of the selected agronomic trait, the strategies to follow may differ. Agronomic traits can be classified as qualitative or quantitative. For qualitative traits, the phenotype is discrete, for example the plastic response of flowering initiation in relation to photoperiod in some rice varieties (Yano *et al.*, 2001) or the seed colour in soybean (Tuteja *et al.*, 2004). These kinds of traits are determined by one or a few genes. This is in contrast to quantitative traits, in which the phenotype varies continuously. The continuous pattern of variation is determined by

Alternative Respiratory Pathways in Higher Plants, First Edition.
Edited by Kapuganti Jagadis Gupta, Luis A.J. Mur and Bhagyalakshmi Neelwarne.

the combined effects of the environment and genetics through various segregating genes, and therefore is under multigenic control (e.g. level and stability of yield; see a review in Barton and Keightley, 2002). FM development for these traits is consequently more challenging than for qualitative traits. Encouragingly, there are several examples of FMs that were developed and applied in plant breeding programmes to assist quantitative trait selection. However, the number of FMs that can be used across populations and species is strongly limited despite *some* efforts (see review in Cardoso and Arnholdt-Schmitt, 2013).

The first critical step for FM development is the identification of candidate genes and sequence polymorphisms that affect protein (enzyme) activity and consequently induce phenotypic variations (functional polymorphisms). Candidate genes for FM development can be identified by high-throughput differential gene expression (eQTL), association mapping, and QTL analysis followed by fine mapping, (bulk) segregation for a trait or by hypothesis-driven research. Hypothesis-driven selection of candidate genes is a targeted approach and is thus a highly promising strategy in molecular plant breeding (Arnholdt-Schmitt, 2005; Collins *et al.*, 2008).

Plant abiotic stress tolerance is one of the most important and complex traits considered in breeding programmes. Adaptive plasticity upon environmental changes influences the stability of plant biomass and consequently, yield production. Plant stress tolerance, as a quantitative multigenic trait, involves the effect of a large set of genes belonging to different signalling and metabolic pathways, hampering the selection of the most appropriate gene(s) for FM development. Good candidate genes are those involved in global cell coordination and decision making of cell fate in plant responses to the environment. *AOX* genes have been proposed and adopted as candidate genes for FM development related to multi-stress tolerance and phenotype plasticity (Arnholdt-Schmitt *et al.*, 2006; Polidoros *et al.*, 2009; Cardoso and Arnholdt-Schmitt, 2013). However, although *AOX* genes could be general candidate markers related to diverse types of abiotic and biotic stress reactions, the role of *AOX* can differ between species and needs to be validated at species as well as at target tissue or cell level depending on the crop and breeding goals (Arnholdt-Schmitt, 2005; Arnholdt-Schmitt *et al.*, 2006; Vanlerberghe, 2013; see also Elliot *et al.*, Chapter 5 in this book).

Alternative oxidase is increasingly becoming a focus of research on stress acclimation and adaptation and seems to play a key role in regulating the process of cell reprogramming by ameliorating metabolic transitions related with the cellular redox state and the flexible carbon balance (Arnholdt-Schmitt *et al.*, 2006; Rasmusson *et al.*, 2009). Clifton *et al.* (2005, 2006) pointed to the importance of this pathway as an early sensoring system for cell programming. Phenotypic changes related to adaptation to environmental changes might be coordinated by AOX, due to its upstream role in biotic and abiotic stress responses (McDonald and Vanlerberghe, 2006; Plaxton and Podestá, 2006; Cardoso and Arnholdt-Schmitt, 2013; Vanlerberghe, 2013). These responses can include morphogenic responses (Fiorani *et al.*, 2005; Ho *et al.*, 2007; Campos *et al.*, 2009; Frederico *et al.*, 2009a; Santos Macedo *et al.*, 2009, 2012). Differential expression of *AOX* genes in genotypes from the same species but with contrasting stress responses provides supporting evidence for a functional role of this gene in stress adaptation (Mhadhbi *et al.*, 2013).

After selecting a suitable candidate gene, the next step for FM development consists of the identification of polymorphisms within the candidate gene sequence that are likely to be functional and associated with phenotypic variation. This includes characterization of alleles and/or copy number variation in genotypes with different degrees of stress tolerance or responses affecting plant phenotypes. After that, the validation of these polymorphisms as markers is needed. Candidate gene-based association studies are commonly used to establish a link between genotypes and phenotypes. These are powerful methods which allow the identification of markers that are significantly linked to traits in natural or breeding populations (Andersen and Lübberstedt, 2003; Neale and Kremer, 2011). However, phenotyping in the field is one of the most laborious and technically challenging steps in molecular plant breeding. Population screening for a desirable trait needs replicates across environmental conditions (Furbank and Tester, 2011). This procedure involves screening of large amounts of replicated samples because the variability in the measured parameters is expected to be high owing to the multi-causal nature of most of the desirable traits and environmental effects.

To overcome these drawbacks, alternative approaches for phenotyping are being considered. Recently, Furbank and Tester (2011) recommended a new approach for breeding named 'deep phenotyping'. Deep phenotyping aims to dissect agronomic traits by examining plant metabolic and physiological processes to elucidate key processes and components that have large effects on the final trait. To carry out deep phenotyping, it is necessary to identify biological material related to the agronomic trait; that is, specific tissues or cells where the main processes determining the final phenotype take place. It is equally important to choose the exact time point at which physiological or biochemical parameters are going to be measured.

Characterization of FMs from candidate gene sequences will be less time consuming and will require fewer samples when phenotyping of the polymorphic genotypes is done in a focused way, that is, by performing deep phenotyping. By identifying the relevant biochemical and/or physiological processes in target tissues/cells – 'deep traits' – the association between them and polymorphic sequences is easier to explore, because fewer samples are required as the process studied is more directly influenced by the candidate gene (i.e. targeted and thus with fewer factors masking the gene effect). Functional polymorphisms that can be used as FM will be much more easily identified than just measuring the final trait, which is influenced by many other factors, thus reducing the degree of robustness of the putative FM.

While this approach might be valid for many candidate genes, it is especially relevant for *AOX*. The central and upstream role that *AOX* has in adaptive metabolism and several biological processes makes its regulation too complex to easily obtain a link with a specific desirable trait. Consequently, identifying a link between the *AOX* gene sequence and a biochemical or metabolic 'deep trait' which highly determines the agronomical trait of interest will make FM development more efficient.

This strategy is being applied recently in several studies related to FM development for *AOX* genes. For example, Santos Macedo *et al.* (2009, 2012) investigated the involvement of *AOX* in olive adventitious rooting for FM development related to the efficiency of this process. Adventitious root formation can be considered a morphological response to stressful treatments which involves cell reprogramming and *de novo* differentiation. That fact leads to the selection of *AOX* as a candidate gene for FM development. For these studies the ring from the basal portion of olive semi-hardwood shoots was taken, where cells are reprogrammed to perform adventitious rooting, a process that is important for efficient, commercially relevant propagation of the trees. Metabolic analyses were performed in the target tissues and demonstrated that phenylpropanoid and/or lignin content could be suitable 'deep traits' for association studies with *AOX* polymorphisms (Santos Macedo *et al.*, 2012).

The appropriateness of *in vitro* culture systems for studying the linkage of *AOX* to a morphological process could recently be confirmed by comparing *AOX* gene transcript accumulation during adventitious root induction in semi-hardwood olive shoots and *in vitro* microshoots. A similar *AOX* gene expression pattern could be found in both systems (C. Noceda and E. Santos Macedo, personal communication), which makes future studies on the functionality of *AOX* gene polymorphisms for efficient adventitious rooting reasonable. Applying the *in vitro* system will make screening much more efficient. Different genotypes can be checked under *in vitro* culture conditions at the same time in a reasonably small space compared to the space necessary for greenhouse plant trials. Additionally, genetic stability and robustness of the polymorphic sites and their effects can easily be screened under these conditions.

In vitro systems can also be applied as a highly efficient tools for 'deep phenotyping' of phenotypic plasticity upon environmental stress. Induction of adventitious organogenesis and somatic embryogenesis (SE) can be rated as examples of phenotypic plasticity responses expressed upon changing environmental conditions in plant material (Pasternak *et al.*, 2002; Zavattieiri *et al.*, 2010). Frederico *et al.* (2009a) have shown that differential *AOX* gene expression is involved in the process of embryo development initiation ('realization') due to the depletion of auxins. Recently, this idea of using SE as a test system for stress behaviour in relation to reactive oxygen specis (ROS) production was adopted and validated for *AOX* gene-transformed transgenic cassava breeding lines by Afuape *et al.* (2013). A discussion on functional studies using transgenes is provided by Kumar and Sathiskumar (Chapter 13.3).

Another example of *in vitro* culture application as a strategy for 'deep phenotyping' for FM development is the use of a primary culture system for *D. carota* (Campos *et al.*, 2009). This system was first established by Steward *et al.* (1952) and consists of inducing a cell programme change in differentiated secondary phloem explants from tap roots in a nutrient media containing cytokinin and auxin, which initiates callus growth. The primary culture system has been applied at diverse temperatures and adopted as a test system for the genetic potential for carrot yield production and to distinguish carrot genotypes (Arnholdt-Schmitt, 1999). The rapid observation of differences in callus growth behaviour between carrot genotypes makes primary cultures a promising system to test the functionality of polymorphisms in *AOX* gene sequences. First results during the initiation of the exponential growth phase (14 days) confirmed the involvement of *AOX* through differential transcript accumulation of *AOX1a* and *AOX2a* (Campos *et al.*, 2009).

FM validation is complicated by heterozygosity and the number of genes within a gene family. Individual genome and metabolic complexities will interfere with the degree of functionality of a polymorphic site. Also, the interplay with endophytes or symbiotic organisms may modify functionality. The proof of causal relationships between SNPs and/or InDels and the final traits is not an easy task. It cannot be assumed that polymorphisms or InDels that have a phenotypic effect in one genotype will have the same effect in another genotype when confronted with defined environmental conditions. This is the reason why traditional breeders first select individual elite plants from which new breeding populations are derived once a polymorphism is identified that associates to a trait. Therefore, analyses for deep phenotype screening must be carried out on individual plants or groups of plants that have a fully or almost identical genetic background (inbred lines, recombinant inbred lines). Functional assessment analyses of *AOX* gene polymorphisms can be done in critical tissues/cells in complementation to *in silico* methodologies by transcript (expression) and enzymatic activity (respiration) analysis using diverse methodologies, such as qPCR and oxygen isotope ratio measurements. However, for breeding it might not be sufficient to point out the direct consequence of an *AOX* polymorphism at the levels of expression and metabolic pathways, but rather it would be important to study *the effect* of the polymorphism at cell and tissue level that finally affects the desired trait of the whole plant and the breeding population.

For this purpose, as a final step in FM development and prior to field trials for its validation, appropriate screening tools to identify the final trait need to be used. Several advanced tools are being developed for this (reviewed in Furbank and Tester, 2011), including calorespirometry (Nogales *et al.*, 2013). This technology has recently been presented as a novel tool for efficient phenotype screening related to temperature responses and related growth potentials (Nogales *et al.*, 2013). Preliminary results point to its great potential to detect functional *AOX* gene polymorphisms for molecular breeding in *D. carota* (see Arnholdt-Schmitt *et al.*, Chapter 14.1).

Conclusion

The development of new 'deep phenotyping' techniques for FM development on *AOX* gene sequences are expected to greatly increase the efficiency of association studies between the candidate FM sequences and the desired phenotype. However, it is critical to perform these studies in the appropriate target tissue/cell at the correct time point. Complementary approaches such as transcript accumulation and *AOX* enzyme activity studies together with appropriate screening tools that identify the target trait will finally make FM development much more focused and less time and effort consuming.

CHAPTER 13.2

AOX gene diversity can affect DNA methylation and genome organization relevant for functional marker development

Carlos Noceda†, Jan T. Svensson*, Amaia Nogales and Birgit Arnholdt-Schmitt
EU Marie Curie Chair, ICAAM - Instituto de Ciências Agrárias e Ambientais Mediterrânicas, Universidade de Évora, Évora, Portugal
**Current address: Nordic Genetic Resource Center, Alnarp, Sweden*
†Current address: Prometeo Project (SENESCYT), CIBE (ESPOL), Guayaquil, Ecuador

In the first decades of molecular genetics, attention was focused on gene encoding parts of DNA, and a number of sequence motifs that control transcriptional activity (*cis*-elements) were found. The natural question then arose; how can cells or organisms with identical DNA have different phenotypes? The studies on gene regulation were born to explore chromatin changes, from chemical marks in DNA (e.g. cytosine methylation) or in its associated proteins (e.g. histone modifications) to structural re-organizations (e.g. copy number variations, chromosome variants). Among all the gene regulatory mechanisms, epigenetic changes have shown to play a crucial role in development and adaptation, including the stress response.

Expanding our understanding on genome control associated with stress responses in species of agronomic interest will have a significant impact on breeding for improved varieties with increased stress tolerance. That is crucial for FM development. Thus, FM developement should take into account that genome regulatory mechanisms may adapt the function of a DNA sequence to distinct and changing conditions, and even may be heritable, such as epigenetic traits. Moreover, many gene regulatory events, including epigenetic ones, are DNA-sequence dependent. Consequently, gene diversity regarding isoforms, allelic polymorphisms or copy heterogeneity and dosage, differentially affects the regulation of a gene, and thereby influences the phenotype constituting a base to focus FM search.

Alternative Respiratory Pathways in Higher Plants, First Edition.
Edited by Kapuganti Jagadis Gupta, Luis A.J. Mur and Bhagyalakshmi Neelwarne.

DNA sequence interacts with DNA methylation

Chromatin remodelling is principally dependent on several inter-connected epigenetic mechanisms (Chapman and Carrington, 2007; Henderson and Jacobsen, 2007; Kasschau *et al.*, 2007; Kouzarides, 2007; Pfluger and Wagner, 2007; Chinnusamy and Zhu, 2009; Law and Jacobsen, 2010) among which DNA methylation is the most clearly dependent on DNA sequence: for example in eukaryotes, cytosines can be (de)methylated depending on nucleotide contexts; also, plant siRNAs from introns can mediate DNA methylation in host genes (Chen *et al.*, 2011). Gene methylation and transcription are complexly interwoven processes and usually a decrease in DNA methylation in the gene body or promoter is correlated with gene up-regulation (Zilberman *et al.*, 2006). This is also frequent for stress-specific genes (Boyko and Kovalchuk, 2011). Consequently, DNA methylation dynamics may be employed by plants to regulate responses to different stresses, leading to an increase of tolerance or adaptation.

As *AOX* is involved in plant stress responses, it might have a central role in plant adaptation to environmental changes (McDonald and Vanlerberghe, 2006; Plaxton and Podestá, 2006; Cardoso and Arnholdt-Schmitt, 2013). *AOX* genes are probably regulated by epigenetic mechanisms and this should be considered in order to better target associated FM development, such as was suggested for *AtAOX1a* methylation via methyltransferase I and RNA-directed DNA methylation (RdDM) pathway (Nogueira *et al.*, 2011). It is easy to assume that there is a differential epigenetic regulation for the distinct members of the *AOX* gene family, since they are inducible by different factors: *AOX1* genes are generally more responsive to stress stimuli whereas *AOX2* genes are more developmental or tissue-specific expressed (reviewed by Vanlerberghe, 2013).

In *Arabidopsis*, a third of the expressed genes and 5% of the promoter regions are methylated (Zhang *et al.*, 2006b). Recently, a survey of 152 methylomes from leaves and inflorescences of distinct *Arabidopsis* accessions (ecotypes) was carried out together with transcriptome analysis (Schmidt *et al.*, 2013). Authors analysed the methylation pattern and transcription of the five *AtAOX* genes in leaves of six ecotypes (Figure 13.1), constituting two geographical groups – Northern Europe and Southern Europe. The public analysis tool allows for a general overview of the data set but does not allow for analysis down to single base level. Both *AOX1a* and *AOX1b* displayed methylation in the promoter region, and methylation was evident in the gene body of *AOX1b* although at a low level but no methylation appeared in *AOX1a*. The expression of *AOX1a* was similar in four ecotypes but lower in Es-O followed by Ann-1, each corresponding to a defined geographical group. No expression was detected for *AOX1b*. In *AOX1c*, no methylation was found in the promoter region, but some methylation was detected in exons 2 and 3; however, this was not consistent amongst the six

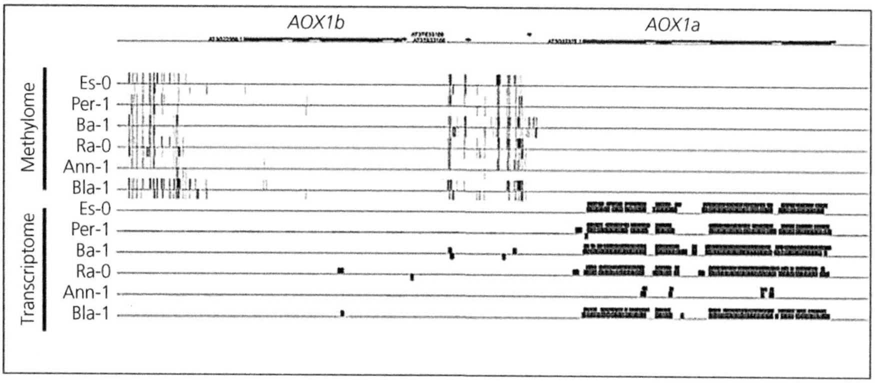

Figure 13.1 Leaf methylome and transcriptome of *Arabidopsis AOX1a* and *AOX1b*.

The picture illustrates a part of chromosome 3 were *AOX1b* and *AOX1a* are located (top line). Below are data from six ecotypes represented on the top half by methylome and bottom half by transcriptome (vertical text). Methylated cytosines in different contexts (CG, CHG, CHH, being H any base) are represented as vertical bars. Bars above the chromosome sequence denote plus strand and bars below the sequence denote minus strand. Transcriptome data from RNA-seq are represented as blocks in the bottom part of the picture. Six ecotypes were analysed: Es-0 (60.19°N, 24.56°E), Per-1 (58.0°N, 56.3°E), Ba-1 (56.45°N, 4.79°E), Ra-0 (46.0°N, 3.3°E), Ann-1 (45.9°N, 6.13°E) and Bla-1 (41.68°N, 2.80°E). Data processed at http://neomorph.salk.edu/1001_epigenomes.html.

ecotypes. *AOX1d* showed no methylation in the promoter region, but possessed methylation in exon 3 in one ecotype (Es-0), and it was moderately expressed in all ecotypes with the exception of Es-0 followed by Ann-1, where only low levels of expression were found. In *AOX2* no methylation was found in the promoter region or in the gene, and no expression was detected. Taken together, a deeper analysis of all 152 methylomes down to deoxynucleotide level could gain valuable information of the methylation pattern and a possible link to differential expression for *AtAOX* genes.

Prolonged exposure to stress could convert an epigenetic modification into a stable (epi)genetic trait of tolerance or resistance (Boyko and Kovalchuk, 2011). While methylated cytosines are highly prone to spontaneous transition mutations, genomic areas with low levels of methylation may be more inclined to chromosomal rearrangements (Chen and Ni, 2006; Boyko *et al.*, 2007; Boyko and Kovalchuk, 2011). Consequently, the change of methylation pattern in a DNA sequence in response to stress may have a significant impact on the rate and type of genetic changes in that sequence, and may lead to the appearance of new alleles in a population. Since genes involved in stress response (like *AOX1* genes) are highly affected by environmental conditions, it is plausible that different stress-induced epigenetic scenarios around those genes bias the type and frequency of mutations in their sequences, making them rich sources of genomic polymorphisms, which could be exploited for FM development.

DNA sequence interacts with genome rearrangements

The genetic variation introduced via stress-triggered genome rearrangements is not completely random (Boyko *et al.*, 2007; DeBolt, 2010). Nevertheless, it is still unclear whether these rearrangements are directed to certain loci or, alternatively, whether certain genome regions are generally more prone to rearrangements due to development and in response to environmental conditions related to both abiotic and biotic stresses. Whatever the case, it is clear that concrete sequences affect genome rearrangements under varying environmental conditions, which has implications for the search of FM candidate sequences. This could be the case of TEs associated to genes. TEs play an important role in genomic rearrangements and may affect gene transcription. TEs have already been detected in several *AOX* genes (*A. thaliana, C. papaya, D. carota, V. vinifera*) (see Cardoso *et al.*, Chapter 12.1). Nevertheless, we have no evidence of structural variants related to *AOX* gene copy number so far. There are, however, known cases in plants, for example copy numbers of concrete alleles have been shown to be an important factor for wheat flowering time and vernalization requirement (Diaz *et al.*, 2012) with both variables regulated by one gene each.

DNA organization may not only affect gene expression, but provide genome protection, depending on environmental conditions or developmental stage, and be tissue- or cell-specific. In somatic cells, homologous recombination (HR) maintains genome stability, whereas in meiotic cells it is responsible for crossing-over and thus generates diversity (Mézard *et al.*, 2007). Many stresses are known to alter the frequency of somatic and meiotic recombination events (Lucht *et al.*, 2002; Kovalchuk *et al.*, 2003; Molinier *et al.*, 2006; Kathiria *et al.*, 2010). In plants, HR may serve as an important mechanism involved in rapid diversification of concrete genomic sequences such as resistance genes (R-genes) in response to stress (Boyko *et al.*, 2007, 2010; DeBolt, 2010).

It is an attractive hypothesis that stress can guide and accelerate plant genome evolution using HR and possibly other DNA repair pathways to trigger locus-specific genome rearrangements. Presumably intraspecific intrafamily gene diversity originated from that type of event, such as is the case of the *AOX* gene family with consequent effects on phenotype. In the cases of post-adaptive (cell-regulated) stress-induced DNA changes, there would be a tendency to a convergence in similar sequences for similar environments. Consequently, the robustness of a FM derived from those sequences may be restricted to these environments. Additionally, if the genomic changes are pre-adaptive (virtually alleatory), irreversible and do not negatively affect fitness, the association of the new allele to similar environmental conditions will be lower, then affecting the validity of a derived candidate FM.

Conclusions and implications for FM development strategies

The sequence diversity in regulatory regions of a gene affects the way in which its expression is modulated by allowing differential action of *trans*-regulatory factors. On the other hand, epigenetic events may affect DNA sequence, which may have consequences at population and evolutionary levels. *AOX* genes, due to their diversity and with differential methylation marks, are likely also subjected to such an interplay between sequence and regulatory mechanisms. In general, the study of regulatory switches of a gene may provide information not only about the stability of the gene activity under a number of conditions, but also about the tendency of the sequence to suffer post-adaptive changes, or pre-adaptive changes conditioned by possible epigenetic scenarios. This could guide the search of allelic polymorphisms.

Polymorphisms in coding sequences may directly affect protein function, but expression regulatory switches are more abundant in non-coding regions. Consequently, the possibility of success in a 'bottom-up' search of a robust FM into alleles known to confer a desirable trait is considerable if the quest is performed on non-coding regulatory sequences, that is enhancer, promoter, introns or untranslated regions. The length of these regions is a factor to take into account, since the number of spanned regulatory elements confers more plasticity to gene expression and thereby more average adaptability to different conditions. Examples for several plant traits of both proposed and developed FMs derived from both coding and regulatory sequences are reviewed by Cardoso and Arnholdt-Schmitt (2013).

Alleles of *AOX* genes from several plant species are being intensively explored (Cardoso *et al.*, 2009, 2011; Costa *et al.*, 2009b; Ferreira *et al.*, 2009; Frederico *et al.*, 2009a, 2009b; Santos Macedo *et al.*, 2009) and a vast amount of new data is in the pipeline (B. Arnholdt-Schmitt, pers. communication; see also Quaresma *et al.*, Chapter 12.3). Further identification of *cis*-regulatory elements for *AOX* genes and the study of the overall involvement of epigenetic regulation for gene expression under distinct conditions will assist identifying appropriate FM candidates for distinct traits.

CHAPTER 13.3

Gene technology applied for *AOX* functionality studies

Sarma Rajeev Kumar and Ramalingam Sathishkumar
Plant Genetic Engineering Laboratory, Department of Biotechnology, Bharathiar University, Coimbatore, India

Most transgenic approaches applied to *AOX* genes aim to better understand *AOX* functionality in terms of 'deep phenotype traits' in metabolism and physiology. Only recently, Afuape *et al.* (2013) reported transgenic *Manihot esculenta* (cassava) transformed with an *AOX* gene from *Arabidopsis* that negatively influenced embryo formation in transgenic lines. This cassava study was conducted to apply breeding strategies specifically to diminish oxidative stress during post-harvest. Till date, all efforts in gene technology have focused only on *AOX1a* (Vanlerberghe *et al.*, 2009). This strongly suggests the importance of expression of *AOX1a* across diverse species and its potential role during abiotic and biotic stress responses. Functional analyses become more complicated as the *AOX* gene family encodes different isoforms having similar/different functions (see Cardoso *et al.*, Chapter 12.1). Moreover, the recent discovery of high sequence polymorphisms within individual genes (Cardoso *et al.*, 2009; Costa *et al.*, 2009b; Ferreira *et al.*, 2009; Frederico *et al.*, 2009b; Santos Macedo *et al.*, 2009; Costa and Svensson, Chapter 12.2) and variability through genetic (e.g. gene copy variation, heterozygosity, DNA methylation) and developmental ploidy changes (e.g. somaclonal variation, gametoclonal variation) make functional studies even more complex. Hence, functional studies should take this entire complex picture into account. *AOX* polymorphic sequences with potentially diverse functions or with efficiency in metabolic regulation could be used for the development of FMs useful to plant breeders and genetic engineers or in native gene substitution to create new breeding material.

Even though biotic and abiotic stresses pose the greatest threat to crop production, conventional selection and traditional breeding techniques have taken a long time to achieve limited progress towards stress resistance. Nowadays, the transgenic approach is a widely used method for crop improvement programmes. Gene discovery and functional genomics have revealed infinite mechanisms and

Alternative Respiratory Pathways in Higher Plants, First Edition.
Edited by Kapuganti Jagadis Gupta, Luis A.J. Mur and Bhagyalakshmi Neelwarne.

led to the identification of many new potential gene families, which could confer adaptation and improved productivity during severe abiotic stresses (Kumar *et al.*, 2012a). These genes can be modified, expressed ectopically, or transferred to interspecies, where they are totally absent (Harish *et al.*, 2013). Hence, the ability to genetically transform major crop species with genes from any biological source (from different kingdoms like plant, animal and microbes) is an extremely powerful tool for molecular breeding. Transgenic plants can be used as a resource for the development of new cultivars or their germplasm as a new source for the variation in breeding programmes. They are also extremely useful as proof-of-concept tools for 'deep phenotyping' (see Nogales *et al.*, Chapter 13.1) to dissect and identify the activity and interplay of novel gene networks during stress and during development, for example seed germination, leaf senescence and flowering. The role of plant breeding in generating new varieties of plants can never be substituted by genetic engineering, but the latter provides an important additional tool for increasing genetic variability. This chapter focuses on the explored functions of *AOX* genes studied through transgenic approaches and also highlights some of the limitations of this methodology.

Novel functions of *AOX* revealed through transgenic technology

A. thaliana or *Nicotiana tabacum* were transformed with *AOX1a,* which is a member of the *AOX* gene family (Vanlerberghe *et al.*, 2009). In both species *AOX1a* was shown to be more stress-responsive with more tissue specific or developmental specific expression patterns compared to other *AOX* genes (Vanlerberghe and McIntosh, 1992, 1994; Saisho *et al.*, 2001a; Mittler *et al.*, 2004; Clifton *et al.*, 2005, 2006). Transgenic tomato lines were established to study the role of *AOX* during fruit ripening. Lines with reduced *AOX1a* (gene was silenced using RNAi) exhibited retarded ripening and down-regulation of many ripening related genes. However, tomato lines over-expressing *AOX1a* accumulated more lycopene (Xu *et al.*, 2012a). Arnholdt-Schmitt *et al.* (2006) and Vanlerberghe (2013) highlighted that *AOX* plays a key role in regulating cell metabolism to adapt external environmental signals as well.

Role of AOX during abiotic stress in different transgenic host systems

Transgenic approaches were used to evaluate the role of *AOX* in plant tolerance during various physiological situations and stresses (Table 13.1). The alternative respiratory pathway of plant mitochondria uncouples respiration from ATP production and may ameliorate plant performance under adverse conditions

Table 13.1 Studies on *AOX* functionalities during various stress/physiological conditions.

Gene source	Methodology	Outcome	Reference
AtAOX1a	Antisense and over-expressing lines of *A. thaliana* lines subjected to low temperature stress	Antisense lines lead to reduced early shoot growth whereas over-expression enhanced early shoot growth.	Fiorani *et al.*, 2005
AtAOX1a		Oxidative stress in antisense lines in presence of KCN while in over expression lines showed no symptoms	Umbach *et al.*, 2005
TaAOX1a	Over-expressing *A.thaliana* subjected to chilling stress	Over-expression lead to delayed expression of the endogenous *AtAOX1a* following shift to 4 °C	Sugie *et al.*, 2006
AtAOX1a	T- DNA lines subjected to low temperature stress	Antioxidant defence genes were induced and malondialdehyde content was lower in knockout lines	Watanabe *et al.*, 2008
NtAOX1a	*AOX1a* knock down and *AOX1a* over-expressing lines subjected to low temperature stress	Knockdown lines showed less cold-induced sugar accumulation and over expressions lineshowed enhanced sugar accumulation.	Wang *et al.*, 2011
AtAOX1a	Over-expression lines subjected to salinity stress in *A. thaliana*	Lower ROS formation, improved growth rates and lower shoot Na^+ content in over-expressing lines	Smith *et al.*, 2009
AtAOX1a	T-DNA insertional line exposed to moderate light under drought stress	Transcripts involved in the synthesis of anthocyanins, chloroplastic and mitochondrial components, cell wall synthesis and sucrose and starch metabolism were altered	Giraud *et al.*, 2008
AtAOX1a *AtAOX1d*	T-DNA insertional line (*AOX*1a knockout) expressing *AOX*1D subjected to anitimycin A inhibition	Inhibition of photosynthesis, increased ROS and amplified membrane leakage and necrosis	Strodtkötter *et al.*, 2009
NtAOX1	Antisense *AOX1a* subjected to P limitation in suspension cells	Knockdown lines showed altered growth, morphology, cellular composition, increased ROS and patterns of respiratory C flow to amino acid synthesis	Parsons *et al.*, 1999
AtAOX1a	*AOX1a* knockout lines under N limitation	*AOX* deficiency altered the levels of sugars and sugar phosphates under low-N stress	Watanabe *et al.*, 2010
AtAOX1a	Over-expression in Cassava	Transgenic lines hindered organised embryogenic structure development	Afuape *et al.*, 2013
NtAOX1a	*AOX1a* silenced lines in suspension culture/leaves treated with SA, peroxides or protein phosphatase inhibitor	Knockdown of *AOX1a* increased susceptibility to PCD induced by cyt pathway dysfunction	Robson and Vanlerberghe, 2002; Vanlerberghe *et al.*, 2002

(*continued*)

Table 13.1 (*continued*)

Gene source	Methodology	Outcome	Reference
NtAOX1a	*AOX1a* knockdown or over-expressed in tobacco prior to pathogen attack	Alterations in *AOX* capacity did not affect the overall response of the plants to systemic disease with or without prior SA treatment	Gilliland *et al.*, 2003
LeAOX1a	*AOX 1a* over-expressed in tomato and petunia prior to virus challenge	Over-expressing lines showed lower levels of tomato spotted wilt virus (TSWV) levels	Ma *et al.*, 2011
NaAOX1a	Plants silenced with *AOX 1a* challenged with different pathogens	Silenced plants showed different metabolite accumulation and defence mechanism based on the pathogens	Zhang *et al.*, 2012
LeAOX1a	*AOX 1a* knockdown lines tomato	Knockdown lines exhibited retarded ripening, reduced carotene accumulation, ethylene production and down-regulation of ripening-associated genes	Xu *et al.*, 2012a
Glycine max	*GUS* was driven under different *AOX* promoter in soybean suspension cells and *A.thaliana*	Promoter activity of the upstream fragments of *AOX2a* and *AOX2b* displayed the same tissue specificity in both systems	Thirkettle-watts *et al.*, 2003
A.thaliana	*GUS* was driven under *AOX 1a* promoter	ABA sensitive elements are present in *AOX* promoter	Giraud *et al.*, 2009

like cold by preventing excess accumulation of reactive oxygen species (ROS) (Wagner *et al.*, 1998). *AOX* competes for electrons with the cytochrome pathway (CP) during stress conditions (Finnegan *et al.*, 2004). Most crops of tropical as well as many of subtropical origin are sensitive to chilling temperatures. The negative impact of ROS during chilling stress is a well-studied phenomenon (Einset *et al.*, 2007). The maintenance of redox balance is crucial because electron input in excess leads to the production of ROS and it is a common phenomenon during stress conditions. It has been postulated that *AOX* may play a significant role in allowing plants to tolerate frost or chilling-induced ROS damage (Purvis and Shewfelt, 1993).

The effect of *AtAOX1a* in response to low temperature stress in *Arabidopsis* was explored by silencing the gene by either knockout or knockdown. Fiorani *et al.* (2005) reported that at 12 °C, *AOX1a* antisense lines lead to reduced early shoot growth in *Arabidopsis,* whereas over-expression enhanced early shoot growth. Over-expression of wheat *AOX1a* in *A. thaliana* delayed the expression of the endogenous *AOX1a* during chilling stress (Sugie *et al.*, 2006). Watanabe *et al.* (2008)

studied the effect of knockout of *AtAOX1a* using T-DNA insertional lines. From the studies of *Arabidopsis* mentioned earlier, it can be concluded that at low temperature, knockout of *AOX1a* lines showed not only enhanced expression of ROS scavengers but also lowered lipid peroxidation levels. Under low temperatures, *AOX1a* transcript was strongly induced in wild-type (WT) plants. The transformed plants were unable to up-regulate *AOX1a*; however, increased cyanide-resistant respiration showed increased uncoupling protein 1 (*UCP1*) expression. It was also observed that lack of *AOX* was linked to a difference in the carbon:nitrogen (C:N) balance and an up-regulation of antioxidant defence system in response to low temperature stress. In contrast to the growth reduction observed in the antisense plants (Fiorani *et al.*, 2005), chilling-related phenotypic changes were not observed in the T-DNA lines (Watanabe *et al.*, 2008). Transgenic rice seedlings over-expressing *AOX1a* exhibited thermotolerance after acute exposure at 41–45 °C for 10 minutes, or chronic exposure at 37 °C for eight days, whereas these high temperature stresses resulted in significant growth inhibition in WT and transgenic plants with antisense *OsAOX1a*. The elevated levels of *AOX* in over-expressing lines are considered to protect several heat-sensitive components of plastids, thus improving the inhibition of shoot growth in rice plants (Murakami and Toriyama, 2008). Therefore, *AOX* not only plays a significant role during low temperature stress, but also provides high temperature tolerance.

Salinity stress has been shown to disturb the cellular redox status leading to mitochondrial dysfunction and increased *AOX* respiration (Borsani *et al.*, 2001). *Arabidopsis* over-expressing *AtAOX1a* showed lower ROS formation, improved growth rates and lower shoot Na^+ content compared with WT under stress conditions. The transgenic lines also displayed lower levels of peroxides than WT and this change might be for exclusion of Na^+ from the shoots (Smith *et al.*, 2009).

Under the combined stresses of moderate light with drought, *AtAOX1a* mutants (T-DNA insertional line) accumulated higher amounts of anthocyanins, O_2^- radicals and an altered transcript level for chloroplastic and mitochondrial components when compared to the WT plants (Giraud *et al.*, 2008). The combined stresses also resulted in accumulation of transcripts related to anthocyanin biosynthesis, cell wall synthesis, various transcription factors, and chloroplastic and mitochondrial components indicating that the effects of the mutation were not only confined to mitochondria but also had an impact on other cell organelles. The *Arabidopsis* T-DNA mutant of *AOX1a* plants did not show any phenotypic changes under normal conditions, but after inhibition of the CP using antimycin A, photosynthesis was affected, which increased ROS formation and membrane leakage (Strodtkotter *et al.*, 2009).

Plant growth and productivity depends to a large extent on the availability of mineral nutrients and, among the macronutrients, nitrogen (N) and phosphorous (P) often limit growth in natural and agricultural settings. There is a drastic change in physiological, morphological, biochemical and molecular mechanisms, which are induced under nutrient stress that will improve the acquisition and use of nutrients. Nutrient stresses are always associated with reduced growth, a physiological response that may have some adaptive advantage (Thomas and Sadras, 2001). The pool of amino acids derived from different carbon intermediates (like 2-oxoglutarate) was reduced in cells lacking *AOX* grown under P limitation (Parsons *et al.*, 1999). Altered levels of amino acids derived from other glycolytic intermediates like phosphoenolpyruvate (PEP) and pyruvate were also observed, indicating disruption of normal metabolism at this regulatory point in respiration. Antisense lines of *AOX1a* magnified ROS generation and restricted carbon metabolism during P-limited growth in a suspension culture of tobacco (Yip and Vanlerberghe, 2001). In another study, knockdown of *AOX1a* suspension culture tobacco lines grown under P or N limiting conditions, enhanced biomass accumulation and caused other nutrient-specific effects on cellular redox and carbon balance (Sieger *et al.*, 2005). An increase of C : N ratio was observed in *A. thaliana* leaves lacking *AOX1a,* after transferring the plants to low temperature (Watanabe *et al.*, 2008). The induction of *AOX* respiration is an important plant metabolic adaptation during P limitation. Hence, *AOX* prevents redirection of C metabolism and excessive generation of free radicals in the mitochondria.

Transgenic tobacco lines over-expressing the *AOX1a* gene accumulated more soluble sugars like glucose and fructose than control plants, while plants with suppressed *AOX* accumulated less sugars (Wang *et al.*, 2011). These studies suggest that *AOX* respiration aids carbon metabolism under different stress conditions and CP alone is not able to compensate for a lack of *AOX*, resulting in accumulation of carbohydrate substrate. As explained above, lack of *AOX* during stress can lead to redirections in carbon metabolism, possibly due to particular bottlenecks in the plant metabolic pathway (Vanlerberghe *et al.*, 2009).

Maxwell *et al.* (1999) reported transgenic tobacco cells with altered levels of *AOX1* playing a crucial role to reduce the formation of ROS. It was found that suppression (by antisense RNA) of *AOX1a* resulted in significantly higher level of ROS compared with WT cells, whereas the over-expression leads to lower ROS abundance during oxidative stress in tobacco. It was also observed that cells over-expressing *AOX* showed lower expression of genes encoding ROS scavenging enzymes like superoxide dismutase (SOD) and glutathione peroxidase (GPx), whereas transcripts encoding catalase and pathogenesis-related protein were significantly higher in cells lacking *AOX*. In another study, the over-expression of *AOX* in transgenic tobacco plants triggered an increased sensitivity to ozone and over-expression resulted in decreased ROS level, which in turn altered the mitochondria defensive to the nuclear signalling pathway that activates ROS scavenging systems (Pasqualini *et al.*, 2007). The reasonable explanation to *AOX* lowering ROS levels is that a second oxidase downstream of the ubiquinone retains upstream electron-transport components in a more oxidized state, lowering free radical generation by over-reduced electron carriers. These findings clearly suggest that *AOX* plays a significant role in lowering ROS formation in plant mitochondria. Afuape *et al.* (2013) tried to make use of this function in cassava breeding on improved post-harvest stress by *AtAOX1a* under 35S promoter in cassava breeding lines. The effect of transgene (*AtAOX1a*) was assessed using stress-inducible somatic embryogenesis as a test system. Somatic embryogenesis has shown to exhibit differential *AOX* gene activities in carrot (Frederico *et al.*, 2009).

Role of *AOX* during biotic stress response

AOX have also been proved equally important during biotic stress using transgenic plants (Cvetkovska and Vanlerberghe, 2013; Garcia *et al.*, 2013). Looking from a physiological point of view, respiratory metabolism in plants and the defence response to biotic stress could be linked at the biochemical level. Pathogen infection enhanced biosynthesis of different aromatic secondary metabolites, such as salicylic acid (SA), phytoalexin, lignin and plays an important role in the host defence response (Bennett and Wallsgrove, 1994). A regulatory role of *AOX* in biosynthesis of phenolic derivatives has been reported by Sircar *et al.* (2012). These compounds are produced from the shikimate pathway and precursors of the shikimate pathway are erythrose-4-phosphate and PEP, which are produced by the respiratory oxidation of glucose (Arcuri *et al.*, 2004). Thus, intermediate products of respiratory carbon metabolism provide the backbone substrates for the biosynthesis of aromatic compounds related to the host defence response.

Lacomme and Roby (1999) reported that *AOX* transcripts were transiently induced by an avirulent strain of *Xanthomonas campestris* in cell suspension culture of *Arabidopsis* and later during infection with a virulent strain, but no detectable *AOX* was found. Gilliland *et al.* (2003) reported that manipulation of *AOX1a* expression neither had an effect on basal susceptibility to tobacco mosaic virus (TMV) nor SA-induced resistance to systemic viral disease in transgenic tobacco lines. Transgenic tobacco plants or suspension cultured cells silenced with *AOX* have increased susceptibility to programmed cell death (PCD) (Robson and Vanlerberghe, 2002; Amirsadeghi *et al.*, 2006). PCD depends on *AOX*, which clearly contributes to the steady-state ROS level by influencing the rate of mitochondrial-generated ROS. In another study using the same system, knockdown of *AOX1a* resulted in increased susceptibility to PCD induced by CP dysfunction or by treatment with SA, H_2O_2 or different protein phosphatase inhibitors (Robson and Vanlerberghe, 2002; Vanlerberghe *et al.*, 2002). Ordog *et al.* (2002) reported that *AOX* is not an essential component of viral disease resistance but may play a role in the hypersensitive response (a form of PCD) during viral infection. Recently, Ma *et al.* (2011) reported that transgenic tomato and petunia lines over-expressing tomato *AOX1a* showed lower levels of tomato spotted wilt virus

(TSWV) than WT, unlike the report by Ordog *et al.* (2002). Nevertheless, it is not clear how *AOX* provides resistance during viral infection.

To conclude, engineering *AOX* appears to have relatively minor consequences during non-stress conditions; however, during stress conditions it has been proved to play a very important role. These results are consistent with the idea that *AOX* respiration has important role(s) under stress conditions (Simons and Lambers, 1999). Taken together, all the findings indicate that plants with reduced or no expression of *AOX1a* have an altered stress response even when the mitochondria are not the primary targets of the stress and they also suggest that *AOX1a* plays a crucial role in cell metabolism by balancing the redox status in the cell.

Limitations of transgenic technology

Though transgenic technology is quite useful in the identification of novel or 'deep' functions of *AOX* during different stress conditions, this strategy also has several drawbacks. A major limitation of plant transgenic technology aiming for a single gene for knockout or over-expression is the multiple number of *AOX* genes (endogenous *AOX* already present in the system) that may have a similar function in the cell, which may obscure any alteration including phenotypic change. In these circumstances, the function of a silenced (single) gene could be compensated by another endogenous isoform. Usually transgenic studies involve cloning and expression of cDNA (not introns) in heterogeneous host systems under the control of constitutive or inducible promoters because intron and untranslated regions are not generally considered for functional studies. The new findings related to the presence of regulatory elements on these regions (e.g. transcription factors, miRNA encoding sites, transposable elements; and see Cardoso *et al.*, Chapter 12.1) and exclusion of such regions could represent an important bottleneck in *AOX* functionality studies.

Role of regulatory elements on *AOX* gene expression

The ability of introns to enhance the expression of genes by intron mediated enhancement (IME) depends upon the sequence and the position of the intron within a gene (Bourdon *et al.*, 2001; Rose, 2002; Rose *et al.*, 2008). The critical feature of IME is that not all introns are capable of enhancing expression and splicing, exon junction complexes, and so on. Reports suggest that introns can increase the expression levels generally between 2- and 10-fold and can even go up to 100-fold in some cases (Maas *et al.*, 1991; Bartlett, *et al.*, 2009). As mentioned earlier, transgenic studies involve cloning and expression of cDNA (not introns) in heterogeneous host systems. In such cases, IME will not have any role and this may lead to low levels of transgene expression by an insufficient recombinant protein. A higher degree of polymorphisms in *AOX* genes are found typically in introns compared to exons and can cover regulative motifs (Cardoso *et al.*, 2009; Ferreira *et al.*, 2009; Santos Macedo *et al.*, 2009; Cardoso *et al.*, 2011; see Cardoso *et al.*, Chapter 12.1). We also presume that IME may play an important role in *AOX*, where many polymorphisms have been reported in UTR and variations in UTR-regulated differential gene expression in olive and carrot (Cardoso *et al.*, 2009, 2011; Santos Macedo *et al.*, 2009). A small but significant fraction of introns are also found to reside within the untranslated regions (5′-UTRs and 3′-UTRs) of expressed sequences (Chung *et al.*, 2006). Hence, the incorporation of UTR region should also be considered while expressing *AOX* cDNA in the heterogeneous host system in future. Another limitation of transgenic studies and its application is the necessity of considering haplotypes and whole genomes as a critical context for regulatory networks in an individual plant. In many cases, a desired effect from a transgene can be found only in selected transgenic lines, but not in all the established transgenic lines. For example, Afuape *et al.* (2013) reported that transgenic cassava lines expressing *AtAOX1a* reduced organized embryogenic structure formation through the reduction of auxin-induced ROS production, which is essential to induce the morphogenic re-differentiation. However, this effect was not found for all transformed breeding lines for reasons that were uncertain. Thus, it is important to recognize that functional studies may depend on individuality and diagnosis. Therefore, it is not only the plant level that should be considered for studies related to function as well as for breeding purposes but also cell and tissue levels.

To date, most of the studies have been focused only on the manipulation of *AOX1a* gene expression and this is not always compensated by changes in expression of other *AOX* genes. For example, knockout of *AOX1a* did not have any impact on the expression of the other four *AOX* gene family members in *A. thaliana* (Umbach *et al.*, 2005; Giraud *et al.*, 2008). The accumulation of *AOX1d* isoform in *AOX1a* mutant in *Arabidopsis*, caused inhibition of photosynthesis and increased ROS resulting in amplified membrane leakage and necrosis when treated with antimycin A. Thus, *AOX1d* was unable to fully compensate the loss of *AOX1A*, when electron flow via the CP was restricted (Strodtkötter *et al.*, 2009).

Future focus

Taking all of these points together, detailed studies should be performed with all *AOX* genes using different host systems to get more insight into their physiological function and crosstalk between different organelles during stress (Figure 13.2). This can be better studied using specific T-DNA insertional mutants for a particular *AOX* isoform. Also, a specific *AOX* gene can be transiently silenced

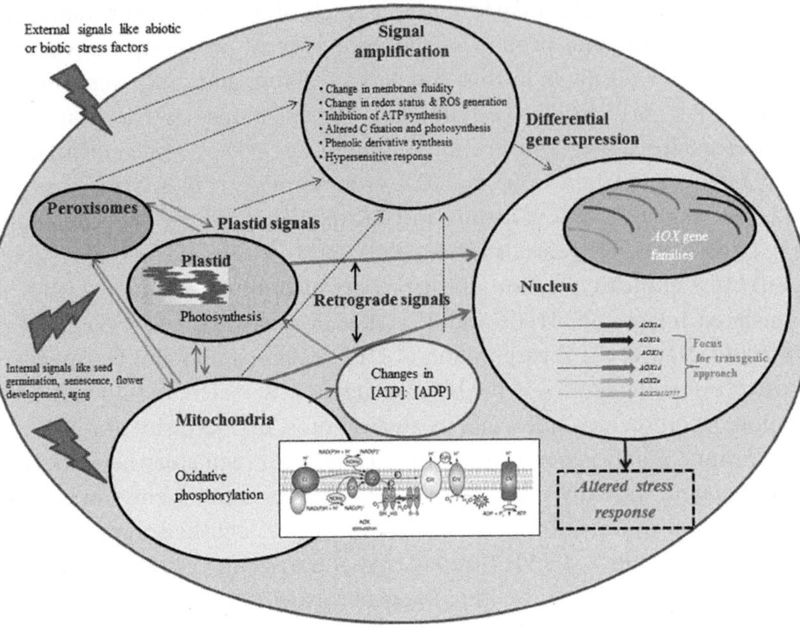

Figure 13.2 Overview of *AOX* signalling during stress and the focus of transgenic technology should be the characterization of *AOX* gene families. Modified with permission from Arnholdt- Schmitt *et al.*, 2006 and Clifton *et al.*, 2006. Dotted arrows, external or internal signal perception, amplification and transmission for altered gene expression; red arrow, retrograde signalling from mitochondria and plastids to nucleus; purple arrow, signalling between mitochondria, plastids and peroxisomes. (*See insert for color representation of the figure.*)

by infiltration or similar methods to study its immediate effect. As already demonstrated, one of the significant effects will be alteration in carbon fixation and photosynthesis. This is because of crosstalk between mitochondria and plastids where the retrograde signals are generated and transmitted to the nucleus for stress response. Recently, Cavalcanti *et al.* (2013) reported that the expression of *AOX1a*, *AOX2b1* and *AOX2b2* showed peculiar spatiotemporal expression patterns after various stress treatments in leguminous plants. However, the role of such 'novel' *AOX2b1* and *AOX2b2* can be explored by developing the transgenic system. This will also aid in the identification of stress responsive pathways and various signalling molecules involved in the crosstalk. This is more important as changes in *AOX* resulted in extramitochondrial metabolism that was more pronounced than mitochondrial metabolism (Fiorani *et al.*, 2005). Giraud *et al.* (2009), reported that the transcription factor abscisic acid insensitive 4 (ABI4) is involved in the retrograde regulation of *AOX1a* in *Arabidopsis* and the promoter is regulated by abscisic acid (ABA). This further confirms the molecular link between retrograde signalling from mitochondrial to plastid as ABI4 has been reported to act downstream of at least two chloroplast retrograde signalling pathways. *AOX* has been already reported as a marker for retrograde signalling response in plants (Suzuki *et al.*, 2011).

Concluding remarks

The main drawback in the present scenario is the missing link between huge amount of molecular data obtained through transgenic technology and its potential application in plant breeding. Nevertheless, transgenic technology will continue to contribute to crop improvement programme, if efforts are directed more towards FM-assisted plant breeding. *AOX* has been proposed as a functional marker for cell reprogramming by Arnholdt-Schmitt *et al.* (2006) and is detailed by Nogales *et al.* (Chapter 13.1). Hence, *AOX* can be put into best use if a dual approach involving genetic transformation and conventional plant breeding go hand in hand.

14

AOX goes risk: A way to application

CHAPTER 14.1

AOX diversity studies stimulate novel tool development for phenotyping: calorespirometry

Birgit Arnholdt-Schmitt[1], Lee D. Hansen[2], Amaia Nogales[1] and Luz Muñoz-Sanhueza[1,*]

[1] *EU Marie Curie Chair, ICAAM - Instituto de Ciências Agrárias e Ambientais Mediterrânicas, Universidade de Évora, Évora, Portugal*

[2] *Department of Chemistry and Biochemistry, Brigham Young University, Provo, Utah, USA*

**Current affliaton: Department of Plant and Environmental Sciences (IPM), Norwegian University of Life Sciences, Ås, Norway*

In plant breeding, cold and heat tolerance are major issues. Variable and extreme temperature conditions are responsible for unstable yield production over a wide range of crops in almost all parts of the world. Field trials to identify temperature tolerant plants are exhaustive in terms of experimental time, personal efforts, space and overall costs. Thus, seed producing companies are highly interested in identifying efficient tools for pre-screening 'deep traits' in plant material. Such tools will aid in narrowing the pool of genotypes for final field screening in the breeding process.

A tool for screening of stress tolerant behaviour can be rated 'efficient' when:

- it can evaluate 'deep traits' that are closely linked to multi-stress tolerance at whole plant level
- measurements can be taken in a simple way during diverse steps of plant development
- upgrading of the technology for higher sample throughput in a short time is possible
- the tool can be adapted to a variety of crops.

Calorespirometry might fulfil all of these criteria and is currently being established as a novel screening tool for traditional and molecular pre-breeding. The first promising results have been achieved for *D. carota* (Nogales *et al.*, 2013).

Alternative Respiratory Pathways in Higher Plants, First Edition.
Edited by Kapuganti Jagadis Gupta, Luis A.J. Mur and Bhagyalakshmi Neelwarne.

Why calorespirometry?

Calorespirometry is the product of respiration rate – the efficiency of carbon incorporation into structural biomass determines the growth rate of plant tissues. Temperature changes thus affect growth properties through respiration. Calorespirometry can be applied to measure respiratory heat and CO_2 rates across a range of temperatures in a technically simple approach, from which growth rates can be calculated (Hansen *et al.*, 2005). The basic concepts and details of calorespirometry for calculating growth potentials in plants as a function of temperature can be found in Hansen *et al.* (2005).

Temperature adaptive growth rates can differ between genotypes and have a significant dependence on AOX and cytochrome oxidase (COX) activities (Hansen *et al.*, 2009). For this reason, studies on the importance of *AOX* gene polymorphisms stimulated the idea to choose calorespirometry as an analytical tool. It was proposed to use calorespirometry to validate the hypothesis that *AOX* gene polymorphisms are closely associated with genotype-specific temperature responses related to growth behaviour.

In contrast to the complex *COX* gene family, *AOX* genes form only a small family (maximum six isozymes; see Cardoso *et al.*, Chapter 12.1). The choice of *AOX* as a source for functional sequences that mark multi-stress behaviour is well justified by recent knowledge of its role in adaptive responses to all kinds of abiotic and biotic environmental stresses (Arnholdt-Schmitt *et al.*, 2006; Arnholdt-Schmitt, 2009; Cardoso and Arnholdt-Schmitt, 2013; Vanlerberghe, 2013; see also Special Issue on *AOX* in *Physiologia Plantarum* 2009, Vol. 137). *COX* genes might also be interesting candidate genes for FM development. However, since *COX* genes are organized in a larger gene family, it will be more complicated to identify the importance of individual or combined polymorphic patterning, since complementing effects between isozymes of the same gene family can be expected to a much higher degree than for *AOX*. In both cases, calorespirometry might serve as a tool to bridge the gap between genomics and phenomics.

FM development for breeding requires close association of a marker sequence in the target gene body or its regulative sequences with a final plant trait, such as yield stability. Only on this basis can the marker sequence aid in marker-assisted plant selection. However, this does not necessarily mean that this association will be based on a (known) biologically causal relationship (Brenner *et al.*, 2013). In physiological or metabolism studies, it is the aim to 'understand' step-by-step the changes that will come through polymorphisms and *how* they contribute to final traits. Biology research is most importantly involved in understanding complexity. Breeding is different; what matters is not to understand *how and why* a gene or polymorphism acts, but instead what matters exclusively is finding out whether there is a correlation or association of a polymorphism with a desired effect in cell or tissue behaviour that finally will positively affect whole plant responses (Arnholdt-Schmitt, 2005). Any important link of a DNA sequence to the target plant trait that allows for efficient selection of promising plant individuals or groups of individuals can help to shorten the breeding process and will thus mean relevant progress. The reason to choose candidate genes for marker development and not neutral marker sequences with no direct functional meaning is just to increase the probability of finding appropriate sequences that show marker characteristics for a trait.

Calorespirometry has been proposed to screen *AOX* polymorphic genotypes for differential effects on growth potentials and temperature behaviour, without the necessity of understanding underlying detailed mechanisms in metabolism for genotypic differences in respiratory heat rate or carbon use efficiency. Calorespirometry can be used to measure the final outcome of relevant complex metabolism changes through a genetic modification in *AOX* gene functionality. However, it can also be applied for novel genotype screening for breeding independent of preceding knowledge on the genotype.

First results confirm the genotype discriminatory power of calorespirometry

The potential of calorespirometry to predict growth potentials and temperature response behaviour for stable yield production was recently studied in carrot (*D. carota*) (Nogales *et al.*, 2013). The results are promising. Genotypes could be clearly distinguished by measuring their growth potential and temperature response. However, it was shown that it is critical to select for the measurements the proper target tissue or cells that are crucial for yield production at plant level. In the case of carrot, the tap root meristem was the most suitable tissue for genotype comparison, since it is responsible for the secondary root growth that determines final yields. Measurements could be performed independently from the age of the carrot plants and the thickness of the tap root.

Perspectives

Calorespirometry is a rapid way to determine the thermal phenotype of target tissues in individual plants. Thus, it can be expected to provide a rapid mean for identifying correlated genes. Research is initiated in running projects to validate calorespirometry as a tool to distinguish phenotypes that are characterized at genome level through different polymorphisms in *AOX* genes. As raised by Arnholdt-Schmitt (2009), the number of functional sites in coding and non-coding *AOX* gene sequences is much higher than expected. A polymorphism can only be claimed to be unimportant when this is definitely shown. Until then we should presume that all polymorphisms can have relevant effects on gene regulation. This challenges screening efforts. High numbers of polymorphisms were identified in *AOX* genes (Cardoso *et al.*, 2009, 2011; Costa *et al.*, 2009b; Ferreira *et al.*, 2009; Santos Macedo *et al.*, 2009). The aim is to identify by calorespirometry the effect of *AOX* polymorphisms in relation to differential growth potentials and to their interaction with environmental stress factors. Temperature variability is one critical component of the environment.

Allelic and copy number variation of the target gene and plastic genomic organization (see Noceda *et al.*, Chapter 13.2) can interfere with the functionality of polymorphisms. For this purpose, genetically at least 'nearly' homogeneous inbred lines with defined polymorphic patterns in their *AOX* genes will be used for screening to avoid allelic and haplotype differences that complicate interpretation of the data. Research is being extended at this stage to include a variety of other important crops, such as cereals. Also, the interaction of genotype functionality with alternative management practices, for example through symbiotic mycorrhizal fungi, will have to be considered (see Mercy *et al.*, Chapter 14.2 and Orfanoudakis *et al.*, Chapter 14.3). Finally, the interaction of plants with endophytes will have to be appraised when the effect of *AOX* gene polymorphism functionality is studied.

CHAPTER 14.2

AOX gene diversity in arbuscular mycorrhizal fungi (AMF) products: a special challenge

Louis Mercy[1], Jan T. Svensson[2,*], Eva Lucic[1], Hélia G. Cardoso[2], Amaia Nogales[2], Matthias Döring[1], Jens Jurgeleit[1], Caroline Schneider[1] and Birgit Arnholdt-Schmitt[2]

[1]*INOQ GmbH, Solkau, Schnega, Germany*

[2]*EU Marie Curie Chair, ICAAM - Instituto de Ciências Agrárias e Ambientais Mediterrânicas, Universidade de Évora, Évora, Portugal*

**Current address: Nordic Genetic Resource Center, Alnarp, Sweden*

A link between plants and fungi: the mycorrhiza

Plants live with a myriad of biotic and abiotic interactions with the soil and environment that determine their growth, productivity and life cycle, whereas the rhizosphere plays an interface role (Jeffries *et al.*, 2003). Arbuscular mycorrhizal fungi (AMF) establish the predominant mutualistic symbiotic relationship within the roots of more than 200 000 species within 85% of plant families (Wang and Qiu, 2006; Smith and Read, 2008). AMF are obligate biotrophs, represented by about 230 species (www.amf-phylogeny.com) found worldwide under a wide range of ecological conditions. The beneficial effects of AMF to the plant host are multiple, as they (i) improve plant growth by a better transfer of inorganic nutrients, especially phosphorus, and water (Smith and Read, 2008); (ii) increase plant pathogen resistance and plant health (Whipps, 2004; Pozo *et al.*, 2009); (iii) boost plant photosynthesis (Quarles, 1999); (iv) stabilize soil by the excretion of a fungal glycoprotein, the glomalin (Rillig and Steinberg, 2002; Bedini *et al.*, 2009); (v) alleviate the impact of abiotic stresses such as salinity (Porras-Soriano *et al.*, 2009), drought (Aroca *et al.*, 2007) and heavy metal (Karimi *et al.*, 2011). In turn, AMF benefits from plants are a habitat in which they can complete their life cycle associated with the uptake of photosynthates.

Many efforts have been conducted to exploit the potential of mycorrhizas, starting with the establishment of mycorrhizal inocula production at various scales. Though several methods are available (*in vivo* production including hydroponic

---

Alternative Respiratory Pathways in Higher Plants, First Edition.
Edited by Kapuganti Jagadis Gupta, Luis A.J. Mur and Bhagyalakshmi Neelwarne.

andaeroponic systems and *in vitro* production in a solid or liquid medium), the average cost of inocula on the market remains prohibitive, thus limiting their use on a large scale (Vosatka *et al.*, 2008; Ijdo *et al.*, 2011; Malusá *et al.*, 2012). In addition, environmental constraints limit mycorrhizal establishment since AMF are usually unable to inhabit soils with high phosphate and nitrogen levels, conditions that are often found in conventional cropping (Smith and Read, 2008). In this case, even a proper inoculation with the best quality inoculum would not always result in a valuable increase in plant health and consequently yield production. Combined with the presence of fake or poor quality mycorrhizal inoculum on the market, farmers may not be convinced, which has adverse consequences for the economy of mycorrhizal fungi (Vosátka *et al.*, 2012). Currently there is no real improvement which allows competition with conventional products and considerable efforts have to be made to produce mycorrhizal inocula containing strongly efficient fungal species in terms of benefits to the plant, as well as in their ability to survive and develop in a refractory environment. Mycorrhizal producers currently face an enormous problem which is related to the high variability of the mycorrhizal inocula produced in terms of effectiveness. In order to contribute to mycorrhizal inocula producers in a breeding perspective, *AOX* was proposed to the scientific community in two COST870 meetings as a target gene for FM development in AMF. The hypothesis that regulation of *AOX* from plants and *AOX* from AMF is connected during the pre-symbiotic phase and plays a crucial role in mycorrhizal colonization was presented (Arnholdt-Schmitt, 2008; Vicente and Arnholdt-Schmitt, 2008). Research based on that hypothesis is now running under an European Project (AGRO-AMF-*AOX* from FP7-PEOPLE-2009-IAPP) which involves a research institution (Universidade de Évora, Portugal) and a private company with long experience in AMF inocula production (INOQ GmbH, Germany).

Some clues on the role of *AOX* in AMF

First data mentioning a possible second respiratory pathway in fungi appeared in the 1930s (Goddard and Smith, 1938). Since these pioneering observations, molecular tools have confirmed the presence of *AOX* genes in the fungal kingdom (Akhter *et al.*, 2003; McDonald and Vanlerberghe, 2006; McDonald, 2008, 2009). Interestingly, while *AOX* genes usually constitute small multigene families in plants, the analysis of 222 fungal genomes currently available (Grigoriev *et al.*, 2012 – http://genome.jgi.doe.gov/), reveals that a majority of fungal species have only one *AOX* gene (79.28%), some possess 2 (9.91%), 3 (1.35%) or no *AOX* gene (9.46%, mainly yeast and/or species having fermentative metabolism). In AMF genomes, the presence of an expressed *AOX* gene is known in *Rhizophagus irregularis* DAOM197198 (Morin *et al.*, http://mycor.nancy.inra.fr/IMGC/GlomusGenome/index3.html) and in *Gigaspora rosea* (Besserer *et al.*, 2009).

Fungal *AOX*s have been shown to play a role in growth regulation and development, resistance, pathogenesis and pathogenicity, and may contribute to fungal ecological fitness (Uribe and Khachatourians, 2008; Ruiz *et al.*, 2011; Grahl *et al.*, 2012; Thomazella *et al.*, 2012; Xu *et al.*, 2012b). In AMF, the role of AOX pathway is fewly explored, but it seems involved in spore germination (Besserer *et al.*, 2009) and may participate in mitochondrial physiological changes in response to root exudates (especially strigolactone), which initiate the pre-symbiotic stage of the fungus (Besserer *et al.*, 2006; Tamasloukht *et al.*, 2003; Besserer *et al.*, 2008). Expression profiles of AMF *AOX* genes have been available since 2011 (Kohler *et al.*, 2011 – GEO DataSets, Accession: GSE29866), but so far it has not been possible to reveal a specific pattern of *AOX* gene expression.

Structural predictions showed that the active *AOX* protein is homodimeric in plants, and monomeric in fungi (Umbach and Siedow, 2000), although dimeric forms are observed for some fungal species (Moore *et al.*, 2013). Unlike plants, fungal *AOX* protein activity cannot be regulated by addition of pyruvate or α-keto acids, but are strongly regulated by purine nucleotides (ADP, AMP, GMP) (Umbach and Siedow, 2000). This regulation pattern in plants depends partly on two well-conserved cysteine residues which are not present in fungi *AOX* (Moore *et al.*, 2013). Indeed, *in silico* analyses highlighted that the encoded protein does not contain any cysteine residues, leading to the conclusion that its regulation is likely different from plant proteins. Small change(s) in *AOX* sequence can also greatly influence the ability of a given organism to interact with its environment and the capacity to adapt to various growth conditions, as was shown for low temperature tolerance of tomato or rice (Abe *et al.*, 2002; Holtzapffel *et al.*, 2003). *R. irregularis AOX* is homologous to classical Zygomycota and Chytridiomycota *AOX* sequences; together these *AOX*s exhibit a higher sequence homology with plant *AOX* compared to almost all other fungal *AOX* (Figure 14.1).

Plants and AMF symbiosis upon stress

Stress response in plants involves numerous signalling and metabolic pathways in which *AOX* plays a central role (Van Aken *et al.*, 2009; Millar *et al.*, 2011). *AOX* gene expression is under the control of a growth regulator, abcisic acid (ABA) (Finkelstein *et al.*, 1998; Choi *et al.*, 2000; Rook *et al.*, 2006; Giraud *et al.*, 2009; Millar *et al.*, 2011; Lynch *et al.*, 2012; Wind *et al.*, 2012). ABA was also demonstrated as a key component for arbuscule formation and functionality within AMF (Herrera-Medina *et al.*, 2007; Martin-Rodriguez *et al.*, 2010, 2011; Aroca *et al.*, 2013). Other compounds such as H_2O_2 can also enhance *AOX* activity and expression (Ho *et al.*, 2008) under various stress conditions such as high salinity (Hasegawa *et al.*, 2000; Liu *et al.*, 2007). Exogenous application of ABA and H_2O_2 increases mycorrhizal development, while SHAM (a known *AOX* inhibitor) has adverse effects on both H_2O_2 content in plants and

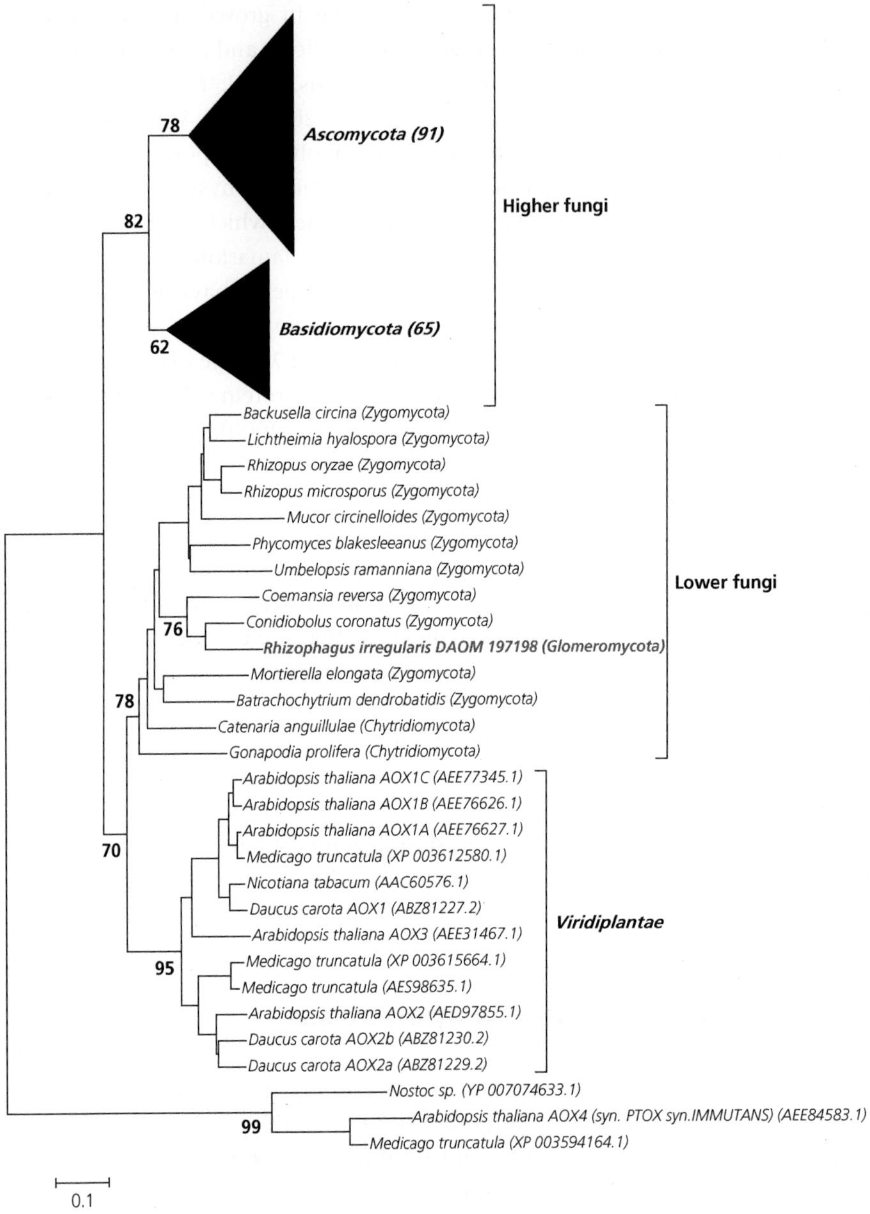

Figure 14.1 Phylogenetic relationships among *AOX* proteins. Complete amino acid sequences were aligned by CLUSTALW and the tree was constructed by the neighbour-joining method using Mega 5.20 (Tamura *et al.*, 2011). p-distances were estimated between all pairs of sequences using the pairwise deletion option. Bootstrap tests were conducted using 1000 replicates, and bootstrap values above 50 and supporting a node of interest are indicated. Accesion numbers are indicated for plant species. All fungal species are obtained from JGI genome portal (Grigoriev *et al.*, 2012 – http://genome.jgi.doe.gov/). The number of sequences from Ascomycota and Basidiomycota are indicated in brackets.

mycorrhizal rate (Liu *et al.*, 2012). It seems clear that the establishment of mycorrhizal symbiosis and its phenotypic variations, as well as mycorrhizal effects on plants, rely on plant stress status and therefore probably strongly involve *AOX*.

Functional marker development in mycorrhiza – a genetic challenge

The use of expressed protein-encoding genes as molecular markers of relevant AMF functions/traits is common and many publications deal with the characterization of new candidates (for review, see Gamper *et al.*, 2010). However, while FM in plant breeding is defined as a sequence motif affecting phenotypic variations (Andersen and Lübberstedt, 2003), in fungal physiology this term refers mainly to the transcript level of the relevant gene rather that to its sequence variations (Gamper *et al.*, 2010). Nonetheless, some studies have also pointed to the molecular diversity for a few genes (Corradi and Sanders, 2006; Corradi *et al.*, 2009). Apart from complex regulations from gene to protein, it can be assumed that polymorphisms in the *AOX* gene sequence can be linked to specific fungal behaviour patterns. The identification of functional polymorphisms will be useful to select AMF strains with the desired traits with direct consequences for plant performance, as shown in plant breeding applications (Arnholdt-Schmitt *et al.*, 2006).

Development of a FM based on gene sequence in AMF poses several difficulties. Foremost, the basic genetics of the *Glomeromycota* is not fully understood, such as ploidy, number of chromosomes, and how genetic exchange and segregation occur. The difficulties in producing a completely annotated and assembled *R. irregularis* genome – the sequencing began in 2004 – demonstrate the complexity of AMF genetics (Martin *et al.*, 2008). A single AMF spore contains hundreds to thousands of genetically different nuclei, and these nuclei may harbour a part of polymorphic genes (Hijri and Sanders, 2005; Marleau *et al.*, 2011; Corradi and Bonfante, 2012; Ehinger *et al.*, 2012; Lin *et al.*, 2014). The overall complexity is further increased by genetic exchange events occurring between different AMF via hyphal fusion (anastomosis) (Croll *et al.*, 2009).

A pilot experiment was conducted to evaluate *R. irregularis AOX* gene (*RiAOX*) polymorphism using a clone sequencing approach on six fungal isolates. The first results revealed the presence of all three plausible types of polymorphisms: between different isolates as expected (inter isolate polymorphism), between spores from one isolate (intra isolate polymorphism), and also evidence of intra spore variability. Group-wise comparison of the different isolates highlighted a relatively low level of polymorphism between five isolates, with only six single nucleotide polymorphisms (SNPs), but the addition of a sixth isolate greatly increased the number of polymorphisms with 41 SNPs. In order to fully characterize variability of the *RiAOX* locus and to detect variants down to 0.5–1%, we are currently conducting ultra-deep sequencing using next generation sequencing technology. These data sets will give us a detailed map of nuclear variants within single spores of several *R. irregularis* isolates.

Perspectives

Further research will compare molecular data to phenological observations (spore production, speed and intensity of mycorrhizal colonization, number and activity of arbuscules, number of vesicles and branching absorptive structures) linked with plant behaviour (survival, growth and yield) in order to determine polymorphic motifs affecting symbiosis and therefore to establish functional groups. At the end, such a marker should allow us to identify isolates or a group of isolates which meet the goals of mycorrhizal producers and farmers. As *AOX* from mycorrhizal fungi are currently poorly studied, this new research field will potentially open a promising way to better understand mycorrhizal functionality and its influence on plant performance.

CHAPTER 14.3

Can *AOX* gene diversity mark herbal tea quality? A proposal

Michail Orfanoudakis[1], Evangelia Sinapidou[2] and Birgit Arnholdt-Schmitt[3]
[1]*Department of Forestry and Management of the Environment and Natural Resources, Forest Soil Lab, Democritus University of Thrace, Orestiada, Greece*
[2]*Department of Agricultural Development, Democritus University of Thrace, Orestiada, Greece*
[3]*EU Marie Curie Chair, ICAAM - Instituto de Ciências Agrárias e Ambientais Mediterrânicas, Universidade de Évora, Évora, Portugal*

The modern way of life dictates a higher need for premium quality natural food products. This has forced the market to switch to products often forgotten but widely used in the past or those still used nowadays in more traditional societies. Products like the herbal tea extracted from *Sideritis* L. could often meet such demands, since they contain chemical constituents exhibiting antioxidant properties (Grzegorczyk *et al.*, 2007). *Sideritis* in particular contains a wide range of phenolic acids, flavonoids, terpenoids, vitamins and tannins with significant antioxidant activity (Bouayed *et al.*, 2007) and is often used in traditional medicine as anti-inflammatory, anti-ulcer, cytostatic, antimicrobial, flu vaccine and stimulant circulatory agents. The effectiveness of herbal products is often affected by the amount of antioxidant substances they contain.

AOX involvement

The prologued characteristics of the *Sideritis* extracts contained in herbal products are essential quality indicators and should therefore be taken into account throughout the production line from breeder to end-product. Genetic diversity as a source of divergence in the phenolic composition and antioxidant properties among *Sideritis* species has been reported in studies comparing species and populations endemic to the Mediterranean region, where over 100 *Sideritis* species of natural and perennial plants are widely distributed (Tunalier *et al.*, 2004). Besides interspecies variation, growing conditions such as light intensity, humidity and temperature are important factors of the constituents of the *Sideritis* extract and their qualities. Differences have been measured in three

Alternative Respiratory Pathways in Higher Plants, First Edition.
Edited by Kapuganti Jagadis Gupta, Luis A.J. Mur and Bhagyalakshmi Neelwarne.

types of phenolics, (namely flavones, hydroxycinnamic acids and phenylethanoid glycosides) in two *Sideritis* species (*S. scardica* and *S. raeseri*) both between the different species, and among infusions prepared from cultivated and wild plants (Petreska *et al.*, 2011).

In terms of breeding, molecular approaches have been proven successful in detecting heterogeneity in gene pools deemed to carry too narrow or even negligible genetic variability such as inbred lines (Gethi *et al.*, 2002). A review of molecular biological studies describing the diversity and regulation of secondary metabolism in medicinal plants, at various levels, revealed that DNA markers can be powerful tools for the investigation of such diversity at interspecies level within the same genus and containing related compounds (Yamazaki, 2002). It is notable that the expression of all structural genes examined was regulated by light. Also, intraspecies variability was reported based on environmental conditions (Daws and Jensen, 2011), while new markers were recently developed for the elucidation of intraspecies differentiation based on retrotransposons, which are known to be activated under stress conditions (Hamon *et al.*, 2011).

Herbal tea quality is strongly related with abiotic stress, in particular drought. *AOX* has been proposed to play a key role in the organization of the efficient acclimation of plants to changing environmental conditions (Arnholdt-Schmitt *et al.*, 2006). Although there are only a few studies dealing with the role of *AOX* during drought stress, increased *AOX* activity has been reported (Bartoli *et al.*, 2005; see also Vanlerberghe *et al.*, Chapter 8).

The development of markers that will highlight differences among *Sideritis* species related to quality traits is imperative in breeding improved cultivars via marker assisted selection (MAS). These quality traits can be either a result of interspecies or intraspecies diversity or growing conditions, namely abiotic stress. FMs such as those based on *AOX* related genes are appealing, especially for industry. These genes are associated with carbohydrate turnover rates and phenolic compound production (Shane *et al.*, 2004; Santos Macedo *et al.*, 2012; Sircar *et al.*, 2012). The existence of reliable DNA markers represents an attractive alternative to less stable morphological and chemotaxonomic markers.

Mycorrhizal symbiosis

Apart from the genetic diversity and variation attributable to growing conditions, variety in growth and flavonoid content was also reported through the introduction of AMF colonization of *Sideritis* (Geneva *et al.*, 2010). There are indications suggestive of an interesting relation between AMF and *AOX* in plants (see Mercy *et al.*, Chapter 14.2), especially for *Sideritis* cultivation where the quality of the extracted products is closely connected with the drought stress the plant was subjected to. Thus, plants in AMF symbiosis have been shown to alleviate oxidative damage and have lower lignification under drought conditions

(Lee *et al.*, 2012), whereas H_2O_2 was significantly higher when AMF was not present in the system.

Furthermore, mycorrhizal symbiosis could occasionally act itself as a stress factor for the host, whose growth response varied from positive to negative as suggested in several pot and field experiments. The reasons for such response variation are both poor phosphorous (P) nutrition and high carbon (C) cost to maintain the symbiosis (Smith and Smith, 2011). The stress to plant growth could be driven by the imbalance from the P uptake via the mycorrhizal versus the root pathway, particularly when P availability is limited as is common in dry Mediterranean soils. Additionally, a high surface soil temperature further increases the C cost of AMF symbiosis due to the disturbance of the external mycelia network, while the plant roots are becoming less able to uptake P. Therefore, mycorrhizal stress on plant growth is increased (Facelli *et al.*, 2010) as the fungus should compensate the low P soil availability. Conclusively, the prologued assumption suggests that AMF application could be an effective approach in the management of *Sideritis* agricultural cultivation with potential benefits to the quality of the final product.

Increased salicylic acid levels have been reported in plants during the first stages of contact with AMF, however these levels are reduced at a later stage when the AMF symbiosis is established (Lendzemo *et al.*, 2007). Salicylic acid is also known to interact with the expression level of *AOX* (Vanlerberghe, 2013). Thus, the interaction of AMF, *AOX* expression and secondary metabolism needs to be considered when AMF treatment will form essential part in farmland production.

Future prospects

The *Sideritis* farmland systems aim at high quality production. Breeding efforts and guided agronomic management practice related to AMF inoculum application are promising means to achieve progress. It is noteworthy that *Sideritis* agrosystems are not industrialized agriculture. Therefore, significant native AMF populations will typically be present in the soil. *Sideritis* breeding for high tea quality might be optimized for efficient plant–AMF interaction through considering both native and applied AMF. The development of FMs from *AOX* genes has the potential to take both factors into account and will be explored in future experimentation.

CHAPTER 14.4

AOX in parasitic nematodes: a matter of lifestyle?

Vera Valadas[1], Margarida Espada[2], Tânia Nobre[1], Manuel Mota[2] and Birgit Arnholdt-Schmitt[1]

[1] *EU Marie Curie Chair, ICAAM - Instituto de Ciências Agrárias e Ambientais Mediterrânicas, Universidade de Évora, Évora, Portugal*

[2] *NemaLab-ICAAM, Departamento de Biologia, Universidade de Évora, Évora, Portugal*

Research on *AOX* in nematodes is scarce. Within our research group, we aim to confirm the existence of alternative oxidase in nematodes and explore its involvement in the animal's metabolism related to its lifestyle (including plant attacks). Here we present our hypothesis that the presence of this gene family in nematodes is a function of their lifestyle.Validation of a role of *AOX* in nematode parasitism might be useful in support of a strategy development against nematode attacks through bio-protection.

Nematodes, commonly named as round worms, are represented in almost every ecological environment from marine to soil, from tropical to polar regions, having different lifestyles, hosts and adaptations, from free-living to parasites (Boucher and Lambshead, 1995; Bongers and Bongers, 1998). Most nematodes are free-living; however, more than 16% (4100 from the 25 000 species described) are plant parasitic. These parasites have a major economic importance and impact in agricultural and forestry ecosystems worldwide, causing significant losses in crop productivity every year (Nicol *et al.*, 2011; Perry and Moens, 2011). Over the decades there has been an effort to understand the mechanisms involved in parasitic–plant host interactions. These parasitic nematodes are capable of degrading and breaking plant cell wall, suppressing and modulating plant defence pathways using, for example antioxidant and detoxifying enzymes, and manipulating plant signalling pathways, like cell regulation and hormone signalling (Davis *et al.*, 2008; Haegeman *et al.*, 2012). Some of these genes have been acquired from bacteria and fungi (Danchin *et al.*, 2010; Whiteman and Gloss, 2010). Indeed, nematodes interact with most different organisms ranging from fungi to bacteria, from protozoans to viruses. Interactions between species,

Alternative Respiratory Pathways in Higher Plants, First Edition.
Edited by Kapuganti Jagadis Gupta, Luis A.J. Mur and Bhagyalakshmi Neelwarne.

and in particularly the close relation between host and symbionts, provide an ongoing source of selection allowing the evolution of different features.

AOX gene homologues in animals have been reported for the first time by McDonald and Vanlerberghe (2004) with its identification in the Mollusca and Chordata. Later, six more animal phyla with *AOX* sequences in their genome were identified: Porifera, Placozoa, Cnidaria, Annellida, Achinoderma and Hemichordate, although it is absent from vertebrates (McDonald *et al.*, 2009). *AOX* has been found in all kingdoms of life except the Archaebacteria, where the existence of *AOX* is still under study (Finnegan *et al.*, 2003, 2004; McDonald, 2008).

Some of these findings were based on *in silico* comparative analysis by sequence similarity searches, thanks to the existing (and continuously rising) set of genomic data that is publicly available. Comparing animal *AOX* sequences with other organisms, several amino acid residues in the central region of the protein that are conserved can be seen that include six iron-binding residues, (McDonald and Vanlerberghe, 2006). In animals, these residues have two of the four iron binding motifs suggesting that they almost certainly encode *AOX* proteins (McDonald and Vanlerberghe, 2004). The endosymbiotic theory proposed for the origin of mitochondria also support the origin of *AOX* in eukaryotic linage (McDonald and Vanlerberghe, 2006).

In Nematoda, *AOX* has been reported in a reduced number of species, but to date no work has consistently searched for this gene family. Previous functional studies in the animal parasitic nematodes *Nippostrongylus brasiliensis, Ascaridia galli* and *Ascaris suum* and in the plant parasitic nematode *Xiphinema index* suggested the existence of an alternative respiratory pathway branching from the cytochrome pathway (CP), which is consistent with the presence of *AOX* (Fry *et al.*, 1983; Paget *et al.*, 1987; Molinari and Miacoli, 1995; Kita *et al.*, 1997). Since then, more than 10 genomes of nematode species with different lifestyles have become available and 20 more will be released soon (Martin *et al.*, 2012; Kumar *et al.*, 2012b). However, until now, only two partial sequences of *AOX* genes have been found in two plant parasitic nematodes: the root-knot nematode, *Meloidogyne hapla* and the root-lesion nematode *Pratylenchus vulnus* (McDonald and Vanlerberghe, 2004; McDonald *et al.*, 2009). These two nematodes are, respectively, sedentary and migratory and are established endoparasites that represent economic losses every year to important crops worldwide (Bird and Kaloshian, 2003). As in plants, fungi and protists, nematode *AOX* seems to have a sparse distribution, shown by the presence of *AOX* in *M. hapla* and its absence in *Caenorhabditis elegans* (McDonald and Vanlerberghe, 2006). The *AOX* sequence from *M. hapla* is still incomplete and thus not fully comparable with other organisms. Nevertheless, it shows the expected conserved features, presenting two of the four iron-binding motifs suggesting that they almost certainly encode for *AOX* protein. A unique feature in protein distinguishes animal *AOX* from the ones found on plants or fungi: the absence of an N-terminal cysteine residue (important for enzyme regulation in plants) and the presence of a unique C-terminal region (McDonald *et al.*, 2009).

The reason for the observed sparse distribution of animal *AOX* is unknown, but it is tempting to hypothesize events of horizontal gene transfer (HGT), that is the transmission of genes between organisms by mechanisms other than vertical inheritance from an ancestor to an offspring. HGT is often assumed in plant-parasitic nematodes to explain a series of genes encoding plant cell wall-degrading or wall-modifying enzymes that exhibit a high similarity to bacteria (e.g. Danchin *et al.*, 2010). *AOX* function seems sufficiently important to resist to selective pressure during evolution. However, the alternative hypothesis of a single *AOX* origin by a most recent common ancestor (MRCA) and several events of gene loss, although to us less likely, cannot yet be discharged. The existing sample size is still highly reduced, but so far only parasitic lifestyle nematodes have (putative) *AOX* transcripts in their genome. Animal parasitism as a lifestyle is inferred to have arisen independently at least six times, and plant parasitism three times within Nematode (Dorris *et al.*, 1999; Holterman *et al.*, 2009). Therefore, if *AOX* is confirmed only for parasitic nematodes, than HGT is the most parsimonious origin for this gene. But then, can *AOX* be an advantageous feature in parasitism?

Little information about how to compare animal *AOX* sequence with those from other kingdoms and what implications this may have to enzyme regulation is available. McDonald and co-workers (2004, 2009) suggested a hypothesis for the functional role of animal *AOX*. Alternative oxidase respiration pathway on animals links to stress response, since it seems that *AOX* promote homeostasis and additional flexibility in metabolisms when reactive oxygen species (ROS) production increases. It is known that parasitic nematodes evolved many strategies during evolution that allow them to survive within their host, usually resulting from a co-evolutionary arms-race. Oxidative stress is caused by higher levels of ROS, that once released into the mitochondrial matrix and cytoplasm leads to perturbations on proteostasis, affecting lipids, membranes and cellular components (Sedensky and Morgan 2006; Rodriguez *et al.*, 2013). One can hypothesized that, since *AOX* can modulate the generation of ROS by preventing the over-reduction of the respiratory chain, nematodes in stress conditions can benefit from the presence of *AOX*, because it promotes homeostasis (McDonald and Vanlerberghe, 2004). Additionally, plants produce nitric oxide (NO) and cyanide (CN) in response to pathogens or other environmental factors, while animals just produce NO as defense signalling (McDonald and Vanlerberghe, 2004; Wendehenne *et al.*, 2001). Both CN and NO function as CP inhibitors (McDonald and Vanlerberghe, 2004). Because *AOX* is resistant to CN and NO, then a parasitic nematode with this alternative respiration pathway would be able to maintain respiration even if its CP is disrupted by plant defenses. In *M. hapla* (a parasitic nematode), *AOX* seems to contribute for its virulence (McDonald and

Vanlerberghe, 2004). Further indications of such a mechanism arise from the parasitic nematode *index.* Although *AOX* has not yet been studied in this organism, previous studies from Molinari and Miacola (1995) observe that its respiratory activity was not inhibited by antimycin but was inhibited by *m*-CLAM (m-chlorobenzhydroxamic acid). This suggests the presence of hydroxamic acid-sensitive terminal oxidases, in agreement with an active *AOX* respiration pathway.

The validation of these hypotheses (or their refutation) requires analyzing *AOX,* both molecularly and biochemically, in an as wide as possible nematodes species to encompass different lifestyles and phylogenetic origins. The data available to date only allows speculating on the importance of this gene family for specific nematode lifestyles.

The experimental tractability of several species of nematodes have promoted their use as models in various research areas, being suitable to study a variety of biological and ecological relevant hypotheses. Their applied importance raises the societal importance of studying these organisms. For example, the serious economic problem caused by the pinewood nematode *Bursaphelenchus xylophilus* worldwide and the spread of the disease in Portugal after 2008 (Valadas *et al.*, 2012a, 2012b) makes relevant the search for a bio-protection strategy against this particular nematode. The recent release of *B. xylophilus* genome (Kikuchi *et al.*, 2011) can greatly contribute to the search for genes putatively involved in the parasitic relation, and in the interaction homeostasis. The role of *AOX* in parasitism can be revealed by studying diverse species across nematode phylogeny, search among them for presence/absence variability (PAV) and relate the homologies found with traits relative to lifestyle. Phylogenetic analysis can be then used to find and test potential examples of HGT, which will highlight the possible functional roles of the gene.

Perspectives

Grasping the role of *AOX* in nematodes represents a new approach that can lead to a deeper understanding of their metabolism and survival mechanisms, as well as to insights into lifestyle pathways and their evolutionary consequences. As parasitic nematodes have significant economic impact worldwide, the applied potential of these studies is also foreseen. Knowing the function of *AOX* and its impact on nematode–plant interaction and nematode survival can aid in finding potential mechanisms that allow their control with minimal side effects.

CHAPTER 14.5

Bacterial *AOX*: a provocative lack of interest!

Cláudia Vicente[1], José Hélio Costa[2] and Birgit Arnholdt-Schmitt[3]

[1] *NemaLab, ICAAM - Instituto de Ciências Agrárias e Ambientais Mediterrânicas, Departamento de Biologia, Universidade de Évora, Évora, Portugal*

[2] *Department of Biochemistry and Molecular Biology, Federal University of Ceara, Fortaleza, Ceara, Brazil*

[3] *EU Marie Curie Chair, ICAAM - Instituto de Ciências Agrárias e Ambientais Mediterrânicas, Universidade de Évora, Évora, Portugal*

Complete bacterial genomes are important research tools of the modern genomics era which contribute to the discovery of new genes, interspecies and intraspecies comparative genomics, and also to the study of the evolutionary events behind bacterial adaptation and speciation (Koonin and Wolf, 2008). The release of the *Novosphingobium aromaticivorans* genome in 2003 enabled the discovery of the first *AOX* homologous gene in bacteria (*NaAOX*), which was found to be functionally active in an *E. coli* mutant deficient in terminal oxidase and highly expressed under microaerobic conditions (Stenmark and Nordlund, 2003). The protein sequence of *NaAOX* was shown to be highly similar to *Arabidopsis AOX*1a (nearly 58% identity) suggesting a possible horizontal gene transfer (HGT) event between both organisms (Stenmark and Nordlund, 2003). Another hypothesis proposed the entrance of ancestral prokaryotic *AOX* into eukaryotes via primary endosymbiosis (McDonald *et al.*, 2003; Atteia *et al.*, 2004). To understand the prokaryotic origins for mitochondrial *AOX* and plastid terminal oxidase nuclear (*PTOX*) genes, also referred to as DOX (di-iron carboxylate quinol oxidase), Finnegan *et al.* (2003) aligned prokaryotic DOX sequences with eukaryotic DOX sequences and observed strong phylogenetic affinities and an overall protein structure conservation, emphasizing endosymbiotic events for the origin of both organelles (mitochondria and plastid). Later, McDonald and Vanlerberghe (2005) assessed the presence and diversity of DOX proteins in the marine microbial community of the Sargasso Sea, which is highly dominated by prokaryotes. Sixty-nine different putative *AOX* proteins were identified widely distributed in Eubacteria (and possibly Archaea), and suggested to be functionally involved in the respiratory O_2 consumption in the oligotrophic conditions of the Sargasso Sea. The N-terminal sequence of bacterial *AOX* was also found to be

Alternative Respiratory Pathways in Higher Plants, First Edition.
Edited by Kapuganti Jagadis Gupta, Luis A.J. Mur and Bhagyalakshmi Neelwarne.

phylogenetically related to *AOX* of higher plants (McDonald and Vanlerberghe, 2005). Once more, *AOX* and PTOX were proposed to have originated from a common prokaryotic ancestral DOX protein, which diverged into Alphaproteobacteria and Cyanobacteria, and dispersed via vertical transmission through the eukaryotic domain in a series of endosymbiotic events, further originating mitochondria and chloroplasts (Finnegan *et al.*, 2004; McDonald and Vanlerberghe, 2006). Interestingly, Dunn *et al.* (2010) found that the *AOX*-encoding gene in *Vibrio fischeri* ES114, a marine bacterium living symbiotically in bobtail squid *Euprymna scolopes*, may be required for a lifestyle outside the host and induced in nitric oxide stress conditions.

By taking advantage of the 6198 Bacteria and 249 Archaea genomes presently available from the Doe Joint Genome Institute (https://img.jgi.doe.gov/cgi-bin/w/main.cgi), it was possible to advance the origin and distribution of *AOX* / PTOX. The data analyses revealed that *AOX* and *PTOX* genes are present in genomes of some bacteria and absent in Archaea. In Bacteria, both genes were detected only in Proteobacteria and Cyanobacteria in an analysis that covered 33 bacterial species (Table 14.1). While the majority of the data supported the endosymbiotic theory with the presence of *AOX* in Proteobacteria and PTOX in Cyanobacteria, an *AOX*, instead of a *PTOX*, gene was found in the Cyanobacteria *Mastigocoleus testarum*. This finding raises doubt about the origin of *AOX* and *PTOX*, or alternatively, taking into account that the endosymbiotic theory is true, *AOX* could be used to reclassify this species as a Proteobacteria since the present classification was based on morphological characteristics according to Lagerheim (1886). Interestingly, the *AOX* and *PTOX* genes were detected only in 4.46 (118/2643) and 11.11% (15/135) of the Proteobacteria and Cyanobacteria, respectively (Table 14.1). This finding suggests that the majority of these Bacteria have lost the *AOX* or *PTOX* genes during evolution. In Proteobacteria, the *AOX* deletion appears to have occurred extensively since it was not detected in Deltaproteobacteria, Epsilonproteobacteria or Zetaproteobacteria and it was identified only in 4.34 (28/644), 3.51 (13/370) and 5.97% (77/1288) of Alphaproteobacteria, Betaproteobacteria and Gammaproteobacteria, respectively.

In the next steps of the *AOX/PTOX* research in Bacteria, it is crucial to clarify why only few bacteria possess *AOX/PTOX* as well as to investigate their functional role. Generally, a single *AOX* or *PTOX* gene was found in Proteobacteria or Cyanobacteria, respectively, although some gene duplications were identified: three *AOX* genes in *O. antarcticus* 238, two *AOX* genes in *Brevundimonas* sp. BAL3 and *Thioalkalivibrio versutus* AL2I as well as three PTOX genes in *Acaryochloris marina* MBIC11017 and two *PTOX* genes in *Acaryochloris* sp CCMEE 5410 and *Oscillatoria* sp. PCC 7112. Thus, while most bacteria deleted *AOX/PTOX*, a few of them duplicated them. It will be of great interest to understand this paradigm. Furthermore, only one *AOX* gene identified in the species *O. antarcticus* 238 revealed the presence of introns (two introns). Curiously, this species also has two other *AOX* genes without introns. This finding was evidence for the

Table 14.1 Distribution of *AOX* and *PTOX* genes in bacteria.

Bacteria	No. of genomes/species	No. of species with *AOX* or PTOX
Firmicutes	1473	—
Actinobacteria	772	—
Cyanobacteria	136	15→*PTOX* 1→*AOX*
Proteobacteria	2643	118→*AOX*
unclassified	259	—
Tenericutes	101	—
Acidobacteria	15	—
Bacteroidetes	365	—
candidate division CD12	1	—
Verrucomicrobia	28	—
Synergistetes	14	—
Chloroflexi	22	—
Aquificae	18	—
Armatimonadetes	1	—
Planctomycetes	10	—
Spirochaetes	115	—
Caldiserica	2	—
Deferribacteres	5	—
Elusimicrobia	3	—
Nitrospirae	4	—
Poribacteria	4	—
Chlamydiae	65	—
Chlorobi	12	—
Deinococcus-Thermus	38	—
Chrysiogenetes	1	—
Dictyoglomi	2	—
candidate division EM 3	1	—
Thermotogae	17	—
Fibrobacteres	2	—
Fusobacteria	39	—
Gemmatimonadetes	2	—
Ignavibacteria	2	—
Lentisphaerae	2	—
Thermodesulfobacteria	5	—

The different kinds of bacteria are listed according their phylogenetic proximity.

The data were obtained from Blast search in the bacteria genomes available in Integrated Microbial Genomes database (https://img.jgi.doe.gov/cgi-bin/w/main.cgi).

O. antarcticus 238 species to be explored as an attractive model for studying *AOX* gene structure and sequence variability in relation to functionality for defined growing conditions.

Another line of investigation in this matter could be the study of *AOX* presence in plant-associated bacteria as a FM for the discovery of potential strains with biotechnology applications, such as cellulose-degrading bacteria or plant growth-promoting bacteria. Plant basal defences are induced upon bacterial invasion, regardless of their phenotype (plant pathogen or plant endophyte). These defences primarily consist of the generation of reactive oxygen species (ROS) that can suppress bacterial invaders (Torres, 2010). Successful bacteria should harbour an extremely efficient and powerful antioxidant defence system to cope with this oxidative stress condition. In this sense – and similar to other organisms (plant, fungi) – bacteria harbouring *AOX* could represent a competitive feature that may contribute to their adaptation to extreme and distinct conditions.

Perspectives

The surprising lack of interest in bacteria *AOX* will certainly be fulfilled as researchers begin to unravel the potential of the *AOX* pathway in the lifestyle of these unicellular microorganisms.

General conclusion

Developing FMs from *AOX* genes for plant robustness and efficient phenotype plasticity linked to yield stability or plant propagation is promising.

For future research on *AOX*-gene-based FM development and development of screening tools in molecular pre-breeding, it will be fundamental to consider a new approach which targets tissues and cells that determine the desired traits at whole plant level. Success of FM development from *AOX* genes will crucially depend on the identification and functional validation of polymorphic sequences as candidates through:

- collection of massive data on *AOX* gene diversity in samples from various environments or stress treatments combined with eco-physiological modelling;
- progress in bioinformatics tool development for *AOX* gene polymorphism discovery with integrated consideration of global and local plastic genome organization factors and mechanisms;
- the definition of appropriate 'deep traits' for phenotype screening of *AOX*-polymorphic genotypes;
- the availability of feasible and rapid screening tools for phenotyping of *AOX*-polymorphic genotypes in association studies;
- the availability of genetic maps that can be integrated with the results of *AOX* polymorphisms and phenotyping.

New knowledge on flexible genome organization during development and environment-induced phenotype plasticity is expected to revolutionize our understanding of plants and will challenge future strategies for molecular breeding. Future breeding should also consider novel traits for crop improvement such as efficient plant interaction with endophytes and symbionts. Additional perspectives for identifying general markers for pre-breeding may further arise from progress in virtual plant design and stress response analyses combined with FM simulation studies.

Alternative Respiratory Pathways in Higher Plants, First Edition.
Edited by Kapuganti Jagadis Gupta, Luis A.J. Mur and Bhagyalakshmi Neelwarne.

FM development from *AOX* genes is supposed to narrow the pool of promising genotypes that enters into the breeding process. Finally, success will depend on the robustness of the marker, and testing of individual plants and breeding population characteristics in field trials under various conditions (years, locations) by species-specific conventional breeding strategies.

Acknowledgements

B.AS acknowledges the financial support of the European Commission and the Foundation of Science and Technology (FCT, Portugal), namely FEDER funds through the Operational Program for competitiveness Factors – COMPETE, and national funds through FCT – Foundation for Science and Technology, under the Strategic Projects PEst-C/AGR/UI0115/2011 and PEst-OE/AGR/UI0115/2014, and the projects FCOMP-01-0124-FEDER-027385 (EXCL/AGR-PRO/0038/2012), FCOMP-01-0124-FEDER-041563 (EXPL/AGR-FOR/1324/2013), FCOMP-01-0124-FEDER-014116 (PTDC/AGR-GPL/111196/2009), FCOMP-01-0124-FEDER-009638 (PTDC/EBB-BIO/099268/2008), and FCOMP-01-0124-FEDER-008819 (PTDC/AGR-GPL/099263/2008). B.AS. and H.C. thank FCT for the support given under the program POPH (Ciência 2007 and 2008: C2008-UE/ICAM/06) and to ICAAM for the support given to H.C. (BPD Uevora ICAAM INCENTIVO AGR UI0115). V.V. is supported by a PostDoc fellowship under the project FP7-SME-2012-315464. T.N. is supported by a Marie Curie fellowship (FP7-PEOPLE-2012-CIG Project Reference 321725) and by the Portuguese Foundation for Science and Technology (SFRH/BCC/52187/2013). The author thanks the editors for the invitation and for giving exceptional space for manuscript creation. She is grateful to Tânia Nobre, Jan T. Svensson, Vera Valadas and Isabel Velada for their engagement to improve final manuscript organization.

References

Abe, F., Saito, K., Miura, K. and Toriyama, K. (2002) A single nucleotide polymorphism in the alternative oxidase gene among rice varieties differing in low temperature tolerance. *FEBS Letters* **527**: 181–185.

Achard, P., Herr, A., Baulcombe, D.C. and Harberd, N.P. (2004) Modulation of floral development by a gibberellin-regulated microRNA. *Development* **131**: 3357–3365.

Affourtit, C., Albury, M.S., Krab, K. and Moore, A.L. (1999) Functional expression of the plant alternative oxidase affects growth of the yeast *Schizosaccharomyces pombe*. *Journal of Biological Chemistry* **274**: 6212–6218.

Afuape, S.O., Sayre, R., Tawanda, Z. and Kahya, S.S. (2013) Transgenic cassava lines carrying heterologous alternative oxidase (*AtAOX1a*) showed impaired quantitative and qualitative response to embryogenesis. *African Journal of Biotechnology* **12**: 4303–4309.

Akhter, S., McDade, H.C., Gorlach, J.M. *et al.* (2003) Role of alternative oxidase gene in pathogenesis of *Cryptococcus neoformans*. *Infection and Immunity* **71**: 5794–5802.

Albury, M.S., Affourtit, C. and Moore, A.L. (1998) A highly conserved glutamate residue (Glu270) is essential for plant alternative oxidase activity. *Journal of Biological Biochemistry* **273**: 30301–30305.

Albury, M.S., Dudley, P., Watts, F.Z. and Moore, A.L. (1996) Targeting the plant alternative oxidase protein to *Schizosaccharomyces pombe* mitochondria confers cyanide-insensitive respiration. *Journal of Biological Chemistry* **271**: 17062–17066.

Albury, M.S., Affourtit. C., Crichton, P.G. and Moore, A.L. (2002) Structure of the plant alternative oxidase: Site-directed mutagenesis provides new information on the active site and membrane topology. *Journal of Biological Biochemistry* **277**: 1190–1194.

Albury, M.S., Elliott, C. and Moore, A.L. (2009) Towards a structural elucidation of the alternative oxidase in plants. *Physiologia Plantarum* **137**: 316–327.

Albury, M.S., Elliott, C. and Moore, A.L. (2010) Ubiquinol-binding site in the alternative oxidase: Mutagenesis reveals features important for substrate binding and inhibition. *Biochimica et Biophysica Acta* **1797**: 1933–1939.

Amirsadeghi, S., Robson, C.A., McDonald, A.E. and Vanlerberghe, G.C. (2006) Changes in plant mitochondrial electron transport alter cellular levels of reactive oxygen species and susceptibility to cell death signaling molecules. *Plant Cell Physiology* **47**: 1509–1519.

Andersen, J.R. and Lübberstedt, T. (2003) Functional markers in plants. *Trends in Plant Science* **8**: 554–560.

Alternative Respiratory Pathways in Higher Plants, First Edition.
Edited by Kapuganti Jagadis Gupta, Luis A.J. Mur and Bhagyalakshmi Neelwarne.

Andersson, M.E. and Nordlund, P. (1999) A revised model of the active site of alternative oxidase. *FEBS Letters* **449**: 17–22.

Ansorge, W.J. (2009) Next-generation DNA sequencing techniques. *New Biotechnology* **25**: 195–203.

Aoufouchi, S., Yélamos, J. and Milstein C. (1996) Nonsense mutations inhibit RNA splicing in a cell-free system: Recognition of mutant codon is independent of protein synthesis. *Cell* **85**: 415–422.

Arcuri, H.A., Canduri, F., Pereira, J.H., *et al.* (2004) Molecular models for shikimate pathway enzymes of *Xylella fastidiosa. Biochemical and Biophysical Research Communications* **320**: 979–991.

Arnholdt-Schmitt, B. (1999) On the physiology of yield production in carrots – Implications for breeding towards nutrient efficiency.*Gartenbauwissenschaft* **64**: 26–32.

Arnholdt-Schmitt, B. (2004) Stress-induced cell reprogramming. A role for global genome regulation? *Plant Physiology* **136**: 2579–2586.

Arnholdt-Schmitt, B. (2005) Efficient cell reprogramming as a target for functional-marker strategies? Towards new perspectives in applied plant-nutrition research. *Journal of Plant Nutrition and Soil Science* **168**: 617–624.

Arnholdt-Schmitt, B. (2008) A novel gene-candidate approach of socio-economic interest? – Breeding on efficient plant genotype – mycorrhiza interaction. COST 870 Meeting 'From production to application of arbuscular mycorrhizal fungi in agricultural systems: a multidisciplinary approach', Denmark, May 27–30, 2008, pp.47–50.

Arnholdt-Schmitt, B. (2009) Alternative oxidase (*AOX*) and stress tolerance – approaching a scientific hypothesis. *Physiologia Plantarum* **137**: 314–315.

Arnholdt-Schmitt, B., Cardoso, H., Campos, M.D. *et al.* (2009) Carrot AOX – a source for potential functional marker candidates to assist breeding on multi-stress tolerance. Invited Talk at the 33rd Carrot Conference, University of California, USA, January 18–21, 2009.

Arnholdt-Schmitt, B., Costa, J.H. and De Melo, D.F. (2006) *AOX* – a functional marker for efficient cell reprogramming under stress? *Trends in Plant Science* **11**: 281–287.

Aroca, R., Porcel, R. and Ruiz-Lozano, J.M. (2007) How does arbuscular mycorrhizal symbiosis regulate root hydraulic properties and plasma membrane aquaporins in *Phaseolus vulgarisunder* drought, cold or salinity stresses? *New Phytologist* **173**: 808–816.

Aroca, R., Ruiz-Lozano, J.M., Zamarreno, A.M. *et al.* (2013) Arbuscular mycorrhizal symbiosis influences strigolactone production under salinity and alleviates salt stress in lettuce plants. *Journal of Plant Physiology* **170**: 47–55.

Atteia, A., Van Lis, R., Van Hellemond, J.J. *et al.* (2004) Identification of prokaryotic homologues indicates an endosymbiotic origin for the alternative oxidases of mitochondria (*AOX*) and chloroplasts (PTOX). *Gene* **330**: 143–148.

Aung, K., Lin, S.-I., Wu, C.-C. *et al.* (2006) pho2, a phosphate overaccumulator, is caused by a nonsense mutation in a microRNA399 target gene. *Plant Physiology* **141**: 1000–1011.

Baek, J.-M., Han, P., Iandolino, A. and Cook, D.R. (2008) Characterization and comparison of intron structure and alternative splicing between *Medicago truncatula, Populus trichocarpa, Arabidopsis* and rice. *Plant Molecular Biology* **67**: 499–510.

Baldi, P. and Brunak, S. (2001) Bioinformatics: The machine learning approach. In *Adaptive Computation and Machine Learning, vol. 2*, pp. 1–476. MIT Press, Cambridge, USA.

Barrett, L.W., Fletcher, S. and Wilton, S.D. (2013) Untranslated gene regions and other non-coding elements. In *Regulation of Eukaryotic Gene Expression* (Barrett, L.W. and Fletcher, S., eds), pp. 1–56. Springer, Basel.

Bartlett, J.G., Snape, J.W. and Harwood, W.A. (2009) Intron-mediated enhancement as a method for increasing transgene expression levels in barley. *Plant Biotechnology Journal* **7**: 856–866.

Bartoli, C.G., Gomez, F., Gergoff, G. *et al.* (2005) Up-regulation of the mitochondrial alternative oxidase pathway enhances photosynthetic electron transport under drought conditions. *Journal of Experimental Botany* **56**: 1269–1276.

Barton, N.H. and Keightley, P.D. (2002) Understanding quantitative genetic variation. *Nature Reviews Genetics* **3**: 11–21.

Bedini, S., Pellegrino, E., Avio, L. *et al.* (2009) Changes in soil aggregation and glomalin-related soil protein content as affected by the arbuscular mycorrhizal fungal species *Glomus mosseae* and *Glomus intraradices*. *Soil Biology and Biochemistry* **41**: 1491–1496.

Bendall, D.S. and Bonner, W.D. (1971) Cyanide-insensitive respiration in plant mitochondria. *Plant Physiology* **47**: 236–245.

Bennett, R.N. and Wallsgrove, R.M. (1994) Secondary metabolites in plant defence mechanisms. *New Phytologist* **127**: 617–633.

Berthold, D.A., Andersson, M.E. and Nordlund, P. (2000) New insight into the structure and function of the alternative oxidase. *Biochimica et Biophysica Acta* **1460**: 241–254.

Besserer, A., Becard, G., Jauneau, A. *et al.* (2008) GR24, a synthetic analog of strigolactones, stimulates the mitosis and growth of the arbuscular mycorrhizal fungus *Gigaspora rosea* by boosting its energy metabolism. *Plant Physiology* **148**: 402–413.

Besserer, A., Becard, G., Roux, C. and Sejalon-Delmas, N. (2009) Role of mitochondria in the response of arbuscular mycorrhizal fungi to strigolactones. *Plant Signaling Behaviour* **4**: 75–77.

Besserer, A., Puech-Pages, V., Kiefer, P. *et al.* (2006) Strigolactones stimulate arbuscular mycorrhizal fungi by activating mitochondria. *PLoS Biology* **4**: e226. DOI: 10.1371/journal.pbio.0040226.

Bhangale, T.R., Stephens, M. and Nickerson, D.A. (2006) Automating resequencing-based detection of insertion-deletion polymorphisms. *Nature Genetics* **38**: 1457–1462.

Bird, D. McK. and Kaloshian, I. (2003) Are roots special? Nematodes have their say. *Physiological and Molecular Plant Pathology* **62**: 115–123.

Bongers, T. and Bongers, M. (1998) Functional diversity of nematodes. *Applied Soil Ecology* **10**: 239–251.

Borecky, J., Nogueira, F.T.S., De Oliveira, K.A.P. *et al.* (2006) The plant energy-dissipating mitochondrial systems: Depicting the genomic structure and the expression profiles of the gene families of uncoupling protein and alternative oxidase in monocots and dicots. *Journal of Experimental Botany* **57**: 849–864.

Borsani, O., Valpuesta, V. and Botella, M.A. (2001) Evidence for a role of salicylic acid in the oxidative damage generated by NaCl and osmotic stress in *Arabidopsis* seedlings. *Plant Physiology* **126**: 1024–1030.

Bouayed, J., Piri, K., Rammal, H. *et al.* (2007) Comparative evaluation of the antioxidant potential of some Iranian medicinal plants. *Food Chemistry* **104**: 364–368.

Boucher, G. and Lambshead, P.J.D. (1995) Ecological biodiversity of marine nematodes in samples from temperate, tropical, and deep-sea regions. *Conservation Biology* **9**: 1594–604.

Bourdon, V., Harvey, A. and Lonsdale, D.M. (2001) Introns and their positions affect the translational activity of mRNA in plant cells. *EMBO Reports* **2**: 394–398.

Boyko, A., Blevins, T., Yao, Y. *et al.* (2010) Transgenerational adaptation of *Arabidopsis* to stress requires DNA methylation and the function of Dicer-like proteins. *PLoS One* **5** (3): e9514. DOI: 10.1371/journal.pone.0009514.

Boyko, A., Kathiria, P., Zemp, F.J. *et al.* (2007) Transgenerational changes in the genome stability and methylation in pathogen-infected plants: (virus-induced plant genome instability). *Nucleic Acids Research* **35**: 1714–1725.

Boyko, A. and Kovalchuk, I. (2011) Genome instability and epigenetic modification—heritable responses to environmental stress? *Current Opinion in Plant Biology* **14**: 260–266.

Brenner, E.A., Beavis, W.D., Andersen, J.R. and Luebberstedt, T. (2013) Prospects and limitations for development and application of functional markers in plants. In *Diagnostics in Plant Breeding* (Lübberstedt, T. and Varshney, R.K., eds), vol. 16: pp. 329–346. Dordrecht, Springer Netherlands.

Breviario, D., Gianì, S., Ponzoni, E. *et al.* (2008) Plant tubulin intronics. *Cell Biology International* **32**: 571–573.

Bromberg, Y. and Rost B. (2007) SNAP: Predict effect of non-synonymous polymorphisms on function. *Nucleic Acids Research* **35**: 3823–3835.

Browning, S.R. and Browning, B.L. (2011) Haplotype phasing: Existing methods and new developments. *Nature Reviews Genetics* **12**: 703–714.

Bureau, T.E. and Wessler, S.R. (1994) Stowaway: A new family of inverted repeat elements associated with the genes of both monocotyledonous and dicotyledonous plants. *The Plant Cell* **6**: 907–916.

Byers, P.H. (2002) Killing the messenger: New insights into nonsense-mediated mRNA decay. *Journal of Clinical Investigation* **109**: 3–6.

Campos, M.D., Cardoso, H.G., Linke, B. *et al.* (2009) Differential expression and co-regulation of carrot *AOX* genes (*Daucus carota*). *Physiologia Plantarum* **137**: 578–591.

Cao, J., Schneeberger, K., Ossowski, S. *et al.* (2011) Whole-genome sequencing of multiple *Arabidopsis thaliana* populations. *Nature Genetics* **43**: 956–963.

Cardoso, H.G. and Arnholdt-Schmitt, A. (2013) Functional marker development across species in selected traits. In *Diagnostics in Plant Breeding* (Lübberstedt, T. and Varshney, R.K., eds) vol. 21: pp. 467–515. Dordrecht, Springer Netherlands.

Cardoso, H., Campos, M.D., Costa, A.R. *et al.* (2009) Carrot alternative oxidase gene *AOX2a* demonstrates allelic and genotypic polymorphisms in intron 3. *Physiologia Plantarum* **137**: 592–608.

Cardoso, H., Campos, M.D., Nothnagel, T. and Arnholdt-Schmitt, B. (2011) Polymorphisms in intron 1 of carrot *AOX2b* – a useful tool to develop a functional marker? *Plant Genetic Resources* **9**: 177–180.

Carlini, D.B., Chen, Y. and Stephan, W. (2001) The relationship between third-codon position nucleotide content, codon bias, mRNA secondary structure and gene expression in the drosophilid alcohol dehydrogenase genes *Adh and Adhr*. *Genetics* **159**: 623–633.

Cavalcanti, J.H.F., Oliveira, G.M., da Cruz Saraiva, K.D. *et al.* (2013) Identification of duplicated and stress-inducible *AOX2b* gene co-expressed with *Aox1* in species of the *Medicago genus* reveals a regulation linked to gene rearrangement in leguminous genomes. *Journal of Plant Physiology*. DOI: org/10.1016/j.jplph.(2013)06.012.

Cenik, C., Chua, H.N., Zhang, H. *et al.* (2011) Genome analysis reveals interplay between 5′UTR introns and nuclear mRNA export for secretory and mitochondrial genes. *PLoS Genetics* **7**: 14.

Chapman, E.J. and Carrington, J.C. (2007) Specialization and evolution of endogenous small RNA pathways. *Nature Reviews Genetics* **8**: 884–896.

Chen, D., Meng, Y., Yuan, C. *et al.* (2011) Plant siRNAs from introns mediate DNA methylation of host genes. *RNA* **17**: 1012–1024.

Chen, Z.J. and Ni, Z. (2006) Mechanisms of genomic rearrangements and gene expression changes in plant polyploids. *Bioessays* **28**: 240–252.

Chinnusamy, V. and Zhu, J.K. (2009) Epigenetic regulation of stress responses in plants. *Current Opinion in Plant Biology* **12**: 133–139.

Chiou, T.-J., Aung, K., Lin, S.-I. *et al.* (2006) Regulation of phosphate homeostasis by microRNA in *Arabidopsis*. *The Plant Cell* **18**: 412–421.

Choi, H., Hong, J., Ha, J. *et al.* (2000) ABFs, a family of ABA-responsive element binding factors. *Journal of Biological Chemistry* **275**: 1723–1730.

Chung, B.Y., Simons, C., Firth, A.E. *et al.* (2006) Effect of 5′UTR introns on gene expression in *Arabidopsis thaliana*. *BMC Genomics* **7**: 120.

Clark, R.M., Schweikert, G., Toomajian, C. *et al.* (2007) Common sequence polymorphisms shaping genetic diversity in *Arabidopsis thaliana*. *Science* **317**: 338–342.

Clifton, R., Lister, R., Parker, K.L. *et al.* (2005) Stress-induced co-expression of alternative respiratory chain components in *Arabidopsis thaliana*. *Plant Molecular Biology* **58**: 193–212.

Clifton, R., Millar, A.H. and Whelan, J. (2006) Alternative oxidases in *Arabidopsis*: A comparative analysis of differential expression in the gene family provides new insights into function of non-phosphorylating bypasses. *Biochimica et Biophysica Acta* **1757**: 730–741.

Collins, N.C, Tardieu, F. and Tuberosa, R. (2008) Quantitative trait loci and crop performance under abiotic stress: Where do we stand? *Plant Physiology* **147**: 469–486.

Considine, M.J., Daley, D.O. and Whelan, J. (2001) The expression of alternative oxidase and uncoupling protein during fruit ripening in mango. *Plant Physiology* **126**: 1619–1629.

Considine, M.J., Holtzapffel, R.C., Day, D.A. *et al.* (2002) Molecular distinction between alternative oxidase from monocots and dicots. *Plant Physiology* **129**: 949–953.

Conti, E. and Izaurralde, E. (2005) Nonsense-mediated mRNA decay: Molecular insights and mechanistic variations across species. *Current Opinion in Cell Biology* **17**: 316–325.

Corradi, N. and Bonfante, P. (2012) The arbuscular mycorrhizal symbiosis: Origin and evolution of a beneficial plant infection. *Plos Pathogens* **8** (4): e1002600.

Corradi, N. and Sanders, I.R. (2006) Evolution of the P-type II ATPase gene family in the fungi and presence of structural genomic changes among isolates of *Glomus intraradices. BMC Evolutionary Biology* **6**: 21.

Corradi, N., Ruffner, B., Croll, D. *et al.* (2009) High-level molecular diversity of copper-zinc superoxide dismutase genes among and within species of arbuscular mycorrhizal fungi. *Applied Environmental Microbiology* **75**: 1970–1978.

Costa, J.H., Cardoso, H.G., Campos, M.D. *et al.* (2009a) *Daucus carota* L – an old model for cell reprogramming gains new importance through a novel expansion pattern of alternative oxidase (*AOX*) genes. *Plant Physiology and Biochemistry* **47**: 753–759.

Costa, J.H., de Melo, D.F., Gouvei, Z. *et al.* (2009b) The alternative oxidase family of *Vitis vinifera* reveals an attractive model to study the importance of genomic design. *Physiologia Plantarum* **137**: 553–565.

Costa, J.H., Mota, E.F., Cambursano, M.V. *et al.* (2010) Stress-induced co-expression of two alternative oxidase (Vu*AOX*1 and 2b) genes in *Vigna unguiculata. Journal of Plant Physiology* **167**: 561–570.

Costa, J.M., Hasenfratz-Sauder, M.P., Pham-Thi, A.T. *et al.* (2004) Identification in *Vigna unguiculata* (L.) Walp. of two cDNAs encoding mitochondrial alternative oxidase orthologous to soybean alternative oxidase genes 2a and 2b. *Plant Science* **167**: 233–239.

Crichton, P.G., Affourtit, C., Albury, M.S. *et al.* (2005) Constitutive activity of *Sauromatum guttatum* alternative oxidase in *Schizosaccharomyces pombe* implicates residues in addition to conserved cysteines in α-keto acid activation. *FEBS Letters* **579**: 331–336.

Crichton, P.G., Albury, M.S., Affourtit, C. and Moore, A.L. (2010) Mutagenesis of the *Sauromatum guttatum* alternative oxidase reveals features important for oxygen binding and catalysis. *Biochimica et Biophysica Acta* **1797**: 732–737.

Cristianini, N. and Shawe-Taylor, J. (2000) *An Introduction to Support Vector Machines and Other Kernel-based Learning Methods*, Cambridge University Press, Cambridge, UK.

Croll, D., Giovannetti, M., Koch, A.M. *et al.* (2009) Nonself vegetative fusion and genetic exchange in the arbuscular mycorrhizal fungus *Glomus* intraradices. *New Phytologist* **181**: 924–937.

Cvetkovska, M. and Vanlerberghe, G.C. (2013) Alternative oxidase impacts the plant response to biotic stress by influencing the mitochondrial generation of reactive oxygen species. *Plant Cell Environment* **36**: 721–32.

Danchin, E.G.J., Rosso, M-N., Vieira, P. *et al.* (2010) Multiple lateral gene transfers and duplications have promoted plant parasitism ability in nematodes. *Proceedings of the National Academy of Sciences of the USA* **107** (41): 17651–17656.

Davis, E.L., Hussey, R.S., Mitchum, M.G. and Baum, T.J. (2008) Parasitism proteins in nematode–plant interactions. *Current Opinion in Plant Biology* **11**: 360–366.

Daws, M.I. and Jensen, M. (2011) Effects of developmental heat sum on fruit traits of clonal lines of *Quercus petraea* grown under controlled conditions. *Plant Growth Regulation* **64**: 203–206.

Day, D.A. and Wiskich, J.T. (1995) Regulation of alternative oxidase activity in higher-plants. *Journal of Bioenergetics and Biomembranes* **27**: 379–385.

DeBolt, S. (2010) Copy number variation shapes genome diversity in *Arabidopsis* over immediate family generational scales. *Genome Biology and Evolution* **2**: 441–453.

Den Dunnen, J.T. and Antonarakis, E. (2001) Nomenclature for the description of human sequence variations. *Human Genetics* **109**: 121–124.

Díaz, A., Zikhali, M., Turner ,A.S. *et al.* (2012) Copy number variation affecting the photoperiod B1 and vernalization-A1 genes is associated with altered flowering time in wheat (*Triticum aestivum*). *PLoS ONE* **7**: e33234. DOI: 10.1371/journal.pone.0033234.

Dinant, M., Baurain, D., Coosemans, N. *et al.* (2001) Characterization of two genes encoding the mitochondrial alternative oxidase in *Chlamydomonas reinhardtii. Current Genetics* **39**: 101–108.

Djajanegara, I., Holtzapffel, R., Finnegan P.M. *et al.* (1999) A single amino acid change in the plant alternative oxidase alters the specificity of organic acid activation. *FEBS Letters* **454**: 220–224.

Dorris, M., De Ley, P. and Blaxter, M.L. (1999) Molecular analysis of nematode diversity and the evolution of parasitism. *Parasitology Today* **15** (5): 188–193.

Doss, C.G.P. and Sethumadhavan, R. (2009) Investigation on the role of nsSNPs in HNPCC genes – a bioinformatics approach. *Journal of Biomedical Science* DOI: 10.1186/1423-0127-16-42.

Duester, G., Jörnvall, H. and Hatfield, G.W. (1986) Intron-dependent evolution of the nucleotide-binding domains within alcohol dehydrogenase and related enzymes. *Nucleic Acids Research* **14**: 1931–1941.

Dunn, A.K., Karr, E.A., Wang, Y. *et al.* (2010) The alternative oxidase (*AOX*) gene in *Vibrio fischeri* is controlled by NsrR and upregulated in response to nitric oxide. *Molecular Microbiology* **77**: 44–55.

Ehinger, M.O., Croll, D., Koch, A.M. and Sanders, I.R. (2012) Significant genetic and phenotypic changes arising from clonal growth of a single spore of an arbuscular mycorrhizal fungus over multiple generations. *The New Phytologist* **196**: 853–861.

Einset, J., Winge, P. and Bones, A. (2007) ROS signaling pathways in chilling stress. *Plant Signaling Behaviour* **2**: 365–367.

Ensembl Plants [website] http://plants.ensembl.org/Multi/Search/New?db=core (accessed 7 November 2014).

Facelli, E., Smith, S.E., Facelli, J.M. *et al.* (2010) Underground friends or enemies: Model plants help to unravel direct and indirect effects of arbuscular mycorrhizal fungi on plant competition. *The New Phytologist* **185**: 1050–1061.

Ferreira, A., Cardoso, H., Macedo, E.S. *et al.* (2009) Intron polymorphism pattern in *AOX1b* of wild St John's Wort (*Hypericum perforatum L*) allows discrimination between individual plants. *Physiologia Plantarum* **137**: 520–531.

Finkelstein, R.R., Wang, M.L., Lynch, T.J. *et al.* (1998) The *Arabidopsis* abscisic acid response locus ABI4 encodes an APETALA 2 domain protein. *The Plant Cell* **10**: 1043–1054.

Finnegan, P.M., Soole, K.L. and Umbach, A.L. (2004) Alternative mitochondrial electron transport proteins in higher plants. In *Plant Mitochondria: From Gene to Function* (Day, D.A., Millar, A.H. and Whelan, J., eds), *Advances in Photosynthesis and Respiration,* vol. 17: pp. 163–230. Kluwer, Dordrecht, The Netherlands

Finnegan, P.M., Umbach, A.L. and Wilce, J.A. (2003) Prokaryotic origins for the mitochondrial alternative oxidase and plastid terminal oxidase nuclear genes. *FEBS Letters* **555**: 425–430.

Fiorani, F., Umbach, A.L. and Siedow, J.N. (2005) The alternative oxidase of plant mitochondria is involved in the acclimation of shoot growth at low temperature. A study of *Arabidopsis AOX1a* transgenic plants. *Plant Physiology* **139**: 1795–1805.

Flynt, A.S. and Lai, E.C. (2008) Biological principles of microRNA-mediated regulation: shared themes amid diversity. *Nature Reviews Genetics* **9**: 831–842.

Frederico, A.M., Campo, M.D., Cardoso, H.G. *et al.* (2009a) Alternative oxidase involvement in *Daucus carota* somatic embryogenesis. *Physiologia Plantarum* **137**: 498–508.

Frederico, A.M., Zavattieri, M.A., Campos, M.D. *et al.* (2009b) The gymnosperm *Pinus pinea* contains both *AOX* gene subfamilies, *AOX1* and *AOX2*. *Physiologia Plantarum* **137**: 566–577.

Fry, M., Bazil, C. and Jenkins, D.C. (1983) A comparison of mitochondrial electron transport in the intestinal parasiticnematodes *Nippostrongylus brasiliensis and Ascaridia galli. Comparative Biochemistry and Physiology* **75B**: 451–460.

Furbank, R.T. and Tester, M. (2011) Phenomics – technologies to relieve the phenotyping bottleneck. *Trends in Plant Science* **16** (12): 635–644.

Gamper, H.A., Van Der Heijden, M.G.A. and Kowalchuk, G.A. (2010) Molecular trait indicators: moving beyond phylogeny in arbuscular mycorrhizal ecology. *New Phytologist* **185**: 67–82.

García, I., Rosas, T., Bejarano, E.R. *et al.* (2013) Transient transcriptional regulation of the CYS-C1 gene and cyanide accumulation upon pathogen infection in the plant immune response. *Plant Physiology* **4**: 2015–2027.

Genetic Information Research Institute (GIRI) [website] www.girinst.org/censor/ (accessed 7 November 2014).

Geneva, M., Hristozkova, M., Yonova, P. *et al.* (2010) Effect of endomycorrhizal colonization with *Glomus intraradices* on growth and antioxidant capacity of *Sideritis scardica* Griseb. *GAPP* **36**: 47–54.

Gethi, J.G., Labate, J.A., Lamkey, K.R. *et al.* (2002) SSR Variation in important U.S. maize inbred lines. *Crop Science* **42**: 951–957.

Gilliland, A., Singh, D.P., Hayward, J.M. *et al.* (2003) Genetic modification of alternative respiration has differential effects on antimycin A-induced versus salicylic acid-induced resistance to tobacco mosaic virus. *Plant Physiology* **132**: 1518–1528.

Giraud, E., Ho, L.H.M., Clifton, R. *et al.* (2008) The absence of alternative oxidase1a in *Arabidopsis* results in acute sensitivity to combined light and drought stress. *Plant Physiology* **147**: 595–610.

Giraud, E., Van Aken, O., Ho L.H. and Whelan J. (2009) The transcription factor ABI4 is a regulator of mitochondrial retrograde expression of Alternative oxidase 1a. *Plant Physiology* **150**: 1–33.

Glomus intraradices DB [website] http://mycor.nancy.inra.fr/IMGC/GlomusGenome/index3.html (accessed 7 November 2014).

Goddard, D.R and Smith, P.E. (1938) Respiratoty block in the dormant spores of *Neurospora tetrasperma. Plant Physiology* **13**: 241–264.

Gordon, D., Abajian, C. and Green, P. (1998) Consed: A graphical tool for sequence finishing. *Genome Research* **8**: 195–202.

Goto, Y., Yue, L., Yokoi, A. *et al.* (2001) A novel single-nucleotide polymorphism in the 3′-untranslated region of the human dihydrofolate reductase gene with enhanced expression. *Clinical Cancer Research* **7**: 1952–1956.

Graça, A., Lynce, I., Marques-Silva, J. and Oliveira, A.L. (2010) Efficient and accurate haplotype inference by combining parsimony and pedigree information. In *Algebraic and Numeric Biology, Lecture notes in Computer Science* (Horimoto, K., Nakatsui, M. and Popov, N. eds), vol. 6479: pp. 38–56. Springer.

Grahl, N., Dinamarco,T.M., Willger, S.D. *et al.* (2012) *Aspergillus fumigatus* mitochondrial electron transport chain mediates oxidative stress homeostasis, hypoxia responses and fungal pathogenesis. *Molecular Microbiology* **84**: 383–399.

Grant, N., Onda, Y., Kakizaki, Y. *et al.* (2009) Two cys or not two cys? That is the question; alternative oxidase in the thermogenic plant sacred lotus. *Plant Physiology* **150**: 987–995.

Grigoriev, I.V., Nordberg, H., Shabalov, I. *et al.* (2012) The genome portal of the Department of Energy Joint Genome Institute. *Nucleic Acids Research* **40**: D26–D32.

Grillo, G., Turi, A., Licciulli, F. *et al.* (2010) UTRdb and UTRsite (RELEASE 2010): A collection of sequences and regulatory motifs of the untranslated regions of eukaryotic mRNAs. *Nucleic Acids Research* **38**: D75–D80.

Grzegorczyk, I., Matkowski, A. and Wysokinska, H. (2007) Antioxidant activity of extracts from *in vitro* cultures of *Salvia officinalis* L. *Food Chemistry* **104**: 536–541.

Haddrill, P.R., Charlesworth, B., Halligan, D.L. and Andolfatto, P. (2005) Patterns of intron sequence evolution in *Drosophila* are dependent upon length and GC content. *Genome Biology* **6**: R67.

Haegeman, A., Mantelin, S., Jones, J.T. and Gheysen, G. (2012) Functional roles of effectors of plant-parasitic nematodes. *Gene* **492**: 19–31.

Hamon, P., Duroy, P.-O., Dubreuil-Tranchant, C. *et al.* (2011) Two novel Ty1-copia retrotransposons isolated from coffee trees can effectively reveal evolutionary relationships in the *Coffea* genus (Rubiaceae). *Molecular Genetics and Genomics* **285**: 447–460.

Hansen, L.D., Criddle, R.S. and Smith, B.N. (2005) Calorespirometry in plant biology. In *Plant Respiration: From Cell to Ecosystem* (Lambers, H. and Ribas-Carbo, M., eds), *Advances in Photosynthesis and Respiration, vol. 18*: pp. 17–30. Springer Netherlands.

Hansen, L.D., Thomas, N.R. and Arnholdt-Schmitt, B. (2009) Temperature responses of substrate carbon conversion efficiencies and growth rates of plant tissues. *Physiologia Plantarum* **137**: 446–458.

Harish, M.C., Dachinamoorthy, P., Balamurugan, S. *et al.* (2013) Enhancement of α-tocopherol content through transgenic and cell suspension culture systems in tobacco. *Acta Physiologiae Plantarum* **4**: 1121–1130.

Hasegawa, P.M., Bressan, R.A., Zhu, J.-K. and Bohnert, H.J. (2000) Plant cellular and molecular responses to high salinity. *Annual Review of Plant Physiology and Plant Molecular Biology* **51**: 463–499.

Henderson, I.R. and Jacobsen, S.E. (2007) Epigenetic inheritance in plants. *Nature* **447**: 418–424.

Herrera-Medina, M.J., Steinkellner, S., Vierheilig, H. *et al.* (2007) Abscisic acid determines arbuscule development and functionality in the tomato arbuscular mycorrhiza. *New Phytologist* **175**: 554–564.

Hijri, M. and Sanders, I.R. (2005) Low gene copy number shows that arbuscular mycorrhizal fungi inherit genetically different nuclei. *Nature* **433**: 160–163.

Ho, L.H.M., Giraud, E., Lister, R. *et al.* (2007) Characterization of the regulatory and expression context of an alternative oxidase gene provides insights into cyanide-insensitive respiration during growth and development. *Plant Physiology* **143**: 1519–1533.

Ho, L.H.M., Giruad, E., Uggalla, V. *et al.* (2008) Identification of regulatory pathways controlling gene expression of stress responsive mitochondrial components in *Arabidopsis*. *Plant Physiology* **147**: 1858–1873.

Holterman, M., Karssen, G., van den Elsen, S. *et al.* (2009) Small subunit rDNA-based phylogeny of the Tylenchida sheds light on relationships among some high-impact plant-parasitic nematodes and the evolution of plant feeding. *Phytopathology* **99**: 227–235.

Holtzapffel, R.C., Castelli, J., Finnegan, P.M. *et al.* (2003) A tomato alternative oxidase protein with altered regulatory properties. *Biochimica et Biophysica Acta* **1606**: 153–162.

Huang, W., Li, L., Myers, J.R. and Marth, G.T. (2012) ART: A next-generation sequencing read simulator. *Bioinformatics* **28**: 593–594.

Huh, W.K. and Kang, S.-O. (1999) Molecular cloning and functional expression of alternative oxidase from *Candida albicans. Journal of Bacteriology* **181**: 4098–4102.

Huh, W.K. and Kang, S.O. (2001) Characterization of the gene family encoding alternative oxidase from *Candida albicans. The Biochemical Journal* **356**: 595–604.

Hull, J., Campino, S., Rowlands, K. *et al.* (2007) Identification of common genetic variation that modulates alternative splicing. *PLoS Genetics* **3**: 10.

Ijdo, M., Cranenbrouck, S. and Declerck, S. (2011) Methods for large-scale production of AM fungi: Past, present, and future. *Mycorrhiza* **21**: 1–16.

IPK Barley Blast Server [website] http://webblast.ipk-gatersleben.de/barley (accessed November 2014).

Ito, Y., Saisho, D., Nakazono, M. *et al.* (1997) Transcript levels of tandem-arranged alternative oxidase genes in rice are increased by low temperature. *Gene* **203**: 121–129.

Jaillon, O., Bouhouche, K., Gout, J.-F. *et al.* (2008) Translational control of intron splicing in eukaryotes. *Nature* **451**: 359–362.

Jeffries, P., Gianinazzi, S. and Perotto, S. (2003) The contribution of arbuscular mycorrhizal fungi in sustainable maintenance of plant health and soil fertility. *Biology and Fertility of Soils* **37**: 1–16.

Jiang, R., Yang, H., Zhou, L. *et al.* (2007) Sequence-based prioritization of nonsynonymous single-nucleotide polymorphisms for the study of disease mutations. *The American Journal of Human Genetics* **81**: 346–360.

Joint Genome Institute [website] https://img.jgi.doe.gov/cgi-bin/w/main.cgi (accessed 7 November 2014).

Karimi, A., Khodaverdiloo, H., Sepehri, M. and Sadaghiani ,M. R. (2011) Arbuscular mycorrhizal fungi and heavy metal contaminated soils. *African Journal of Microbiology Research* **5**: 1571–1576.

Kasschau, K.D., Fahlgren, N., Chapman, E.J. *et al.* (2007) Genome-wide profiling and analysis of *Arabidopsis* siRNAs. *PLoS Biology* **5**: e57. DOI: 10.1371/journal.pbio.0050057.

Kathiria, P., Sidler, C., Golubov, A. *et al.* (2010) Tobacco mosaic virus infection results in an increase in recombination frequency and resistance to viral, bacterial, and fungal pathogens in the progeny of infected tobacco plants. *Plant Physiology* **153**: 1859–1870.

Kawase, T., Akatsuka, Y., Torikai, H. *et al.* (2007) Alternative splicing due to an intronic SNP in HMSD generates a novel minor histocompatibility antigen. *Blood* **110**: 1055–1063.

Kikuchi, T., Cotton, J.A., Dalzell, J.J. *et al.* (2011) Genomic insights into the origin of parasitism in the emerging plant pathogen *Bursaphelenchus xylophilus. PLoS Pathogens* **7**: e1002219. DOI: 10.1371/journal.ppat.1002219.

Kimchi-Sarfaty, C., Oh. J.M., Kim, I.-W. *et al.* (2007) A 'silent' polymorphism in the MDR1 gene changes substrate specificity. *Science* **315**: 525–528.

Kita, K., Hirawake, H. and Takamiya, S. (1997) Cytochomes in the respiratory chain of helminth mitochondria.*International Journal ofParasitology* **27**: 617–630.

Kohany, O., Gentles, A.J., Hankus, L. and Jurka, J. (2006) Annotation, submission and screening of repetitive elements in Repbase: RepbaseSubmitter and Censor. *BMC Bioinformatics* **7**: 474.

Kohler, A., Tisserant, E., Martin, F. *et al.* (2011) The transcriptome of the arbuscular mycorrhizal fungus Glomus intraradices. GEO DataSets, Accession: GSE29866. [website] http://www.ncbi.nlm.nih.gov/geo/query/acc.cgi?acc=GSE29866 (accessed 7 November 2014).

Komar, A.A. (2007) SNPs, silent but not invisible. *Science* **315**: 466–467.

Koonin, E.V, and Wolf, Y.I. (2008) Genomics of bacteria and archaea: the emerging dynamic view of the prokaryotic world. *Nucleic Acids Research* **36**: 6688–6719.

Kouzarides, T. (2007) Chromatin modifications and their function. *Cell* **128**: 693–705.

Kovalchuk, I., Kovalchuk, O., Kalck, V. *et al.* (2003) Pathogen-induced systemic plant signal triggers DNA rearrangements. *Nature* **423**: 760–762.

Kuersten, S. and Goodwin, E.B. (2003) The power of the 3′ UTR: Translational control and development. *Nature Reviews Genetics* **4**: 626–637.

Kumar, P., Henikoff, S. and Ng, P.C. (2009) Predicting the effects of coding non-synonymous variants on protein function using the SIFT algorithm. *Nature Protocols* **4**: 1073–1081.

Kumar, S.R., Sivalingam, A., Selvaraj, D. *et al.* (2012a) Isolation and characterization of cold inducible genes in carrot by suppression subtractive hybridization. *Biologia Plantarum* **57**: 97–104.

Kumar, S., Schiffer ,P.H. and Blaxter, M. (2012b) 959 Nematode genomes: A semantic wiki for coordinating sequencing projects. *Nucleic Acids Research* **40**: D1295–D1300.

Lacomme, C. and Roby, D. (1999) Identification of new early markers of the hypersensitive response in *Arabidopsis thaliana*. *FEBS letters* **459**: 149–53.

Lagerheim, G. (1886) Note sur le *Mastigocoleus*, nouveau genre des algues marines de l'ordre des Phycochromacées. *Notarisia* **1**: 65–69.

Lambert, J.-C., Luedecking-Zimmer, E., Merrot, S. *et al.* (2003) Association of 3′-UTR polymorphisms of the oxidised LDL receptor 1 (OLR1) gene with Alzheimer's disease. *Journal of Medical Genetics* **40**: 424–430.

Law, J.A. and Jacobsen, S.E. (2010) Establishing, maintaining and modifying DNA methylation patterns in plants and animals. *Nature Reviews Genetics* **11**: 204–220.

Lee, B.-R., Muneer, S., Jung, W.-J. *et al.* (2012) Mycorrhizal colonization alleviates drought-induced oxidative damage and lignification in the leaves of drought-stressed perennial ryegrass (*Lolium perenne*). *Physiologia Plantarum* **145**: 440–449.

Lendzemo, V.W., Kuyper, T.W., Matusova, R. *et al.* (2007) Colonization by arbuscular mycorrhizal fungi of sorghum leads to reduced germination and subsequent attachment and emergence of *Striga hermonthica*. *Plant Signaling & Behaviour* **2**: 58–62.

Levenshtein, V.I. (1966) Binary codes capable of correcting deletions, insertions, and reversals. *Soviet Physics Doklady* **10**: 707–710.

Li, S.-C., Shiau, C.-K. and Lin, W. (2008) Vir-Mir db: Prediction of viral microRNA candidate hairpins. *Nucleic Acids Research* **36**: D184–D189.

Lin, K., Limpens, E., Zhang, Z. *et al.* (2014) Single nucleus genome sequencing reveals high similarity among nuclei of an endomycorrhizal fungus. *PLoS Genetics* DOI: 10.1371/journal.pgen.1004078.

Liu, H.Q., Che, Y.M., Zhao, F.G. *et al.* (2012) Hydrogen peroxide participates symbiosis between AM fungi and tobacco plants. *Acta Ecologica Sinica* **32**: 6085–6091.

Liu, Y., Wu, R., Wan, Q. *et al.* (2007) Glucose-6-phosphate dehydrogenase plays a pivotal role in nitric oxide-involved defense against oxidative stress under salt stress in red kidney bean roots. *Plant Cell Physiology* **48**: 511–522.

Louro, R., Nakaya, H.I., Amaral, P.P. *et al.* (2007) Androgen responsive intronic non-coding RNAs. *BMC Biology* **5**: 4.

Lübberstedt, T. and Varshney, R.K. (eds) (2013) *Diagnostics in Plant Breeding* pp. 521. Dordrecht, Springer Netherlands, doi: 10.1007/978-94-007-5687-8.

Lucht, J.M., Mauch-Mani, B., Steiner H.-Y. *et al.* (2002) Pathogen stress increase somatic recombination frequency in *Arabidopsis*. *Nature Genetics* **30**: 311–314.

Lynch, T., Erikson. B.J. and Finkelstein, R.R. (2012) Direct interactions of ABA-insensitive (ABI) clade protein phosphatase (PP)2Cs with calcium-dependent protein kinases and ABA response element-binding bZIPs may contribute to turning off ABA response. *Plant Molecular Biology* **80**: 647–658.

Ma, H., Song, C., Borth, W. *et al.* (2011) Modified expression of alternative oxidase in transgenic tomato and petunia affects the level of tomato spotted wilt virus resistance. *BMC Biotechnology* **11**: 96–102.

Maas, C., Laufs, J., Grant ,S. *et al.* (1991) The combination of a novel stimulatory element in the first exon of the maize Shrunken-1 gene with the following intron 1 enhances reporter gene expression up to 1000-fold. *Plant Molecular Biology* **16**: 199–207.

Macko-Podgorni, A., Nowicka, A., Grzebelus, E. *et al.* (2013) DcSto: Carrot Stowaway-like elements are abundant, diverse, and polymorphic. *Genetica* **141**: 255–267.

Mallory, A.C., Reinhart, B.J., Jones-Rhoades, M.W. *et al.* (2004) MicroRNA control of PHABULOSA in leaf development: importance of pairing to the microRNA 5′ region. *The European Molecular Biology Organization Journal* **23**: 3356–3364.

Malusá, E., Sas-Paszt, L. and Ciesielska, J. (2012) Technologies for beneficial microorganisms inocula used as biofertilizers. *The Scientific World Journal* **2012**: 491206 (http: //dx.doi.org/10.1100/2012/491206).

Mao, L., Wood, T.C., Yu, Y. *et al.* (2000) Rice transposable elements: A survey of 73,000 sequence-tagged-connectors. *Genome Research* **10**: 982–990.

Marleau, J., Dalpe, Y., St-Arnaud, M. and Hijri, M. (2011) Spore development and nuclear inheritance in arbuscular mycorrhizal fungi. *BMC Evolutionary Biology* **11**: 91.

Marth, G.T., Korf, I., Yandell, M.D. *et al.* 1999. A general approach to single-nucleotide polymorphism discovery. *Nature Genetics* **23**: 452–456.

Martin, F., Gianinazzi-Pearson, V., Hijri, M. *et al.*(2008) The long hard road to a completed *Glomus intraradices* genome. *New Phytology* **180**: 747–750.

Martin, J., Abubucker, S., Heizer, E. *et al.* (2012) Nematode.net update 2011: addition of data sets and tools featuring next-generation sequencing data. *Nucleic Acids* **40** (D1): D720–D728.

Martin-Rodrigez, J.A., Leon-Morcillo, R., Vierheilig, H. *et al.* (2010) Mycorrhization of the notabilis and sitiens tomato mutants in relation to abscisic acid and ethylene content. *Journal of Plant Physiology* **167**: 606–613.

Martín-Rodríguez, J., León-Morcillo, R., Vierheilig H. *et al.* (2011) Ethylene-dependent/ethylene independent ABA regulation of tomato plants colonized by arbuscular mycorrhiza fungi. *New Phytologist* **190**: 193–205.

Mattick, J.S. and Gagen, M.J. (2001) The evolution of controlled multitasked gene networks: The role of introns and other noncoding RNAs in the development of complex organisms. *Molecular Biology and Evolution* **18**: 1611–1630.

Matukumalli, L.K., Grefenstette, J.J., Hyten, D.L. *et al.* (2006) Application of machine learning in SNP discovery. *BMC Bioinformatics* **7**: 4.

Maxwell, D.P., Wang, Y. and McIntosh, L. (1999) The alternative oxidase lowers mitochondrial reactive oxygen production in plant cells. *Proceedings of the National Academy of Sciences of the USA* **96**: 8271–8276.

McDonald, A.E. (2008) Alternative oxidase: An inter-kingdom perspective on the function and regulation of this broadly distributed 'cyanide resistant' terminal oxidase. *Functional Plant Biology* **35**: 535–552.

McDonald, A.E. (2009) Alternative oxidase: What information can protein sequence comparisons give us? *Physiologia Plantarum* **137**: 328–341.

McDonald, A.E. and Vanlerberghe, G.C. (2004) Branched mitochondrial electron transport in the Animalia: Presence of alternative oxidase in several animal phyla. *IUBMB Life* **56**: 333–341.

McDonald, A.E. and Vanlerberghe, G.C. (2005) Alternative oxidase and plastoquinol terminal oxidase in marine prokaryotes of the Sargasso Sea. *Gene* **349**: 15–24.

McDonald, A.E. and Vanlerberghe, G.C. (2006) Origins, evolutionary history, and taxonomic distribution of alternative oxidase and plastoquinol terminal oxidase. *Comparative Biochemistry and Physiology-Part D: Genomics and Proteomics* **1**: 357–364.

McDonald, A.E., Amirsadeghi, S. and Vanlerberghe, G.C. (2003) Prokaryotic orthologues of mitochondrial alternative oxidase and plastid terminal oxidase. *Plant Molecular Biology* **53**: 865–876.

McDonald, A.E., Vanlerberghe, G.C. and Staples, J.F. (2009) Alternative oxidase in animals: unique characteristics and taxonomic distribution. *The Journal of Experimental Biology* **212**: 2627–2634.

Mézard, C., Vignard, J., Drouaud, J. and Mercier, R. (2007) The road to crossovers: Plants have their say. *Trends in Genetics* **23**: 91–99.

Mhadhbi, H., Fotopoulos, V., Mylona, P.V. *et al.* (2013) Alternative oxidase 1 (*AOX*1) gene expression in roots of *Medicago truncatula* is a genotype-specific component of salt stress tolerance. *Journal of Plant Physiology* **170**: 111–114.

Mignone, F., Gissi, C., Liuni, S. and Pesole, G. (2002) Untranslated regions of mRNAs. *Genome Biology* **3** (3): reviews0004.1–reviews0004.10.

Millar, A.H., Whelan, J., Soole, K.L. and Day, D.A. (2011) Organization and regulation of mitochondrial respiration in plants. *Annual Review of Plant Biology* **62**: 79–104.

Mittler, R., Vanderauwera ,S., Gollery, M. and Van Breusegem, F. (2004) Reactive oxygen gene network of plants. *Trends in Plant Science* **9**: 490–498.

Molinari, S. and Miacola, C. (1995) Respiration and terminal oxidases of *Xyphimena index. Nematologia Mediterranea* **23**: 359–365.

Molinier, J., Ries, G., Zipfel, C. and Hohn, B. (2006) Transgenerational memory of stress in plants. *Nature* **442**: 1046–1049.

Moore, A.L. and Albury, M.S. (2008) Further insights into the structure of the alternative oxidase: From plants to parasites. *Biochemical Society Transactions* **36**: 1022–1026.

Moore, A.L., Carré, J.E., Affourtit, C. *et al.* (2008) Compelling EPR evidence that the alternative oxidase is a diiron carboxylate protein. *Biochimica et Biophysica Acta* **1777**: 327–330.

Moore, A.L., Shiba, T., Young, L. *et al.* (2013) Unraveling the heater: New insights into the structure of the alternative oxidase. *Annual Review of Plant Biology* **64**: 637–63.

Morello, L., Gianì, S., Troina, F. and Breviario, D. (2011) Testing the IMEter on rice introns and other aspects of intron-mediated enhancement of gene expression. *Journal of Experimental Botany* **62**: 533–544.

Morin, E., Martin, F. and Uroz, S. MycorWeb. *Glomus* intraradices DB. [website] mastered by the Ecogenomics of Interactions Lab, INRA Nancy, Champenoux. http://mycor.nancy.inra.fr/IMGC/GlomusGenome/index3.html (accessed 7 November 2014).

Murakami, Y. and Toriyama, K. (2008) Enhanced high temperature tolerance in transgenic rice seedlings with elevated levels of alternative oxidase, *OsAOX1a. Plant Biotechnology* **25**: 361–364.

Nakamura, K., Sakamoto, K., Kido, Y. *et al.* (2005) Mutational analysis of the *Trypanosoma vivax* alternative oxidase: the E(X)6Y motif is conserved in both mitochondrial alternative oxidase and plastid terminal oxidase and is indispensable for enzyme activity. *Biochemical and Biophysical Research Communications* **334**: 593–600.

Nakaya, H.I., Amaral, P.P., Louro, R. *et al.* (2007) Genome mapping and expression analyses of human intronic noncoding RNAs reveal tissue-specific patterns and enrichment in genes related to regulation of transcription. *Genome Biology* **8**: R43.

Neale, D.B. and Kremer, A. (2011) Forest tree genomics: growing resources and applications. *Nature Reviews Genetics* **12**: 111–122.

Neale, D.B. and Savolainen O. (2004) Association genetics of complex traits in conifers. *Trends in Plant Science* **9**: 325–330.

Needleman, S.B. and Wunsch, C.D. 1970. A general method applicable to the search for similarities in the amino acid sequence of two proteins. *Journal of Molecular Biology* **48**: 443–453.

Neimanis, K., Staples, J.F., Hüner, N.P. and McDonald, A.E. (2013) Identification, expression, and taxonomic distribution of alternative oxidases in non-angiosperm plants. *Gene* **526**: 275–286.

Ner-Gaon, H., Leviatan, N., Rubin, E. and Fluhr, R. (2007) Comparative cross-species alternative splicing in plants. *Plant Physiology* **144**: 1632–1641.

Ng, S., Giraud, E., Duncan, O. *et al.* (2013) Cyclin-dependent kinase E1 (CDKE1) provides a cellular switch in plants between growth and stress responses. *Journal of Biological Chemistry* **288**: 3449–59.

Ngamphiw, C., Kulawonganunchai, S., Assawamakin, A. *et al.* (2008) VarDetect: A nucleotide sequence variation exploratory tool. *BMC Bioinformatics* **9** (Suppl12): S9, DOI: 10.1186/1471-2105-9-S12-S9.

Nickerson, D.A., Tobe, V.O. and Taylor, S.L. (1997) PolyPhred: Automating the detection and genotyping of single nucleotide substitutions using fluorescence-based resequencing. *Nucleic Acids Research* **25**: 2745–2751.

Nicol, J.M., Turner, S.J.,Coyne, D.L. *et al.* (2011) Current nematode threats to world agriculture. In *Genomics and Molecular Genetics of Plant–Nematode Interactions*, (Jones, J.T., Gheysen, G. and Fenoll, C.. eds), vol. 2: pp. 21–44. Heidelberg, Springer.

Nogales, A, Muñoz-Sanhueza, L., Hansen, L.D. and Arnholdt-Schmitt, B. (2013) Calorespirometry: a new tool for phenotyping carrot genotypes for cold tolerance and functional marker development. *Abstract Book of the 36th Carrot Conference*, pp. 12–13, Madison, Wisconsin, USA.

Nogales, A., Muñoz-Sanhueza, L., Hansen, L.D. and Arnholdt-Schmitt, B. (2013) Calorespirometry as a tool for studying temperature response in carrot (*Daucus carota* L.). *Engineering in Life Sciences* **13** (6): 541–548.

Nogueira, F.T.S., Sassaki, F.T. and Maia, I.G. (2011) *Arabidopsis thaliana* uncoupling proteins (AtUCPs): Insights into gene expression during development and stress response and epigenetic regulation. *Journal of Bioenergetics and Biomembranes* **43**: 71–79.

Noh, S.-J., Lee, K., Paik, H. and Hur, C.-G. (2006) TISA: tissue-specific alternative splicing in human and mouse genes. *DNA Research* **13**: 229–243.

Novelli, G., Borgiani, P., Mango, R. *et al.* (2007) Further evidence that polymorphisms of the OLR1 gene are associated with susceptibility to coronary artery disease and myocardial infarction. *Nutrition, Metabolism and Cardiovascular Diseases* **17**: 7–8.

Olsson, U., Sirijovski, N. and Hansson, M. (2004) Characterization of eight barley xantha-f mutants deficient in magnesium chelatase. *Plant Physiology and Biochemistry* **42**: 557–564.

Ordog, S.H., Higgins, V.J. and Vanlerberghe, G.C. (2002) Mitochondrial alternative oxidase is not a critical component of plant viral resistance but may play a role in the hypersensitive response. *Plant Physiology* **129**: 1858–1865.

Ouyang, S. and Buell, C.R. (2004) The TIGR Plant Repeat Databases: A collective resource for the identification of repetitive sequences in plants. *Nucleic Acids Research* **32**: D360–D363.

Paget, T.A., Fry, M. and Lloyd, D. (1987) Hydrogen peroxide production in uncoupled mitochondria of the parasitic nematode worm *Nippostrongylus brasiliensis*. *Biochemical Journal* **243**: 589–595.

Parsons, H.L., Yip, J.Y.H. and Vanlerberghe, G.C. (1999) Increased respiratory restriction during phosphate-limited growth in transgenic tobacco cells lacking alternative oxidase. *Plant Physiology* **121**: 1309–1320.

Pasqualini, S., Paolocci, F., Borgogni, A. *et al.* (2007) The overexpression of an alternative oxidase gene triggers ozone sensitivity in tobacco plants. *Plant Cell Environment* **30**: 1545–1556.

Pasternak, T.P., Prinsen, E., Ayaydin, F. *et al.* (2002) The role of auxin, pH, and stress in the activation of embryogenic cell division in leaf protoplast-derived cells of alfalfa. *Plant Physiology* **129**: 1807–1819.

Perry, R.N. and Moens, M. (2011) Introduction to plant-parasitic nematodes; modes of parasitism. In *Genomics and Molecular Genetics of Plant–Nematode Interactions* (Jones, J.T., Gheysen, L. and Fenoll, C., eds), vol. 1: pp. 3–20.Heidelberg, Springer.

Petreska, J., Stefova, M., Ferreres, F. *et al.* (2011) Dietary burden of phenolics per serving of 'Mountain Tea' (*Sideritis*) from Macedonia and correlation to antioxidant activity. *NPC* **6**: 1305–1314.

Pfluger, J. and Wagner, D. (2007) Histone modifications and dynamic regulation of genome accessibility in plants. *Current Opinion in Plant Biology* **10**: 645–652.

Plant Repeat Database [website] http://plantrepeats.plantbiology.msu.edu/search.html (accessed 7 November 2014).

Plaxton, W.C. and Podestá, F. (2006) The functional organization and control of plant respiration. *Critical Reviews in Plant Sciences* **25**: 159–198.

Polidoros, A.N., Mylona, P.V, Pasentsis, K. *et al.* (2005) The maize alternative oxidase 1a (*AOX*1a) gene is regulated by signals related to oxidative stress. *Redox Report: Communications in Free Radical Research* **10**: 71–78.

Polidoros, A.N., Mylona, P.V. and Arnholdt-Schmitt, B. (2009) *AOX* gene structure, transcript variation and expression in plants. *Physiologia Plantarum* **137**: 342–353.

Porras-Soriano, A., Soriano-Martín ,M.L., Porras-Piedra, A. and Azcón, R. (2009) Arbuscular mycorrhizal fungi increased growth, nutrient uptake and tolerance to salinity in olive trees undernursery conditions. *Journal of Plant Physiology* **166** (13): 1350–1359.

Pozo, M.J., Verhage, A., Garcıa-Andrade, J. *et al.* (2009) Priming plant defence against pathogens by arbuscular mycorrhizal fungi. In *Mycorrhizas – Functional Processes and Ecological Impact* (Azcon-Aguilar, C., Barea, J.M., Gianinazzi ,S. and Gianinazzi-Pearson, V., eds), vol. 9: pp. 123–136. Berlin, Springer Verlag .

Purvis, A.C. and Shewfelt, R.L. (1993) Does the alternative pathway ameliorate chilling injury in sensitive plant tissues? *Physiologia Plantarum* **88**: 712–718.

Quarles, W. (1999) Plant disease biocontrol and VAM fungi. *IPM Practitioner* **21**: 1–9.

Quinlan, J.R. (1993) *C4.5: Programs for Machine Learning.* In Morgan Kaufmann (Eds) Kluwer Academic Publishers, *San Mateo, California* **1**: 235–240.

Rambaldi, D. and Ciccarelli, F.D. (2009) FancyGene: Dynamic visualization of gene structures and protein domain architectures on genomic loci. *Bioinformatics* **25**: 2281–2282.

Rana, T.M. (2007) Illuminating the silence: understanding the structure and function of small RNAs. *Nature Reviews Molecular Cell Biology* **8**: 23–36.

Rasmusson, A.G., Fernie, A.R. and Van Dongen, J.T. (2009) Alternative oxidase: A defence against metabolic fluctuations? *Physiologia Plantarum* **137**: 371–382.

Rhoades, M.W., Reinhart, B.J., Lim, L.P. *et al.* (2002) Prediction of plant microRNA targets. *Cell* **110**: 513–20.

Rhoads, D.M. and McIntosh, L. (1991) Isolation and characterization of a cDNA clone encoding an alternative oxidase protein of *Sauromatum guttatum* (Schott). *Proceedings of the National Academy of Sciences of the USA* **88**: 2122–2126.

Rhoads, D.M., Umbach, A.L., Sweet, C.R. *et al.* (1998) Regulation of the cyanide-resistant alternative oxidase of plant mitochondria – identification of the cysteine residue involved in alpha-keto acid stimulation and intersubunit disulfide bond formation. *Journal of Biological Chemistry* **273**: 30750–30756.

Rillig, M.C. and Steinberg, P.D. (2002) Glomalin production by an arbuscular mycorrhizal fungus: a mechanism of habitat modification? *Soil Biology and Biochemistry* **34**: 1371–1374.

Robson, C.A. and Vanlerberghe, G.C. (2002) Transgenic plant cells lacking mitochondrial alternative oxidase have increased susceptibility to mitochondria-dependent and -independent pathways of programmed cell death. *Plant Physiology* **129**: 1908–1920.

Rodriguez, M., Snoek, L.B., De Bono, M. and Kammenga, J.E. (2013) Worms under stress: *C. elegans* stress response and its relevance to complex human disease and aging. *Trends in Genetics* **29**: 367–74.

Rook, F., Hadingham, S.A., Li, Y. and Bevan, M.W. (2006) Sugar and ABA response pathways and the control of gene expression. *Plant Cell Environment* **29**: 426–434.

Rose, A.B. (2002) Requirements for intron-mediated enhancement of gene expression in *Arabidopsis*. *RNA New York NY* **8**: 1444–1453.

Rose, A.B. (2008) Intron-mediated regulation of gene expression. *Current Topics in Microbiology and Immunology* **326**: 277–290.

Rose, A.B., Elfersi, T., Parra, G. and Korf, I. (2008) Promoter-proximal introns in *Arabidopsis thaliana* are enriched in dispersed signals that elevate gene expression. *The Plant Cell* **20**: 543–551.

Ruiz, O.H., Gonzalez, A., Almeida, A.J. *et al.* (2011) Alternative oxidase mediates pathogen resistance in *Paracoccidioides brasiliensis* infection. *PLoS Neglected Tropical Diseases* **5** (10): e1353. DOI: 10.1371/journal.pntd.0001353.

Saisho, D., Nakazono, M., Lee, K.H. *et al.* (2001a) The gene for alternative oxidase-2 (*AOX2*) from *Arabidopsis thaliana* consists of five exons unlike other *AOX* genes and is transcribed at an early stage during germination. *Genes & Genetic Systems* **76**: 89–97.

Saisho, D., Nakazono, M., Tsutsumi, N. and Hirai, A. (2001b) ATP synthesis inhibitors as well as respiratory inhibitors increase steady-state level of alternative oxidase mRNA in *Arabidopsis thaliana. Journal of Plant Physiology* **158**: 241–245.

Saisho, D., Nambara, E., Naito, S. *et al.* (1997) Characterization of the gene family for alternative oxidase from *Arabidopsis thaliana. Plant Molecular Biology* **35**: 585–596.

Sakajo, S., Minagawa, N., Komiyama, T. and Yoshimoto, A. (1991) Molecular cloning of cDNA for antimycin A-inducible mRNA and its role in cyanide-resistant respiration in *Hansenula anomala. Biochimica et Biophysica Acta* **1090**: 102–108.

SALK Epigenomic Diversity [website] http://neomorph.salk.edu/1001_epigenomes.html (accessed 7 November 2014).

Sanger, F., Nicklen, S. and Coulson, A.R. (1977) DNA sequencing with chain-terminating inhibitors. *Proceedings of the National Academy of Sciences of the USA* **74**: 5463–5467.

Santos Macedo, E., Cardoso, H.G., Hernández, A. *et al.* (2009) Physiologic responses and gene diversity indicate olive alternative oxidase as a potential source for markers involved in efficient adventitious root induction. *Physiologia Plantarum* **137**: 532–552.

Santos Macedo, E.S., Sircar, D., Cardoso, H.G. *et al.* (2012) Involvement of alternative oxidase (*AOX*) in adventitious rooting of *Olea europaea* L. microshoots is linked to adaptive phenylpropanoid and lignin metabolism. *Plant Cell Reports* **31**: 1581–1590.

Sattler, S.E., Saathoff, A.J., Haas, E.J. *et al.* (2009) A nonsense mutation in a cinnamyl alcohol dehydrogenase gene is responsible for the *Sorghum* brown midrib6 phenotype. *Plant Physiology* **150**: 584–595.

Schmidt, R.J., Schultz, M.D., Urich, M.A. *et al.* (2013) Patterns of population epigenomic diversity.*Nature* **495**: 193–198.

Sedensky, M.M. and Morgan, P.G. (2006) Mitochondrial respiration and reactive oxygen species in *C. elegans.Experimental Gerontology* **41**: 957–967.

Seli, E., Yaba, A., Guzeloglu-Kayisli, O. and Lalioti, M.D. (2008) Alternative splicing of the mouse embryonic poly (A) binding protein (Epab) mRNA is regulated by an exonic splicing enhancer: a model for post transcriptional control of gene expression in the oocyte. *Molecular Human Reproduction* **14**: 393–398.

Shabalina, S.A. and Spiridonov, N.A. (2004) The mammalian transcriptome and the function of non-coding DNA sequences. *Genome Biology* **5**: 105.

Shane, M.W., Cramer, M.D., Funayama-Noguchi, S. *et al.* (2004) Developmental physiology of cluster-root carboxylate synthesis and exudation in harsh hakea. Expression of phosphoenolpyruvate carboxylase and the alternative oxidase. *Plant Physiology* **135**: 549–560.

Shiba, T., Kido, Y., Sakamoto, K. *et al.* (2013) Structure of the trypanosome cyanide-insensitive alternative oxidase. *Proceedings of the National Academy of Sciences of USA* **110**: 4580–4585.

Siedow, J.N., Umbach, A.L. and Moore, A.L. (1995) The active site of the cyanide-resistant oxidase from plant mitochondria contains a binuclear iron center. *FEBS Letters* **362**: 10–14.

Sieger, S.M., Kristensen, B.K., Robson, C.A. *et al.* (2005) The role of alternative oxidase in modulating carbon use efficiency and growth during macronutrient stress in tobacco cells. *Journal of Experimental Botany* **56**: 1499–1515.

Simons, B.H. and Lambers, H. (1999) The alternative oxidase: is it a respiratory pathway allowing a plant to cope with stress? In *Plant Responses to Environmental Stress: From Phytohormones to Gene Reorganization* (Lerner, H.R., Simons, B.H. and Lambers, H., eds), pp. 265–286. Marcel Dekker Inc., New York, USA.

Sircar, D., Cardoso, H.G., Mukherjee, C. *et al.* (2012) Alternative oxidase (*AOX*) and phenolic metabolism in methyl jasmonate-treated hairy root cultures of *Daucus carota* L. *Journal of Plant Physiology* **169**: 657–63.

Smith, C.A., Melino, V.J., Sweetman, C. and Soole, K.L. (2009) Manipulation of alternative oxidase can influence salt tolerance in *Arabidopsis thaliana. Physiologia Plantarum* **137**: 459–472.

Smith, S.E. and Read, D.J. (2008) *Mycorrhizal Symbiosis*, 3rd edn. Academic Press, London, ISBN-10: 0123705266.

Smith, S.E. and Smith, F.A. (2011) Roles of arbuscular mycorrhizas in plant nutrition and growth: New paradigms from cellular to ecosystem scales. *Annual Review of Plant Biology* **62**: 227–250.

Smith, T.F. and Waterman, M.S. (1981) Identification of common molecular subsequences. *Journal of Molecular Biology* **147**: 195–197.

Stenmark, P. and Nordlund, P. (2003) A prokaryotic alternative oxidase present in the bacterium *Novosphingobium aromaticivorans. FEBS Letters* **552**: 189–192.

Steward, F.C., Caplin, S.M. and Millar, F.K.(1952) Investigations on growth and metabolism of plant cells. I. New techniques for the investigation of metabolism, nutrition and growth in undifferentiated cells. *Annals of Botany* **16**: 57–77.

Strodtkötter, I., Padmasree, K., Dinakar, C. *et al.* (2009) Induction of the *AOX1D* isoform of alternative oxidase in *A. thaliana* T-DNA insertion lines lacking isoform *AOX1A* is insufficient to optimize photosynthesis when treated with antimycin A. *Molecular Plant* **2**: 284–297.

Sugie, A., Naydenov, N., Mizuno, N. *et al.* (2006) Overexpression of wheat alternative oxidase gene *Waox1a* alters respiration capacity and response to reactive oxygen species under low temperature in transgenic *Arabidopsis. Genes Genetic Systems* **81**: 349–354.

Suzuki, N., Koussevitzky, S., Mittler, R. and Miller, G. (2011) ROS and redox signalling in the response of plants to abiotic stress. *Plant Cell Environment* **35**: 259–70.

Takata, M., Kiyohara, A., Takasu, A. *et al.* (2007) Rice transposable elements are characterized by various methylation environments in the genome. *BMC Genomics* **8**: 469.

Takumi, S., Tomioka, M., Eto, K. *et al.* (2002) Characterization of two non-homoeologous nuclear genes encoding mitochondrial alternative oxidase in common wheat. *Genes genetic systems* **77**: 81–88.

Tamasloukht M, Séjalon-Delmas N, Kluever, A. *et al.* (2003) Root factors induce mitochondrial related gene expression and fungal respiration during the developmental switch from asymbiosis to presymbiosis in the arbuscular mycorrhizal fungus *Gigaspora rosea. Plant Physiology* **131**: 1468–1478.

Tamura, K., Peterson, D., Peterson, N. *et al.* (2011) MEGA5: Molecular Evolutionary Genetics Analysis using Maximum Likelihood, Evolutionary Distance, and Maximum Parsimony Methods. *Molecular Biology and Evolution* **28**: 2731–2739.

The Arabidopsis Genome Iniative (2000) Analysis of the genome sequence of the flowering plant *Arabidopsis thaliana. Nature* **408**: 796–815.

Thirkettle-Watts, D., McCabe, T.C., Clifton, R. *et al.* (2003) Analysis of the alternative oxidase promoters from soybean. *Plant Physiology* **133**: 1158–1169.

Thomas, H. and Sadras, V.O. (2001) The capture and gratuitous disposal of resources by plants. *Functional Ecology* **15**: 3–12.

Thomazella, D.P., Teixeira, P.J., Oliveira, H.C. *et al.* (2012) The hemibiotrophic cacao pathogen *Moniliophthora perniciosa* depends on a mitochondrial alternative oxidase for biotrophic development. *New Phytologist* **194**: 1025–1034.

TIGR Plant Repeat Databases [website] http://plantrepeats.plantbiology.msu.edu/search.html (accessed 7 November 2014).

Torres, M.A. (2010) ROS in biotic interactions. *Physiologia Plantarum* **138**: 414–429.

Tunalier, Z., Kosar, M., Ozturk, N. *et al.* (2004) Antioxidant properties and phenolic composition of *Sideritis* species. *Chemistry of Natural Compounds* **40** (3): 206–210.

Tuteja, J.H., Clough, S.J., Chan, W.C. and Vodkin, L.O. (2004) Tissue-specific gene silencing mediated by a naturally occurring chalcone synthase gene cluster in *Glycine max*. *Plant Cell* **16**: 819–835.

Umbach, A.L., Fiorani, F. and Siedow, J.N. (2005) Characterization of transformed *Arabidopsis* with altered alternative oxidase levels and analysis of effects on reactive oxygen species in tissue. *Plant Physiology* **39**: 1806–1820.

Umbach, A.L., Gonzalez-Meler, M.A., Sweet, C.R. and Siedow, J.N. (2002) Activation of the plant mitochondrial alternative oxidase: insights from site-directed mutagenesis. *Biochimica et Biophysica Acta* **1554**: 118–128.

Umbach, A.L., Ng, V.S. and Siedow, J.N. (2006) Regulation of plant alternative oxidase activity: a tale of two cysteines. *Biochimica et Biophysica Acta* **1757**: 135–42.

Umbach, A.L. and Siedow, J.N. (2000) The cyanide-resistant alternative oxidases from the fungi *Pichia stipitis* and *Neurospora crassa* are monomeric and lack regulatory features of the plant enzyme. *Archives of Biochemistry and Biophysics* **378**: 234–245.

Uribe, D. and Khachatourians, G.G. (2008) Identification and characterization of an alternative oxidase in the entomopathogenic fungus *Metarhizium anisopliae*. *Canadian Journal of Microbiology* **54**: 119–127.

Valadas, V., Laranjo, M., Mota, M. and Oliveira, S. (2012a) Molecular characterization of Portuguese populations of *Bursaphelenchus xylophilus* using cytochrome b and cellulase genes. *Journal of Helminthology* **16**: 1–10.

Valadas, V., Laranjo, M., Barbosa, P. *et al.* (2012b) The pinewood nematode, *Bursaphelenchus xylophilus,* in Portugal: Possible introductions and spread routes of a serious biological invasion revealed by molecular methods. *Nematology* **14**: 899–911.

Valentine, C.R. (1998) The association of nonsense codons with exon skipping. *Mutation Research* **411**: 87–117.

Van Aken, O., Giraud, E., Clifton, R. and Whelan, J. (2009) Alternative oxidase: A target and regulator of stress responses. *Physiologia Plantarum* **137**: 354–361.

Vanlerberghe, G.C (2013) Alternative oxidase: A mitochondrial respiratory pathway to maintain metabolic and signaling homeostasis during abiotic and biotic stress in plants. *International Journal of Molecular Sciences* **14**: 6805–6847.

Vanlerberghe, G.C., Cvetkovska, M. and Wang, J. (2009) Is the maintenance of homeostatic mitochondrial signaling during stress a physiological role for alternative oxidase? *Physiologia Plantarum* **137**: 392–406.

Vanlerberghe, G.C. and McIntosh, L. (1992) Lower growth temperature increases alternative pathway capacity and alternative oxidase protein in tobacco. *Plant Physiology* **100**: 115–119.

Vanlerberghe, G.C. and McIntosh, L. (1994) Mitochondrial electron transport regulation of nuclear gene expression. Studies with the alternative oxidase gene of tobacco. *Plant Physiology* **105**: 867–874.

Vanlerberghe, G.C., Mcintosh, L. and Yip, J. (1998) Molecular localization of a redox-modulated process regulating plant mitochondrial electron transport. *Plant Cell* **10**: 1551–1560.

Vanlerberghe, G.C., Robson, C.A. and Yip, J.Y.H. (2002) Induction of mitochondrial alternative oxidase in response to a cell signal pathway down-regulating the cytochrome pathway prevents programmed cell death. *Plant Physiology* **129**: 1829–1842.

Vicente, C. and Arnholdt-Schmitt, B. (2008) Characterization of Mediterranean AM: Initiation of a novel functional marker approach. COST 870 Meeting 'From production to application of arbuscular mycorrhizal fungi in agricultural systems: a multidisciplinary approach' Mycorrhiza application in sustainable agriculture and natural systems, Thessaloniki, September, pp.45–48.

Vinogradov, A.E. (2005) Noncoding DNA, isochores and gene expression: Nucleosome formation potential.*Nucleic Acids Research* **33** (2): 559–563.

Vosátka, M., Albrechtova, J. and Patten, R. (2008) The international market development for mycorrhizal technology. In *Mycorrhiza* (Varma, A., ed.), pp. 419–438. Berlin, Springer.

Vosátka, M., Látr, A., Gianinazzi, S. and Albrechtová, J. (2012) Development of arbuscular mycorrhizal biotechnology and industry: current achievements and bottlenecks. *Symbiosis* **58**: 29–37.

Wagner, A.M., Wagner, M.J. and Moore, A.L. (1998) *In vivo* ubiquinone reduction levels during thermogenesis in araceae. *Plant Physiology* **117**: 1501–1506.

Wang, B. and Qiu, Y.L. (2006) Phylogenetic distribution and evolution of mycorrhizas in land plants. *Mycorrhiza* **16**: 299–363.

Wang, J., Rajakulendran, N., Amirsadeghi, S. and Vanlerberghe, G.C. (2011) Impact of mitochondrial alternative oxidase expression on the response of *Nicotiana tabacum* to cold temperature. *Physiologia Plantarum* **142**: 339–351.

Wang, J.-F., Zhou, H., Chen, Y.-Q. *et al.* (2004) Identification of 20 microRNAs from Oryza sativa. *Nucleic Acids Research* **32**: 1688–1695.

Wang, X., Zhao, X., Zhu, J. and Wu, W. (2005) Genome-wide investigation of intron length polymorphisms and their potential as molecular markers in rice (*Oryza sativa* L.). *DNA Research* **12**: 417–427.

Watanabe, C.K., Hachiya, T., Terashima, I. and Noguchi, K. (2008) The lack of alternative oxidase at low temperature leads to a disruption of the balance in carbon and nitrogen metabolism, and to an up-regulation of antioxidant defence systems in *Arabidopsis thaliana* leaves. *Plant, Cell and Environment* **31**: 1190–1202.

Watanabe, C.K., Hachiya, T., Takahara, K. *et al.* (2010) Effects of AOX1a deficiency on plant growth, gene expression of respiratory components and metabolic profile under low-nitrogen stress in *Arabidopsis thaliana. Plant and Cell Physiology* **51**: 810–822.

Weckx, S., Del-Favero, J., Rademakers, R. *et al.* (2005) novoSNP, a novel computational tool for sequence variation discovery. *Genome Research* **15**: 436–442.

Wendehenne, D., Pugin, A., Klessig, D.F. and Durner, J. (2001) Nitric oxide: Comparative synthesis and signaling in animal and plant cells. *Trends in Plant Science* **6** (4): 177–183.

Wessler, S.R. (1988) Phenotypic diversity mediated by the maize transposable elements *Ac* and *Spm. Science* **242**: 399–405.

Wessler, S.R., Bureau, T.E. and White, S.E. (1995) LTR-retrotransposons and MITEs: Important players in the evolution of plant genomes. *Current Opinion in Genetics Development* **5**: 814–821.

Whelan, J., Millar, A.H. and Day, D.A. (1996) The alternative oxidase is encoded in a multigene family in soybean. *Planta* **198**: 197–201.

Whipps, J.M. (2004) Prospects and limitations for mycorrhizas in biocontrol of root pathogens. *Canadian Journal of Botany* **82**: 1198–1227.

Whiteman, N.K. and Gloss, A.D. (2010) Parasitology: Nematode debt to bacteria. *Nature* **468**: 641–642.

Wind J.J., Peviani A., Snel B. *et al.* (2012) ABI4: versatile activator and repressor. *Trends of Plant Science* **19**: 1360–1385.

Wu, J. and Jiang, R. (2013) Prediction of deleterious nonsynonymous single-nucleotide polymorphisms for human diseases. *The Scientific World Journal* **675851**. DOI: 10.1155/2013/675851. Epub 2013 Jan 30.

Xie, X., Lu, J., Kulbokas, E.J. *et al.* (2005) Systematic discovery of regulatory motifs in human promoters and 3′-UTRs by comparison of several mammals. *Nature* **434**: 338–345.

Xing, D. and Li, Q.Q. (2011) Alternative polyadenylation and gene expression regulation in plants. *Wiley interdisciplinary reviews RNA* **2**: 445–58.

Xu, F., Yuan, S., Zhang, D.-W. *et al.* (2012a) The role of alternative oxidase in tomato fruit ripening and its regulatory interaction with ethylene. *Journal of Experimental Botany* DOI 10.1093/jxb/ers226.

Xu, T., Yao, F., Liang, W.-S. *et al.* (2012b) Involvement of alternative oxidase in the regulation of growth, development, and resistance to oxidative stress of *Sclerotinia sclerotiorum*. *Journal of Microbiology* **50**: 594–602.

Xu, Z. and Ramakrishna, W. (2008) Retrotransposon insertion polymorphisms in six rice genes and their evolutionary history. *Gene* **412**: 50–58.

Yamazaki, M. (2002) Molecular biological studies on diversity of secondary metabolism in medicinal plants and application to the production in transgenic plants. *Yakugaku zasshi Journal of the Pharmaceutical Society of Japan* **122**: 47–56.

Yano, M., Kojima, S., Takahashi, Y. *et al.* (2001) Genetic control of flowering time in rice, a short-day plant. *Plant Physiology* **127**: 1425–1429.

Yip, J.Y.H. and Vanlerberghe, G.C. (2001) Mitochondrial alternative oxidase acts to dampen the generation of active oxygen species during a period of rapid respiration induced to support a high rate of nutrient uptake. *Physiologia Plantarum* **112**: 327–333.

Zavattieri, M.A., Frederico, A.M., Lima, M. *et al.* (2010) Induction of somatic embryogenesis as an example of stress-related plant reactions. *Electronic Journal of Biotechnology* **13**: 1–9.

Zerjal, T., Joets, J., Alix, K. *et al.* (2009) Contrasting evolutionary patterns and target specificities among three Tourist-like MITE families in the maize genome. *Plant Molecular Biology* **71**: 99–114.

Zhang, B., Pan, X., Cobb, G.P. and Anderson, T.A. (2006a) Plant microRNA: A small regulatory molecule with big impact. *Developmental Biology* **289**: 3–16.

Zhang, X., Yasaki, J., Sundaresan, A. *et al.* (2006b) Genome-wide high resolution mapping and functional analysis of DNA methylation in *Arabidopsis*. *Cell* **126**: 1189–1201.

Zhang, Z. and Gerstein, M. (2003) Patterns of nucleotide substitution, insertion and deletion in the human genome inferred from pseudogenes. *Nucleic Acids Research* **31**: 5338–5348.

Zhao, J., Hyman, L. and Moore, C. (1999) Formation of mRNA 3′ ends in eukaryotes: Mechanism, regulation, and interrelationships with other steps in mRNA synthesis. *Microbiology and Molecular Biology Reviews* **63**: 405–445.

Zilberman, D., Gehring, M., Tran, R.K. *et al.* (2006) Genome-wide analysis of *Arabidopsis thaliana* DNA methylation uncovers an interdependence between methylation and transcription. *Nature Genetics* **39**: 61–69.

SECTION C
Protocols

Contents

CHAPTER 15

Technical protocol for mitochondria isolation for different studies

Renate Horn
Institut für Biowissenschaften, Abteilung Pflanzengenetik, Universität Rostock, Rostock, Germany

Introduction

Recently, the interest in mitochondria, especially with regard to electron partitioning between the cytochrome *c* pathway and the alternative pathway has considerably increased as the role of mitochondria in biotic and abiotic stress is increasingly revealed (Pastore *et al.*, 2007; Suzuki *et al.*, 2012; Cvetkovska and Vanlerberghe, 2013; Vanlerberghe, 2013). The alternative respiratory pathway seems to lower reactive oxygen species (ROS) as well as reactive nitrogen species (RNS) by reducing the electron flow from the electron transport chain to oxygen and nitrite in the cytochrome pathway (Cvetkovska and Vanlerberghe, 2012). In addition, it becomes clear that the alternative oxidase (AOX) is also involved in various developmental processes like fruit ripening (Xu *et al.*, 2012), adventitious rooting (Santos Macedo *et al.*, 2012) or thermogenesis (Zhu *et al.*, 2011).

Although mitochondria purification has been well established for a number of model plants like *Arabidopsis* (Giegé *et al.*, 2003; Keech *et al.*, 2005; Sweetlove *et al.*, 2007), peas (Moore *et al.*, 1993; Rödiger *et al.*, 2010), potato (Considine *et al.*, 2003) and rice (Bardel *et al.*, 2002; Huang *et al.*, 2009), it still presents a challenge for other plant species, which might require complex purification steps, as for *Medicago sativa* (Dubinin *et al.*, 2011). Here, after three differential centrifugation steps, a density centrifugation followed using first a continuous Percoll gradient between 15% and 55% and then two three-step Percoll gradients (14%, 26%, 45%; 18%, 23%, 40%). For isolation of mitochondria from wheat seedlings, complex purification steps also involve isopycnic centrifugation, but in self-generating density gradients, consisting of 0.5 M sucrose and 28% (v/v) Percoll, combined with a linear gradient of 0–10% PVP-40 (Soccio *et al.*, 2010). However, simpler protocols were also successfully applied for wheat using linear Percoll gradients from 2% to 60% (Goldstein *et al.*, 1980).

Alternative Respiratory Pathways in Higher Plants, First Edition.
Edited by Kapuganti Jagadis Gupta, Luis A.J. Mur and Bhagyalakshmi Neelwarne.

For maize seedlings, mitochondria were isolated after two differential centrifugation steps followed by a sucrose density gradient (Leaver *et al.*, 1983; Prasad *et al.*, 1994).

For studies on thermogenesis, mitochondria needed to be isolated from the spadix of the skunk cabbage (*Symplocarpus renifolius*) to investigate the role of AOX and plant uncoupling mitochondrial proteins (PUMPS) in dissipating chemical energy into metabolic heat (Onda *et al.*, 2007). In this case, mitochondria from the florets were purified using a three-step centrifugation, which included a three-step Percoll gradient (27%, 45%, 60%). Mitochondria were recovered from the 27%/45% interphase.

Although most procedures for isolating mitochondria rely on differential centrifugations followed by a density gradient (isopycnic) centrifugation, alternative options that do not address size and density are available (Eubel *et al.*, 2007). Free-flow electrophoresis based on surface charge can assist conventional, centrifugation-based techniques by providing a different means to separate plastids and peroxisomes from mitochondria (Eubel *et al.*, 2007). For purification of soybean mitochondria from roots and hypocotyls, a QProteome Mitochondrial Isolation kit (Qiagen, Hilden, Germany) was used after the differential centrifugation steps to obtain intact mitochondria without cytosolic contaminations (Komatsu *et al.*, 2011).

However, the purification of mitochondria from non-model plant species as well as *in planta* investigations (Cvetkovska and Vanlerberghe, 2012) have become increasingly important to understanding the role of the mitochondria in the homeostasis of nitrite oxide (NO) and ROS, in which the alternative oxidase plays a central role (Gupta *et al.*, 2012).

In this chapter, general aspects for isolating mitochondria as well as a specific protocol developed for sunflower is presented, which should allow the development of a method for mitochondria purification for other species with slight modifications, for example we also used the method for the isolation of wheat mitochondria. Using etiolated seedlings and the buffers described by Goldstein *et al.* (1980), we performed discontinuous Percoll gradient (13%/21%/45%) and successfully purified wheat mitochondria.

General aspects of plant mitochondria isolation

Choice of starting plant material

Depending on the research focus, the choice of the starting plant material might decide the success of the research. If the species is defined, the next question is what part of the plant is to be used for the extraction, for example leaves, etiolated shoots, roots, flowers or anthers. For small plants like *Arabidopsis thaliana* or *H. petiolaris,* callus or cell suspension cultures represent an alternative. However, the choice of plant material will put different demands on the extraction buffer to be used. In addition, whether the plants are grown under a day/night cycle or can be cultivated in the dark (etiolated), under sterile conditions on solid or liquid media, on soil or on vermiculite need to be considered. The advantage of growing plants in the dark is that the chloroplasts are not yet fully developed; instead yellow etioplasts, containing neither chlorophyll nor starch with lighter buoyant density, are present in the cells. The use of liquid media is of special interest if mitochondria from roots are investigated, for example in drought experiments using polyethylene glycol (Fulda *et al.*, 2011).

Methods for the disruption of the plant material

The use of a blender can be recommended for the disruption of various plant tissues. The time has to be short (2 × 3 seconds) at low speed to avoid damaging the mitochondria by shearing forces. Depending on the starting material, the plant tissue has to be cut in advance with a scalpel. If etiolated dicot seedlings are used, this is normally not necessary. However, for monocot leaves it is essential to cut them into 1 cm pieces with a razor blade before placing them into the blender as otherwise within a short time the blades have to be freed using a scalpel. Other methods for disruption like a mortar and pestle can be used, especially for smaller amounts of plant material, but the blender can be universally used for plants like sunflower (Leipner and Horn, 2002) and potato (Lössl *et al.*, 1999) as well as for wheat.

Extraction buffer

The extraction buffer requires in principle the following components:

(a) Buffer substance (e.g. Tris/HCl or potassium phosphate buffer), which should have a high buffering capacity as disrupting the plant cell can release acid compounds, especially from the vacuoles that might cause a considerable drop in pH. Establishing a mitochondria isolation protocol for a new species, it is recommended to test the pH after extraction. The pH should be around 7.2 and 7.8;

(b) An osmoticum like mannit or sucrose to avoid disruption of the mitochondria due to osmotic changes;

(c) Reducing agents as β-mercaptoethanol or dithiothreitol (DTT) and if necessary ascorbic acid in addition. This is required to avoid immediate browning of the extraction solution due to oxidation processes and the oxidation of phenolic compounds. The amounts might vary according to the demand and requires working under a hood when using ß-mercaptoethanol (which is nevertheless the most recommended);

(d) Polyvinylpyrrolidone (PVP), which binds phenolic compounds and is available in two forms: soluble PVP-40 and as an insoluble powder, known as Polyclar AT (Serva). Using soluble PVP-40 makes the solution more viscous and can never be totally eliminated during purification. The use of the insoluble powder Polyclar AT is recommended. This reacts during the first round of extraction, eliminating the deleterious effects of phenolic compounds released by the disrupted cells and is mostly discarded with the first centrifugation step or at the latest in the density gradient step;

(e) Ethylene glycol bis(2-aminoethylether)-tetra acetic acid (EGTA) preferentially used when a higher selectivity for Ca^{+} over Mg^{+} is required; or

(f) Ethylene diamine tetra acetic acid (EDTA) to chelate released metal ions like Ca^{+} and Mg^{+} and thereby reducing activities of enzymes (proteases and phospholipases) released from the disrupted cells and various organelles;

(g) Cysteine for providing additional free sulfhydryl groups and thereby reducing the deleterious oxidation of sulfhydryl groups in mitochondrial proteins;

(h) Bovine serum albumin (BSA fraction V, fatty acid free) stabilizes protein complexes and binds free fatty acids that would be deleterious to membranes.

Differential centrifugation

The first round for purifying plant mitochondria is in general a differential centrifugation. In the first centrifugation step with low *g* debris, nuclei and any denser material, for example the insoluble Polyclar AT are also separated from the mitochondria, which remain in the supernatant. Discarding the pellet, the second centrifugation step allows the sedimentation of the mitochondria, separating them from all material with a lighter buoyant density.

The supernatant of this centrifugation step is carefully poured off and the mitochondria pellet is resuspended in 2 ml of a suspension buffer with a soft paint brush. This suspension buffer may contain DNAase to eliminate any DNA attached to the membranes in case mitochondrial DNA will be isolated. Mitochondria suspension should only be handled with cut-off tips to avoid shearing forces. This represents the crude mitochondria preparation, which requires further purification over density gradients.

Purification using density gradients

Frequently either Percoll (sterile, GE Health Care) or sucrose are used for purification of mitochondria on density gradients. These gradients can be continuous or discontinuous consisting of different steps. The recommendation is to pour stepwise gradients with three steps using different percentages of Percoll and Corex tubes (with adapter). Compared to sucrose, Percoll is inert and not sticky, but it is expensive. The percentages of the different steps have to be optimized for different plant species. However, using steps of 14%, 21% and 45% represents a good starting point. Using discontinuous versus continuous gradients has the advantage that the mitochondria are concentrated in a sharper band. Pouring discontinuous gradients requires patience but pouring the first two steps is much easier if the step with the middle percentage is filled in first and the heavier one is carried out as an under layer using a long Pasteur pipette. This considerably speeds up the procedure of pouring gradients and only leaves one step to be done as an over layer; this has to be carried out carefully so that a sharp line can still be seen between the steps. Centrifugation should result in a very light white (opaque) mitochondria band. Sometimes chloroplasts can have the same buoyant density as the mitochondria, which makes purification more difficult. One solution for this can be to use plant material grown in the dark (etiolated) so that the chloroplasts have not yet differentiated and are present as etioplasts, which have lighter buoyant density. Another solution is to reduce the starch content in the chloroplasts by harvesting the plant material immediately after the dark phase and thereby reducing the amount of starch in the chloroplasts to a minimum. Another option frequently used to reduce contamination by chloroplasts is to repeat the earlier steps before differential centrifugation. Using callus or cell suspension cultures as the starting material can also help to circumvent this problem.

Final washing

In the final washing steps the mitochondria have to be freed from the remaining Percoll or sucrose and resuspended in a small volume of buffer appropriate for the measurements or treatments that follow.

Optimising the mitochondria purification protocol

For a number of plant species, the protocols for isolating mitochondria are well developed, for example for peas, potato, but also for *Arabidopsis*. However, establishing a protocol for a new species of interest will require optimization of the individual steps. Purification of mitochondria can be followed by visual control of the different steps using microscopy technology (light and electron microscopy) combined with cytochemical staining methods (e.g. Mito Tracker), immunofluorescence and verification of organelle integrity (Agrawal *et al.*, 2011). However, this requires equipment that is relatively expensive and might not be directly available. The easiest way to follow the purification of a mitochondria preparations is to measure enzyme activities specific for the different organelles in the cell (Agrawal *et al.*, 2011), for example cytochrome-*c* oxidase as marker for the mitochondria (Prasad *et al.*, 1994) or NADH-cytochrome *c* reductase (Bergman *et al.*, 1980) for endoplasmatic reticulum (inhibition of the mitochondrial form by antimycin A). Another biochemical method would be to use specific antibodies for the organelles, for example α and β subunits of the F_1F_0 ATPase for mitochondria (Luethy *et al.*, 1993) or cytochrome oxidase II and isocitrate dehydrogenase (Agrisera, Vännäs, Sweden) to verify the state of purification of the mitochondria or contamination by chloroplasts using antibodies against the ribulose-1,5-biphosphate carboxylase/oxygenase or chlorophyll fluorescence or its measurement (Rödiger *et al.*, 2010).

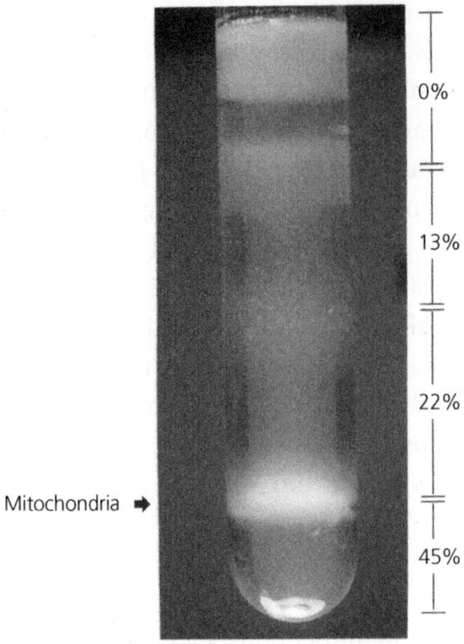

Figure 15.1 Purification of sunflower mitochondria on a discontinuous Percoll density gradient (13%, 22% and 45%) after centrifugation for 15 min, 14 000*g*. (*See insert for color representation of the figure.*)

Specific protocol for isolation of sunflower mitochondria as a basic protocol

A simple, straightforward method for isolating mitochondria from etiolated sunflower seedlings is presented here, which represents a slightly modified protocol of the one used by Horn *et al.* (1991) and Köhler *et al.* (1991) and does not require, apart from the Waring Blender, the Corex tubes (including adapters and rack) and soft paint brushes, anything, that would not be present in a regular laboratory. This protocol can be a good starting point to develop a protocol for any other plant species, probably only requiring slight modifications (Figure 15.1).

This protocol was successfully used for mitochondria preparation from sunflowers to measure the mitochondrial respiratory activity, the capacity of the alternative pathway and the involvement of the alternative pathway in cytoplasmic male sterile lines (ANL1, ANL2, GIG1, MAX1 and PET2) and the corresponding maintainer lines (Leipner and Horn, 2002). It was also applied to isolate mitochondria for radioactively labelling mitochondrial encoded proteins with ^{35}S-methionine via in organello translation (Horn *et al.*, 1996; Horn, 2002) and for mitochondrial DNA extractions (Horn and Friedt, 1999).

Solutions for the mitochondria isolation

All steps are performed on ice using ice-cold solutions.

Extraction buffer

50 mM Tris/HCl
0.3 M Mannit
1 mM EGTA
1 mM $MgCl_2$ (0.5 M stock solution)
adjust pH to 7.4 (about 3 ml HCl 32% per l)
autoclave, cool down and store in refrigerator
add freshly per 100 ml
0.1 *g* BSA fraction V
0.1 *g* cysteine (basic)
140 µl mercaptoethanol or 0.031 *g* DTT (2 mM)
0.5 *g* PVP insoluble
stored on ice for at least 30 min before use (2 tablets protease inhibitor cocktail (cOmplete ULTRA tablets, Roche)/500 ml might be dissolved in it for proteomic studies – requires time!)

Suspension buffer

50 mM Tris/HCl
0.3 M Mannit
10 mM $MgCl_2$ (0.5 M stock solution)
adjust pH with HCl to 7.4
autoclave, cool down and store in refrigerator

Resuspension buffer (RB) 2× (EDTA)

0.6 M Mannit
20 mM Tricine
20 mM EDTA
adjust pH with 20% KOH to 7.2
autoclave, cool down and store in refrigerator

Washing buffer

50 mM Tris/HCl
0.3 M Mannit
1 mM EGTA
1 mM $MgCl_2$ (0.5 M stock solution)
adjust pH to 7.4 (about 3 ml HCl 32% per l)
autoclave, cool down and store in refrigerator

Percoll (GE Health Care)

Percoll gradient (6 gradients, in 30 ml Corex tubes)

Percentage	Volume	Percoll	ddH_2O	RB 2× (EDTA)
13%	6 ml	6.5 ml	18.5 ml	25 ml
22%	10 ml	16.5 ml	21.0 ml	37.5 ml
45%	5 ml	16.95 ml	1.95 ml	18.75 ml

Fill each layer of percentage with long, sterile plugged Pasteur pipettes into the Corex-tube:
22% first layer to be filled in
45% to be filled in as sublayer (go with the tip to the bottom of the Corex tube)
13% as an over layer
!Use Corex tubes only with adapters in a superspeed centrifuge, e.g. Sorvall RC 6 Plus, SS34 rotor!

Resuspension buffer 1× (EGTA)

0.3 M Mannit
10 mM Tricine
10 mM EGTA
adjust pH with 20% KOH to 7.2
autoclave, cool down and store in refrigerator

Isolation procedure for sunflower mitochondria

For isolating six different sunflower lines 2 l extraction buffer are required. The etiolated 12- to 14-day-old sunflower seedlings are first measured. The weight should be around 15–20 g/line. Using much higher or lower amounts of plant material leads to reduced yields and, for more starting material, less pure mitochondria. Isolating mitochondria for respiratory activity or *in organello* translation, 0.1% (w/v) BSA is freshly added to the extraction buffer as described, but for analysing mitochondrial proteins via polyacrylamide gel electrophoresis this is omitted and instead protease inhibitor cocktail (cOmplete ULTRA tablets, Roche) should be added.

Disruption

- Fill Waring Blender with seedlings, add about 120 ml extraction buffer, blend 2 × 3 seconds at low speed, filter through six layers of cotton gauze (or Miracloth [Calbiochem] or cheese cloth) using a broad funnel into 500 ml centrifugation tubes
- re-extract 2 × 3 seconds at low speed, filter through six layers of new cotton gauze, pool fractions

Differential centrifugation

- 1. centrifugation 5 min, 2 600 *g*, 4 °C
- pour supernatant into new tubes, discard pellet; **pellet → debris, nuclei**

- 2. centrifugation 20 min, 13 000*g*, 4 °C
- discard supernatant; **pellet → mitochondria**
- add 2 ml suspension buffer to the pellets, dissolve carefully with a soft (horsehair) paintbrush size 8 or 9, use only cut tips for pipetting!

Percoll density gradient centrifugation

Layer the mitochondria suspension on top of the density gradient

- centrifugation 15 min, 14 000*g*, no brakes, 4 °C
- take the mitochondria layer (between 22%/45% with a sterile Pasteur glass pipette)
- transfer to 30 ml Corex tubes, fill up with washing buffer
- centrifugation 10 min, 16 000*g*, 4 °C
- suck up the pellet with a long Pasteur glass pipette
- transfer into an Eppendorf tube, fill up with 1× resuspension buffer EGTA (½ tablet proteinase inhibitor/20 ml for proteomic studies, otherwise omit)
- centrifugation 5 min, full speed in an Eppendorf centrifuge at 4 °C, suck away the supernatant with extra thinly stretched Pasteur glass pipette
- resuspend pellet in 1× resuspension buffer EGTA
- use immediately for respiratory activity measurements or in organello translation or store at −80 °C for proteomic analysis.

Appliances (sterile for *in organello* translation)

Cotton gauze (medical supply), six layers each

6 centrifugation tubes (500 ml) (place piece of aluminium foil between the lid and the cup for the autoclaving procedure)

6 funnel with large outlets

12 Corex tubes (30 ml), autoclave in tulip glasses (Weck), place cotton gauze at the bottom to avoid breakage of the Corex tubes

6 centrifugation tubes (30 ml)

6 soft (horsehair) paint brushes size 8 or 9 (sterile: wrap in total in aluminium foil, autoclave)

Glass of 1.5 ml Eppendorf tubes

Boxes of yellow tips

Boxes of blue tips (normal and cut ones)

Pasteur glass pipettes (short, long, thinly stretched ones), in tulip glasses with cotton gauze at the bottom (plugged, sterile)

1 rack for Corex tubes

2 styropor ice buckets filled with ice

Set of variable pipettes (10 µl, 100 µl, 1000 µl)

Pipetting aid for Pasteur pipettes

Conclusions

Purification of mitochondria from non-green tissue like etiolated seedlings, roots, callus or cell suspension culture is still the best option for isolating mitochondria with minimal contamination by plastids. However, this will not address all research fields for alternative oxidase and for some scientific questions, the simultaneous isolation of chloroplast and mitochondria from the same plant tissue might be required. This was successfully achieved with pea leaves by Rödiger *et al.* (2010). For comprehensive analysis of the mitochondrial proteome under different conditions, address the different proteins. Computational addressing by the novel GelMap software package of mitochondrial protein complexes after separation by two-dimensional blue native/sodium dodecyl sulfate-polyacrylamide gel electrophoresis in combination with mass spectrometry represents an interesting option, which will give a more detailed insight into the differences between the mitochondrial proteomes of various plant species (Klodmann *et al.*, 2011).

With the increasing interest in the role of the alternative oxidase and the alternative pathway in biotic and abiotic stress reactions, purification of mitochondria from different plant species will be a challenging task in the future.

Acknowledgements

The research leading to the development of the protocol for the isolation of mitochondria was funded by the Deutsche Forschungsgemeinschaft.

References

Agrawal, K. G., Bourguignon, J., Rolland, N. *et al.* (2011) Plant organelle proteomics: Collaborating for optimal cell function. *Mass Spectrometry Reviews* **30**: 772–853.

Bardel, J., Louwagie, M., Jaquinod, M. *et al.* (2002) A survey of the plant mitochondrial proteome in relation to development. *Proteomics* **2**: 880–898.

Bergman, A., Gardeström, P. and Ericson, I. (1980) Method to obtain a chlorophyll-free preparation of intact mitochondria from spinach leaves. *Plant Physiology* **66**: 442–445.

Considine, M.J., Goodman, M., Echtay, K.S. *et al.* (2003) Superoxide stimulates a proton leak in potato mitochondria that is related to the activity of uncoupling protein. *Journal of Biological Chemistry* **25**: 22298–22302.

Cvetkovska, M. and Vanlerberghe, G.C. (2012) Alternative oxidase modulates leaf mitochondrial concentrations of superoxide and nitric oxide. *New Phytologist* **195**: 32–39.

Cvetkovska, M. and Vanlerberghe, G.C. (2013) Alternative oxidase impacts the plant response to biotic stress by influencing the mitochondrial generation of reactive oxygen species. *Plant, Cell and Environment* **36**: 721–732.

Dubinin, J., Braun, H.-P., Schmitz, U. and Colditz, F. (2011) The mitochondrial proteome of the model legume *Medicago truncatula*. *Biochimica et Biophysica Acta* **1814**: 1658–1668.

Eubel, H., Lee, C.P., Kuo, J. *et al.* (2007) Free-flow electrophoresis for purification of plant mitochondria by surface charge. *The Plant Journal* **52**: 583–594.

Fulda, S., Mikkat, S., Stegmann, H. and Horn, R. (2011) Physiology and proteomics of drought stress acclimation in sunflower (*Helianthus annuus* L.). *Plant Biology* **13**: 632–642.

Giegé, P., Heazlewood, J.L., Roessner-Tunali, U. *et al.* (2003) Enzymes of glycolysis are functionally associated with the mitochondrion in *Arabidopsis* cells. *The Plant Cell* **15**: 2140–2151.

Goldstein, A.H., Anderson, J.O. and McDaniel, R.G. (1980) Cyanide-insensitive and cyanide-sensitive O2 uptake in wheat. 1. Gradient-purified mitochondria. *Plant Physiology* **66**: 488–493.

Gupta, K.J., Igamberdiev, A.U. and Mur, L.A.J. (2012) NOS and ROS homeostasis in mitochondria: A central role for alternative oxidase. *New Phytologist* **195**: 1–3.

Horn, R. (2002) Molecular diversity of male sterility inducing and male fertile cytoplasms in the genus *Helianthus*. *Theoretical and Applied Genetics* **104**: 562–570.

Horn, R. and Friedt, W. (1999) CMS sources in sunflower: different origin but same mechanism? *Theoretical and Applied Genetics* **98**: 195–201.

Horn, R., Hustedt, J.E.G., Horstmeyer, A. *et al.* (1996) The CMS-associated 16 kDa protein encoded by *orfH522* is also present in other male sterile cytoplasms of sunflower. *Plant Molecular Biology* **30**: 523–538.

Horn, R., Köhler, R.H. and Zetsche, K. (1991) A mitochondrial 16 kDa protein is associated with cytoplasmic male sterility in sunflower. *Plant Molecular Biology* **7**: 29–36.

Huang, S., Taylor, N.L., Narsai, R. *et al.* (2009) Experimental analysis of the rice mitochondrial proteome, its biogenesis, and heterogeneity. *Plant Physiology* **149**: 719–734.

Keech, O., Dizengremel, P. and Gardeström, P. (2005) Preparation of leaf mitochondria from *Arabidopsis thaliana*. *Physiologia Plantarum* **124**: 403–409.

Köhler, R.H., Horn, R., Lössl, A. and Zetsche, K. (1991) Cytoplasmic male sterility in sunflower is correlated with the co-transcription of a new open reading frame with the *atpA* gene. *Molecular and General Genetics* **227**: 369–376.

Klodmann, J., Senkler, M., Rode, C. and Braun, H.-P. (2011) Defining the protein complex proteome of plant mitochondria. *Plant Physiology* **157**: 587–598.

Komatsu, S., Yamamoto, A., Nakamura, T. *et al.* (2011) Comprehensive analysis of mitochondria in roots and hypocotyls of soybean under flooding stress using proteomics and metabolomics techniques. *Journal of Proteome Research* **10**: 3993–4004.

Leaver, C., Hack, E. and Forde, B.G. (1983) Protein synthesis by isolated plant mitochondria. *Methods in Enzymology* **97**: 476–484.

Leipner, J. and Horn, R. (2002) Nuclear and cytoplasmic differences in the mitochondrial respiration and protein expression of CMS and maintainer lines of sunflower. *Euphytica* **123**: 411–419.

Lössl, A., Adler, N., Horn, R. *et al.* (1999) Chondriome type characterization of potato: Mt α, β, γ, δ, ε and novel plastid-mitochondrial configurations in somatic hybrids. *Theoretical and Applied Genetics* **99**: 1–10.

Luethy, M.H., Horak, A. and Elthon T.E. (1993) Monoclonal antibodies to the α- and β-subunits of the plant mitochondrial F1-ATPase. *Plant Physiology* **101**: 931–937.

Moore, A., Gemel, J. and Randall, D.D. (1993) The regulation of pyruvate dehydrogenase activity in pea leaf mitochondria (the effect of respiration and oxidative phosphorylation). *Plant Physiology* **103**: 1431–1435.

Onda, Y., Kato, Y., Abe, Y. *et al.* (2007) Pyruvate-sensitive AOX exists as a non-covalently associated dimer in the homeothermic spadix of the skunk cabbage, *Symplocarpus renifolius*. *FEBBS Letters* **581**: 5852–5858.

Pastore, D., Trono, D., Laus, M.N. *et al.* (2007) Possible plant mitochondria involvement in cell adaptation to drought stress. A case study: durum wheat mitochondria. *Journal of Experimental Botany* **58**: 195–210.

Prasad, T.K., Anderson, M.D. and Stewart, C.R. (1994) Acclimation, hydrogen peroxide, and abscisic acid protect mitochondria against irreversible chilling injury in maize seedlings. *Plant Physiology* **105**: 619–627.

Rödiger, A., Baudisch, B. and Klösgen, R.-B. (2010) Simultaneous isolation of intact mitochondria and chloroplast from a single pulping of plant tissue. *Journal of Plant Physiology* **167**: 620–624.

Santos Macedo, E., Sircar, D., Cardoso, H.G. *et al.* (2012) Involvement of alternative oxidase (AOX) in adentitious rooting of *Olea europaea* L. microshoots is linked to adaptive phenylpropanoid and lignin metabolism. *Plant Cell Reports* **31**: 1581–1590.

Soccio, M., Laus, M.N., Spera, G.P. *et al.* (2010) Mitochondrial proline oxidation is affected by hyperosmotic stress in durum wheat seedlings. *Annals of Applied Biology* **157**: 1–11.

Sweetlove, L.J., Taylor, N.L. and Leaver, C.J. (2007) Isolation of intact, functional mitochondria from the model plant *Arabidopsis thaliana.* Mitochondria: Practical Protocols. *In Methods in Molecular Biology (Leister, D. and Kerrmann, J.M., eds)*, vol. **372**, pp. 125–136.

Suzuki, N., Koussevitzky, S., Mittler, R. and Miller, G. (2012) ROS and redox signalling in the response of plants to abiotic stress. *Plant, Cell and Environment* **35**: 259–270.

Xu, F., Yuan, S., Zhang, D.W. *et al.* (2012) The role of alternative oxidase in tomato fruit ripening and its regulator interaction with ethylene. *Journal of Experimental Botany* **15**: 5705–5716.

Vanlerberghe, G.C. (2013) Alternative oxidase: A mitochondrial respiratory pathway to maintain metabolic and signaling homeostasis during abiotic and biotic stress in plants. *International Journal of Molecular Science* **14**: 6805–6847.

Zhu, Y., Lu, J., Wang, J. *et al.* (2011) Regulation of thermogenesis in plants: The interaction of alternative oxidase and plant uncoupling mitochondrial protein. *Journal of International Plant Biology* **53**: 7–13.

CHAPTER 16

Simultaneous isolation of root and leaf mitochondria from *Arabidopsis*

Kapuganti Jagadis Gupta[1,*] and Ralph Ewald[2]
[1] *Department of Plant Sciences, University of Oxford, Oxford, UK*
[2] *Institut für Biowissenschaften, Abteilung Pflanzengenetik, Universität Rostock, Rostock, Germany*
**Current address: National Institute of Plant Genome Research, Aruna Asaf Ali Road, New Delhi, India*

Introduction

In all aerobic organisms mitochondria generate ATP via oxidative phosphorylation. Mitochondria are involved not only in energy production but also in various processes such as generation of reactive oxygen species, participation in retrograde signalling, nitric oxide production, calcium regulation, programmed cell death, involvement in photorespiration, providing carbon skeletons for amino acid biosynthesis, and thermogenesis (Kowaltowski, 2000). For physiological, biochemical and proteomic studies it is often important to isolate uncontaminated, physiologically active and intact mitochondria. For bulky tissues such as potato tubers and cauliflower and for larger crop plants such as tobacco, pea, soybean or etiolated seedlings it is possible to get a good yield of mitochondria. However, for tiny model plants like *Arabidopsis* it is very difficult to get sufficient quantities of mitochondria for various studies. Moreover, for comparative studies it is very important to isolate leaf and root mitochondria.

Due to the lack of chlorophyll, root mitochondria isolation is often an easy task. Leaf mitochondria isolation has the advantage that higher amounts of tissues can be obtained from the plants in comparison to root material, but chlorophyll contamination can be a problem. Here we describe how to isolate root mitochondria with sufficiently high yields, and how to obtain chlorophyll-free leaf mitochondria simultaneously.

Materials

It is essential that you consult the appropriate Material Safety Data Sheets and your institution's Environmental Health and Safety Office for proper handling of the equipment and hazardous materials used in this protocol.

Alternative Respiratory Pathways in Higher Plants, First Edition.
Edited by Kapuganti Jagadis Gupta, Luis A.J. Mur and Bhagyalakshmi Neelwarne.

Reagents

Extraction buffer

0.3 M sucrose
100 mM HEPES pH 7.4
0.1% (w/v) Bovine serum albumin (BSA)
0.6% PVP (w/v)
1 mM EDTA
2 mM $MgCl_2$
4 mM cysteine
1 mM KH_2PO_4
Ready-to-use protease cocktail ('Complete', Roche, Mannheim, Germany, 1 tablet in 100 ml medium)
50 mM sodium ascorbate (for leaf mitochondria)

Suspension buffer

20 mM HEPES pH 7.4
0.3 M sucrose
2 mM $MgCl_2$
1 mM EDTA
0.1 mM KH_2PO_4

Gradient buffer for root mitochondria

Percoll (Sigma-Aldrich, Munich Germany)
0.25 M sucrose
25 mM HEPES pH 7.4

Equipment

Clips
Centrifuge Sorvall R6 Plus
Centrifuge tubes (250 ml; 50 ml)
Cheese cloth
Miracloth (Calbiochem, Darmstadt, Germany)
Oxygen electrode
Petri dishes
Paintbrush
Pasture pipette
Razor blades
Rotors, Sorval SLA-1500, rotor, Sigma 12156-H
Syringe with needle
Ultra Turrax IKA (Janke and Kunkel, Germany)

Method

1 Growth of *Arabidopsis* plants: Sterilize the seeds with 70% ethanol and 10% sodium hypochlorite and rinse several times with autoclaved distilled water. In order to avoid contamination carry on the sterilization procedure in a laminar air flow chamber. Prepare square Petri dishes that contain Murashige and Skoog basal medium (MS) (Duchefa Biochemie) (Murashige and Skoog, 1962) containing 1% sucrose and 1% agar. Close the plates with Leukopor tape (BSN Medical). With a small sterile syringe place the seeds in a linear manner on the agar plates as shown in Figure 16.1. Keep the plates in vertically in a Percival growth chamber (12/12 h) light/dark cycle, 22/18 °C, photosynthetic active photon flux density about 150 µmol $m^{-2} s^{-1}$). Plants will be ready in two to three weeks.

2 All mitochondrial isolation steps should be carried out at 4 °C. It is very important to release intact mitochondria from the tissues without mechanically disturbing them. Due to the rigid cell walls it is very difficult to rupture plant cells. So the roots and leaves need to be chopped up with a sharp razor blade into approximately 0.5 cm slices, 2 g of slices per 10 ml of solution placed in a 50 ml plastic measuring cylinder and the tissue ground using an Ultra Turrax. This instrument is much more efficient than mixers using this method is much higher than for other extraction methods.

3 Homogenize the filtrate using one layer of Miracloth and four layers of nylon mesh (80–100 µm).

4 Centrifuge the filtrate at 2000 ***g*** for 10 min. Discard the pellet.

5 Centrifuge the supernatant at 12000 ***g*** for 30 min. Discard the supernatant.

6 The pellet should be removed by passing a soft paint brush over the pellet or by repeatedly rinsing the pellet with a small volume of medium using a pasture pipette. The pellet has to be finally suspended in 2 ml of suspension buffer.

7 For root mitochondria, place the mitochondrial suspension on the discontinuous Percoll gradient (Figure 16.1). Required concentrations of gradients should be prepared by mixing specific concentrations of Percoll in a gradient buffer solution (according to Vanlerberghe *et al.*, 1995; Nishimura *et al.*, 1982). More specifically, the first layer (from below) contains 3 ml of 60% Percoll (v/v) and then overlay with 4 ml 45% (v/v) and then overlay with 4 l of 28% (v/v) Percoll and then on the top with 4 ml of 5% (v/v) Percoll. The Percoll can be loaded gently with pasture pipette at a 40° angle. Preparation of gradient solution can be carried out using gradient mixture or by smooth pipetting. Gradients can be made a day before isolation and can be stored in refrigerator at 4 °C.

8 The mitochondrial fraction appears at the interface between the 45% and 28% (v/v). Gently remove the layer with a pasture pipette and place it in a 50 ml centrifuge tube that contains 15 ml of suspension buffer, and centrifuge at

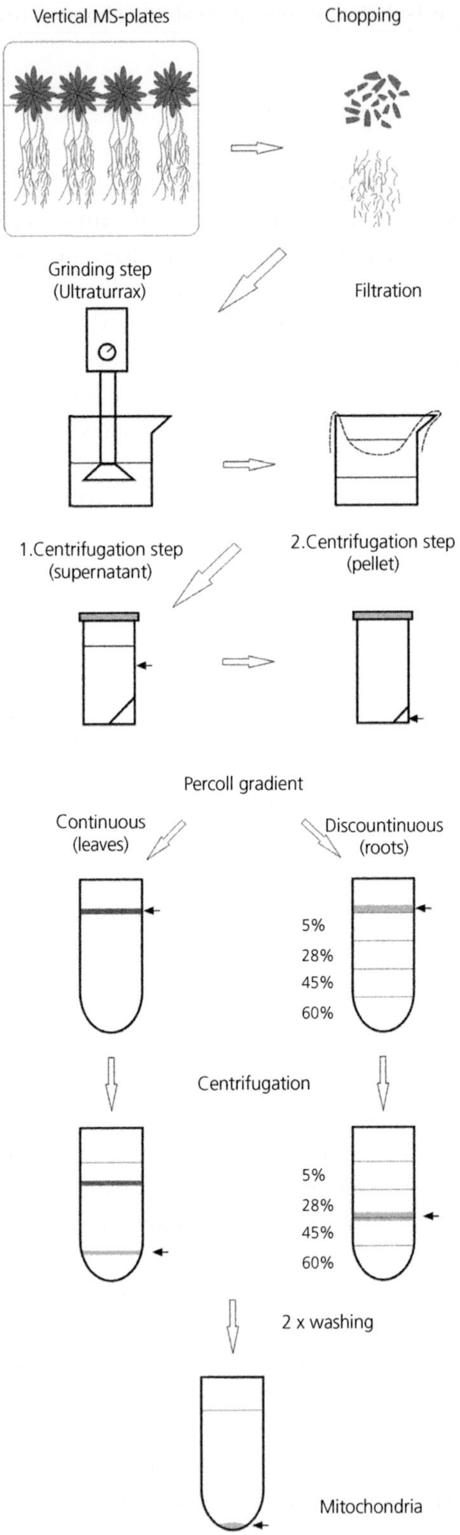

Figure 16.1 Stepwise procedure for mitochondria isolation procedure from roots and leaves of *Arabidopsis* axenic cultures.

18000*g* for 15 min. Discard the supernatant, resuspend the pellet in 15 ml suspension buffer and centrifuge again at 18000*g* for 15 min.

9 Finally a yellowish brown pellet containing the root mitochondria can be seen at the bottom of the tube.

10 For leaf mitochondria prepare a continuous Percoll gradient by centrifuging 30 ml of 50% Percoll for 30–40 min at 40000*g* (Keech *et al.*, 2005). The pellet from the second step centrifugation (refer to Figure 16.1) has to be placed on the top of the continuous Percoll gradient and should be centrifuged for 20 min at 15000*g*.

11 A Whitish band of leaf mitochondria can be seen at the bottom of the centrifuge tube.The band has to be aspired with suspension buffer as described for root mitochondria and centrifuged twice at 18000*g* for 15 min.

Methods to check activity and integrity of the mitochondria

1 Monitoring mitochondrial activity: state 3/state 4 ratio is referred to as the respiratory control ratio (P:O). State 3 respiration means ADP enhanced respiration. State 4 means respiration in the complete absence of the ATP synthesis. State 4 can be achieved by adding the ATP synthase inhibitor oligomycin (1 µg ml^{-1}). Oxygen uptake measurements for checking state 3/state 4 ratio can be done using oxygen electrode or by using the Microx TX2 oxygen sensing device (PreSens Precision Sensing). A respiratory control ratio (P:O) ratio of 3 means that mitochondria are well coupled, values lower than this indicate that they are only loosely coupled and that membranes are eventually damaged.

2 Peroxisomal contamination can be checked by adding 1 mM H_2O_2 to the mitochondrial suspension. Rates of oxygen evolution are proportional to the peroxisomal content.

3 Cytosolic contamination can be checked by measuring a cytosolic marker such as PEPC (phosphoenolpyruvate carboxylase) activity.

4 Thylakoid contamination can be checked by measuring the chlorophyll content in leaf mitochondria.

5 Western blots can be done by using antibodies against various marker enzymes of subcellular compartments, for example against peroxysomal protein KAT2 (3-ketoacyl-CoA thiolase-2) for checking peroxysomal contamination. Chloroplast contamination can be checked by antibody against large subunit of Rubisco (Duncan *et al.*, 2011).

Discussion

For various studies, metabolically active, well-coupled mitochondria are essential. Here we optimized the protocol based on information from various protocols (Vanlerberghe *et al.*, 1995; Nishimura *et al.*, 1982; Sweetlove *et al.*, 2007; Keech *et al.*, 2005). By our method, mitochondria can be isolated from leaves and roots simultaneously by following same steps until Percoll gradient separation. Our Ultra Turrax method of rupturing the tissues gives more mitochondria from less tissue. This is especially important for limited tissues such as roots. If the plants are grown on soil there is a possibility that bacteria are also isolated with the mitochondria. Microbial contaminations can be avoided by growing plants on axenic cultures. Growing plants on vertical plates has advantage over horizontal planes in root harvesting. The protocol described here is well suited for all physiological and biochemical studies.

Acknowledgements

We thank Werner Kaiser and Hermann Bauwe for providing laboratory facilities. Werner Kaiser for the suggestion of using Ultra Turrax. K.J.G. thanks Maria Stiomenova for introducing the mitochondrial technique. Authors thank Abir U. Igamberdiev and Werner Kaiser for critical reading of the manuscript.

Notes

1 It is important to fill sterile vertical Petri dishes with medium up to 60% of volume.
2 Do not place the seeds on the Petri dishes until the medium is completely cool.
3 Clean glass and plastic ware with distilled water to avoid any contamination with detergent.
4 Percoll should be carefully removed otherwise mitochondria become extensively contaminated.
5 It is very important to calibrate the oxygen electrodes before measuring respiration for determination of the respiratory control ratio.
6 All buffers should be freshly prepared.

References

Duncan, O., Taylor, N.L., Carrie, C. *et al.* (2011) Multiple lines of evidence localize signaling, morphology, and lipid biosynthesis machinery to the mitochondrial outer membrane of Arabidopsis. *Plant Physiology* **157** (3):1093–1113.

Keech, O., Dizengremel, P. and Gardeström, P. (2005) Preparation of leaf mitochondria from *Arabidopsis thaliana*. *Physiologia Plantarum* **124**: 403–409.

Kowaltowski, A.J. (2000) Alternative mitochondrial functions in cell physiopathology: beyond ATP production. *Brazilian Journal of Medical and Biological Research* **33** (2): 241–250.

Murashige, T. and Skoog, F. (1962) A revised medium for rapid growth and bioassays with tobacco cultures. *Physiologia Plantarum* **15**: 473–497.

Nishimura, M., Douce, R. and Akazawa, T. (1982) Isolation and characterization of metabolically competent mitochondria from spinach leaf protoplasts. *Plant Physiology* **669**: 916–920.

Sweetlove, L.J., Taylor, N.L. and Leaver, C.J. (2007) Isolation of intact, functional mitochondria from the model plant *Arabidopsis thaliana. Methods in Molecular Biology* **372**: 125–136.

Vanlerberghe, G.C., Day, D.A., Wiskich, J.T. *et al.* (1995) Alternative oxidase activity in tobacco leaf mitochondria. Dependence on tricarboxylic acid cycle-mediated redox regulation and pyruvate activation. *Plant Physiology* **109**: 353–361.

Index

Alternative Respiratory Pathways in Higher Plants, First Edition.
Edited by Kapuganti Jagadis Gupta, Luis A.J. Mur and Bhagyalakshmi Neelwarne.

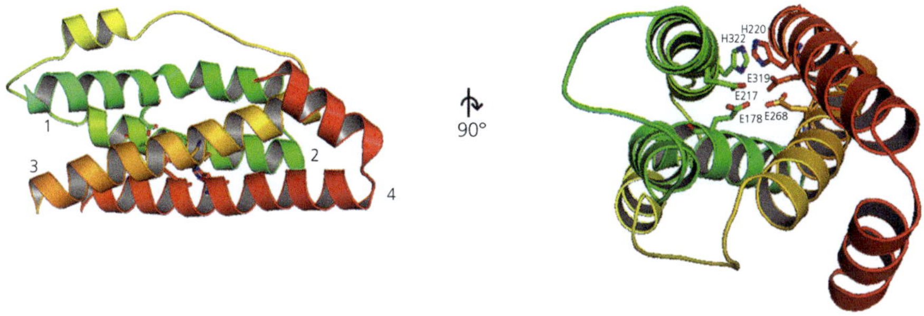

Figure 5.1 A modified version of the 1999 *AOX* model, indicating iron-binding residues (right, as per Table 5.2) within the four helix bundle (left, numbers indicate helices 1–4).

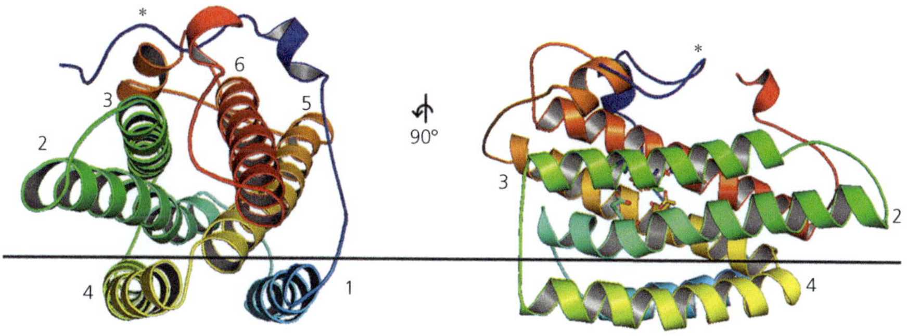

Figure 5.2 The monomeric *S. guttatum* homology model based on TAO (3VV9). On the left, all six helices are labelled and on the right only the nearest are labelled. *, the location of the conserved Cys 122 residue in the unstructured N-terminal region shown in dark blue (see text for details); black line, the approximate placement of the membrane with respect to the protein. The image on the right is the 90° anticlockwise rotation of the image on the left.

Alternative Respiratory Pathways in Higher Plants, First Edition.
Edited by Kapuganti Jagadis Gupta, Luis A.J. Mur and Bhagyalakshmi Neelwarne.

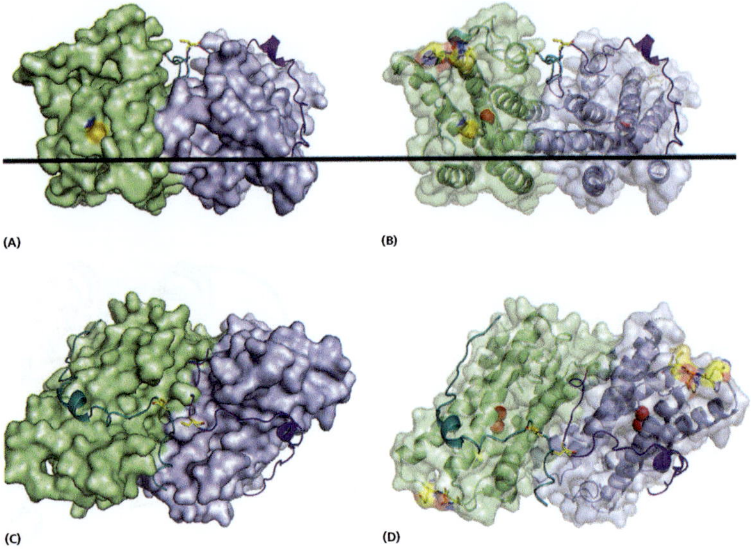

Figure 5.3 The dimeric *S. guttatum* homology model based on TAO (3VV9) is shown here embedded into the inner surface of the inner membrane (see Box 5.1 for further details) with helices 1 and 4, as in Figure 5.2) lying approximately 5 Å below the lipid/solvent interface (solid line). A. Surface representation of the plant *AOX* showing N-terminal extension and location of Cys 122. B. As A but surface rendered transparent (40%) and showing helices and Fe atoms. C. As A but looped 90° . D. As C but surface rendered transparent (40%) and showing helices and Fe atoms. The yellow and red sticks indicate the position of the QDC motif.

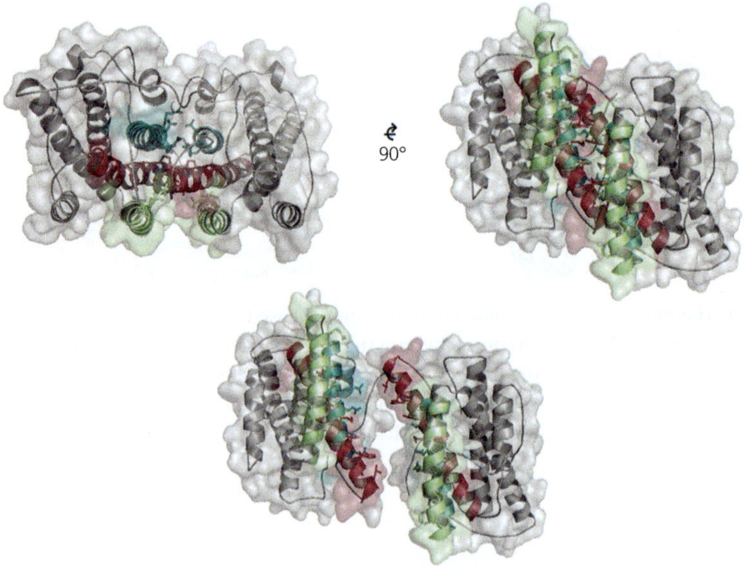

Figure 5.4 A graphical representation of the potential dimer interface, showing conserved residues on helices 2 (red), 3 (teal) and 4 (pale green) as listed in Table 5.3. The top two images represent the whole dimeric model both parallel (top left) and perpendicular (right) to the membrane, whilst the bottom image shows the monomeric models separated artificially to show the extent of tessellation between the two units which overlap rather than lying flush to one another.

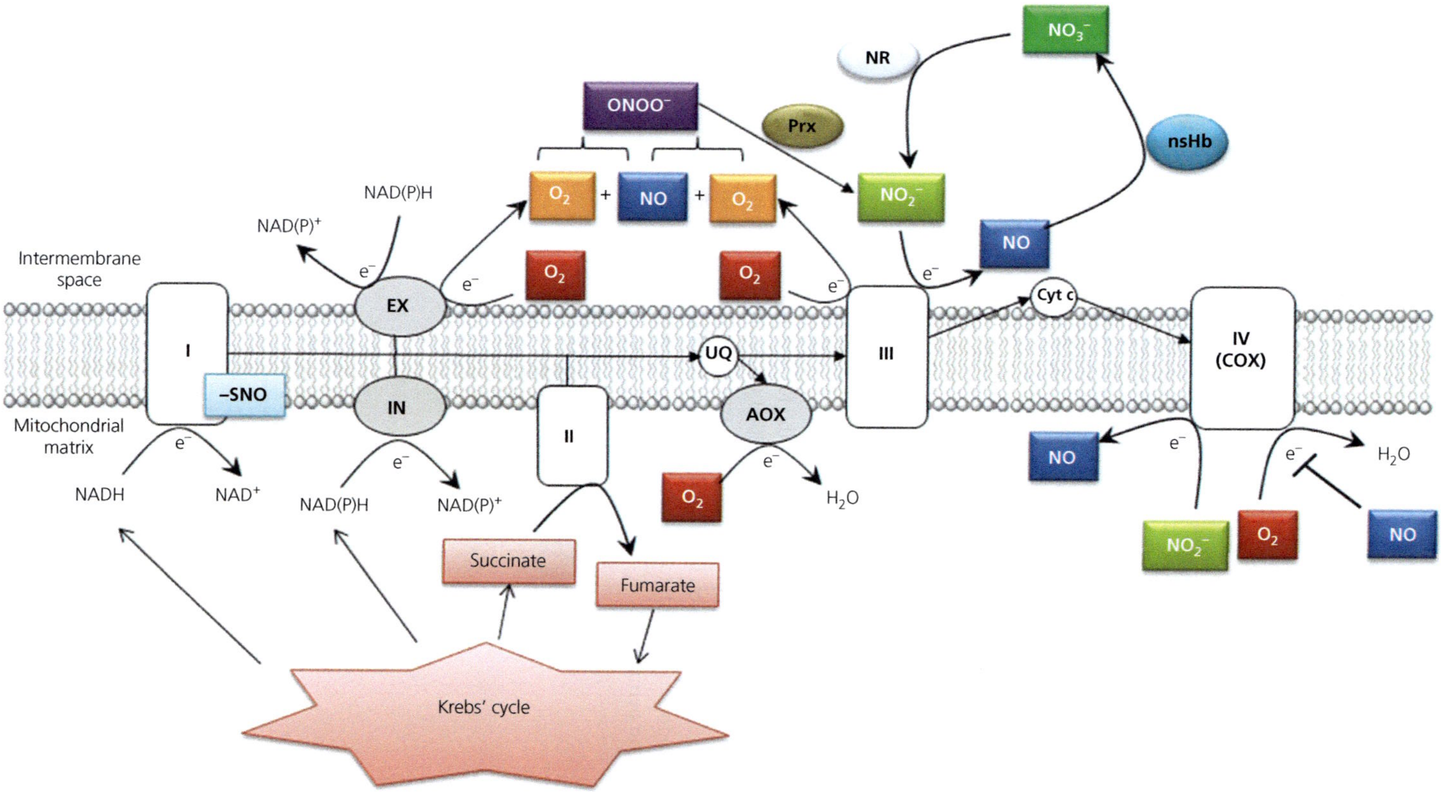

Figure 6.1 Schematic model for the maintenance of NO homeostasis by plant mitochondria. Nitrate (NO_3^-) is reduced to nitrite (NO_2^-) by cytosolic nitrate reductase (NR). NO_2^- is then reduced to NO by cytochrome *c* oxidase (COX) or complex III of mitochondrial respiratory chain. At physiological levels NO causes reversible inhibition of COX and can also lead to *S*-nitrosylation of complex I. The resulting restriction of electron flux through the cytochrome pathway stimulates production of superoxide (O_2^-) by external NAD(P)H dehydrogenases (EX) and complex III. These enzymes then contribute to NO degradation because NO promptly reacts with O_2^- to produce peroxynitrite ($ONOO^-$). Conversely, alternative oxidase (*AOX*) allows mitochondrial electron flow in the presence of NO and decreases electron leakage and NO consumption. $ONOO^-$ can be metabolised back to NO_2^- by peroxiredoxins (Prx) and NO can also be metabolised to NO_3^- by cytosolic class 1 non-symbiotic haemoglobins (nsHb), closing the cycle.

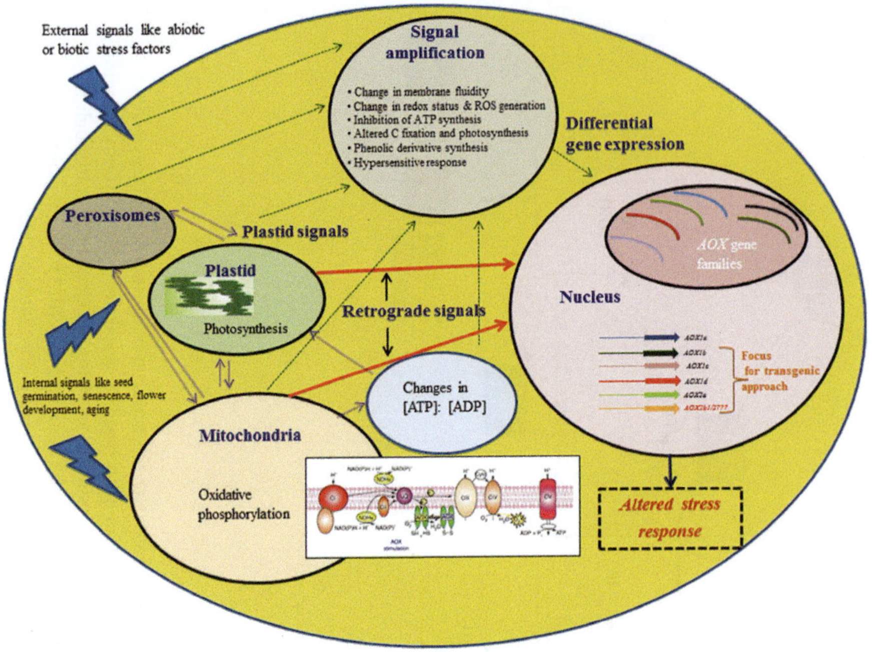

Figure 13.2 Overview of *AOX* signalling during stress and the focus of transgenic technology should be the characterization of *AOX* gene families. Modified with permission from Arnholdt-Schmitt *et al.*, 2006 and Clifton *et al.*, 2006. Dotted arrows, external or internal signal perception, amplification and transmission for altered gene expression; red arrow, retrograde signalling from mitochondria and plastids to nucleus; purple arrow, signalling between mitochondria, plastids and peroxisomes.

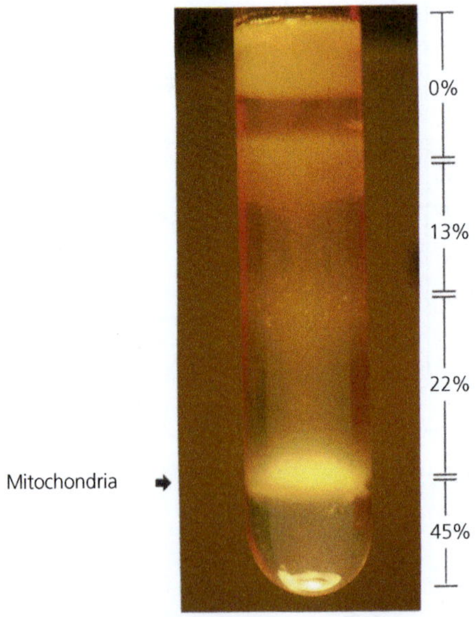

Figure 15.1 Purification of sunflower mitochondria on a discontinuous Percoll density gradient (13%, 22% and 45%) after centrifugation for 15 min, 14 000 g.

Printed and bound by CPI Group (UK) Ltd, Croydon, CR0 4YY

30/09/2026

14973388-0001